An Introduction to Pain and its Relation to Nervous System Disorders

Cover Image Legends

Clockwise:

1) **Synaptic plasticity in the dorsal horn:** the primary mechanism underlying pathological pain.

2) **A cannabis leaf:** Cannabinoids mediate anti-nociception when binding to their receptors, widespread in both the peripheral and the central nervous system.

3) **Adult rat Dorsal Root Ganglion (DRG) nociceptors labelled with anti-ephrinB2 (x20).** Slide prepared by A Battaglia. Image observed with a Leica (DMRB) epifluorescence microscope and acquired with a Hamamatsu camera by A Battaglia. EphBs and their ligand ephrinsB have been shown to play an important role in the pathophysiology of pain.

4) **Healthy and damaged nerves as found in Multiple Sclerosis:** an autoimmune disease leading to loss of myelin around axons and axon degeneration.

5) **Quiescent microglia.** Microglia are a type of glial cell performing phagocytic and immune functions in the central nervous system (CNS); microglia activates following insult to the CNS and to the Peripheral Nervous System leading to a pro-inflammatory response with pathological effects.

6) **Brain made of pills:** A representation of placebo, the 'healing ritual' that can modify our brains and our perception of pain.

7) **A human chromosome** showing details of chromatin and DNA wrapped around histone proteins, whose modifications have an important role in epigenetic regulation of chronic pain.

All images (except 3) drawn for this book by scientific illustrator Dr Claudia Stocker, Vivid Biology, www.vividbiology.com / claudia.stocker@gmail.com.

An Introduction to Pain and its Relation to Nervous System Disorders

EDITED BY

Anna A. Battaglia

WILEY Blackwell

Library of Congress Cataloging-in-Publication data applied for

9781118455913 (Hardback)
9781118455975 (Paperback)

A catalogue record for this book is available from the British Library.

Wiley also publishes its books in a variety of electronic formats. Some content that appears in print may not be available in electronic books.

Cover image: 1-2 and 4-7: © 2016 Claudia Stocker, Vivid Biology, www.vividbiology.com / claudia.stocker@gmail.com

Set in 9.5/13pt Meridien by SPi Global, Pondicherry, India
Printed and bound in Singapore by Markono Print Media Pte Ltd

1 2016

To Marco, Niccolò, Leonardo and Stefano, for supporting me all the way through this project and for accepting a wife and a mum busier than usual; to my parents Elio and Marisa, for all their love and support; to my sister Arabella, who has experienced pain all her life, in the hope that this book will give her some answers, if not relief; and, last but not least, to my brother Erik, whose creativity and talent with words and notes I will never match.

I would also like to dedicate this book to the memory of my university professor and mentor, prof Aldo Fasolo who unexpectedly left us in November 2014; his endless curiosity for the world, love for life, interest in people, integrity and passion for Neuroscience had a long-lasting influence on my approach to work and on my relationship with the students.

Contents

List of contributors

Martina Amanzio
Assistant Professor of Psychobiology in the Department of Psychology,
University of Torino, Italy

Amaria Baghdadli
Professor of child and adolescent psychiatry, and head of the Child and Adolescent
Psychiatry Department, CHRU of Montpellier, France

Anna A. Battaglia
Senior Lecturer in the Faculty of Life Sciences and Medicine, King's College London, UK

Massimiliano Beltramo
Director of Research at the Institut National de la Recherche Agronomique, Physiologie de
la Reproduction et des Comportements, France

Maeve Campbell
Housing Monitoring and Support Officer for Local Government, UK

Kallol Ray Chaudhuri
Professor of Movement Disorders and Director of the National Parkinson Foundation Centre
of Excellence, Kings College London, UK

Luana Colloca
Associate Professor, Department of Pain Translational Symptom Science, University of Maryland
School of Nursing and Dept of Anesthesiology, University of Maryland School of Medicine, USA

Franziska Denk
Research Assistant at the Wolfson Center for Age-related Diseases, Institute of Psychiatry,
Psychology and Neuroscience (IoPPN), King's College London, UK

Amandine Dubois
Lecturer in Psychology at the Faculte des Lettres et Sciences Humaines, Universite de Bretagne
Occidentale, France

Damien Finniss
Associate Professor at the Pain Management Research Institute, University of Sydney and
Royal North Shore Hospital, Sydney; also at Griffith University, Australia

Isabella Gavazzi

Lecturer in Physiology at the Wolfson Centre for Age-related Diseases, Institute of Psychiatry, Psychology and Neuroscience (IoPPN), King's College London, UK

Segzi Goksan

Department of Paediatrics, John Radcliffe Hospital, University of Oxford, UK

Joan Hester

Lead Clinician and Consultant in Pain Medicine, Anaesthetics Department, King's College Hospital, UK

Adam P. Horin

National Institutes of Health, Department of Bioethics, USA

Angela Iannitelli

Psychiatrist, Psychoanalyst and Professor in the Department of Human Studies, Università degli Studi dell'Aquila, Italy

Miriam Kunz

Assistant Professor, Rosalind Franklin Fellow, Gerontology Section, Department of General Practice, University Medical Center Groningen, Netherlands

Alban Latrémolière

Research Fellow at the Boston Children's Hospital and Harvard Medical School, MA

Stefan Lautenbacher

Professor of Physiological Psychology, University of Bamberg, Germany

Marzia Malcangio

Professor of Neuropharmacology, Wolfson Centre for Age-related Diseases, Institute of Psychiatry, Psychology and Neuroscience (IoPPN), King's College London, UK

Lance M. McCracken

Professor of Behavioural Medicine, Health Psychology Section, Psychology Department, Institute of Psychiatry, Psychology and Neuroscience (IoPPN), King's College London, UK

Stephen B. McMahon

Sherrington Professor of Physiology, Director of the London Pain Consortium and Head of the Pain group in Wolfson Centre for Age-related Diseases, Institute of Psychiatry, Psychology and Neuroscience (IoPPN), King's College London, UK

Michele Messmer Uccelli

Director of Healthcare Professional and Client Programs, Italian Multiple Sclerosis Society, Italy

Massieh Moayedi
Honorary Research Fellow, Department of Neuroscience, Physiology, and Pharmacology, University College London, UK

Fiona Moultrie
Academic Clinical Fellow, Department of Paediatrics, Children's Hospital, John Radcliffe Hospital, University of Oxford, UK

Louise S. C. Nicol
Wolfson Centre for Age-related Diseases, Institute of Psychiatry, Psychology and Neuroscience (IoPPN), King's College London, UK

Elizabeth A. Old
Wolfson Centre for Age-related Diseases, Institute of Psychiatry, Psychology and Neuroscience (IoPPN), King's College London, UK

Ravi Poorun
DPhil Student, Nuffield Department of Clinical Neurosciences, John Radcliffe Hospital, University of Oxford, UK

Cecile Rattaz
Psychologist, PhD, Autism Resources Center, Child and Adolescent Psychiatry Department, CHRU of Montpellier, France

Tim V. Salomons
Assistant Professor of Neuroscience at the School of Psychology and Clinical Language Sciences, Pain Emotion and Cognition Laboratory, University of Reading, UK

Rebeccah Slater
Associate Professor of Paediatric Neuroimaging, Wellcome Trust Career Development Fellow and Fellow of Green Templeton College, Department of Paediatrics, John Radcliffe Hospital, University of Oxford, UK

Elisaveta Sokolov
The Royal Marsden Hospital, UK

Claudio Solaro
Neurologist, Department of Head and Neck, Neurology Unit, Italy

Claudia Stocker
Scientific Illustrator and Graphic Designer, Vivid Biology, UK

Matthew Thakur
Wolfson Centre for Age-Related Diseases, King's College and Science Divison, Wellcome Trust, UK

Paola Tirassa

Headgroup in the Institute of Cellular Biology and Neurobiology (IBCN), National Council of Research (CNR), Italy

Andrew J. Todd

Professor at the Institute of Neuroscience and Psychology (Spinal Cord Group), University of Glasgow, UK

Loraine Williams

Project Officer in Education, UK

Panagiotis Zis

National Parkinson Foundation Centre of Excellence, Kings College London, UK

Foreword

Maeve had an unpleasant forceps delivery when her third child was born; the intervention was traumatic and painful at the time and after delivery the pain did not go away – it radiated into her thigh, across her perineum, it raged away and affected her recovery, her self-esteem, her care of her child. And no one believed her, it was inexplicable, there were no physical signs of damage, no major nerve had been severed, it did not fit any diagnostic category, therefore it was not real. Bitterness, loss of self-confidence, loss of hope, litigation ensued, but still the pain persisted.

Is this scenario explained by new research into the neurobiology of pain? Is brain imaging making it easier for us to understand it? Every small step forward adds a small piece to the jigsaw; but there is a long way to go. She could not wait, her life lay in tatters with smouldering anger and resentment.

Hilary Mantel writes eloquently about her own chronic pain from endometriosis. She rebukes doctors for not listening and finds that the best therapy is an empathic and supportive relationship with a caring doctor who may not have any answers but who listens, cajoles, explains as much as he or she understands and tries to wear the shoes of the pain sufferer. Maeve found a similar pathway four years after her pain started and now, 14 years later, she leads a normal life – still with a small amount of medication that she dare not stop, and with an annual review. It was not injections or sophisticated gadgetry that helped her, but a listening ear and guidance from an educated pain professional.

Loraine suffered a whiplash injury 20 years ago; she had persistent tingling and pain in her arm and hand but no demonstrable nerve injury. In the course of her rehabilitation she was lucky enough to be referred to a pain management programme. She continues to see her physiotherapist. As a result, she returned to full-time work and manages her children without assistance. It could have been very different.

Wilbur is 32 years old and moves slowly; he has a constant back pain. He has been successfully treated for aplastic anaemia with chemotherapy and no longer needs any blood transfusions. Despite his miraculous 'cure', he appears angry, as the back pain is 'ruling my life'. He has served several spells in prison and has been deserted by family and friends because of outbursts of uncontrolled anger. His gold front teeth flash as he relaxes into an engaging smile and talks about his fears. A scan of his back shows stable marrow changes but no pathology to account for his pain. His back movements are reduced and painful in all directions, pain being exacerbated by anxiety and anxiety causing more pain. He is very willing to engage with a psychologist, to learn how to control his anger and to go to the gym. He seems to need permission to do so. Resolve to change the course of your life needs support and nurture.

There are many patients with chronic pain who have not managed to lead a fulfilling life and who continue to look for a cure. Injections, surgical operations and more operations often make the situation worse. There is always the hope that this time it will be different; if there is no hope there is despair, and this features highly in chronic pain sufferers. It is hard for attending health professionals to manage such a complex condition as chronic pain. We may not be able to alter the level of pain, but we can alter life around the pain, provide hope where there is despair, laughter instead of tears, relaxation instead of anxiety, and we can encourage physical activity that may have seemed an impossible task. We can alleviate only some of the suffering that chronic pain causes in one's life.

This book looks at some very needy populations in intractable pain. Pain within a serious neurological disorder such as Parkinson's disease or multiple sclerosis leads to a double dose of suffering; pain in dementia can be very hard to treat, as the normal distraction of performing simple tasks is often absent; pain in the neonate and in the autistic adult needs a whole new level of understanding. Pain, depression and anxiety are recurring themes in patients with chronic pain. We cannot remove the pain, so we must address the things we can help; depression and anxiety can very definitely be helped by counselling, mindfulness and other psychological techniques, and through the judicious use of drugs. There is increasing awareness that opioids do not work well in the management of chronic non-cancerous pain and may cause more harm than good. Injection therapies and neuromodulation, in highly selected cases, have a small but important part to play in pain management.

The science of pain has moved way ahead of the techniques available for managing pain, which get fewer and fewer as research shows that any one technique or drug works well in only 15–30 per cent of people. New pain pathways and receptors are being discovered in the nervous system, neuroimmune mechanisms are shown to have an important role, genetics plays its part, plasticity is all important; but there is no single answer for any individual, and there is more complexity for the doctor to deal with. We must not lose sight of the person who suffers. Science cannot explain suffering but it can add insights into the causes of suffering.

Despite these scientific advances, pain medicine remains an art; but the insights from scientific discovery serve to further our understanding of the origin and perpetuation of chronic pain. Helping to alter thinking and behavioural patterns in an individual requires devoted input from doctors, nurses, psychologists, physiotherapists, the pain sufferer and his or her family and friends. When pain management is well done, it is remarkably effective.

Joan Hester
27 February 2015

Acknowledgements

I wish to thank all the authors for their brilliant and thoughtful chapters: without all of your contributions this book would not have been possible. I know that taking time out of your research has not been easy and I do appreciate that you have been able to do it in the end. In particular, thank you to Isabella Gavazzi and Steve B. McMahon, with whom I started my scientific journey here in London at King's College. Dr Gavazzi has supported me all these years, and it is also thanks to her that the neuroscience module called Perspectives on Pain and Nervous System Disorders, which we both organise, is such a success with the students.

I must thank both Joan Hester and her two chronic pain patients, Maeve Campbell and Loraine Williams: they have been great in helping me to set up a very successful Introduction to the Pain Scenario, which I organise for medical students at King's College London. Maeve's and Loraine's interviews with Dr Hester in front of 350 students have been one of the best ways to tell everyone what being in pain means and how profoundly it affects a person's life. I am grateful that they all agreed to contribute to this book and to share – this time with the readers – both their experiences with patients and those of being a patient.

The invitation to write this book came to me courtesy of Nicky McGirr, then publishing editor in life sciences at Wiley, who gave me the strength and motivation to put forward a proposal. I look forward to sending her a copy of the book. Thank you, Nicky, for reaching out and for your support and encouragement. I know you are not at Wiley any more but I hope your new career is going well.

I would also like to thank Celia Carden, the development editor who has been in charge of this project from the start and for most of the time I've spent working on it; she has always been there for me when I needed help and advice or when things got difficult. Thank you also to Fiona Seymour, senior project editor, for her useful advice and help, and to Audrie Tan, the project editor who has taken the relay from Ms Seymour and has been really helpful and supportive in the final stages of the book's preparation. Thank you to Jenny Cossham and Janine Maer, members of Wiley's staff, for their support. Thank you to Manuela Tecusan, the book's copy-editor, for her great swork.

Last but not least, thank you to my creative scientific illustrator Claudia Stocker, who is responsible for the great book cover.

Notes on authors

Dr Anna A. Battaglia, BSc and MSc Biological Sc; BSc (Hons) Psychology, PhD Neuroendocrinology; PGCAP, PGDip Academic Practice

Senior Lecturer, Faculty of Life Sciences and Medicine, Department of Anatomy/ Neuroscience, Division of Bioscience Education, King's College London; Associate Lecturer, Faculty of Science, Open University, UK

Dr Anna A. Battaglia is a neurobiologist with international research experience. She holds a degree in Biological Sciences from the University of Turin and a PhD in Neuroendocrinology from the University of Milan; in the United Kingdom she obtained a BSc in Psychology from the Open University. She has worked in Nobel Prize winner Rita Levi-Montalcini's lab in Rome for post-lauream training; then in Paris, both at the Institut des Neurosciences (Université Pierre et Marie Curie) and at the Collège de France; then at the Stazione Zoologica A. Dohrn in Naples; and she has been a visiting research fellow at Stanford University, where she worked on glutamatergic receptors. Since 2001 she came to London, where she started as a Wellcome Trust Research Fellow at King's College, working on research projects on the spinal cord modulation of nociceptive stimuli and on chronic pain conditions; her research led to important discoveries in the field. More recently, as a senior lecturer, she has been involved in educational projects and leads pain and nervous system disorders and behavioural science modules. Passionate about students' education, she embarked on a Master in Academic Practice educational pathway and was granted first a postgraduate certificate and then a postgraduate diploma from King's Learning Institute, which will be completed with a pedagogical piece of research on critical thinking from the students' perspectives. Dr Battaglia is also an associate lecturer in the Faculty of Science at the Open University since 2007, where she teaches two courses – 'Biological Psychology: Exploring the Brain' and 'Science of the Mind: Investigating Mental Health' – to mature students, students with disabilities and students in prison. Dr Battaglia is a STEM (Science, Technology, Engineering and Mathematics) ambassador in the UK.

Dr Battaglia lives in St Albans with her husband and her three sons.

Dr Joan Hester, MBBS, FRCA, MSc, FFPMRCA

Consultant in Pain Medicine, King's College Hospital, and St Christopher's Hospice, UK

Dr Joan Hester has worked in the UK National Health Service for over 40 years. For over 25 years she was consultant in anaesthesia and pain management at Eastbourne, where she co-founded St Wilfrid's Hospice, and she was also medical director of Eastbourne Hospitals until moving to London in 2004. Joan has worked as consultant in pain medicine at King's College Hospital, London and at St Christopher's Hospice, Sydenham – the first specialist palliative care unit in the United Kingdom, founded by the late Dame Cicely Saunders. She is a past president of the British Pain Society. Her work in promoting the understanding of pain and education in pain management has earned her a national and international reputation. She has also been a founding and a board member of the Faculty of Pain Medicine at the Royal College of Anaesthetists. Joan is primarily a front-line clinician who treats patients with many kinds of pain, particularly complex cancer pain. She has contributed to a wide number of publications on pain-related topics, particularly on the long-term endocrine effects of opioids and on interventional cancer pain management.

Introduction

Anna A. Battaglia

The worst pain a man can suffer: to have insight into much and power over nothing.

Herodotus (5th century BC)

Why pain? Pain is fascinating, it is about our physical and mental existence; it is about the subjectivity of our sensory experiences and the impossibility to fully share them with our fellow human beings, so it is also about the loneliness and the silence we feel when intractable chronic pain syndromes grip us.

What is pain?

Defining pain is elusive; philosophers and scientists have endlessly tried to produce a satisfactory definition. Plato (428/7–348/7 BC) thought that pain is a sensation and corresponds to the illness itself, while one of the authors of the Hippocratic corpus (roughly contemporary with Plato) acknowledged that pain happens to the body, while suffering happens to the whole person. The famous ancient doctor Galen (AD 129–c. 216) and the medieval philosopher Ibn Sina, known as Avicenna (980–1038), in contrast with Aristotle (384–322 BC), both believed that the brain was the main organ for the perception of pain, which was considered a sensation opposite to pleasure. Galen was cogently aware that pain is useless to the person in pain, while Ibn Sina, in defining and describing the nature of pain in five pages of his medical treatise *The Canon*, was ahead of his times when he suggested that 'the true cause of pain was a change of the physical condition of an organ whether there was an injury present or not' (quoted in Tashani and Johnson, 2010: 00). In this he anticipated the International Association of Pain (IASP)'s modern definition, according to which pain is an 'unpleasant sensory and emotional experience associated with actual or potential tissue damage, or described in terms of such damage' (Merskey and Bogduk, 1994: III, 3). It has been argued that it is time for a review of this definition. In 2011 A. Wright has written a paper criticising

An Introduction to Pain and its Relation to Nervous System Disorders, First Edition. Edited by Anna A. Battaglia.

the IASP's definition of pain and highlighting the difficulty of the task, which involves giving 'an objective grounding for the definition of a subjective experience' (Wright, 2011: 00). Wright notes how the subjectivity of the phenomenal experience we humans call pain is shown by the huge number of the pain descriptors used in attempting to communicate the quality or character of an individual's perception of his/her pain; and he acknowledges at the same time that it is impossible to appreciate someone else's particular painful experience. He then argues that, given this impossibility, the only way to define pain in a satisfactory manner is to refer to its evolutionary role and to see pain 'as an unpleasant sensation that has evolved to motivate behaviour which avoids or minimises tissue damage, or promotes recovery' (Wright, 2011: 00). Wright also recalls the older debate about the IASP's pain definition (Anand and Craig, 1996); this relying on self-report seems to exclude many categories of people who do not posses the ability to communicate effectively – such as infants, people with learning disabilities, people with forms of dementia. Many chapters in this book address these issues.

In the *Stanford Encyclopedia of Philosophy*, Murat (2013) argues that there is an ambiguity in the conception of pain: we seem to be able to ascribe a sort of objectivity when we report feeling pain in a specific body location and we treat pains like physical objects; on the other hand, pain is also defined as being a subjective experience, which seems to exist only if there is someone to feel it. Again, this double character makes the distinction between objective and subjective quite fuzzy.

Will the new imaging techniques be able to solve this apparent incongruence? In an intellectually stimulating issue of the *Journal of Consciousness Studies* that is all dedicated to pain, Camporesi et al. (2011) discuss the topic. They ask whether it is really the case that the new techniques do really allow us to 'see' the pain in others or to 'know' whether others are in pain; the legal implications of these possibilities are also discussed. In the present book, Chapter 8 ('Brain Imaging in Experimental Pain') presents the state of the art in this field.

This book is about this phenomenal experience that we call pain. Pain accompanies us when we are young, both in health and in disease; we can find it hard to communicate to others if we are affected by neurodevelopmental disorders; pain is also inextricably linked to our mental health, as adults experiencing depression are aware of the very complicated interplay between the two forms of suffering; and, finally, when we became old and are possibly affected by forms of dementia or other disorders of the nervous system, pain will also inevitably be part of our life.

Aims

This book does not want to be a comprehensive textbook on pain: there are already plenty of incredibly good resources in the literature – one for each of the themes in *Wall and Melzack's Textbook on Pain* (McMahon et al., 2014).

The idea behind this publication stems from the perceived need to give health professionals, neuroscience undergraduates and medical students in their preclinical years an easily accessible resource that should take them through the most recent advances in pain research, giving them examples from different areas of enquiry; this is in order to foster an appreciation of the importance of a multimodal approach to studying and possibly treating complex pain syndromes in humans.

Research advances have been chosen so as to focus mostly on chronic pain; this is due to unmet clinical needs in the treatment of persistent pain states, for which there are not many successful pharmacological remedies regardless of the wealth of knowledge in the basic neurobiology of pain pathways and mechanisms. It is suggested that 20 per cent of the adult population worldwide suffers from pain and 10 per cent are newly diagnosed with chronic pain each year (Goldberg and McGee, 2011). This is a huge social and economic problem: the cost per annum is €200 billion in Europe and $150 billion in the United States (Tracey and Bushnell, 2009). The fact that it is not possible to measure objectively the level of pain is a hindrance to effective clinical trials designed to test for new drugs. Another area of interest in the book is the study of placebo effects, which goes beyond their role in clinical trials and has opened an area of research that has the potential to revolutionise the way we understand pharmacological treatments. Moreover, J. Mogil has recently published a study that highlights recent failures of clinical trials of novel analgesics to treat neuropathic pain (Tuttle et al., 2015); analysing data from 84 clinical trials conducted from 1990 to 2013, his group found out that in the United States placebo responses are steadily increasing. This makes it harder to prove that a drug has an advantage over placebo.

Other sections in the book address pain both in relation to the lifecycle (e.g., pain in neonates and infants) and in relation to nervous system disorders. This field is huge and the book by no means aims to cover all the available knowledge. Only a few examples of pain in nervous system disorders will be offered – in particular, pain in autism, in Alzheimer's disease, in Parkinson's and in multiple sclerosis is addressed; the psychology of pain and the co-morbidity of pain with depression are also extensively reviewed. Overall, these areas of research into pain are given less prominence in textbooks and other publications on pain, and we wanted to fill the gap in the educational resources available to students and practitioners in the health professions.

The book should prompt readers to become critical and independent thinkers and should motivate them to do further reading on the topics discussed. As in any other scientific publication, the researchers' approach and their critical point of view to the subject matter will be evidently based on the theoretical framework that guides their research. Each researcher, clinician, psychologist and psychoanalyst contributing to the book has a different cultural and scientific background and a different goal in her/his investigation of pain. Seemingly working in parallel worlds with no apparent connection with one another, nonetheless they all pursue the same overarching aim, which is to better understand human pain and humans in pain and, in the end, to relieve human beings from their suffering.

There is considerable expertise at King's College London – including at the Institute of Psychiatry, Psychology and Neuroscience (IoPPN – in research both on pain and on a number of neurological and psychiatric disorders; hence many of the present contributors belong to this academic community in the United Kingdom. Other valued contributions are from Italy, France, Germany and the United States; they all give an international flavour to the book.

The next part of this Introduction will summarise each individual contribution so as to give an overview of the book's general aims – which are to provide an account of recent advances in the neurobiology of pain, to discuss what it is currently possible to know about the role of our brain in the perception of pain and, finally, to review pain in the lifecycle and in some very common nervous system disorders.

Summary of contributions

Section I Neurobiology of pain: Recent advances

Chapter 1 contains an up-to-date overview of the anatomy of pain pathways. A. Todd covers brilliantly the fine details of the spinal cord role in the integration of nociceptive stimuli. It is clear from this chapter that the local spinal cord micro-circuitry is extremely complex and, despite decades of research, we still have a limited knowledge and understanding of it. What is certain is that the dorsal horn of the spinal cord is pivotal to the modulation of normal and pathological pain states.

In Chapter 2 A. Latremolière discusses extensively and with rare clarity one of the most complex issues in pain research: the plasticity of pain pathways. Mechanisms of nociception are also here briefly covered.[1] This chapter comprehensively covers recent advances in the understanding of the mechanisms of central sensitisation, which underlie chronic pain states, notoriously extremely difficult to treat from a pharmacological point of view. The most important message is that, once the nociceptive system is in a state of sensitisation, it is no longer coupled to the environment. If on the one hand this can have an evolutionary advantage by allowing the healing of damaged tissue, on the other hand, 'if central sensitisation occurs, or is maintained in the absence of tissue damage, this becomes a pathological and maladaptive plasticity of the nociceptive system'. The author says in conclusion: 'Central sensitisation can be triggered by many different mechanisms, which can be schematically classed into two broad categories: a gain in excitability and a loss of inhibitory controls.'

In Chapter 3 M. Thakur and S. McMahon give an overview of symptoms and pathology in neuropathic pain, which is a chronic pain state among the hardest to treat. Quantitative sensory tests currently used to characterise symptoms and signs of neuropathic pain are presented. The interest of this chapter lies in the extensive critical evaluation of animal models of neuropathic pain used in basic research in the hope of finding mechanisms that could become the target of pharmacological intervention. Peripheral and central pathological mechanisms are discussed in the final part of the chapter without detailed coverage of a particular aspect of neuropathic pain, namely the importance of neuroimmune interactions.

Recent advances in neuroimmune interactions in neuropathic pain are the topic of Chapter 4, by E. Old, L. Nicol and M. Malcangio. The authors describe in the first place the role of microglia, the resident macrophages of the central nervous system (CNS). The activation of microglia and astrocytes is presented, then glial pro-nociceptive mediators are taken

---

[1] Many chapters in this book contain an overview of the mechanisms of nociception. This helps the authors to frame the subsequent discussion; therefore I have decided not to modify these sections. I hope the reader will understand and accept some degree of overlap.

into consideration one by one. The authors summarise their chapter by saying: 'Both microglia and astrocytes contribute substantially to the development and maintenance of the pain-associated behaviours that arise as the result of peripheral nerve injury. Under neuropathic conditions microglial response within the spinal cord is associated with the development of pain; inhibitors of the activation of these cells attenuate pain behaviours.' Most importantly, there are pilot clinical studies suggesting that 'some aspects of the preclinical findings approximate the human condition and that microglial and astrocyte targets represent new therapeutic avenues for the treatment of neuropathic pain.'

An important role in the individual variation to pain perception is played by genetic and epigenetic mechanisms. Pain sensation seems to be partly heritable, and Chapter 5, by F. Denk and S. McMahon, gives an overview of genetics and epigenetics in pain. The authors adopt a critical stance towards the current approaches to the study of pain genetics, highlighting for example the limitations of both candidate gene and genome-wide association studies (GWAS). One very recent area of research is the study of epigenetics in chronic pain, which is really in its infancy but holds the possibility of allowing us to understand how environmental influences can have a persistent impact on nociceptive pathways and on pain conditions developed later on.

The two chapters that follow have been chosen as examples, among many others, of the study of molecules involved in modulating nociception and molecules that have a role in mechanisms underlying central sensitisation. For an extensive overview of the many more molecules implicated in these mechanisms, see *Wall and Melzack's Textbook of Pain* (McMahon et al., 2014).

Cannabinoids are a classic example of molecules known since antiquity for their pharmacological analgesic properties. In Chapter 6 M. Beltramo gives an extensive overview of their role in analgesia. An historical approach is taken to tell the reader about cannabis use throughout centuries and about the discovery of the endocannabinoid system and of the synthesis and degradation of cannabinoid receptor intracellular signalling. The analgesic effects of cannabinoid agonists, both in human and in animal models, have validated the role of cannabinoids in analgesia. What is difficult to understand, though, is that clinical trials have mostly failed, as they showed many undesirable side effects and a general lack of efficacy. The complexity of the cannabinoid system is considered the main cause of this failure. A major section of the chapter is on cannabinoids and nociception and describes the distribution of the different components of the cannabinoid system in order to make us understand its role in the modulation of nociception. In a nutshell, it seems that CB1 is quite widespread in the brain, while CB2 is present in the dorsal horn of the spinal cord and in dorsal root ganglia. The body of evidence for cannabinoids antinociceptive effects is then reviewed concluding with some controversial recent findings in relation to the possible pronociceptive effect of cannabinoids agonists. In general it seems that cannabinoids are a better target for chronic rather than acute pain. In the final reflection section the author points out that a less well known fact has a very important role in the failure of clinical trials: the importance of the constitutive activity of the native receptors: 'some CB2 ligands could behave as protean agonists, which means they could be agonists, antagonists or inverse agonists, pending on the constitutive activity of the receptor'. The author is nonetheless hopeful that the antinociceptive properties of cannabinoids will be harnessed in the future.

The involvement of the ephrin/EhpB system in the spinal processing of noxious stimuli is a relatively recent area of research, and one full of promises in terms of potential therapeutic role. In Chapter 7 I. Gavazzi explains the newly discovered role of EphB1 receptors in the onset and maintenance of central sensitisation in inflammatory, neuropathic and cancer pain. The author gives a brief introduction to the system represented by the membrane-bound EphB receptor and their ligands, the ephrins, to familiarise the reader with a very complex system, mostly known in the literature for its role in neural development. At the beginning of the year 2000 a new role started to emerge: the involvement of these receptors in excitatory synaptogenesis and in phenomena of synaptic plasticity. The author's team then discovered that, in animal models of pain, the EphB modulates the synaptic function at spinal cord level, changing the sensitivity to noxious input. The next section is, then, a recall of the experimental evidence gathered during a few years and constitutes a very useful example for new generations of students if they wish to understand how science works. Moreover, these experiments were successfully replicated in many other laboratories around the world, which further validated the first discovery that EphB receptors in the spinal cord contribute to increased sensitivity to both thermal and mechanical stimulation. Chimeric molecules have enabled the author's group to show that ephrinB2–Fc sensitise the animals mainly to thermal stimuli and that the blocking of EphB receptor activation prevented the development of thermal and mechanical hyperalgesia in models of persistent pain. Later the use of EphB1 knock-out mice in models of persistent pain showed reduced mechanical and thermal hyperalgesia and spontaneous pain. Hypothesised mechanisms are then discussed that show the important role of EphB receptors in regulating N-metyl-D-aspartate (NMDA) receptors. The author explains in a comprehensive manner how these early studies led to a wealth of experiments, which demonstrate a far greater role of these molecules in many different types of pain such as visceral and cancer pain. The latest discovery demonstrates a role of the EphB receptors in morphine tolerance. The potential for translation is then discussed in the final section of the chapter.

As this section is called Neurobiology of pain: recent advances, I could not have failed to mention the very recent (4th December 2015) discovery by Prof J Wood and colleagues (UCL Medicine) who found out that Nav1.7 deletion incredibly has profound effects on gene expression, leading to an upregulation of enkephalin precursor Penk mRNA and met-enkephalin protein, a natural opioid peptide, in sensory neurons. In 2006 it was found that loss-of-function mutations in the SCN9A gene encoding voltage-gated sodium channel Nav1.7 cause congenital insensitivity to pain (CIP) in humans and mice (see Chapter 5). A good number of studies have shown that contrary to expectations, many potent selective antagonists of Nav1.7 are weak analgesics. J Wood and his group have thus realized that the "….application of the opioid antagonist naloxone potentiates noxious peripheral input into the spinal cord and dramatically reduces analgesia in both female and male Nav1.7-null mutant mice, as well as in a human Nav1.7-null mutant. These data suggest that Nav1.7 channel blockers alone may not replicate the analgesic phenotype of null mutant humans and mice, but may be potentiated with exogenous opioids"(Minett et al, 2015, pg 1). The 39-year old woman with the rare mutation in Nav 1.7 who was given naloxone felt pain for the first time in her life.

Section II Pain in the brain

An exciting avenue in pain research is the use of functional magnetic resonance imaging (fMRI). Chapter 8, by M. Massieh and T. Salomons, gives an extensive overview, both from a technical and from a philosophical angle, of how the use of fMRI has powerfully changed our ideas of pain perception in humans, allowing us to explore how the human brain shapes the experience of pain. Particularly exciting is the section debating the existence of the so called 'pain matrix', widely reported as being the collection of brain areas specifically devoted to integrating nociceptive signals from the periphery so as to give rise to the pain experience; its existence has been recently questioned, after clever experiments aimed at challenging this dogma in the neurobiology of pain. It has been demonstrated that, in the authors' words, 'salient (but non-painful) stimuli in other sensory modalities … can elicit patterns of activation remarkably similar to those generated in response to nociceptive stimuli' and that 'pain matrix stimulation' is neither 'sufficient' nor 'necessary' 'for the experience of pain'. Thus it seems that we are far away from being able to 'see pain' and to 'objectively' define the human pain experience.

L. Colloca, A. Horin and D. Finniss give both an historical and a contemporary overview of placebo effects in Chapter 9. They go over important areas of debate in placebo effects research, which are not always directly linked to placebo in analgesia but are relevant to understand the complexity of the phenomenon. Psychological mechanisms of verbal suggestions, conditioning and social learning can trigger all sorts of expectations. Interestingly, defining a true placebo effect can be quite difficult; its potential challenges and pitfalls are examined. Studies of pain, which have formed the basis of many placebo experiments, are then described. The chapter devotes an important section to the neurobiology of placebo analgesia, focusing on the history of experiments in the field and on recent advances.

L. McCracken provides expert insight into the relationship between psychology and pain in Chapter 10. The author clearly stated, from the beginning, that it is very important to go beyond a dichotomous approach of pain that regards it as either physical or psychological, and he argues against the use of phrases such as 'psychogenic pain', which he considers useless to the patient. If human pain perception is represented as being affected by a range of factors that are biological, social and psychological, adopting such an integrated view allows for psychological treatment to have a role in the relief of pain. Acknowledging the different perspectives that one can take on the phenomenon of pain in making sense of it, the particular point of view of this chapter is to examine the human experience of it as a psychosocial problem. The author examines the weak link existing between pathophysiology and the occurrence of pain; moreover, he provides an interesting example of pain that appears without a noxious stimulus represented by the nocebo effect. Extremely interesting is the section on psychological flexibility as a new model of pain and suffering: this section shows that the role this flexibility has appears more important than the traditional pain management strategies for eliciting positive outcomes. Psychosocial approaches to the treatment of pain are reviewed, acceptance and commitment therapy proposing a direct role for psychological flexibility in relation to treatment outcomes.

Section III Pain in the lifecycle and in nervous system disorders

F. Moultrie, S. Goksan, R. Poorun and R. Slater address the topic of pain in neonates and infants in Chapter 11. The authors consider the different measures used to assess pain in this population, describe how pain is managed and, very interestingly, demonstrate how early pain experiences may have long-lasting consequences on later life. Ethical considerations are also taken into account. For obvious reasons, health professional cannot rely on self-report in the case of the painful experiences of infants and neonates. Incredibly, only recently has it been acknowledged that the lack of verbal skills does not necessarily entail an absence of perception of pain. A similar issue presents itself to those who deal with young or adult people with autism spectrum disorders (ASD) (see Chapter 13). Most tools devised to assess pain in infants take into account both physiological responses (e.g., changes in heart rate), and behavioural ones (e.g., changes in facial expression), and the variable success of these tools is due to their lack of nociceptive specificity. These difficulties in assessing pain in a younger population have an obvious impact on infant pain management. Pain experiences in early life and their long-term effects are then considered in the chapter; it seems that, among other effects, the one on cognition is quite powerful (e.g., structural MRI shows a decrease in both white matter and subcortical grey matter). The chapter ends with a philosophical analysis of why pain in neonates and infants has been neglected for such a long time.

C. Rattaz, A. Dubois and A. Baghdadli examine pain in ASD in Chapter 12. I would like to remind the reader that there is very limited literature on this topic, which requires more attention both from research and from the clinic. Very recently, in February 2015, the British Academy of Childhood Disability has published a survey stating that children with disabilities such autism are being failed and left in pain for too long. In view of these circumstances, Chapter 12 is particularly welcome in our book. The authors discuss the many instances where people with ASD are likely to experience pain due to their challenging behaviours, other medical syndromes (e.g., co-morbidity with epilepsy) or life conditions (e.g., prolonged inactivity or lack of dental care). The authors highlight the difficulty of assessing and treating people with ASD; it is acknowledged that these people might suffer alterations of the pain experience at many different levels (i.e., neurochemical, sensory and cognitive). Moreover, due to the unique sociocommunicative impairments of people with ASD, clinicians face challenges similar to those discussed in Chapter 11, where the authors describe how difficult it has been – historically, in the clinic – to recognise that neonates and infants do perceive pain regardless their ability to communicate this particular experience. As a result of their communication impairment, people with ASD have been considered to be less sensitive to pain than others, or even altogether insensitive. The reasons why this idea has been challenged and is no longer supported are described and discussed in the chapter. Main theories about pain in autism are then considered. A very recent approach is the study of sensory abnormalities in ASD. Cognitive explanations are also taken into account, which show the 'relationship between pain expression and the level of cognitive development'. Interesting to note here is the fact that children with disabilities lack the normal inhibitory control of behavioural and emotional reactions to noxious stimuli – that is, the control shown by normally developing children. So the still unanswered question is how to assess people with ASD. Difficulty in managing pain in this population stems from (and reflects) the difficulty of answering this question.

In Chapter 13 A. Iannitelli and P. Tirassa give a fascinating overview of the co-morbidity of pain and depression. In the first place, the authors discuss the problems that the famous Cartesian mind–body dualism created for research on pain and for psychiatry in general; they adopt a psychoanalytic point of view that tries to overcome the Cartesian perspective. The authors provide a comprehensive philosophical and historical account of how human pain, discomfort and unhappiness have been described over the centuries. It is then stated that a psychoanalytical approach 'is the best lens to understand the painful depression and depressive attribute pain: two sides of the same coin'. A brief history of the *Diagnostic and Statistical Manual of Mental Disorders* (DSM) is presented before the authors address the issue of the re-evaluation of somatic symptoms and related disorders in the new 2013 edition – DSM-V. The authors ask: 'Is the reality of pain dependent on the medical diagnosis or, better, on an organic cause?' In their view, 'the DSM … shows the need to refer to a medical pathophysiology explanation in order to understand the symptomatology'. The authors continue by admitting: 'There is a further distance to be covered between the pain perceived by patients and the nosographic classification. Moreover, there is a prejudice to be overcome, in psychiatry as well as in medical science: pain is not an accessory to human life.' The complex interrelationship between different forms of chronic pain and depression is then discussed. The chapter proposes an analysis of the concept of mental pain that, again, advocates for the need to go beyond Cartesian dualism. Both depression and chronic pain have been shown to alter the brain, and this is discussed in a section devoted to neuronal plasticity. The hypothesis that chronic pain and mood disorders represent a maladaptation of the reward system is also considered. The authors conclude their chapter from a pharmacological point of view, by presenting the use of antidepressants as analgesic, to further confirm the tight link between the two sets of disorders.

In Chapter 14 C. Solaro and M. Messmer Uccelli discuss the topic of pain in patients affected by multiple sclerosis (MS). Pain is considered to account for 30 per cent of all symptomatic treatments. The authors give an overview of the pathophysiology of all major pain symptoms in MS. They explain that 'MS pain is thought to be a type of central pain due to demyelinating lesions in areas involved in pain perception.' Various sections in the chapter address various types of pain in MS. The first section is devoted to the treatment of central neuropathic pain in MS; it follows an overview of musculoskeletal pain and secondary pain. Very detailed and informative tables accompany the chapter, informing the reader about the many studies on the prevalence of pain in MS; the medication used for central neuropathic pain in MS; and all the double-blind randomised placebo-controlled trials on cannabinoids for neuropathic pain in MS.

In Chapter 15 M. Kunz and S. Lautenbacher give an overview of pain perception in dementia, with a particular focus on Alzheimer's disease. The authors provide an exhaustive discussion of the challenges posed by the clinical assessment of pain perception in people with dementia. There is a strong concern that people in this population may suffer in silence, their pain going untreated. The chapter covers both experimental and clinical perspectives.

In Chapter 16 M. Amanzio discusses the role of cognitive impairment in placebo and nocebo effects, highlighting how the loss of placebo mechanisms may require increased therapeutic dosage. This is a research area in its infancy, and very few publications exist on the topic. The chapter describes the difficulties of people affected by dementia, in particular by Alzheimer's disease, in communicating their pain – difficulties that leave them at risk of untreated pain. The role of cognitive–evaluative areas of the brain (such as the prefrontal cortex) in the placebo response is discussed. The disruption of these areas in Alzheimer's disease has important consequences for the patient's ability to experience a placebo effect.

P. Zis, E. Sokolov and K. R. Chaudhury conclude this section by providing an extensive overview of pain in Parkinson's disease (PD) in Chapter 17. The chapter recognises that, in patients suffering from PD, pain can even overshadow the motor symptoms and is often the first symptom in the development of motor PD. The authors then examine the epidemiology and classification systems of pain in PD; this is followed by a typology of pain in PD that makes a distinction between pain of central origin and secondary pain. The treatment of the co-morbidity of pain and motor symptoms in PD is complicated. The role of levodopa, dopamine agonists and non-dopaminergic alternatives such as deep brain stimulation are analysed in detail; opioids are increasingly recognised as a useful tool in managing pain in PD.

> We all must die. But if I can save him from days of torture, that is what I feel is my great and ever new privilege.
>
> Pain is a more terrible lord of mankind than even death himself.
>
> *Albert Schweitzer (1875–1965; Nobel Peace Prize Laureate in 1952)*

References

Anand, K. and Craig, K. D. (1996) New perspectives on the definition of pain. *Pain* 67 (1): 3–6.

Camporesi, S., Bottalico, B. and Zamboni, G. 2011. Can we 'see' Pain? *Journal of Consciousness Studies* 18 (9–10): 257–76.

Clegg, R. 2015. Disabled children 'left in pain'. At http://www.bbc.co.uk/news/disability-31312000 (accessed 3 October 2015).

Goldberg, D. S. and McGee, S. J. 2011. Pain as a global public health priority. *BMC Public Health* 11 (770). doi: 10.1186/1471-2458-11-770.

McMahon, S., Koltenzburg, M., Tracey, I. and Turk, D. C. 2014. *Wall and Melzack's Textbook of Pain*, 6th edn. Philadelphia, PA: Elsevier.

Merskey, H. and Bogduk, N., eds. 1994. *Classification of Chronic Pain*, 2nd edn. Seattle, WA: IASP Task Force on Taxonomy, IASP Press.

Minett, M. S., Pereira V., Sikandar, S., Matsuyama, A., Lolignier, S., Kanellopoulos, A. H., Mancini, F., Iannetti, G. D., Bogdanov, Y. B., Santana-Varela, S., Millet, Q. Baskozos, G., MacAllister, R., Cox, J. J., Zhao, J. and Wood, J. N. (2015) Endogenous opioids contribute to insensitivity to pain in humans and mice lacking sodium channel Nav1.7. *Nature Communications* 6: 8967.

Murat, A. 2013. Pain. In *The Stanford Encyclopedia of Philosophy*, ed. by E. N. Zalta. At http://plato.stanford.edu/archives/spr2013/entries/pain (accessed 2 October 2015).

Tashani, O. A., and Johnson, M. I. 2010. Avicenna's concept of pain. *The Libyan Journal of Medicine* 5. doi:10.3402/ljm.v5i0.5253.

Tracey, I. and Bushnell, M. C. 2009. How neuroimaging studies have challenged us to rethink: Is chronic pain a disease? *Journal of Pain* 10 (11): 1113–20.

Tuttle, H., Tohyama, S., Ramssay, T., Kimmelman, J., Schweinhardt, P., Bennett G. J. and Mogil, J. S. 2015.Increasing placebo responses over time in US clinical trials of neuropathic pain. *Pain*. doi: 10.1097/j.pain.0000000000000333.

Wright, A. 2013. A criticism of the IASP's definition of pain. *Journal of Consciousness Studies* 18 (9–10): 19–44.

SECTION I
Neurobiology of pain: Recent advances

CHAPTER 1
Anatomy of pain pathways

Andrew J. Todd

Basic organisation of spinal pain pathways

Primary afferent axons belonging to the somatosensory system can respond to a range of mechanical, thermal and chemical stimuli. Many of these afferents are activated by stimuli that damage (or threaten to damage) tissues, and these are known as nociceptors. Primary afferents that innervate the limbs and the trunk enter the spinal cord through the dorsal roots and form excitatory (glutamatergic) synapses with neurons in the dorsal horn. The dorsal horn contains a large number of neurons, the great majority of which have axons that arborise locally and remain in the spinal cord; these are known as interneurons and are involved in the local processing of sensory information. In addition, the dorsal horn contains projection cells – that is, neurons with axons that enter the white matter and travel rostrally to the brain. The axons of these cells are grouped into a number of different ascending tracts. The final neuronal component consists of descending axons, which originate from cells in the brain (particularly the brainstem) and terminate diffusely within the dorsal horn.

Rexed (1952) divided the grey matter of the dorsal horn into six parallel laminae (numbered from dorsal to ventral), and this scheme is widely used, for example to describe the arborisation of primary afferents and the distribution of different populations of spinal neurons (Figure 1.1). The dorsal horn is somatotopically organised, the body being mapped in a bidimensional pattern onto the rostrocaudal and mediolateral axes. The other, dorsoventral dimension is arranged in a modality-specific pattern, as will be described later.

Fifty years ago, Melzack and Wall (1965) proposed that neurons within the superficial dorsal horn could 'gate' the inputs from nociceptors and thus modify the perception of pain. This theory attracted a great deal of interest, and there have been numerous attempts to unravel the neuronal circuitry that underlies pain processing in the spinal cord. It turns out that this circuitry is highly complex, and we still have only a limited understanding of it. Nonetheless, it is clear that this region is very important for modulating pain in both normal and pathological states. The superficial dorsal horn is also likely to provide important targets

An Introduction to Pain and its Relation to Nervous System Disorders, First Edition. Edited by Anna A. Battaglia.
© 2016 John Wiley & Sons, Ltd. Published 2016 by John Wiley & Sons, Ltd.

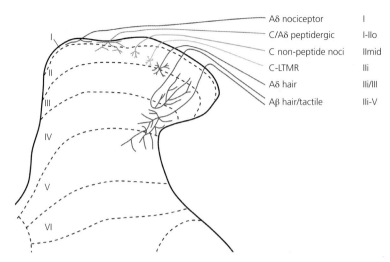

Aδ nociceptor	I
C/Aδ peptidergic	I–IIo
C non-peptide noci	IImid
C-LTMR	IIi
Aδ hair	IIi/III
Aβ hair/tactile	IIi–V

Figure 1.1 Rexed's laminae and primary afferent inputs. The diagram shows the main areas targeted by the central arbors of different types of primary afferent, apart from proprioceptors. These have been superimposed on Rexed's laminar scheme, as applied to the rat dorsal horn. Note that Woodbury and Koerber (2003) have described another class of myelinated nociceptor, with axonal arbors that extend throughout laminae I–V, and this has not been included. Source: Todd, 2010. Reproduced with permission of Nature Publishing Group.

for new drugs designed to treat pain, as it contains a wealth of receptors and signalling molecules. This chapter will summarise some of the available information about the anatomy of spinal somatosensory pathways, placing particular emphasis on the organisation of neuronal populations and their synaptic connections.

Primary afferent input to the dorsal horn

Primary afferents have their cell bodies in the dorsal root ganglia and an axon that bifurcates, sending one branch to a peripheral target and the other to the spinal cord. They can be classified according to a series of criteria. On the basis of their peripheral targets, they can be divided into afferents that innervate the skin, the muscles, the viscera and so on. Secondly, they can be characterised in terms of the strength and type of the adequate stimulus (e.g., hot/cold nociceptors, low-threshold mechanoreceptors). They also vary in axonal diameter – which is related to conduction velocity – and in whether or not they are myelinated: very fine afferents are unmyelinated and are known as C fibres, while the remainder can be divided into large-diameter (Aβ) and small-diameter (Aδ) myelinated fibres. Finally, there are various neurochemical markers, for example neuropeptides, that can be used to distinguish functional populations. These parameters are interrelated; for example, the majority of C and Aδ fibres are nociceptors or thermoreceptors, whereas most Aβs are low-threshold mechanoreceptors (LTMRs) that respond to touch or hair movement. However, this is not an absolute distinction and there are LTMRs among both the Aδ and the C fibre classes, while some nociceptors conduct action potentials in the Aβ range.

Although the focus of this chapter is on pain, the LTMRs are relevant because in many pathological pain states touch can elicit pain (tactile allodynia), and at least a part of this effect is mediated by the low-threshold afferents (Campbell et al., 1988).

The laminar distribution pattern of the main classes of primary afferent is shown in Figure 1.1.

Termination of nociceptors within the dorsal horn

Myelinated nociceptors (mostly Aδ fibres) convey 'fast' pain, whereas the nociceptive C fibres underlie 'slow' pain. The central projections of Aδ nociceptors have been demonstrated through intra-axonal labelling techniques. Many of these fibres terminate in a compact distribution in lamina I and in the outermost part of lamina II, while some arborise diffusely throughout laminae I–V (Light and Perl, 1979, Woodbury and Koerber, 2003). The central projections of myelinated afferents have also been studied through bulk-labelling after the injection of cholera toxin B subunit (CTb) into peripheral nerves. However, although this technique reveals projections to lamina I, some Aδ afferents (e.g., those terminating throughout laminae I–V) are not labelled.

Because of the small size of C fibres, there have been few studies of the central projections of individual afferents of this type (Sugiura, Lee and Perl, 1986). Conveniently, there are neurochemical markers that can be used to recognise different functional populations. C nociceptors can be divided into those that contain neuropeptides – the peptidergic group – and those that do not – the non-peptidergic group. All of the peptidergic afferents appear to contain calcitonin gene-related peptide (CGRP), which is only found in primary afferents in the dorsal horn. In addition, they can express a variety of other peptides, including substance P. Such afferents, which innervate most tissues of the body (including the skin), can be identified in anatomical experiments through immunocytochemistry for CGRP, although this approach does not distinguish between C fibres and peptidergic Aδ nociceptors (Lawson, Crepps and Perl, 1997). Most peptidergic afferents project to lamina I and the outer part of lamina II (IIo), but some send branches to deeper laminae (III–V) (Todd, 2010).

The non-peptidergic C nociceptors have been identified by their ability to bind the plant lectin IB4, although this property is not restricted to the non-peptidergic C afferents. More recently they have been shown to express mas-related G protein-coupled receptor D (MRGD) (Zylka, Rice and Anderson, 2005). These afferents innervate the skin and terminate more superficially than cutaneous peptidergic fibres. Their central arbors occupy a narrow band in the middle part of lamina II.

Central terminations of low threshold mechanoreceptors

Studies of intra-axonally injected Aβ afferents have shown that these terminate in the deep laminae, with specific patterns for the various types of tactile and hair follicle afferent. Low-threshold mechanoreceptive Aδ fibres correspond to down-hair (D-hair) afferents, and these have a more restricted projection, to the inner half of lamina II (IIi) and the dorsal part of lamina III. Recent studies have identified neurochemical/genetic markers for these afferents in the mouse, and these markers have confirmed the distribution patterns that were previously reported from intra-axonal injection studies (Abraira and Ginty, 2013). Some C fibres respond to tactile stimuli (C-LTMRs), and these have been shown to terminate in lamina IIi (Seal et al., 2009).

Vesicular glutamate transporters and primary afferents

All primary afferents use glutamate as their principal fast transmitter, and this is concentrated into synaptic vesicles by a family of three vesicular glutamate transporters (VGLUT1–3). These are differentially distributed among the primary afferents. All myelinated LTMRs express VGLUT1 and form the main source of VGLUT1 in the dorsal horn, although some originates from descending corticospinal axons (Todd et al., 2003). In contrast, Aδ nociceptors possess VGLUT2.

Both peptidergic and non-peptidergic C nociceptors express VGLUT2 (Brumovsky, Watanabe and Hokfelt, 2007), but the level of protein detected with immunocytochemistry in their central terminals is generally very low. The C-LTMRs are unique, in that they express VGLUT3 (Seal et al., 2009).

Receptors expressed by primary afferents

Primary afferents can express a wide variety of ligand-gated ionotropic, metabotropic and tyrosine kinase receptors (Todd and Koerber, 2012). For example, peptidergic and non-peptidergic C nociceptors have different growth factor receptors: trkA and RET, respectively. Glutamate receptors of the NMDA and AMPA type are widely expressed, as are $GABA_A$ and $GABA_B$ receptors, while non-peptidergic nociceptors possess purinergic $P2X_3$ receptors. There are several receptors for neuropeptides, including μ, δ and κ opioid receptors. In addition, TRP channels (TRPV1, TRPA1, TRPM8), which transduce thermal and chemical signals, are found on the peripheral and central terminals of certain afferents.

Ultrastructure of primary afferent central terminals

While most central terminals of primary afferents form relatively simple synaptic arrangements, others form complex arrangements known as synaptic glomeruli (Ribeiro-da-Silva and Coimbra, 1982) (Figure 1.2). Two different types of synaptic glomerulus have been identified, and these involve the central terminals of non-peptidergic C nociceptors and Aδ D-hair afferents. Both types have a central axonal bouton (the primary afferent terminal), which is surrounded by dendritic profiles (mainly dendritic spines) and peripheral (GABAergic) axons (Todd, 1996). Some of the dendritic spines that originate from GABAergic interneurons also possess synaptic vesicles. Glomeruli are sites of complex synaptic interactions, in which the primary afferent bouton receives axoaxonic (inhibitory) synapses from the peripheral axons (and in some cases from vesicle-containing dendrites) and forms axodendritic (excitatory) synapses with the various dendritic profiles. There are also triadic synapses: for example, a peripheral axon can form a synapse with the central axon and also with a dendritic spine that receives a synapse from the central axon. Since all glomerular central axons are of primary afferent origin, they provide a convenient way of identifying primary afferent boutons in the dorsal horn. However, as stated above, many primary afferents have simpler endings, forming axodendritic synapses onto a small number of dendritic shafts or spines and receiving axoaxonic synapses.

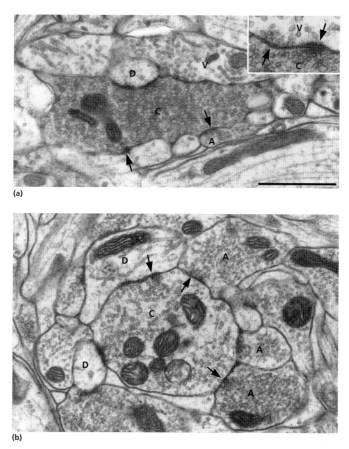

Figure 1.2 Two types of synaptic glomerulus in the superficial dorsal horn of the rat. Some primary afferents form the central axons of synaptic glomeruli, and examples are shown here. **(a)** The central axon of a type I glomerulus (C). These are derived from non-peptidergic C nociceptors and are typically surrounded by vesicle-containing dendrites (V), dendrites that lack vesicles (D), and a single peripheral axon (A). The peripheral axon is presynaptic at an axoaxonic synapse with the central bouton. The central bouton forms synapses with both types of dendrite and can receive dendroaxonic synapses from the vesicle-containing dendrites. In some cases reciprocal axodendritic synapses are present, as shown in the inset. **(b)** The central ending of a type II glomerulus (C). These are thought to originate from Aδ D-hair endings, and are presynaptic to dendrites (D), most of which lack vesicles. These glomeruli are often surrounded by several peripheral axons (A), which are presynaptic to the C bouton and can form triadic synapses involving the C bouton and a dendrite. Arrows in both parts indicate synapses. Scale bar: 1 μm. Source: Todd, 1996. Reproduced with permission of John Wiley & Sons.

Projection neurons and ascending tracts

Projection neurons are present in relatively large numbers in lamina I and are scattered throughout the deeper laminae of the dorsal horn (III–VI); but they are essentially absent from lamina II (Figure 1.3). The lamina I projection cells, together with many of those in laminae III–VI, have axons that cross the midline and enter the ventral or lateral funiculus on the opposite side, before ascending to the brain. The axons of these cells travel to several brain regions – including the thalamus, the periaqueductal grey matter (PAG), the lateral parabrachial area

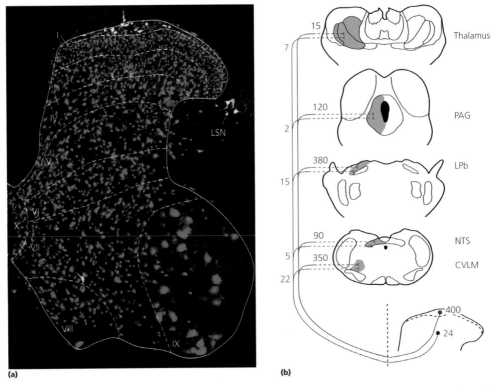

(a)   **(b)**

Figure 1.3 Anterolateral tract projection neurons in the rat lumbar enlargement. **(a)** A transverse section from the L4 segment of a rat that had received injections of cholera toxin B subunit (CTb) into the caudal ventrolateral medulla (CVLM) and Fluorogold into the lateral parabrachial area (LPb). The section was immunostained to reveal CTb (red), Fluorogold (green) and the neuronal marker NeuN (blue). Tracer injections into these two sites can label virtually all of the projection neurons in lamina I, as well as scattered cells throughout the deep dorsal horn (laminae III–VI). Note that some cells have taken up both tracers, and therefore appear yellow. **(b)** Quantitative data from the rat lumbar spinal cord showing the approximate numbers of ALT projection neurons in laminae I and III, and the numbers that can be retrogradely labelled from each target. There are approximately 400 ALT neurons in lamina I (~5 per cent of the total neurons in this laminae), and ~24 of these cells in lamina III. LSN = lateral spinal nucleus; PAG = periaqueductal grey matter; NTS = nucleus of the tractus solitarius. Source: In Handbook of Clinical Neurology, 3rd series: Pain, ed. by F. Cervero and T. S. Jensen, vol. 81. Reproduction with permission of Elsevier..

(LPb), and some nuclei in the caudal brainstem. Individual cells can have axons that project to several of these targets, and the projection is often referred to as the 'anterolateral tract' (ALT). Quantitative studies in the rat have shown that there are around 400 ALT neurons in lamina I on each side of the L4 segment and that virtually all of these neurons project as far as the LPb, one third reaching the PAG. However, somewhat surprisingly, less than 5 per cent of those in the lumbar cord project to the thalamus. Spinothalamic lamina I neurons are much more common in the cervical enlargement of the rat and in both lumbar and cervical enlargements of the cat and of the monkey (Todd, 2010). The ALT cells in deeper laminae of the dorsal horn are less numerous. For example, there are around twenty of these cells in lamina III, but these have projection targets similar to those of the lamina I cells.

In the rat, around 80 per cent of lamina I projection neurons, as well as the ALT neurons in lamina III, express the neurokinin 1 receptor (NK1r). This is a target for substance P, which is released from nociceptive primary afferents, and internalisation of the receptor following ligand binding can be seen after noxious stimulation (Mantyh et al., 1995). Not surprisingly, ablation of these cells with substance P conjugated to the cytotoxin saporin leads to a dramatic reduction in hyperalgesia in both inflammatory and neuropathic pain states (Nichols et al., 1999). Among the lamina I ALT cells that lack the NK1r we have identified a population of very large (giant) cells, which have an extremely high density of excitatory and inhibitory synapses that coat their cell bodies and dendrites (Polgár et al., 2008). These cells are very sparse (they represent about 3 per cent of all lamina I projection neurons) but have extensive dendritic trees that are widely distributed throughout the lamina.

Electrophysiological studies have demonstrated that virtually all lamina I projection neurons in the rat are activated by noxious stimuli, while some are also activated by innocuous mechanical or thermal stimuli (Andrew, 2009; Bester et al., 2000). In addition, some of these cells are likely to respond to pruritic stimuli. Most lamina I projection neurons have dendrites that remain within the lamina. Various morphological types (e.g. fusiform, pyramidal, multipolar) have been identified, and it has been suggested that these represent specific functional populations and that pyramidal cells in the cat respond to thermal, rather than noxious stimuli (Han, Zhang and Craig, 1998). However, this hypothesis remains controversial, since pyramidally shaped lamina I projection neurons in the rat express the transcription factor Fos in response to noxious stimuli (Todd et al., 2002). In addition, these cells account for around one third of lamina I projection cells in the rat and, as stated above, virtually all of these respond to noxious stimuli.

The ALT is thought to be responsible for the perception of various stimuli as pain or as itch, as well as for thermal sensation. However, it is not the only ascending tract to originate from the dorsal horn. The spinocervical tract and the postsynaptic dorsal column (PSDC) pathway both arise from cells located in the deep dorsal horn, mainly lamina IV (Abraira and Ginty, 2013; Brown, 1981). Much less is known about the functional role of either of these pathways.

Dorsal horn interneurons

Interneurons account for around 95 per cent of the neurons in lamina I, virtually all of those in lamina II, and an unknown proportion of those in the deeper laminae (Todd, 2010). Numerous anatomical studies have been carried out on interneurons in laminae I–III, which were labelled either with the Golgi technique or during electrophysiological recording experiments carried out *in vivo* or *in vitro*. These studies have shown that the interneurons almost invariably give rise to axonal boutons within the same segment, often forming complex local axonal arbors. These arbors are generally found in the lamina that contains the cell body, but they often extend into adjacent laminae. However, it is also clear that many interneurons in laminae I–III also give off long intersegmental branches, although the function of the latter is not yet understood (Todd and Koerber, 2012).

There are undoubtedly several different functional populations of interneurons in this region, and these are likely to perform distinct tasks. It is very important to be able to define these populations, so that we may be able to explore their locations in the synaptic circuits of the dorsal horn and to investigate their functions, for example by selectively ablating or silencing them.

Classification of interneurons in laminae I–III
Inhibitory and excitatory interneurons

A fundamental distinction can be made between inhibitory and excitatory interneurons. The inhibitory interneurons use GABA and/or glycine as their major fast transmitter(s), and both amino acids can be revealed in neuronal cell bodies with immunocytochemistry. Quantitative studies in rat and mouse have shown that GABA is present in around 25–30 per cent of neurons in laminae I–II and in around 40 per cent of neurons in lamina III (Polgar et al., 2013a). Since the main role of GABA is to act as an inhibitory transmitter, it is very likely that these cells are all GABAergic interneurons. Glycine is also enriched in some neurons; but, because glycine has other roles apart from neurotransmission, it is not certain that all glycine-enriched neurons are glycinergic (Zeilhofer et al., 2005). Nonetheless, the vast majority of glycine-enriched neurons in laminae I–III were also found to be GABA-immunoreactive. This is significant for two reasons: first, it indicates that essentially all inhibitory interneurons can be revealed with GABA antibodies; and, secondly, it suggests that many of these interneurons can co-release GABA and glycine. However, despite evidence that GABA and glycine are co-localised in both interneuron cell bodies and their axon terminals, it appears that co-transmission is relatively rare in the adult spinal cord, possibly because of differential distribution of the postsynaptic receptors (Zeilhofer, Wildner and Yevenes, 2012).

All of the neurons in this region that lack GABA immunoreactivity are likely to be excitatory, glutamatergic neurons. These neurons include projection cells (in laminae I and III), but most are excitatory interneurons. Although glutamate can be revealed with immunocytochemistry, this has not proved to be a useful technique for identifying cell bodies of glutamatergic neurons, presumably because metabolic (i.e. non-transmitter) levels of glutamate are relatively high and this makes it difficult to distinguish them from 'transmitter glutamate'. Until around ten years ago it was difficult to identify glutamatergic neurons in anatomical studies. However, the discovery of the VGLUTs has provided very useful markers for identifying these cells. Unfortunately the levels of VGLUTs in cell bodies are too low for immunocytochemical detection, although their mRNAs can be revealed with in situ hybridisation (Malet et al., 2013). VGLUTs are concentrated in axon terminals and, where axons of individual neurons can be identified, they can be tested for the presence of the different VGLUT proteins (Maxwell et al., 2007, Todd et al., 2003, Yasaka et al., 2010). The results of these different approaches have shown that VGLUT2 is the major vesicular transporter used by glutamatergic neurons in laminae I–III, although some neurons in lamina III express low levels of VGLUT3.

We know something about the roles of inhibitory interneurons from studies in which antagonists acting at GABA$_A$ or glycine receptors have been applied intrathecally (Yaksh, 1989). Although some GABAergic and glycinergic axons in the superficial dorsal horn originate in the brainstem (Antal et al., 1996), most are thought to belong to local interneurons, and this means that the effects of intrathecal antagonists are most likely to result from blocking transmission by local inhibitory interneurons. Studies of this type have suggested that these neurons are involved in limiting the extent and severity of pain in response to noxious stimuli, in preventing spontaneous pain and in avoiding non-noxious stimuli (e.g., touch) from being perceived as painful (Sandkuhler, 2009). There is also evidence that one role of inhibitory interneurons in the superficial laminae is to suppress itch, for example when a counterstimulus such as scratching is applied (Ross et al., 2010).

Much less is known about the roles of excitatory interneurons, but these roles may include the transmission of primary afferent information across laminar borders. For example, it is thought that excitatory interneurons provide a polysynaptic pathway through which tactile afferents (which terminate in laminae IIi–VI) can activate nociceptive projection neurons in lamina I (Lu et al., 2013, Torsney and MacDermott, 2006). This pathway may normally be closed by local inhibitory mechanisms, but it could open up in pathological states, thus allowing these cells to acquire low-threshold inputs. This phenomenon may contribute to tactile allodynia.

Morphological classification

Elsewhere in the central nervous system (CNS), the morphology of neurons has often allowed them to be assigned to distinct populations; this approach has therefore been used extensively in the spinal dorsal horn. Most anatomical studies have concentrated on lamina II and, and while the early reports were based on Golgi staining, more recent investigations have examined neurons that were labelled during electrophysiological recording. The most widely accepted classification scheme was that developed by Grudt and Perl (2002), who identified four main morphological types of lamina II neuron: islet, vertical, radial and central cells (Figure 1.4a). Islet cells have dendritic trees that are highly elongated in the rostrocaudal axis; vertical cells generally have a dorsally located soma and a cone-shaped, ventrally directed dendritic arborisation; radial cells have short, radiating dendrites; while central cells are somewhat similar to islet cells but have much more restricted dendritic trees. However, although these cells have been identified in several other studies, a limitation of this approach is that many cells (typically ~30 per cent) cannot be assigned to any of these classes.

Yasaka et al. (2010) recorded from a large sample of lamina II neurons and related the morphology of the recorded neurons to their transmitter phenotype by immunostaining for VGLUT2 and for the vesicular GABA transporter (VGAT) in sections that contained the axons of these cells. They found that all islet cells were inhibitory – that is, had VGAT-immunoreactive axons – while radial cells and most vertical cells were excitatory. However, cells that were classified as central could be either excitatory or inhibitory interneurons; and there were many 'unclassified' cells among both transmitter types. These results suggest that, while certain functional populations can be identified on the basis of morphological criteria, many cells cannot be classified in this way. In addition, care is needed when using this approach: for example, some cells that resemble glutamatergic vertical cells are in fact inhibitory interneurons (Maxwell et al., 2007, Yasaka et al., 2010).

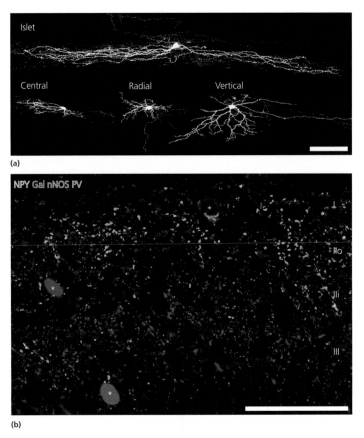

Figure 1.4 Schemes for classifying interneurons in the superficial dorsal horn. **(a)** Confocal images of four lamina II neurons, recorded in parasagittal spinal cord slices from young adult rats. Neurobiotin in the pipette allowed labelling with fluorescent avidin after whole-cell recording. The cells correspond to the four main classes recognised by Grudt and Perl (2002). **(b)** Four non-overlapping populations can be recognised among the inhibitory interneurons in laminae I–III of the rat dorsal horn on the basis of expression of neuropeptide Y (NPY), galanin (Gal), neuronal nitric oxide synthase (nNOS) and parvalbumin (PV). This confocal image shows a single optical plane through a transverse section of rat lumbar spinal cord that had been reacted with antibodies to each of these substances. A single cell of each type is present, and these cells are indicated through asterisks. Approximate positions of laminae are shown. Scale bars: 100 µm (a), 50 µm (b). Sources: (a) Yasaka 2010.

Neurochemical classification

Neurons in laminae I–III of the dorsal horn express a wide variety of potential neurochemical markers: neuropeptides, neuropeptide receptors, calcium-binding proteins and miscellaneous other proteins – for example, neuronal nitric oxide synthase (nNOS) and the γ isoform of protein kinase C (PKCγ). It has been found that many of these substances are restricted to limited numbers of neurons with distinctive laminar distributions, and in some cases to either excitatory or inhibitory interneurons (Todd and Koerber, 2012). For example, among the neuropeptides, it has been reported that somatostatin, neurotensin, substance P, gastrin-releasing

Table 1.1 Expression of various neuropeptides and proteins by interneurons in laminae I–III of the dorsal horn. Some of the neuropeptides and proteins that are selectively distributed among inhibitory and excitatory interneurons in laminae I–III. Most of these are described in the text. The neurokinin 3 (NK3) receptor is expressed by specific types of inhibitory and excitatory interneuron, while the NPY Y1 is found on certain types of excitatory interneuron: neuronal nitric oxide synthase (nNOS) and protein kinase Cγ (PKCγ).

	Inhibitory (GABAergic)	Excitatory (glutamatergic)
Neuropeptides	Neuropeptide Y Galanin Enkephalin Dynorphin	Somatostatin Neurotensin Neurokinin B Substance P Gastrin-releasing peptide Enkephalin Dynorphin
Neuropeptide receptors	sst2A NK3	NK1 NK3 MOR-1 NPY Y1
Other proteins	nNOS Parvalbumin	nNOS Calbindin Calretinin Parvalbumin PKCγ

peptide and neurokinin B (NKB) are expressed by excitatory interneurons, neuropeptide Y (NPY) and galanin by inhibitory interneurons, and the opioid peptides dynorphin and enkephalin by both types (see Table 1.1). An additional complication here is that some of these peptides (substance P, somatostatin and galanin) are also present in peptidergic primary afferents. However, since all peptidergic afferents are thought to contain CGRP in the rat, this feature can be used to distinguish the axons of these interneurons from primary afferents in this species.

Neurochemical populations among the inhibitory interneurons

We have recently identified four non-overlapping populations among the inhibitory interneurons in laminae I–III of the rat, on the basis of expression of NPY, galanin, nNOS and parvalbumin (Polgar et al., 2013b) (Figure 1.4b). The galanin cells also contain the opioid peptide dynorphin, although this is additionally expressed by some excitatory interneurons in the dorsal horn (Sardella et al., 2011). Between them, these four populations account for at least half of the inhibitory interneurons in laminae I–II and for a smaller proportion of those in lamina III. A similar pattern has been observed in the mouse, except that in that case there is a moderate degree of overlap between nNOS and galanin/dynorphin populations (Iwagaki et al., 2013). There are differences in the distribution of these cell types, since the galanin/dynorphin-containing cells are concentrated in laminae I–IIo and the parvalbumin cells in laminae IIi–III, while the other two classes are present throughout laminae I–III.

We found that there were differences among these populations in their responses to noxious stimulation, as determined by expression of the transcription factor Fos (Hunt, Pini and Evan, 1987) or by the phosphorylation of extracellular signal-regulated kinases (ERKs) (Ji et al., 1999). Many of the galanin and NPY-expressing cells responded to noxious heat, as well as to the intradermal injection of capsaicin or formalin (Polgar et al., 2013b). Although nNOS-containing cells up-regulated Fos after noxious heat or formalin injection, they did not do so after injection of capsaicin. In contrast, the parvalbumin-positive cells were not apparently activated by any of these noxious stimuli.

Quantitative studies have shown that around 50 per cent of the inhibitory interneurons in laminae I-II possess the somatostatin receptor sst_{2A}. This appears to be the only receptor for somatostatin that is expressed by dorsal horn neurons; and it is virtually restricted to the inhibitory cells, which makes it a very convenient marker. In addition, there is evidence that the receptor is functional, since the application of somatostatin results in hyperpolarisation of these cells (Iwagaki et al., 2013, Yasaka et al., 2010). Interestingly, in both rat and mouse, nearly all of the galanin/dynorphin and nNOS-containing inhibitory interneurons were found to be included in the sst_{2A}-expressing population, while most NPY- and all parvalbumin-containing cells lacked sst_{2A}.

Two further pieces of evidence suggest distinct functions for these neurochemical classes of inhibitory interneurons. Hughes et al. (2012) have shown that parvalbumin-immunoreactive axons, presumably derived from the parvalbumin-containing inhibitory interneurons in laminae II–III, are closely associated with the central terminals of myelinated low-threshold mechanoreceptive afferents, with which they form axoaxonic synapses. Since the parvalbumin cells appear to receive inputs from the same types of afferent, their role may be to generate feedback presynaptic inhibition, and they are likely to be involved in maintaining tactile acuity. Ross et al. (2010) have reported that mice lacking the transcription factor Bhlhb5 develop exaggerated itch, but an almost completely normal pain phenotype. This was shown to be associated with loss of certain inhibitory interneurons from the superficial dorsal horn. We have recently found that the cells that are absent in the Bhlhb5 knock-out mouse correspond to the nNOS- and galanin/dynorphin-expressing populations, which between them account for around two thirds of sst_{2A}-expressing cells and for around one third of all inhibitory interneurons (Kardon et al., 2014). This loss appeared to be selective, since the remaining one third of sst_{2A}-expressing cells were still present and the numbers of sst_{2A}-negative inhibitory interneurons were unchanged. This suggests that either or both of the nNOS- and galanin/dynorphin-populations of inhibitory interneurons are involved in the suppression of itch. Since both types can be activated by noxious stimuli (Polgar et al., 2013b), both may be responsible for the scratch-mediated inhibition of itch.

Although in some cases individual neurochemical markers can be used to define discrete neuronal populations, caution is needed in interpreting immunocytochemical results. For example, both nNOS and dynorphin are found in some excitatory interneurons as well as in inhibitory cells. However, in each case the expression of sst_{2A} can be used to distinguish between these cell types, as the receptor is present on virtually all the inhibitory cells but on none of the excitatory ones. It is likely that in many cases a combinatorial approach will be necessary in order to identify populations of interneurons in this region.

Neurochemical populations among the excitatory interneurons

Considerably less is known about the organisation of neurochemical populations among the excitatory interneurons. Many of those in lamina II contain somatostatin: these are distributed throughout the lamina and include some vertical and radial cells (Yasaka et al., 2010). Two different peptides, neurokinin B and neurotensin, are found in mutually exclusive populations of excitatory interneurons in lamina IIi–III. It is known that there is overlap between the somatostatin cells and those that contain NKB or enkephalin, but not between the somatostatin and neurotensin cells. Another peptide– gastrin-releasing peptide (GRP) – is found in presumed excitatory interneurons in lamina II, and these are thought to be involved in itch pathways (Mishra and Hoon, 2013). However, the relation of this group to the other populations mentioned above is not yet known.

Although the NK1r is present on many ALT projection neurons in laminae I and III, it is also expressed by excitatory interneurons, particularly in lamina I. Antibodies directed against the μ-opioid receptor MOR-1 label a distinctive population of excitatory interneurons in the middle part of lamina II, but little is known about the extent to which these cells overlap with the peptide-containing classes described above.

Another protein that has been investigated in several studies is PKCγ, which is expressed by around 30 per cent of neurons in lamina IIi (Peirs et al., 2014), together with scattered cells dorsal and ventral to this. These neurons, which are innervated by myelinated low-threshold mechanoreceptors, have been implicated in the development of neuropathic pain (Lu et al., 2013, Malmberg et al., 1997). It is known that the PKCγ cells overlap extensively with those that express somatostatin, neurotensin or NKB.

Two calcium-binding proteins are present in large numbers of cells in laminae I–III, and most of these are excitatory interneurons. Some overlap between peptide-containing and calbindin-containing cells has been demonstrated.

Descending pathways

The superficial dorsal horn receives descending inputs from various brain areas such as the cortex, the ventromedial medulla (including the raphe nuclei), and the locus coeruleus and surrounding areas of the pons. There has been particular interest in the role of descending monoaminergic systems, because there is much evidence that these play a role in modulating pain transmission at the spinal level. The serotoninergic pathway originates in the nucleus raphe magnus and terminates throughout the dorsal horn, but at particularly high density in laminae I–IIo. Norepinephrinergic axons from the locus coeruleus have a generally similar pattern of termination. Because the monoamine transmitters are thought to act through volume transmission, it is probably more important to know about the distribution of their receptors than about exactly where the axons terminate. However, although several monoamine receptors have been identified in the dorsal horn, we still know relatively little about their expression by specific cell types.

It has been shown that some of the descending axons originating in the ventromedial medulla are GABAergic and/or glycinergic (Antal et al., 1996), but the functions of this pathway are not yet known.

Neuronal circuits

Information about the anatomical organisation of synaptic circuits in the dorsal horn has come from two main approaches. Immunocytochemical studies have been able to demonstrate contacts (and in many cases synapses) between axons belonging to one neurochemically defined population and the dendrites or cell bodies of retrogradely labelled projection neurons or of interneuron populations. Electrophysiological studies involving paired recordings have revealed functional synapses between different cell types. Interestingly, with the latter approach it has been reported that only around 10 per cent of randomly selected pairs of interneurons are synaptically linked, which suggests that there may be fairly selective patterns of synaptic connectivity in this region (Lu and Perl, 2003). The following sections discuss synaptic inputs to projection neurons, interneurons and primary afferent terminals. A diagram summarising some of these synaptic circuits is shown in Figure 1.5.

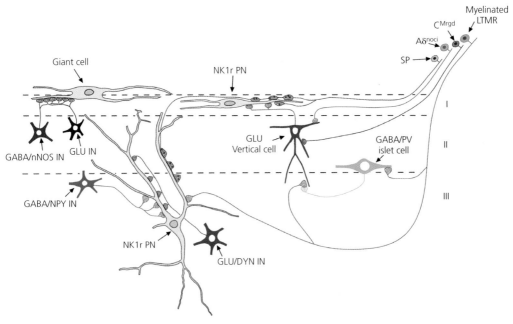

Figure 1.5 Neuronal circuits in the superficial dorsal horn. This diagram shows some of the synaptic connections that have been identified in laminae I–III of the rodent dorsal horn. Three ALT projection neurons are indicated: a lamina I giant cell and projection neurons (PN) in laminae I and III that express the neurokinin 1 receptor (NK1r). Both lamina I and lamina III NK1r-expressing cells receive numerous synapses from peptidergic primary afferents that contain subtance P (SP), and these account for around a half of their excitatory synaptic input. The lamina I NK1r+ PN receives excitatory synapses from glutamatergic (GLU) vertical cells in lamina II, which are thought to be innervated by Aδ nociceptors (Aδnoci), non-peptidergic C fibre nociceptors (CMrgd) and myelinated low-threshold mechanoreceptors (LTMR). The myelinated LTMRs also innervate GABAergic islet cells that contain parvalbumin (PV), and they receive axoaxonic synapses from these interneurons. The lamina III PNs are selectively innervated by two distinct classes of interneuron: inhibitory cells that express neuropeptide Y (NPY) and excitatory (glutamatergic) cells that express dynorphin. The giant lamina I projection neurons seem to receive little or no direct primary afferent input but are densely innervated by excitatory and inhibitory interneurons. Many of the latter contain neuronal nitric oxide synthase (nNOS).

Synaptic inputs to projection neurons
Inputs from primary afferents

The large NK1r-expressing ALT projection neurons in laminae I and III are densely innervated by substance P-containing primary afferents, and the latter account for over half of the glutamatergic synapses on these cells (Polgár, Al Ghamdi and Todd, 2010, Baseer et al., 2012). These cells therefore receive a powerful direct synaptic input from peptidergic nociceptors, although the peripheral targets of the afferents (e.g., skin, viscera) are not yet known. Since the lamina III cells have extensive dendritic trees that pass through several laminae, they could in principle be innervated by a variety of different types of primary afferent. However, while they receive some synapses from myelinated LTMRs, they seem to have little or no input from non-peptidergic C nociceptors (Todd, 2010).

Unlike these two groups of NK1r-expressing ALT cells, the giant lamina I projection neurons seem to receive little if any monosynaptic primary afferent input (Polgár et al., 2008).

Inputs from interneurons

The NK1r-expressing ALT cells in laminae I and III, and the giant lamina I cells, all receive numerous synapses from VGLUT2-immunoreactive boutons, and most of these probably originate from local excitatory interneurons. Some of those on the lamina I projection cells are presumably derived from the axons of lamina II vertical cells, which are known to innervate these cells (Cordero-Erausquin et al., 2009, Lu and Perl, 2005). We have recently shown that nearly 60 per cent of the non-primary glutamatergic synapses on the lamina III projection neurons contain preprodynorphin, and are therefore likely to be derived from local dynorphin-containing excitatory interneurons (Baseer et al., 2012). The fact that preprodynorphin is found in between 5 and 7 per cent of the VGLUT2$^+$ boutons in this region indicates a highly selective targeting of the projection cells by these interneurons.

There is also evidence that inhibitory interneurons innervate the projection cells in a very selective manner (Todd, 2010). The lamina III ALT cells are densely innervated by axons that contain NPY and GABA: over 30 per cent of the GABAergic boutons that synapse with these cells are NPY-immunoreactive, by comparison with less than 15 per cent among the general population of GABAergic boutons in laminae I–II. In contrast, the giant lamina I projection neurons are preferentially innervated by inhibitory interneurons that express nNOS.

Synaptic inputs to interneurons
Inputs from primary afferents

Electrophysiological studies in slice preparations have demonstrated that vertical and radial cells in lamina II can receive monosynaptic input from Aδ and C fibres and that this includes input from C fibres that possess TRPV1 and TRPA1 (Grudt and Perl, 2002, Yasaka et al., 2007, Uta et al., 2010). Although some of the myelinated primary afferent input to vertical cells is from Aδ nociceptors, it is also likely that these cells are innervated by myelinated LTMRs, because their dendrites often extend ventrally into the region where these afferents terminate. In support of this suggestion, we have recently found numerous contacts involving the central terminals of myelinated low-threshold afferents and dendritic spines of vertical cells in laminae IIi–III (Yasaka et al., 2014). Electrophysiological studies have also shown that islet cells receive monosynaptic input from C fibres but not from myelinated primary afferents. PKCγ-expressing excitatory interneurons in lamina IIi are directly innervated by Aβ primary afferents, which are presumably LTMRs (Lu et al., 2013).

Inputs from other interneurons

The interconnections between interneurons in lamina II have been examined in a series of elegant experiments in which pairs of neurons have been recorded simultaneously in spinal cord slices (Lu et al., 2013, Lu and Perl, 2003, 2005, Zheng, Lu and Perl, 2010), and these studies have identified a number of consistent patterns. Islet cells were found to be synaptically connected to another population of inhibitory interneurons, defined by expression of green fluorescent protein (GFP) under the control of the prion promoter (PrP-GFP mouse line). In some cases this connection was reciprocal: either cell could cause an inhibitory post-synaptic current (IPSC) in the other. It has since been shown that the PrP-GFP cells correspond to those that contain nNOS and/or galanin (Iwagaki et al., 2013). It was reported that these two types of inhibitory interneuron were presynaptic to different classes of excitatory interneuron in lamina II: the PrP-GFP cells to vertical cells, and islet cells to a class defined as 'transient central' cells. An excitatory circuit was identified in which transient central cells were presynaptic to vertical cells, which in turn innervated lamina I projection neurons. Finally it was shown that transient central cells could receive from Aβ afferents a disynaptic input that was transmitted by PKCγ cells in lamina IIi.

While these studies have provided remarkable insights into the synaptic connections between interneurons, some caution is needed when interpreting their findings. First, it is not known whether the populations described in these studies are really homogeneous. Secondly, it is very likely that each class of interneuron receives input from, and provides output to, several other neuronal populations. This means that attempting to track the flow of information through circuits involving several neurons that are connected in series becomes very difficult.

Synaptic inputs to primary afferents

Primary afferent boutons often receive axoaxonic synapses from GABAergic neurons that mediate presynaptic inhibition. Various types of afferent differ in the extent to which they receive axoaxonic synapses, and also in the source of these synapses. For example, peptidergic afferents have few if any axoaxonic synapses, while the latter are found in moderate numbers on non-peptidergic C nociceptors (Figure 1.2a) and in large numbers on myelinated LTMRs, in particular on the central terminals of Aδ D-hair afferents (Ribeiro-da-Silva, Tagari and Cuello, 1989, Ribeiro-da-Silva and Coimbra, 1982, Maxwell and Rethelyi, 1987) (Figure 1.2b). The cells giving rise to axoaxonic synapses on C fibres and Aδ D-hair afferents are likely to belong to different populations, as axons belonging to the former are enriched with GABA but not with glycine, while axons belonging to the latter often contain high levels of both amino acids (Todd, 1996). Hughes et al. (2012) have recently identified the parvalbumin-containing inhibitory interneurons (which correspond to a subset of the islet cells) as a source of axoaxonic synapses on the central terminals of the Aδ D-hair afferents, since around 80 per cent of parvalbumin axons in lamina IIi were associated with these afferents.

Future directions

Although our understanding of the neuronal organisation and synaptic circuitry of the dorsal horn has increased dramatically over recent years, there are still major gaps in our knowledge.

Classification of neurons into functional populations is an essential prerequisite for unravelling neuronal circuits, and some progress has been made in the identification of neurochemical populations among the inhibitory interneurons. This scheme needs to be expanded to include those cells that have not already been assigned to classes. For the excitatory interneurons in laminae I–III, it is clear that there are numerous neurochemical markers that are expressed by subsets of these cells, and future studies will need to clarify their interrelationships, for example by looking for mutually exclusive expression patterns. In addition, this information will have to be related to the morphological classes that have been identified among the excitatory interneurons (e.g. vertical and radial cells). Since many of these markers can be detected in axon terminals, it should be possible to identify patterns of the synaptic input that goes from excitatory interneurons to different classes of projection neuron and interneuron. It will also be important to define discrete populations among the projection neurons in lamina I, for example to identify those that are responsible for conveying different types of stimulus that are perceived as itch (Davidson et al., 2007).

While some progress has been made in defining synaptic circuitry, a vast amount remains to be done. Much of what we know concerns the synaptic inputs to projection neurons, which are relatively easy to investigate with conventional anatomical techniques, as they generally have few dendritic spines and a high density of synapses on their dendritic shafts and cell bodies. We need far more information on the synaptic inputs to interneurons, but this will depend on our ability to define functional populations among these cells. It will also be vital to have quantitative data in order to assess the relative strengths of different synaptic inputs. Another important outstanding issue is that of the origin of the axoaxonic synapses on nociceptors. It appears that the presynaptic inhibition of primary afferents is performed by distinct populations of inhibitory interneurons in the spinal cord (Hughes et al., 2005, Hughes et al., 2012). Identifying those that innervate non-peptidergic C nociceptors could reveal novel targets for analgesic drugs and may provide insight into pathological pain states.

One advantage of the neurochemical approach is that it provides genetic and pharmacological targets that can be used to investigate the functions of specific neuronal populations. For example, the intrathecal administration of saporin conjugated to peptides can be used to destroy cells that express the peptide receptors. If these receptors are associated with particular neuronal populations, the resulting behavioural outcomes can provide insight into the function of these cells (Wiley and Lappi, 2003). The neurochemical approach can also identify genetic markers that can be used, for example, to activate cells with the help of optogenetics or to silence or ablate them with toxins (Johansson et al., 2010).

References

Abraira, V. E. and Ginty, D. D. 2013. The sensory neurons of touch. *Neuron* 79: 618–39.

Andrew, D. 2009. Sensitization of lamina I spinoparabrachial neurons parallels heat hyperalgesia in the chronic constriction injury model of neuropathic pain. *Journal of Physiology* 587: 2005–17.

Antal, M., Petko, M., Polgár, E., Heizmann, C. W. and Storm-Mathisen, J. 1996. Direct evidence of an extensive GABAergic innervation of the spinal dorsal horn by fibres descending from the rostral ventromedial medulla. *Neuroscience* 73: 509–18.

Baseer, N., Polgár, E., Watanabe, M., Furuta, T., Kaneko, T. and Todd, A. J. 2012. Projection neurons in lamina III of the rat spinal cord are selectively innervated by local dynorphin-containing excitatory neurons. *Journal of Neuroscience* 32: 11854–63.

Bester, H., Chapman, V., Besson, J. M. and Bernard, J. F. 2000. Physiological properties of the lamina I spinoparabrachial neurons in the rat. *Journal of Neurophysiology* 83: 2239–59.

Brown, A. G. 1981. *Organization in the Spinal Cord: The Anatomy and Physiology of Identified Neurons*. Berlin: Springer.

Brumovsky, P., Watanabe, M. and Hokfelt, T. 2007. Expression of the vesicular glutamate transporters-1 and -2 in adult mouse dorsal root ganglia and spinal cord and their regulation by nerve injury. *Neuroscience* 147: 469–90.

Campbell, J. N., Raja, S. N., Meyer, R. A. and Mackinnon, S. E. 1988. Myelinated afferents signal the hyperalgesia associated with nerve injury. *Pain* 32: 89–94.

Cordero-Erausquin, M., Allard, S., Dolique, T., Bachand, K., Ribeiro-Da-Silva, A. and de Koninck, Y. 2009. Dorsal horn neurons presynaptic to lamina I spinoparabrachial neurons revealed by transynaptic labeling. *Journal of Comparative Neurology* 517: 601–15.

Davidson, S., Zhang, X., Yoon, C. H., Khasabov, S. G., Simone, D. A. and Giesler, G. J., Jr. 2007. The itch-producing agents histamine and cowhage activate separate populations of primate spinothalamic tract neurons. *Journal of Neuroscience* 27: 10007–14.

Grudt, T. J. and Perl, E. R. 2002. Correlations between neuronal morphology and electrophysiological features in the rodent superficial dorsal horn. *Journal of Physiology* 540: 189–207.

Han, Z. S., Zhang, E. T. and Craig, A. D. 1998. Nociceptive and thermoreceptive lamina I neurons are anatomically distinct. *Nature Neuroscience* 1: 218–25.

Hughes, D. I., Sikander, S., Kinnon, C. M., Boyle, K. A., Watanabe, M., Callister, R. J. and Graham, B. A. 2012. Morphological, neurochemical and electrophysiological features of parvalbumin-expressing cells: A likely source of axo-axonic inputs in the mouse spinal dorsal horn. *Journal of Physiology* 590: 3927–51.

Hughes, D. I., Mackie, M., Nagy, G. G., Riddell, J. S., Maxwell, D. J., Szabo, G., Erdelyi, F., Veress, G., Szucs, P., Antal, M. and Todd, A. J. 2005. P boutons in lamina IX of the rodent spinal cord express high levels of glutamic acid decarboxylase-65 and originate from cells in deep medial dorsal horn. *Proceedings of the National Academy of Sciences of the United States of America* 102: 9038–43.

Hunt, S. P., Pini, A. and Evan, G. 1987. Induction of c-fos-like protein in spinal cord neurons following sensory stimulation. *Nature* 328: 632–4.

Iwagaki, N., Garzillo, F., Polgár, E., Riddell, J. S. and Todd, A. J. 2013. Neurochemical characterisation of lamina II inhibitory interneurons that express GFP in the PrP-GFP mouse. *Molecular Pain* 9: 56. doi: 10.1186/1744-8069-9-56.

Ji, R. R., Baba, H., Brenner, G. J. and Woolf, C. J. 1999. Nociceptive-specific activation of ERK in spinal neurons contributes to pain hypersensitivity. *Nature Neuroscience* 2: 1114–9.

Johansson, T., Broll, I., Frenz, T., Hemmers, S., Becher, B., Zeilhofer, H. U. and Buch, T. 2010. Building a zoo of mice for genetic analyses: A comprehensive protocol for the rapid generation of BAC transgenic mice. *Genesis* 48: 264–80.

Kardon, A., Polgár, E., Hachisuka, E., Snyder, L., Cameron, D., Savage, S., Cai, X, Karnup, S., Fan, C., Hemenway, G. et al. 2014. Dynorphin acts as a neuromodulator to inhibit itch in the dorsal horn of the spinal cord. *Neuron* 82(3): 573–586.

Lawson, S. N., Crepps, B. A. and Perl, E. R. 1997. Relationship of substance P to afferent characteristics of dorsal root ganglion neurones in guinea-pig. *Journal of Physiology* 505 (Part 1), 177–91.

Light, A. R. and Perl, E. R. 1979. Spinal termination of functionally identified primary afferent neurons with slowly conducting myelinated fibers. *Journal of Comparative Neurology* 186: 133–50.

Lu, Y. and Perl, E. R. 2003. A specific inhibitory pathway between substantia gelatinosa neurons receiving direct C-fiber input. *Journal of Neuroscience* 23: 8752–8.

Lu, Y. and Perl, E. R. 2005. Modular organization of excitatory circuits between neurons of the spinal superficial dorsal horn (laminae I and II). *Journal of Neuroscience* 25: 3900–7.

Lu, Y., Dong, H., Gao, Y., Gong, Y., Ren, Y., Gu, N., Zhou, S., Xia, N., Sun, Y. Y., Ji, R. R. and Xiong, L. 2013. A feed-forward spinal cord glycinergic neural circuit gates mechanical allodynia. *Journal of Clinical Investigation* 123: 4050–62.

Malet, M., Vieytes, C. A., Lundgren, K. H., Seal, R. P., Tomasella, E., Seroogy, K. B., Hokfelt, T., Gebhart, G. F. and Brumovsky, P. R. 2013. Transcript expression of vesicular glutamate transporters in lumbar dorsal root ganglia and the spinal cord of mice: Effects of peripheral axotomy or hindpaw inflammation. *Neuroscience* 248C, 95–111.

Malmberg, A. B., Chen, C., Tonegawa, S. and Basbaum, A. I. 1997. Preserved acute pain and reduced neuropathic pain in mice lacking PKCgamma. *Science* 278: 279–83.

Mantyh, P. W., Demaster, E., Malhotra, A., Ghilardi, J. R., Rogers, S. D., Mantyh, C. R., Liu, H., Basbaum, A. I., Vigna, S. R., Maggio, J. E., et al. 1995. Receptor endocytosis and dendrite reshaping in spinal neurons after somatosensory stimulation. *Science* 268: 1629–32.

Maxwell, D. J. and Rethelyi, M. 1987. Ultrastructure and synaptic connections of cutaneous afferent fibres in the spinal cord. *Trends in Neurosciences* 10: 117–23.

Maxwell, D. J., Belle, M. D., Cheunsuang, O., Stewart, A. and Morris, R. 2007. Morphology of inhibitory and excitatory interneurons in superficial laminae of the rat dorsal horn. *Journal of Physiology* 584: 521–33.

Melzack, R. and Wall, P. D. 1965. Pain mechanisms: A new theory. *Science* 150: 971–9.

Mishra, S. K. and Hoon, M. A. 2013. The cells and circuitry for itch responses in mice. *Science* 340: 968–71.

Nichols, M. L., Allen, B. J., Rogers, S. D., Ghilardi, J. R., Honore, P., Luger, N. M., Finke, M. P., Li, J., Lappi, D. A., Simone, D. A. and Mantyh, P. W. 1999. Transmission of chronic nociception by spinal neurons expressing the substance P receptor. *Science* 286: 1558–61.

Peirs, C., Patil, S., Bouali-Benazzouz, R., Artola, A., Landry, M. and Dallel, R. 2014. Protein kinase C gamma interneurons in the rat medullary dorsal horn: Distribution and synaptic inputs to these neurons, and subcellular localization of the enzyme. *Journal of Comparative Neurology* 522: 393–413.

Polgár, E., Al Ghamdi, K. S. and Todd, A. J. 2010. Two populations of neurokinin 1 receptor-expressing projection neurons in lamina I of the rat spinal cord that differ in AMPA receptor subunit composition and density of excitatory synaptic input. *Neuroscience* 167: 1192–204.

Polgár, E., Durrieux, C., Hughes, D. I. and Todd, A. J. 2013. A quantitative study of inhibitory interneurons in laminae i-iii of the mouse spinal dorsal horn. *PLoS One* 8: e78309.

Polgár, E., Al-Khater, K. M., Shehab, S., Watanabe, M. and Todd, A. J. 2008. Large projection neurons in lamina I of the rat spinal cord that lack the neurokinin 1 receptor are densely innervated by VGLUT2-containing axons and possess GluR4-containing AMPA receptors. *Journal of Neuroscience* 28: 13150–60.

Polgár, E., Sardella, T. C., Tiong, S. Y., Locke, S., Watanabe, M. and Todd, A. J. 2013. Functional differences between neurochemically defined populations of inhibitory interneurons in the rat spinal dorsal horn. *Pain* 154: 2606–15.

Rexed, B. 1952. The cytoarchitectonic organization of the spinal cord in the cat. *Journal of Comparative Neurology* 96: 414–95.

Ribeiro-Da-Silva, A. and Coimbra, A. 1982. Two types of synaptic glomeruli and their distribution in laminae I–III of the rat spinal cord. *Journal of Comparative Neurology* 209: 176–86.

Ribeiro-Da-Silva, A., Tagari, P. and Cuello, A. C. 1989. Morphological characterization of substance P-like immunoreactive glomeruli in the superficial dorsal horn of the rat spinal cord and trigeminal subnucleus caudalis: A quantitative study. *Journal of Comparative Neurology* 281: 497–15.

Ross, S. E., Mardinly, A. R., Mccord, A. E., Zurawski, J., Cohen, S., Jung, C., Hu, L., Mok, S. I., Shah, A., Savner, E. M., Tolias, C., Corfas, R., Chen, S., Inquimbert, P., Xu, Y., Mcinnes, R. R., Rice, F. L., Corfas, G., Ma, Q., Woolf, C. J. and Greenberg, M. E. 2010. Loss of inhibitory interneurons in the dorsal spinal cord and elevated itch in Bhlhb5 mutant mice. *Neuron* 65: 886–98.

Sandkuhler, J. 2009. Models and mechanisms of hyperalgesia and allodynia. *Physiological Reviews* 89: 707–58.

Sardella, T. C., Polgár, E., Garzillo, F., Furuta, T., Kaneko, T., Watanabe, M. and Todd, A. J. 2011. Dynorphin is expressed primarily by GABAergic neurons that contain galanin in the rat dorsal horn. *Molecular Pain* 7: 76. doi: 10.1186/1744-8069-7-76.

Seal, R. P., Wang, X., Guan, Y., Raja, S. N., Woodbury, C. J., Basbaum, A. I. and Edwards, R. H. 2009. Injury-induced mechanical hypersensitivity requires C-low threshold mechanoreceptors. *Nature* 462: 651–5.

Sugiura, Y., Lee, C. L. and Perl, E. R. 1986. Central projections of identified, unmyelinated (C) afferent fibers innervating mammalian skin. *Science* 234: 358–61.

Todd, A. J. 1996. GABA and glycine in synaptic glomeruli of the rat spinal dorsal horn. *European Journal of Neuroscience* 8: 2492–8.

Todd, A. J. 2006. Spinal cord processing: Anatomy and neurochemistry of the dorsal horn. In *Handbook of Clinical Neurology*, 3rd series: *Pain*, ed. by F. Cervero and T. S. Jensen, vol. 81. Elsevier: New York, pp. 61–76.

Todd, A. J. 2010. Neuronal circuitry for pain processing in the dorsal horn. *Nature Reviews of Neuroscience* 11: 823–36.

Todd, A. J. and Koerber, H. R. 2012. Neuroanatomical substrates of spinal nociception. In *Wall and Melzack's Textbook of Pain*, ed. by S. McMahon, M. Koltzenburg, I. Tracey, and D. C. Turk, 6th edn. Edinburgh: Elsevier, pp. 73–90.

Todd, A. J., Puskar, Z., Spike, R. C., Hughes, C., Watt, C. and Forrest, L. 2002. Projection neurons in lamina I of rat spinal cord with the neurokinin 1 receptor are selectively innervated by substance P-containing afferents and respond to noxious stimulation. *Journal of Neuroscience* 22: 4103–13.

Todd, A. J., Hughes, D. I., Polgar, E., Nagy, G. G., Mackie, M., Ottersen, O. P. and Maxwell, D. J. 2003. The expression of vesicular glutamate transporters VGLUT1 and VGLUT2 in neurochemically defined axonal populations in the rat spinal cord with emphasis on the dorsal horn. *European Journal of Neuroscience* 17: 13–27.

Torsney, C. and MacDermott, A. B. 2006. Disinhibition opens the gate to pathological pain signaling in superficial neurokinin 1 receptor-expressing neurons in rat spinal cord. *Journal of Neuroscience* 26: 1833–43.

Uta, D., Furue, H., Pickering, A. E., Rashid, M. H., Mizuguchi-Takase, H., Katafuchi, T., Imoto, K. and Yoshimura, M. 2010. TRPA1-expressing primary afferents synapse with a morphologically identified subclass of substantia gelatinosa neurons in the adult rat spinal cord. *European Journal of Neuroscience* 31: 1960–73.

Wiley, R. G. and Lappi, D. A. 2003. Targeted toxins in pain. *Advanced Drug Delivery Reviews* 55: 1043–54.

Woodbury, C. J. and Koerber, H. R. 2003. Widespread projections from myelinated nociceptors throughout the substantia gelatinosa provide novel insights into neonatal hypersensitivity. *Journal of Neuroscience* 23: 601–10.

Yaksh, T. L. 1989. Behavioral and autonomic correlates of the tactile evoked allodynia produced by spinal glycine inhibition: Effects of modulatory receptor systems and excitatory amino acid antagonists. *Pain* 37: 111–23.

Yasaka, T., Tiong, S. Y. X., Hughes, D. I., Riddell, J. S. and Todd, A. J. 2010. Populations of inhibitory and excitatory interneurons in lamina II of the adult rat spinal dorsal horn revealed by a combined electrophysiological and anatomical approach. *Pain* 151: 475–488.

Yasaka, T., Tiong, S. Y. X., Polgár, E., Wanabe, M., Kumamoto, E., Riddell, J. S. and Todd, A. J. 2014. A putative relay circuit providing low-threshold mechanoreceptive input to lamina I projection neurons via vertical cells in lamina II of the rat dorsal horn. *Molecular Pain* 10: 3. doi: 10.1186/1744-8069-10-3.

Yasaka, T., Kato, G., Furue, H., Rashid, M. H., Sonohata, M., Tamae, A., Murata, Y., Masuko, S. and Yoshimura, M. 2007. Cell-type-specific excitatory and inhibitory circuits involving primary afferents in the substantia gelatinosa of the rat spinal dorsal horn in vitro. *Journal of Physiology* 581: 603–18.

Zeilhofer, H. U., Wildner, H. and Yevenes, G. E. 2012. Fast synaptic inhibition in spinal sensory processing and pain control. *Physiological Reviews* 92: 193–235.

Zeilhofer, H. U., Studler, B., Arabadzisz, D., Schweizer, C., Ahmadi, S., Layh, B., Bosl, M. R. and Fritschy, J. M. 2005. Glycinergic neurons expressing enhanced green fluorescent protein in bacterial artificial chromosome transgenic mice. *Journal of Comparative Neurology* 482: 123–41.

Zheng, J., Lu, Y. and Perl, E. R. 2010. Inhibitory neurones of the spinal substantia gelatinosa mediate interaction of signals from primary afferents. *Journal of Physiology* 588: 2065–75.

Zylka, M. J., Rice, F. L. and Anderson, D. J. 2005. Topographically distinct epidermal nociceptive circuits revealed by axonal tracers targeted to Mrgprd. *Neuron* 45: 17–25.

CHAPTER 2

Spinal plasticity of the nociceptive system: The role of central sensitisation in chronic pain states

Alban Latrémolière

Introduction

Pain is critical to adequately protecting an organism from injuries. An illustration of this principle is the 'classic' bite or burning of the tongue when drinking hot coffee or tea after a visit to the dentist in which local anaesthetics were used. With no proper feedback from our environment we cannot stop damage to ourselves until it is too late and a tissue lesion has already occurred. A more dramatic example is provided by patients who suffer from hereditary sensory and autonomic neuropathies that make them chronically insensitive to pain (Indo et al., 1996; Cox et al., 2006; Cox et al., 2010). In the course of their lives, these patients undergo massive traumas mostly without realising it, endangering their well-being and general health, as fractures go unchecked and open wounds untreated (Bennett and Woods, 2014) – something that can become life-threatening.

At the other end of the spectrum there are congenital conditions that render individuals hypersensitive to pain or put them under constant pain, which severely deteriorates the quality of their life (Bennett and Woods, 2014). Chronic pain hypersensitivity syndromes can have various aetiologies, from peripheral nerve trauma to viral infection; and they affect millions of people worldwide (Breivik et al., 2006; Torrance et al., 2006; Van Hecke et al., 2014). These syndromes occur around an estimated median age of fifty and, while they affect both genders, they are more prevalent in women when all chronic pain syndromes are pooled together (Breivik et al., 2006; Torrance et al., 2006). As life expectancies have significantly been extended in the past few decades, chronic pain syndromes typically affect working-age populations and represent therefore a major public health issue.

Synaptic plasticity within the spinal cord, the first integration centre of the pain information, can significantly alter the nature of the stimuli perceived. More than a simple anatomical relay, the spinal cord is the site of a very dynamic integration of somatosensory inputs. In normal conditions noxious and innocuous stimuli are processed through distinct pathways. This separation is in part due to an extremely complex and highly adaptive network within the spinal cord. The functional state of a spinal nociceptive neuron in particular is the result

An Introduction to Pain and its Relation to Nervous System Disorders, First Edition. Edited by Anna A. Battaglia.
© 2016 John Wiley & Sons, Ltd. Published 2016 by John Wiley & Sons, Ltd.

of the balance between numerous excitatory and inhibitory inputs. A change in this balance will in turn cause numerous changes in the functional properties of the spinal nociceptive neurons, thereby altering how these neurons respond not only to noxious stimuli, but also to previously sub-threshold stimuli. For example, a sustained intense noxious stimulation triggers a state of hyperexcitability of the spinal neuronal circuitry known as activity-dependent central sensitisation (Woolf, 1983; Latrémolière and Woolf, 2009; Woolf, 2010). When in a state of central sensitisation, spinal nociceptive neurons display one or all of the following features: an increase in excitability and response to noxious stimuli; an enlargement of their receptor field; and an ability to respond to innocuous stimuli, which means that they shift from a nociceptive-specific state to a functional state where they are activated non-selectively by a wide variety of stimuli. The phenomenon of central sensitisation has provided a mechanistic explanation for many of the temporal, spatial and threshold changes in pain sensibility in various chronic clinical pain settings. Therefore, if we want to develop new and more effective therapeutic strategies for treating chronic pain symptoms, it is critical that we understand the triggers and mechanisms responsible for this switch in the somatosensory system from the physiological state, in which the sensory experiences evoked by low-intensity stimuli (innocuous sensations) and by noxious stimuli (pain) are distinct and separate, to a dysfunctional hypersensitive system in which this discrimination is lost.

In this chapter we will first explain how a noxious stimulus is transmitted from the periphery to the central nervous system (CNS) and how this information can be modified at the spinal level through central sensitisation. We will discuss the cellular basis of this phenomenon and its consequences for pain perception. We will then describe the molecular triggers for central sensitisation in physiological states and the mechanisms responsible for the shift in the excitatory/inhibitory balance that occurs. In a third part we will examine how central sensitisation plays a fundamental role in the development and maintenance of pain hypersensitivity in two chronic pain syndromes: chronic inflammatory pain and neuropathic pain. Finally in the last section we will briefly discuss the possible involvement of this state in other diseases associated with abnormal pain symptoms, and will conclude with a short discussion about how it is conceivable that various perturbations within the CNS promote a central sensitisation state.

Nociceptive pain and nociception

Nociceptive pain is the physiological sensation caused by the activation of neural pathways by stimuli of sufficient intensity to potentially lead to tissue damage (for example a burn, a bite or excessive pressure). The detection per se of these stimuli is named 'nociception' (from the Latin *nocere*, 'to do harm/be harmful' and *cepi < capere*, 'to catch, receive') and is a critical protective process that helps prevent injury by generating both a reflex withdrawal from the stimulus and an unpleasant sensation (pain) that promotes avoidance of any further contact with such stimuli. Nociception occurs through highly specialised sensory fibres called nociceptors whose cell bodies are located in the trigeminal ganglia (TG) in the cephalic region and in the dorsal root ganglia (DRG) in extracephalic regions. The sensory neurons detecting innocuous mechanical and thermal stimuli are also located in the TGs and DRGs. All sensory neurons have a characteristic pseudo-unipolar morphology. Upon leaving the soma, the same

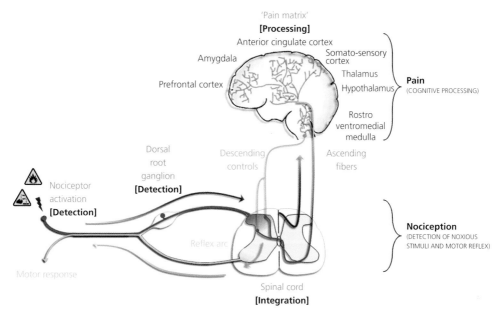

Figure 2.1 Nociception and pain. Nociception corresponds to the detection of a potentially harmful stimulus and the reflex motor response to withdraw a body region from the source of potential danger. Noxious stimuli are detected by specialized nerve fibres called nociceptors, whose cell bodies are located in the dorsal root ganglia. These fibres have a pseudo-unipolar morphology, where one branch of the axon goes to the periphery and the other contacts the second-order neurons in the spinal cord. The withdrawal motor response is triggered through a reflex arc in the spinal cord (marked in yellow). The excitability state of the projection nociceptive spinal neurons can be modulated by local (spinal) and supraspinal controls (the descending controls shown in green, which originate from the midbrain). As a result, the information sent to the brain for processing through the ascending fibres is the result of the integration of many signals that occur within the spinal circuitry. In the brain, several structures, collectively named the 'pain matrix', are involved in the cognitive processing of the nociceptive information that will eventually lead to the pain sensation – a complex experience, subjective in nature. The integration of the sensory signals in the dorsal horn of the spinal cord defines the nature of the nociceptive information sent to the brain, ultimately determining how much pain we will feel or for how long.

axon forms two branches, describing a 'T' shape. One branch innervates the peripheral tissues, where it detects environmental stimuli, while the other branch contacts the neurons within the CNS in order to transmit the neural information (Figure 2.1).

Nociceptors can be divided into two main classes on the basis of their size: Aδ nociceptors and C fibres nociceptors. Aδ nociceptors have a medium-diameter cell body and produce a thinly myelinated axon, allowing for a relatively fast transmission of the neural information (4–30 m/s). They are modality-specific (monomodal), so their activation leads to a well-localised and well-defined 'fast' pain sensation (Julius and Basbaum, 2001). C fibres are unmyelinated, slow-conducting (0.5–2 m/s) sensory fibres that originate in small-diameter neurons (Julius and Basbaum, 2001). They represent 60–90 per cent of all sensory neurons, depending on the species, and can be activated by stimuli of various modalities (thermal, mechanical or chemical); for this reason they are referred to as polymodal (Julius and Basbaum, 2001). Their activation leads to a 'slow' and diffuse pain sensation. Nociceptive C fibres can further be divided into two major classes on the basis of their cellular content: peptidergic and non-peptidergic. Peptidergic

C fibres produce substance P (SP) and calcitonin-related gene peptide (CGRP), two neuropeptides that trigger a massive vasodilatation at the periphery when they are released upon nociceptor stimulation (Holzer, 1988) and cause long-term depolarisation of the second-order neurons when they are released in the spinal cord. Peptidergic C fibres are sensitive to nerve growth factor (NGF), which binds to its tropomyosin receptor kinase A (TrkA) receptor; notably they express the ion channel – the transient receptor potential cation channel member 1 (TRPV1) – which is activated by capsaicin (the active compound found in red hot chili peppers), by acid pH and by heat (above 44°C). These peptidergic fibres are critical to the transduction of noxious heat stimuli (Cavanaugh et al., 2009), which explains why their activation by capsaicin causes a burning pain sensation. Non-peptidergic C fibres also express specific factors such as the ligand-gated ion channel purinergic receptor P2X3 or the G-protein-coupled receptor MRGPRD and are dependent on glia-derived neurotrophic factor (GDNF) (Molliver et al., 1997). Their role is not fully understood, but evidence suggests that they participate in the transduction of mechanical noxious stimuli (Cavanaugh et al., 2009). Recent advances in gene expression profiling from single-cell RNA sequencing have allowed to further define subclasses of sensory neurons on the basis of the pattern of the major genes they express (Usoskin et al., 2014).

The first relay for the transmission nociceptive information is the dorsal horn of the spinal cord, which displays a laminar organisation, mostly based on the fibres that contact the spinal neurons (Figure 2.2). Schematically, lamina I is contacted by both Aδ nociceptors and peptidergic C fibres – both nociceptors and fibres that detect innocuous temperatures (4 to 44°C). Lamina II is a region extremely rich in interneurons that can be further divided into two subregions. The outer region, named lamina IIo, is contacted mostly by peptidergic C fibres, whereas the inner region, called lamina IIi, is almost exclusively contacted by non-peptidergic C fibres. A subpopulation of excitatory interneurons that express the protein kinase C gamma (PKCγ) isoform (Malmberg et al., 1997) are located in the most ventral part of lamina IIi and are the target of non-nociceptive fibres only (Neumann et al., 2008). Laminae III and IV do not receive inputs from nociceptive fibres, but mostly from large-diameter myelinated fibres that convey fine tactile information (Aβ fibres). Finally lamina V, populated by convergent neurons (also named wide dynamic-range neurons), is contacted by Aβ fibres, Aδ nociceptors, and some – but only a few – peptidergic C fibres (Figure 2.2). There are relatively few ascending projection neurons that transfer the nociceptive information to the brain, and these are located mainly in laminae I and V (Gauriau and Bernard, 2002), although a few projection neurons have also been identified in lamina III (Todd, 2010). These projection neurons are under numerous modulatory inputs from local excitatory and inhibitory interneurons scattered throughout the dorsal spinal cord with a high density in lamina II (the region dominated by nociceptor input), as well as from descending controls originating in supraspinal structures (Millan, 2002). The fine anatomical organisation of the connectivity network within the spinal cord is not yet fully elucidated. See A Todd in this book for a comprehensive description and analysis of the anatomy of the spinal somatosensory pathways with particular emphasis on neuronal populations and their synaptic connections. What is clear, however, is that the signals sent by the projection neurons that will ultimately determine in the brain how much pain we feel in response to what stimuli and for how long are the result of the integration of these local and distant excitatory and inhibitory synaptic

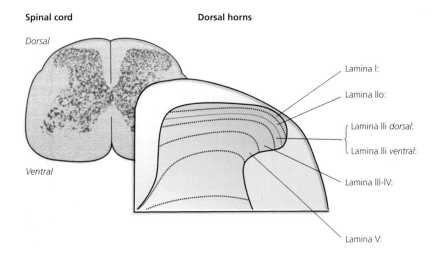

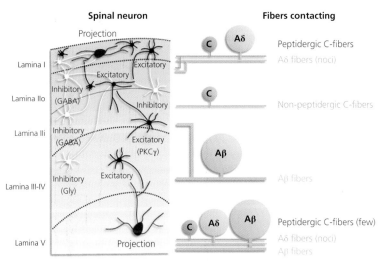

Figure 2.2 Laminar organization of the dorsal horn of the spinal cord, types of neurons and contacting sensory fibres. **Top Panel**: anatomy of the spinal cord at the lumbar level. In blue are the cell somata (bodies) that form the 'grey matter', surrounded by ascending and descending fibres tracks (the 'white matter'). The dorsal horns are organized in lamina, on the basis of their fine anatomy (density and size of neurons) and the fibres that contact them. Laminar boundaries are marked by dotted lines. **Bottom Panel**: there are few projection nociceptive neurons, and they are located mainly in laminae I and V (projection neurons are shown in red). Lamina II, the region dominated by nociceptor input, consists of excitatory glutamatergic (back) and inhibitory GABAergic (white) interneurons. Interneurons that can modulate the projection nociceptive neurons are also found in laminae III–IV, notably the glycinergic inhibitory interneurons. Laminae I and IIo are contacted by peptidergic C fibre nociceptors (in red) and by Aδ nociceptors (in blue), whereas the dorsal part of lamina IIi is only contacted by non-peptidergic C fibres (in purple). The excitatory interneurons of the ventral part of lamina IIi that specifically express the PKCγ isoform are contacted by a small contingent of Aβ fibres (in green), whereas most of these fibres project into laminae III–IV (non-nociceptive lamina). Projection neurons of lamina V (known as 'dynamic wide range' or 'convergent' neurons) are contacted by a small number of peptidergic C fibres, but mostly by Aδ and Aβ fibres.

inputs (Millan, 2002; Torsney and MacDermott, 2006). In other words, the pain we experience is not solely defined by the sensory fibres that are activated but ultimately by the nature of the integrated signal emitted by the spinal nociceptive projection neurons. The state of excitability of the spinal projection neuron is therefore critical to a proper discrimination of noxious stimuli. Altering this state will change how peripheral signals are integrated and will ultimately determine which stimuli are considered painful and which ones are not.

Synaptic plasticity of the nociceptive system

A fundamental characteristic of the nociceptive system is that it can be sensitised, notably after repeated or sustained noxious stimulations or in the presence of pro-inflammatory factors. The sensitisation of nociceptors is called peripheral sensitisation (Gold and Gebhart, 2010; Basbaum et al., 2009) and plays a major role in external injuries where there is a local inflammation at the wound site (Basbaum et al., 2009). Numerous immune cells and the injured epithelial cells release in concert pro-inflammatory factors such as NGF, serotonin (5-HT), prostaglandins, bradykinin, histamine, adenosine tri-phosphate (ATP) and protons (tissue acidosis) collectively referred to as an inflammatory soup (Kessler et al., 1992), which sensitise the nociceptors and lower their activation threshold (Woolf and Ma, 2007). The sensitised nociceptive fibres contribute to symptoms such as heat hyperalgesia, an increased pain response to a noxious heat stimulus; heat allodynia, when an innocuous temperature causes a sensation of pain; and hypersensitivity to pressure applied directly at the site of the wound (Woolf and Ma, 2007). This increased sensitivity is restricted to nociceptors that are in contact with the inflammatory agents. For example, the intradermal injection of NGF at low doses causes heat and pressure hyperalgesia, in line with the effects of nociceptive C fibres sensitisation, whereas fine tactile sensations mediated by Aβ fibres are not affected (Dyck et al., 1997).

The enhancement in the functional status of nociceptive neurons can occur directly, at the spinal cord level, and is then referred to as central sensitisation (Latrémolière and Woolf, 2009, 2010). This form of plasticity is prominent in projection nociceptive neurons of lamina I in the dorsal horn of the spinal cord (Cook, Woolf and Wall, 1986; Woolf and King, 1990; Dougherty and Willis, 1992; Lin et al., 1997; Mantyh et al., 1997; Lin et al., 1999) but has also been detected in projection neurons located in lamina V (Pezet et al., 2008; Roh et al., 2008) and in the spinal nucleus *pars caudalis* (Sp5c) (Burstein and Jakubowski, 2004). Central sensitisation is characterised by three main functional features: an increase in the excitability of spinal nociceptive neurons and in their response to noxious stimuli; the enlargement of their receptor field; and their ability to respond to innocuous stimuli (Latrémolière and Woolf, 2009). The receptor field of a nociceptive spinal neuron is defined by the sensory inputs that can produce a significant action potential output from this cell. In other words it corresponds to the sensory fibres whose activation will systemically lead to the generation of a pain signal from the nociceptive neuron and will therefore define the area of peripheral tissue 'probed' by the spinal nociceptive neuron. In normal conditions, the nociceptive-specific neurons of lamina I are mostly activated by large mono- and polysynaptic inputs from nociceptors (Dahlhaus, Ruscheweyh and Sandkuhler, 2005; Torsney and MacDermott, 2006), but they also receive inputs from nociceptors outside of their receptive field, as well as polysynaptic inputs from low-threshold afferents (Aβ fibres; Torsney and MacDermott, 2006). These additional inputs cause a small amplitude response from the cell, but they are insufficient to drive a functional output on their own. They represent structural but functionally silent connections between the nociceptive neurons and the sensory fibres (Figure 2.3). Their recruitment during central sensitisation is the substrate for receptive field plasticity and provides a mechanistic explanation for pain symptoms such as secondary hyperalgesia and mechanical allodynia (Latrémolière and Woolf, 2009). Secondary hyperalgesia is defined by an increased pain sensitivity in territories adjacent to the site of injury. This phenomenon is not caused by the local diffusion of inflammatory mediators, as it persists even after local blood flow is blocked (Lamotte et al., 1991). Blocking the nociceptive fibres located outside the injury site abolishes this symptom, indicating that it is mediated through the recruitment of novel nociceptors (Figure 2.3). Similarly, mechanical allodynia, defined by the ability of innocuous stimuli such as a light touch or a gentle brush of the skin to cause pain, corresponds to the recruitment of polysynaptic inputs from Aβ fibres that are now sufficient to activate potentials through the nociceptive neurons (Torebjork, Lundberg and Lamotte, 1992; Koltzenburg, Torebjork and Wahren, 1994; Woolf and Doubell, 1994; Ziegler et al., 1999) (Figure 2.3). Although mechanical allodynia is one of the major abnormal pain sensations experienced in chronic pain diseases, it also occurs after minor and transient injuries, where it serves a protective purpose. For example, the development of mechanical allodynia in the irradiated area after sunburn is caused by the central sensitisation of spinal nociceptive neurons (Gustorff et al., 2004; Gustorff et al., 2013). This hypersensitivity to such low-threshold stimuli forces the organism to adapt its global behaviour so as to fully isolate and protect the injured area until recovery. When recovery has occurred, the nociceptive system returns to its original baseline levels, where high threshold stimuli are again required for causing nociceptive pain.

Homosynaptic versus heterosynaptic potentiation

Because the activation of Aβ fibres does not normally trigger central sensitisation (Woolf, 1983) but can activate nociceptive neurons once they have been sensitised (Thompson, Woolf and Sivilotti, 1993), this represents a form of heterosynaptic potentiation (Ji et al., 2003; Latrémolière and Woolf, 2009), where activity in one synapse enhances activity in previously non-activated synapses, typically by 'sensitising' the entire neuron (Figure 2.4). This contrasts with the phenomenon of homosynaptic potentiation, which corresponds to an use-dependent facilitation of one synapse evoked by the activation of that same synapse, and is referred to as long-term potentiation (LTP) (Bliss and Collingridge, 1993) (Figure 2.4). At the spinal cord level, homosynaptic facilitation is observed in lamina I projection neurons after C fibres stimulation (Ikeda et al., 2003; Ikeda et al., 2006; Drdla and Sandkuhler, 2008). While LTP between C fibres nociceptors and projection nociceptive neurons provides a mechanistic explanation for behavioural manifestations such as hyperalgesia, it cannot account for the expansion of nociceptive receptor fields or for the recruitment of innocuous fibres, two major features of central sensitisation (Figure 2.4). The mechanisms underlying the heterosynaptic potentiation observed during central sensitisation are not yet understood. They could involve intrinsic properties of spinal nociceptive projection neurons or be the consequence of specific anatomical characteristics of

Normal states:

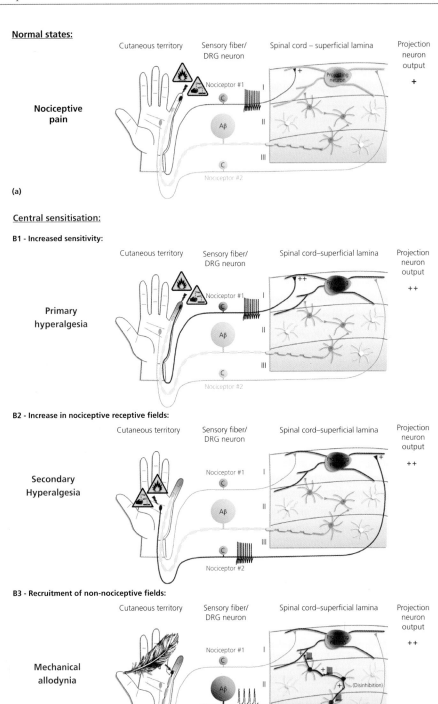

(a)

Central sensitisation:

B1 - Increased sensitivity:

B2 - Increase in nociceptive receptive fields:

B3 - Recruitment of non-nociceptive fields:

(b)

the spinal nociceptive circuitry (Latrémolière and Woolf, 2009, 2010; Sandkuhler, 2010; Naka, Gruber-Schoffnegger and Sandkuhler, 2013). At a more integrated level, the fact that spinal nociceptive neurons are prone to developing heterosynaptic facilitation when stimulated represents a fundamental difference from the homosynaptic facilitation exhibited by most of the other neurons of the CNS – like the LTP produced by hippocampal neurons, which is thought to underlie memory formation (McGaugh, 2000; Goosens and Maren, 2002; Collingridge, Isaac and Wang, 2004; Malenka and Bear, 2004). Memory is a convergent neural operation where the storage of defined information is the consequence of coincident temporal and spatial association between specific stimuli. This convergent phenomenon is critical for the establishment of memory engrams: each specific stimulus needs to be encoded in its context, together with other stimuli that occurred at the same time. For one combination of events – all encoded by specific synapses – one unique engram is formed. In contrast, central sensitisation represents a divergent neural operation, where a conditioning nociceptive input triggers diffuse changes such that any other unstimulated synaptic inputs can now generate pain.

To be triggered, central sensitisation requires input from many C fibres nociceptors over tens of seconds; a single stimulus, such as a pinch, is insufficient (Woolf, 1983). Whereas peripheral tissue injury is not necessary to trigger central sensitisation, in effect the degree of noxious stimulation required almost always produces frank tissue injury. When such injury occurs, the role of the nociceptive system, now sensitised, is perhaps equally important as its preventive detection function. Central sensitisation enhances the overall defense role of the nociceptive system by placing the organism into a hyperalert state where any stimulus applied on the site of injury or in its close vicinity is perceived as a potential threat and triggers an avoidance reflex. Central sensitisation therefore represents an additional level of protection for an injured organism, one that is activated when the first line of defense – the detection of potentially harmful stimuli – has failed and damage has occurred.

Figure 2.3 Characteristics of a nociceptive neuron in a state of central sensitisation. **(a)** Schematic example of a projection neuron of lamina I (in red) contacted by a C fibre nociceptor #1 (in red) that can activate the nociceptive neuron when stimulated on its innervation territory (last digit). This nociceptor contributes to the receptive field of the projection neuron. C-fibre nociceptor #2 (in orange) innervates another cutaneous territory (middle of the hand) and contacts the projection neuron. Stimulation of nociceptor #2 is insufficient to activate the projection neuron and does not contribute to its receptive field. The Aβ myelinated fibre (in green) projects to lamina III; its stimulation does not activate the projection neuron. The activation of the projection nociceptive neuron is specific to noxious stimuli that activate the nociceptor 1. **(b)** Three main characteristics define a neuron in a state of central sensitisation: an increased sensitivity to nociceptive stimuli, an enlargement of the receptive fields and the recruitment of non-nociceptive fibres. **B1**: The stimulation of the nociceptor #1 now strongly activates the projection neuron in a state of central sensitisation (in bright red), causing primary hyperalgesia (a noxious stimulus causes increased pain response). **B2**: The activation of nociceptor #2 is now sufficient to activate the sensitized projection neuron, which has therefore 'extended' its receptive field. Because nociceptor #2 is outside of the normal receptive field of the projection neuron, the pain caused by its activation is referred to as secondary hyperalgesia. **B3**: The activation of the Aβ fibre by an innocuous stimulus can activate the projection neuron through a polysynaptic pathway from lamina III to lamina I (shown in red). This polysynaptic pathway is revealed through the phenomenon of disinhibition and causes allodynia (when an innocuous stimulus, such as a light touch represented by the feather, causes pain). In this state the projection nociceptive neuron is therefore not 'pain-specific': both noxious and innocuous stimuli activate it, ultimately causing pain.

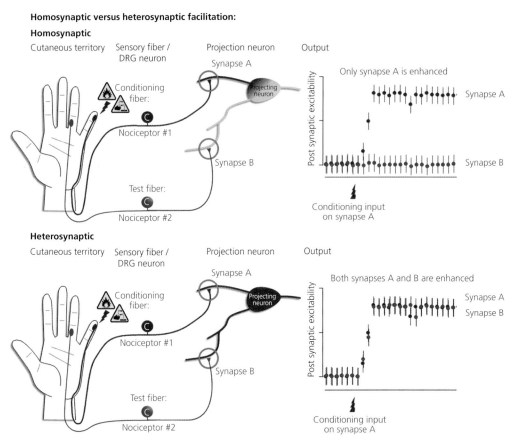

Figure 2.4 Schematic representation of a homosynaptic versus a heterosynaptic potentiation. **Top panel**: homosynaptic potentiation is a form of use-dependent facilitation where the conditioning fibre (nociceptor #1, in red) potentiates the strength of the same synapse (synapse A). The non-conditioned synapse (synapse B) is not potentiated, so the stimulation of the test fibre (nociceptor #2, in blue) does not change the output of the projection neuron. This type of potentiation is commonly called 'long-term potentiation' (LTP). LTP-like homosynaptic potentiation can contribute to primary hyperalgesia: nociceptor #1 causes more activation upon stimulation. **Bottom panel**: heterosynaptic potentiation represents a form of activity-dependent facilitation in which activity in the conditioning fibre (nociceptor #1, in red) potentiates all synapses of the neuron (synapses A and B). Stimulation of the test fibre (nociceptor #2, in blue), which has not been stimulated before, can drive an increased output from the projection neuron. Heterosynaptic potentiation is responsible for the major sensory manifestations of central sensitisation: pain in response to low-threshold afferents (allodynia) and spread of pain sensitivity to non-injured areas (secondary hyperalgesia).

Mechanisms of central sensitisation

Central sensitisation can be triggered by a strong gain in nociceptive inputs or by a reduction of the endogenous inhibitory brake on the spinal nociceptive neurons. Typically any intense and sustained input from C-nociceptors is sufficient for triggering central sensitisation. In experimental settings this can be provoked by a sustained hot stimulation (49°C for 10s) or by selectively activating nociceptive fibres without causing actual peripheral damage – for example by recruiting C fibres through electrical stimulation (1 Hz for 10–20s) (Woolf and Thompson, 1991), or through chemical activation by irritant compounds such as allyl isothiocyanate (mustard oil) and formalin, which both act through the TRPA1 channel (Jordt et al., 2004; McNamara et al., 2007), as well as by capsaicin (Simone, Baumann and Lamotte, 1989; LaMotte et al., 1991), which activates TRPV1 channels (Caterina et al., 1997). Triggering central sensitisation by a nociceptive barrage is referred to as activity-dependent central sensitisation.

Key excitatory neurotransmitters and receptors

Glutamate is the main transmitter of primary afferent neurons and binds to several receptors on postsynaptic neurons in the dorsal horn of spinal cord, including ionotropic amino-3-hydroxy-5-methyl-4-isoxazole propionate (AMPA), N-methyl-D-aspartate (NMDA) and several G-protein-coupled glutamate receptor subtypes (mGluR). In the superficial lamina of the dorsal horn the three types of receptors are present in almost every synapse. Typically AMPA receptors (AMPARs) and NMDA receptors (NMDARs) are present at the core of the synapse, and metabotropic receptors sit at the periphery (Alvarez et al., 2000). In basal conditions, excitatory neurons of lamina I/II appear to express mostly the low Ca^{2+}-permeable GluR2 AMPAR subunits, whereas the inhibitory interneurons appear to preferentially express the highly Ca^{2+}-permeable GluR1 AMPAR subunits (Kerr, Maxwell and Todd, 1998; Tong and MacDermott, 2006). NMDARs are expressed in virtually all glutamatergic synapses in the dorsal horns of the spinal cord (Antal et al., 2008) and constitute a complex formed by two mandatory NR1 subunits associated with two NR2A, NR2B or NR2C subunits (Stephenson, Cousins and Kenny, 2008). The distribution of NMDAR heteromers is relatively lamina-specific, which suggests distinct physiological roles for the different subtypes: NR2A subunits are heavily found in laminae III–IV (Nagy, Watanabe, et al., 2004); NR2B are highly expressed in laminae I–II (Momiyama, 2000; Nagy, Watanabe, et al., 2004); and NR2D are mostly found in lamina I neurons (Hildebrand et al., 2014). Next to these ionotropic receptors sit the slower-acting metabotropic glutamate receptors. Most of the eight existing isoforms are expressed in the spinal cord (Valerio et al., 1997; Berthele et al., 1999; Alvarez et al., 2000), but group I mGluRs (mGluR1 and 5), coupled with $G\alpha q$-proteins, have a prominent role in pain processing and central sensitisation (Neugebauer, Chen and Willis, 1999; Karim, Wang and Gereau, 2001; Price et al., 2007).

At baseline states the NMDAR channel is blocked by a magnesium (Mg^{2+}) ion sitting in the receptor pore (Mayer, Westbrook and Guthrie, 1984). Acute pain responses caused by the detection of intense thermal, mechanical or noxious chemicals (nociceptive pain) do not require the activation of the NMDARs (South et al., 2003), which suggests that they are mostly mediated by the activation of the AMPARs. Activation of the NMDARs is, however, essential to any form of synaptic plasticity in spinal nociceptive neurons. Blocking the NMDARs through pharmacological (Woolf and Thompson, 1991; Ma and Woolf, 1995) or genetic tools (South et al., 2003) fully prevents any form of synaptic plasticity of the nociceptive system. NMDAR blockade by Mg^{2+} is voltage-dependent and can only be removed when the membrane potential is sufficiently depolarised (Mayer et al., 1984). Such depolarisation typically occurs after a very intense or prolonged noxious stimulus and involves the activation of the AMPARs through glutamate effects as well as through other key neurotransmitters.

Most of the nociceptive fibres that contact projection spinal neurons express the neuropeptide SP (Todd, 2010). Noxious stimuli trigger the release of SP (Afrah et al., 2002), which binds to the neurokinin-1 (NK1) G-protein–coupled receptor expressed by the spinothalamic, spinoparabrachial, and spino-PAG neurons – the major populations of spinal projection neurons that are essential to the development of central sensitisation (Mantyh et al., 1997; Khasabov et al., 2002; Willis, 2002). The binding of SP to its receptor causes a long-lasting membrane depolarisation (Henry, 1976), thereby allowing the temporal summation of C fibre-evoked synaptic potentials (Dougherty and Willis, 1991; Xu, Dalsgaard and Wiesenfeld-Hallin, 1992a, 1992b). In addition to SP, both CGRP and brain-derived neurotrophic factor (BDNF), also synthesised by small-diameter sensory neurons, can further promote a temporal summation of postsynaptic excitatory potentials upon nociceptive stimulations (Zhou and Rush, 1996; Kerr et al., 1999; Thompson et al., 1999; Balkowiec and Katz, 2000; Heppenstall and Lewin, 2001) (Figure 2.5). This sustained release of glutamate and neuropeptides by nociceptors at a low frequency (0.3–5 Hz) increases the postsynaptic excitability of the nociceptive neuron, a phenomenon known as 'wind-up' (Mendell and Wall, 1965). Such stimulation paradigm is within the physiological range of the activity of C fibre nociceptors upon noxious stimulation (Iggo, 1960; Kress et al., 1992), although they can transiently exhibit a peak of their frequency discharge (greater than 10 Hz; Kress et al., 1992). The gradual increase in membrane potential observed during wind-up allows a temporal and a spatial summation of the excitatory inputs onto the nociceptive neuron. This eventually leads to sufficient membrane depolarisation to remove the Mg^{2+} from NMDAR, which activates this channel (Davies and Lodge, 1987; Dickenson and Sullivan, 1987).

Entry of calcium and signaling cascades

In this state, glutamate binding to NMDARs causes the channel to open, allowing a massive sodium and Ca^{2+} influx into the cell; this leads to a long-lasting membrane depolarisation (Dingledine, 1983). Whereas wind-up is critical to the establishment of activity-dependent central sensitisation, it does not by itself correspond to a state of generalised excitability, but rather represents the excitability state of a neuron during the conditioning paradigm by nociceptors activity. This activation of NMDARs represents a major boost in synaptic strength and is the key trigger for activity-dependent plasticity of the spinal nociceptive neurons (Woolf and Thompson, 1991).

Heightened intracellular levels of Ca^{2+} are critical to initiating activity-dependent central sensitisation (Ikeda et al., 2003). Most of this increase is due to the massive entry of calcium from the extracellular milieu through NMDARs (MacDermott et al., 1986), AMPARs (Hartmann et al., 2004) and voltage-gated calcium channels (Coderre and Melzack, 1992), but also, to some extent, from the endoplasmic reticulum (Fagni et al., 2000) upon activation

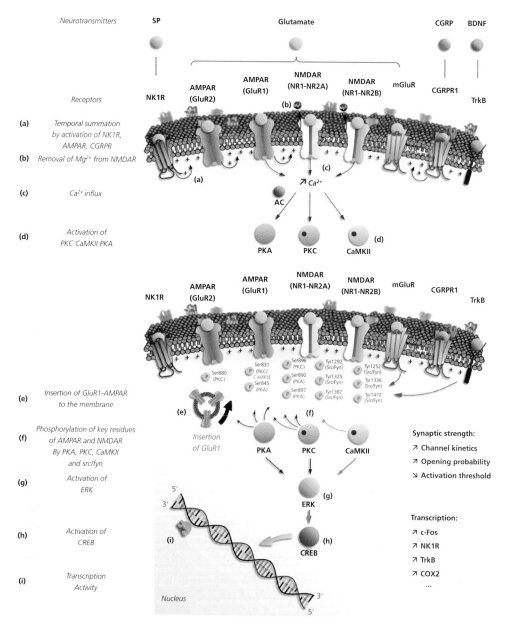

Figure 2.5 Schematic representation of the neurotransmitters, receptors and cellular cascades involved in central sensitisation. **Top panel**: neurotransmitters, receptors and cellular cascades involved in central sensitisation. **(a)** a barrage of activity in the nociceptors releases glutamate, SP and CGRP that bind to their receptors and cause a gradual increase in membrane potential, allowing a temporal and a spatial summation of the excitatory inputs onto the nociceptive neuron. **(b)** This leads to sufficient membrane depolarization to remove the Mg^{2+} from NMDAR, which activates this channel and results in **(c)** a rapid increase of $[Ca^{2+}]i$. **(d)** Rise in intracellular calcium activates PKC, CaMKII and PKA (through adenylyl cyclase (AC) activation). **Bottom panel**: **(e)** PKA activation promotes the insertion of Ca^{2+}-permeable GluR1-containing AMPARs into the membrane from vesicles stored under the synapse, further promoting Ca^{2+} entry into the cell. **(f)** PKA, PKC and CaMKII phosphorylate key residues of AMPAR and NMDAR NR1 subunits. NR2A and NR2B NMDAR subunits are phosphorylated by Src/Fyn kinases. These phosphorylations change the threshold and activation kinetics of NMDA and AMPA receptors, boosting synaptic efficacy. **(g)** PKA, PKC and CaMKII converge to activate the ERK pathway, which leads to the activation of CREB **(h)** and other transcription factors that drive the expression of several genes, including c-Fos, NK1, TrkB, and Cox-2, so as to produce a long-lasting strengthening of the synapse **(i)**.

of the phosphatidylinositol-3-kinase (PI3K) pathway (Pezet et al., 2002) triggered by group 1 metabotropic glutamate receptors (Guo et al., 2004).

The rise of intracellular levels of calcium leads to the activation of the calcium-dependent protein kinase C (PKC) and of the calcium–calmodulin-dependent protein kinase II (CAMKII) that phosphorylate NMDA and AMPA glutamate receptors on key residues (Carvalho, Duarte and Carvalho, 2000; Chen and Roche, 2007) (Figure 2.5). PKC phosphorylates NMDAR subunit NR1 on residue Ser896 (Leonard and Hell, 1997; Tingley et al., 1997; Carvalho et al., 2000), AMPAR subunit GluR2 on residue Ser880, and AMPAR subunit GluR1 on residue Ser831 – this residue being also the target of CAMKII (Carvalho et al., 2000). The phosphorylation of those serine residues leads to dramatic changes of the activation threshold, channel kinetics, and voltage dependence properties of the AMPARs and NMDARs, causing an increase of their activity and an overall postsynaptic hyperexcitability of the nociceptive neuron (Yu and Salter, 1999; Fang et al., 2002; Zou, Lin and Willis, 2002; Fang, Wu, Ling, et al., 2003; Fang, Wu, Zhang, et al., 2003; Sun, Lawand and Willis, 2003; Brenner et al., 2004; Sun et al., 2004; Zou, Lin and Willis, 2004; Jones and Sorkin, 2005; Ultenius et al., 2006; Liu et al., 2008). Once activated, PCK also reduces the Mg^{2+} block of NMDARs and increases their probability of channel opening, so that there are higher chances of NMDAR activation upon future glutamate release (Chen and Huang, 1992) – a mechanism that could explain why established central sensitisation can be maintained through a low activity of C fibres that normally cannot trigger this state (Koltzenburg, Lundberg and Torebjork, 1992).

The nociceptor-induced elevation in intracellular Ca^{2+} also activates adenylyl cyclases 1 and 8, whose cAMP production in turn activates protein kinase A (PKA) (Zou et al., 2002; Sun et al., 2003; Sun et al., 2004; Wei et al., 2006). PKA phosphorylates the NMDAR subunit NR1 on residues Ser890 and Ser897 (Leonard and Hell, 1997; Tingley et al., 1997), which increases the response of these receptors to glutamate (Chen and Huang, 1992; Raman, Tong and Jahr, 1996). NMDAR excitability can also be modulated through the phosphorylation of the NR2A and NR2B subunits, which are heavily expressed in the spinal cord by nonreceptor tyrosine kinases such as Src or Fyn. This leads to an increase in the probability of receptor opening as well as to changes in its conductance properties (Guo et al., 2002; Guo et al., 2004; Salter and Kalia, 2004; Chen and Roche, 2007) and prevents their endocytosis (Chen and Roche, 2007), further strengthening synaptic excitability (Figure 2.5).

In addition to phosphorylation of the receptors, which affects their intrinsic properties, central sensitisation also causes structural changes of the synapse. At basal states neurons in the superficial lamina of the dorsal horn mostly express GluR2 AMPAR subunits, which are relatively low Ca^{2+}-permeable. After nociceptive stimulation there is a rapid increase in the trafficking of GluR1 AMPAR subunits into the synapse (Figure 2.5). The phosphorylation of the AMPAR GluR1 residue Ser845 by PKA, together with the Ser831 by CAMKII, promotes the insertion of GluR1 into the synapse (Esteban et al., 2003; Galan, Laird and Cervero, 2004), which causes an increase in synaptic strength (Banke et al., 2000; Fang, Wu, Zhang, et al., 2003), as well as larger Ca^{2+} influx upon stimulation.

Stimuli that cause central sensitisation also cause a massive activation of extracellular signal-regulated kinase (ERK) in nociceptive neurons (Fang, Wu, Zhang, et al., 2003;

Kawasaki et al., 2004; Slack et al., 2005; Wei et al., 2006; Walker et al., 2007). This activation of the ERK pathway is the convergence point of cellular pathways of several activated receptors such as the NMDAR, group I mGluR, TrkB, NK1, or CGRP1R (Ji et al., 1999; Karim et al., 2001; Hu and Gereau, 2003; Lever et al., 2003). Once phosphorylated, ERK activates the cAMP response element-binding protein (CREB) (Shaywitz and Greenberg, 1999) that will drive the expression of several genes, including c-Fos, NK1, TrkB, and Cox-2 (Ji et al., 2002), thereby promoting the maintenance of the hyperexcitability state (Figure 2.5). These transcriptional changes might also contribute to the fact that the hyperexcitability caused by central sensitisation strongly outlasts the duration of the nociceptive stimulation, to an extent that is not likely to be explained solely by the phosphorylation of receptors (Simone et al., 1989; Woolf and King, 1990; Ji et al., 2003). In addition, they indicate that a nociceptive neuron in a state of central sensitisation has the cellular machinery ready for long-term structural changes.

Taken together, these changes caused by a barrage of nociceptive inputs produce dramatic functional alterations in the nociceptive neurons. In normal conditions, however, they are reversible. Phosphatases dephosphorylate receptors and ion channels so as to reset their activity to baseline levels; the increased trafficking of receptors to the membrane reverses through endocytosis; and, in time, the nociceptive synapses return to their basal state. Activity-dependent central sensitisation is therefore a transient phenomenon.

Loss of inhibitory signals (disinhibition)

The other major way to trigger a state of central sensitisation in spinal nociceptive neurons is by removing the inhibitory brake that is normally exerted on them. The nociceptive neurons of the spinal cord are indeed under strong inhibitory controls that originate locally (segmental inhibition) or distally (descending inhibition) (Roberts, Beyer and Komisaruk, 1986; Dickenson, Chapman and Green, 1997). Segmental inhibition is constituted by inhibitory GABAergic and glycinergic interneurons located in the dorsal horns (Sivilotti and Nistri, 1991) that drive a very high tonic inhibition of local nociceptive neurons; but their activity is also modulated by sensory inputs (Sivilotti and Woolf, 1994). GABAergic interneurons are predominant in superficial laminae (I–II; Tamamaki et al., 2003), where they release GABA locally, whereas glycinergic neurons have their soma mostly located in deeper laminae (III and lower; Todd, 1990) but are nonetheless critical for modulating lamina I projection neurons (Chery and De Koninck, 1999; Zeilhofer, Wildner and Yevenes, 2012). Due to their anatomical localisation, GABAergic interneurons are mostly contacted by nociceptive fibres (Alvarez, Kavookjian and Light, 1992) and glycinergic interneurons by large myelinated fibres (Todd, 1990; Watson, 2004). Upon release, GABA can act through three receptors: the ionotropic GABAA receptor and GABAC receptor and the metabotropic GABAB receptor (Bowery, Hudson and Price, 1987; Bormann, 2000). The chloride-permeable GABAAR is the prominent class of GABA receptors found in the dorsal horns of the spinal cord (Bohlhalter et al., 1996), where their activation causes strong analgesia (Knabl et al., 2008; Knabl et al., 2009). Similarly, the binding of glycine to its ionotropic receptor (glyR) causes the opening of this channel, which in turn allows a massive entry of chloride into the cell, leading to its inhibition (Chery and De Koninck, 1999).

A loss of segmental spinal inhibitory inputs leads to the phenomenon known as disinhibition, where dorsal horn neurons become more susceptible to activation through both nociceptive and non-nociceptive inputs (Figure 2.3). Disinhibition is a critical mechanism for triggering and maintaining central sensitisation (Yaksh, 1989; Sivilotti and Woolf, 1994; Malan, Mata and Porreca, 2002; Baba et al., 2003; Torsney and MacDermott, 2006). Indeed, an intrathecal injection of biccuculine (antagonists of GABAAR), strychnine (antagonists of GlyR) or phaclofen (antagonists of GABABR), or the genetic ablation or silencing of GlyR2-expressing spinal cord neurons in naïve animals induces spontaneous pain-like behaviours and causes mechanical allodynia to light touch stimuli (Beyer, Roberts and Komisaruk, 1985; Roberts et al., 1986; Sivilotti and Woolf, 1994; Loomis et al., 2001; Malan et al., 2002; Foster et al., 2015). When the inhibitory GABA/glycine inhibition tone is removed, this allows a polysynaptic activation of lamina I projection neurons by large-diameter morphine-insensitive fibres (Sherman and Loomis, 1994) – a process that involves the PKCγ excitatory interneurons (Miraucourt, Dallel and Voisin, 2007) located in lamina IIi and contacted by Aβ fibres (Neumann et al., 2008). Punctuate mechanical allodynia can therefore be triggered without any structural changes in the spinal cord, only by reducing the normal inhibition of the spinal nociceptive projection neurons.

Partial disinhibition also occurs during the establishment of activity-dependent central sensitisation, in which the activation of PKC causes a reduction in the sensitivity of GABA and glycine receptors to their ligands and a consequent reduction of their inhibitory tone on the nociceptive neurons (Lin, Peng and Willis, 1996).

The dorsal horns of the spinal cord are also contacted by several supraspinal projections that can modulate the excitability of the nociceptive system (Millan, 2002; Gebhart, 2004). In baseline conditions, these descending controls are mostly driving a tonic inhibition of the nociceptive neurons; but their activation upon nociceptive stimuli can either inhibit or amplify the excitability of the spinal circuitry (Gebhart, 2004). The neurons of the locus coeruleus that project into the spinal cord are noradrenergic (Kwiat and Basbaum, 1992; Millan, 1999). Their activation by noxious stimuli causes the release of norepinephrine, which in turn causes analgesia (Yaksh, 1985; Sullivan, Dashwood and Dickenson, 1987; Men and Matsui, 1994a) through its binding to α2-adrenergic receptors (Yaksh, 1985). Descending neurons of the raphe magnus produce serotonin (Willis et al., 1984; Rivot, Calvino and Besson, 1987) but can also express GABA, dynorphin and enkephalin (Millan, 2002). Descending controls from the raphe magnus nucleus are strongly activated by nociceptive stimuli (Puig, Rivot and Besson, 1992; Taguchi and Suzuki, 1992; Men and Matsui, 1994b). Once activated, they cause an inhibition of the spinal nociceptive neurons (Fields et al., 1977) through the release of serotonin – as well as through the co-release of other neurotransmitters. The inhibitory effects observed after the intrathecal injection of serotonin are mostly caused by the activation of Gi/o-coupled 5-HT$_{1A}$R expressed at the soma of nociceptive neurons (Bardin, Lavarenne, et al., 2000; Bardin, Schmidt, et al., 2000), whereas the activation of the Gi/o-coupled 5-HT$_{1B/1D}$R, expressed at the presynaptic level (Potrebic et al., 2003) presumably reduces the release of neurotransmitters from sensory fibre terminals (Carmichael, Charlton and Dostrovsky, 2008). 5-HT$_3$ receptors, the only serotonin receptors that are ligand-gated ion channels, cause a membrane depolarisation upon activation (Maricq et al., 1991). At the spinal cord level, 5-HT$_3$R is expressed in sensory fibres terminals (Hamon et al., 1989) and in various nociceptive neurons, both excitatory and inhibitory (Laporte et al., 1996; Morales, Battenberg and Bloom, 1998; Tsuchiya, Yamazaki and Hori, 1999). Serotonin released in the spinal cord can therefore either inhibit or facilitate nociceptive transmission. The net effect of serotonin will eventually depend on which cellular targets are contacted, the types of receptors expressed, and the overall state of excitability of the raphe magnus neurons, which can itself be modulated by numerous brain areas (Heinricher et al., 2009; Rodriguez-Gaztelumendi et al., 2014).

It clearly appears on the whole that the nociceptive neurons are finely regulated by both inhibitory and excitatory controls. A loss of these controls, or a shift in a balance in favour of an increased excitability, would facilitate the transition of nociceptive neurons into a state of central sensitisation.

Central sensitisation in pathological states

Central sensitisation represents an additional level of protection for an injured organism. This phenomenon is triggered when the first line of defense, designed to detect potential harmful stimuli, has failed and damage has occurred. Once the nociceptive neurons of the spinal cord are in a state of central sensitisation, they no longer transmit information that reflects with precision the nature and intensity of environmental stimuli. Instead they transform any stimuli they receive into alert messages and convey a message that is uncoupled with its environment. When there is a transient injury, this represents an extremely efficient way to maintain a whole body area protected from any stimuli until its full recovery.

If this state of hypersensitivity is maintained over extended periods of time or is abnormally triggered, the resulting pain is no longer protective. This can happen when the source of the nociceptive input does not resolve, when the nociceptive system itself is damaged, or in any condition where there is an unbalance in the excitatory/inhibitory modulatory systems of the nociceptive system that affects spinal circuitry. Such an exaggerated hypersensitive state represents a veritable burden when it is maintained chronically. Then pain is not a protective alarm system any more, but becomes an ongoing nuisance for the organism and isolates it from its environment. In various disease states these changes have become permanent, literally transforming the first nociceptive relay into a new, hyperactive synapse that cannot discriminate and determine the nature of the information. Several key features of central sensitisation are present in pathologies with chronic pain symptoms such as neuropathic pain, chronic inflammatory diseases (rheumatoid arthritis, osteoarthritis), complex regional pain syndrome (CRPS), pancreatitis, whiplash injury, fibromyalgia, low-back pain or migraine (Woolf, 2010).

Because several redundant factors can trigger and maintain central sensitisation, it is perhaps not surprising that many changes that affect the spinal cord circuitry can cause chronic pain hypersensitivity. NMDAR (Seltzer et al., 1991; Mao et al., 1993; Cheng et al., 2008; Qu et al., 2009), AMPAR (Lu et al., 2008; Park et al., 2008), group I mGluR (Neugebauer, Lucke and Schaible, 1994; Young et al., 1997; Dogrul et al., 2000; Fundytus et al., 2002; Adwanikar, Karim and Gereau, 2004; de Novellis et al., 2004; Zhu et al., 2004; Giles, Trezise and King, 2007), group II–III mGluR (Simmons et al., 2002; Chen and Pan, 2005; Goudet et al., 2008; Zhang,

Chen and Pan, 2009), SP and CGRP (Abbadie et al., 1996; Abbadie et al., 1997; Lee and Kim, 2007) have all been shown to participate in the development and maintenance of central sensitisation in various preclinical models of chronic pain, through the same molecular and cellular mechanisms involved in activity-dependent central sensitisation. The following sections will describe mechanisms that are unique to chronic pain states due to either chronic inflammatory or peripheral nerve injury with regard to the development or maintenance of central sensitisation.

Chronic inflammatory pain

Chronic inflammatory pain is due to the ongoing activation of nociceptors when the source of the inflammation does not resolve. At the periphery, this persistent inflammation can result from the organism's incapacity to heal from an actual injury or from the development of an auto-immune reaction that will constantly produce and activate additional immune factors targeted against the organism's own cells. Inflammation causes peripheral sensitisation, which lowers the activation threshold of nociceptors and is often associated with hyperalgesia. However, it is the ongoing stimulation of nociceptive fibres together with additional pathways within the CNS that is responsible for the development of a chronic central sensitisation state and associated symptoms such as secondary hyperalgesia and mechanical allodynia. In what follows we will present the changes that cause either a gain in excitability or a loss of inhibition in the nociceptive system, which together lead to the maintenance of an abnormal hyperexcitable state.

Gain in excitability

Persisting peripheral inflammation causes several types of sensory neurons to dramatically change their transcription and translation pattern (Rodriguez Parkitna et al., 2006). In addition to the increase in transduction sensitivity and membrane excitability in nociceptive fibres, some sensory neurons undergo a phenotypic switch. Large-diameter neurons begin to express SP and BDNF (Neumann et al., 1996; Mannion et al., 1999), two peptides critical to the wind-up phenomenon. As a result, an activation of these fibres by innocuous stimuli leads to the release of these neurotransmitters, which could trigger and maintain central sensitisation (Indo et al., 1996; Neumann et al., 1996; Baba, Doubell and Woolf, 1999) (Figure 2.6).

A central pathway important for the development of inflammatory pain hypersensitivity involves the induction of cyclo-oxygenase-2 (COX2) in dorsal horn neurons. COX2 drives the production of prostaglandin E2 (PGE2) (Vasko, Campbell and Waite, 1994; Samad et al., 2001), which binds to its EP2 receptor located on dorsal horn neurons to potentiate both AMPAR and NMDAR currents and to activate non-selective cation channels that depolarise the potential membrane of nociceptive neurons (Baba et al., 2001). Genetic deletion of COX-2 only in neurons strongly attenuates the development of mechanical allodynia without altering heat hyperalgesia (Vardeh et al., 2009), which suggests an important role of this pathway in the development of central sensitisation in chronic inflammatory pain (Figure 2.6).

Under normal conditions microglia are the only immunocompetent cells of the nervous system, and they remain in a quiescent state while surveying the local milieu (DeLeo and Yezierski, 2001;

Central sensitisation in pathological settings – Inflammatory pain:

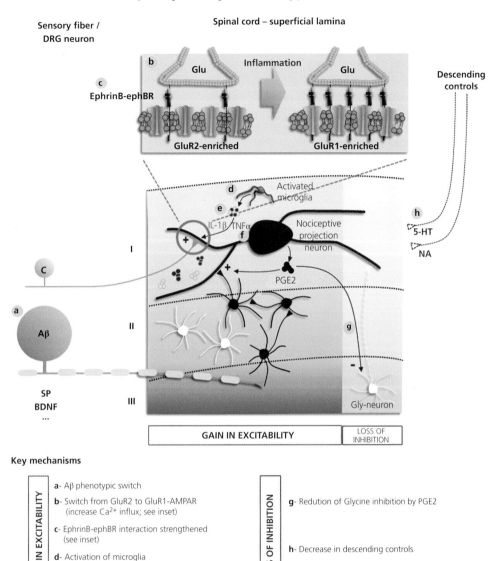

Figure 2.6 Schematic representation of the changes contributing to central sensitisation in the superficial lamina in chronic inflammatory states. **Gain in excitability**: Large DRG neurons undergo a phenotypic switch and **(a)** start to express SP and BDNF, so stimulation of these afferents acquires the capacity to generate long-term postsynaptic depolarization. **(b)** Neurons start to express GluR1-containing AMPAR at their synapse (see inset), which results in an increase of the Ca²⁺ influx on their activation. **(c)** The presynaptic element of the nociceptive synapse is also structurally strengthened though the EphrinB–ephBR interaction (see inset). The kinase activity of EphBR increases NMDAR activity, which potentiates synaptic activity. **(d)** Microglial cells are activated and release factors such as pro-inflammatory cytokines IL-1β and TNF-α, which **(e)** contribute to the development of central sensitisation by enhancing excitatory and reducing inhibitory currents. **(f)** Induction of cyclooxygenase-2 (Cox-2) in dorsal horn neurons drives the production of prostaglandin E2 (PGE2), which potentiate AMPAR and NMDAR currents, boosting post-synaptic excitability. **Loss of inhibition**: **(g)** PGE2 produced by dorsal horn neurons reduces glycine-mediated currents. **(h)** Inhibitory 5-HTergic and NAergic descending controls are diminished. Projection lamina I neuron is shown in red. Excitatory interneurons are in black (the polysynaptic activation pathway from Aβ is highlighted in red) and the inhibitory interneurons are in white. Laminae are marked I–III and separated by dotted lines.

Watkins, Milligan and Maier, 2001; Nimmerjahn, Kirchhoff and Helmchen, 2005). After peripheral inflammation these cells rapidly change their shape, their function and the factors they release (Watkins et al., 2001). In particular, the activation of the p38 MAPK pathway (Svensson, Hua, et al., 2003; Svensson, Marsala, et al., 2003) leads to the synthesis and release of pro-inflammatory factors, among them IL-1β and TNF-α, which contribute to the development of central sensitisation by enhancing excitatory currents (Honore et al., 2000; Raghavendra, Tanga and DeLeo, 2004; Kawasaki et al., 2008; Gruber-Schoffnegger et al., 2013) and by activating or maintaining Cox-2 activity (Figure 2.6). Blockade of microglial activation through the intrathecal administration of minocycline or through the administration of IL-1β receptor antagonist (IL-1ra) reduces the development of pain behaviours, notably thermal hyperalgesia and mechanical allodynia (Hua et al., 2005; Honore et al., 2006; Schreiber, Beitz and Wilcox, 2008; Zhang et al., 2008), whereas the intrathecal injection of IL-1β, but not of TNF-α is sufficient to cause mechanical allodynia (Reeve et al., 2000).

Peripheral inflammation causes a critical rearrangement of the excitatory synapses of the neurons of the superficial lamina that increases the strength of nociceptive transmission (Larsson and Broman, 2008; Vikman, Rycroft and Christie, 2008; Park et al., 2009). The majority of the AMPARs expressed by these neurons usually contain the relatively Ca^{2+}-impermeable GluR2 subunit (Petralia et al., 1997; Nagy, Al-Ayyan, et al., 2004). In peripheral inflammation states there is a shift from GluR2 to GluR1-containing AMPARs expressed at the membrane (Larsson and Broman, 2008; Vikman et al., 2008; Park et al., 2009). This shift is orchestrated by the action of PKA and PKC (Park et al., 2009; Kopach et al., 2013). The activation of the PKA is dependent on the TNF-a (Choi et al., 2010), which is probably produced and released by activated microglial cells (Sawada, Suzumura and Marunouchi, 1992; Raghavendra et al., 2004), and the activation of the PKC is triggered by the Ca^{2+} influx through NMDARs (Park et al., 2009). Once activated, the PKC phosphorylates the GluR2 subunit, thereby reducing its affinity for the scaffolding protein GRIP-1, which is needed to cluster the AMPARs at the synapse (Dong et al., 1997; Hirai, 2001). At the same time, the increase in intracellular Ca^{2+} activates the protein PICK-1 (Chung et al., 2000; Hanley and Henley, 2005; Park et al., 2009), which promotes the endocytosis of GluR2-containing AMPARs that removes them from the membrane (Matsuda, Mikawa and Hirai, 1999; Chung et al., 2000; Lin and Huganir, 2007; Terashima et al., 2008; Park et al., 2009). The reinsertion of the endocytosis vesicle requires the action of the N-ethylmaleimide-sensitive fusion protein (NSF) (Nishimune et al., 1998; Song et al., 1998; Noel et al., 1999; Huang et al., 2005), but the expression of this protein is strongly reduced in the spinal cord during inflammation (Katano et al., 2008). This further favours the insertion of GluR1-containing AMPARs at the synapse – but also at extrasynaptic sites of tonically active lamina II neurons (Kopach et al., 2011) – by PKA (Yang et al., 2011). Once established, GluR1-containing AMPARs are maintained at the membrane as long as there is synaptic activity (Ehlers et al., 2007; Man, Sekine-Aizawa and Huganir, 2007). With this expression pattern of AMPAR subunits, glutamate binding elicits a massive entry of Ca^{2+} into the cell, to levels comparable with NMDAR activation (Luo et al., 2008), which contributes to triggering and maintaining central sensitisation (Figure 2.6). This structural strengthening of the spinal excitatory synapse is also mediated through the rearrangement of the subcellular organisation of metabotropic glutamate receptors, where group I mGluR5 are inserted into the membrane and mGluR1 are clustered closer to the synapse (Pitcher, Ribeiro-Da-Silva and Coderre, 2007). The resulting increased activation of mGluR promotes the excitability of NMDAR through the phosphorylation of its NR2B subunit, which is mediated in part by the activation of the Src cascade (Guo et al., 2002; Guo et al., 2004). Specific blockading of the Src cascade significantly attenuates signs of central sensitisation like mechanical allodynia without affecting nociceptive pain (Guo et al., 2002; Liu et al., 2008).

The presynaptic element of the nociceptive synapse is also structurally strengthened, notably though the EphrinB–EphBR interaction mostly present between C fibres and laminae I–II neurons (Figure 2.6). EphBRs are receptor tyrosine kinases only expressed by postsynaptic neurons, whereas EphrinB is anchored to the presynaptic membrane (Kullander and Klein, 2002). The kinase activity of EphBRs is critical to maintaining the clustering of NMDARs at the synapse (Dalva et al., 2000). The stimulation of EphBRs increases NMDAR activity, which potentiates the Ca^{2+} influx into the neuron and causes the phosphorylation of CREB through the activation Src (Takasu et al., 2002). The activation of EphBRs in naïve animals leads to a facilitation of C fibre input onto nociceptive neurons and to an increased sensitivity to heat, but does not cause mechanical allodynia (Battaglia et al., 2003). Inhibition of this pathway can prevent and reverse pain hypersensitivity in preclinical pain models of inflammatory pain without altering nociceptive pain (Battaglia et al., 2003; Slack et al., 2008). (For an extensive review of the experiments on the EphrinB/EphBR system and the spinal processing of noxious stimuli, see Chapter 7 in this volume.)

The overall hyperexcitability of the nociceptive neurons is further enhanced by an increase in descending facilitation, which is mediated notably through BDNF (Urban and Gebhart, 1999, Guo et al., 2006).

Disinhibition

The production of prostaglandin E2 by COX2 in neurons of the spinal cord selectively blocks glycinergic receptors containing α3 subunits (Ahmadi et al., 2002; Muller, Heinke and Sandkuhler, 2003; Harvey et al., 2004), thereby reducing the glycinergic-mediated inhibitory tonus on nociceptive neurons and promoting the phenomenon of disinhibition (Zeilhofer and Zeilhofer, 2008). In addition to this reduction in segmental inhibition, peripheral inflammation causes, over time, a severe loss of the descending inhibition normally mediated by NE and 5-HTergic fibres (Wei, Dubner and Ren, 1999) (Figure 2.6).

Neuropathic pain

Peripheral neuropathic pain is an extremely complex disease that can occur after lesions to the somatosensory system (Treede et al., 2008). The origin of the nerve lesion can be diverse: mechanical trauma (e.g., section, compression, crush), metabolic diseases (e.g., diabetes), neurotoxic chemicals, infection (e.g., HIV) or tumours. The nociceptive system itself is damaged and no longer processes sensory information adequately (Costigan, Scholz and Woolf, 2009; von Hehn, Baron and Woolf, 2012). As a result, there is a loss of tactile fine sensory perception (part of the so-called 'negative symptom'), but also the apparition of pain hypersensitivity and spontaneous pain attacks – positive symptoms that can persist indefinitely (Costigan, Scholz et al., 2009; for a detailed overview of neuropathic symptoms and pathology, see Chapter 3 in this volume). Central sensitisation plays a critical role in many aspects of neuropathic pain pathophysiology (Woolf, 2010). Key symptoms like mechanical allodynia and cold allodynia are mostly generated centrally, by nociceptive neurons in a chronic state of central sensitisation (Latrémolière and Woolf, 2009; Woolf, 2010). Upon peripheral nerve injury there are considerable changes that contribute to the development of abnormal pain sensitivity. In the following sections we will highlight those that can directly trigger or maintain a state of central sensitisation.

Gain in excitability

Nerve injury caused by a physical trauma – for example, nerve section or compression – causes a massive release of glutamate by the injured fibres into the spinal cord during the first hour after axotomy (Inquimbert et al., 2012). This amount of glutamate, once released, binds and activates AMPARs and NMDARs, which can trigger the cascades leading to central sensitisation. Pre-treatment with a NMDAR blocker such as ketamine or MK801 can prevent the establishment of pain hypersensitivity in various animal models of neuropathic pain (Smith et al., 1994; Kim et al., 1997; Munglani et al., 1999; Shields, Eckert, 3rd and Basbaum, 2003), and blocking the effects of an early axotomy-mediated glutamate release is sufficient to attenuate the development of hypersensitivity for several weeks (Munglani et al., 1999). Neither a preemptive nerve blocking through bupivacaine-loaded microspheres nor analgesic treatment with pregabalin (which blocks α2-δ1 voltage-dependent calcium channels: Field et al., 2006) prevents the development of neuropathic pain (Suter et al., 2003; Yang et al., 2014). This indicates that, whereas NMDAR activation is critical to initiating central sensitisation after peripheral nerve trauma, neuronal activity of the sensory fibres is not necessarily required.

After peripheral nerve injury, both injured and non-injured sensory neurons in the dorsal root ganglion of the damaged nerve exhibit a massive change in transcription, which alters their membrane properties, their growth and their transmitter function (Fukuoka et al., 2001; Costigan et al., 2002; Xiao et al., 2002; Obata et al., 2003; Obata et al., 2004). In particular, there is a considerable production of BDNF, which enhances nociceptive transmission both presynaptically, by improving vesicular release through the activation of the Src and NMDARs, and postsynaptically, by enhancing NMDAR response through phosphorylation of its NR2B subunit (Kerr et al., 1999; Geng et al., 2010). At the spinal cord level, the BDNF activates the ERK cascade in nociceptive neurons of the superficial laminae, which drive the transcriptional activity required for long-term synaptic consolidation (Zhou et al., 2008). Enzymes responsible for determining the intracellular levels of the pteridine (6R)-L-erythro-5,6,7,8-tetrahydrobiopterin (BH4), an essential co-factor for aromatic amino acid hydroxylases (tyrosine and tryptophan hydroxylases) and nitric oxide synthase, are up-regulated in injured neurons (Tegeder et al., 2006), but also in macrophages that infiltrate the injured nerve and the DRGs (Latrémolière et al., 2015). An elevation in BH4 levels triggers an increased nitric oxide (NO) production (Tayeh and Marletta, 1989), which in turn sensitises ion channels such as TRPV1 and TRPA1 by nitrosylation (Miyamoto et al., 2009); these channels could mediate in part the NO-mediated thermal pain hypersensitivity caused by elevated BH4 levels only in sensory neurons (Latrémolière and Costigan, 2011; Latrémolière et al., 2015). In addition, a lowered activation threshold for TRPV1 can lead to an increased probability that the channels would open at temperatures within the physiological range (Waxman et al., 1999), which can cause abnormal spontaneous neuronal activity (Hitomi et al., 2012).

Activity of the sensory fibres that is not triggered by external stimuli is known as ectopic activity and probably contributes to maintaining established central sensitisation (Koltzenburg et al., 1992; Devor, 2009). Neuropathic pain is often associated with two types of spontaneous pain symptoms: an ongoing 'burning' type of pain that could be mediated by the ectopic (and ongoing) activity of nociceptive C fibres (Baron, 2006); and a paroxysmal, 'electric' type of pain attacks that are thought to be caused by a burst of activity in larger, faster conducting sensory fibres (Baron, Binder and Wasner, 2010). In normal states, both myelinated and non-myelinated fibres exhibit very low tonic activity (Tanelian and MacIver, 1991). After peripheral nerve injury, however, they start displaying a high spontaneous discharge rate of action potentials. This ectopic activity can occur in large injured sensory fibres (Devor, 2009; Djouhri et al., 2012), in small-diameter injured C fibres (Seltzer and Devor, 1979; Omana-Zapata et al., 1997; Pan, Eisenach and Chen, 1999) or in small-diameter non-injured C fibres in close vicinity to damaged fibres (Wu et al., 2002; Djouhri et al., 2006; Djouhri et al., 2012) (Figure 2.7). Peripheral nerve blockade totally abolishes spontaneous pain symptoms in patients with neuropathic pain, which confirms its peripheral origin (Haroutounian et al., 2014). The precise mechanisms responsible for the genesis of ectopic activity are not yet fully understood, but evidence suggests a role for depolarising sodium channels ('Na(v)s') and for hyperpolarising potassium channels ('K(v)s'), which are both critical to neuronal excitability. The two TTX-resistant Na(v)1.8 and Na(v)1.9 exhibit changes in their axonal transport after nerve injury, so that there is an accumulation of these channels at the site of injury. This triggers local action potentials that can propagate to nearby fibres (Novakovic et al., 1998; Gold et al., 2003). Potassium channels, which are critical to maintaining normal resting membrane potential (Kang and Kim, 2006; Dobler et al., 2007), are heavily down-regulated after nerve injury (Everill and Kocsis, 1999), which significantly depolarises the cell membrane (Tulleuda et al., 2011) and allows a spontaneous firing of sensory neurons (Acosta et al., 2014). Ectopic activity in the injured nerve has been described in early and chronic phases of neuropathic pain. The administration of anticonvulsants (gabapentin, lamotrigine) reduces mechanical allodynia (Field et al., 1999; Christensen et al., 2001), possibly in part by disrupting the maintenance of central sensitisation (Koltzenburg et al., 1992; Rogawski and Loscher, 2004).

The activation of genetic programs designed to promote intrinsic axonal growth in injured neurons in order to reinnervate peripheral targets also gives the opportunity for rearrangements within the spinal cord. It has been suggested that myelinated Aβ fibres, normally confined within laminae III–IV, make direct contact with nociceptive neurons located in laminae I–II (this is the so-called 'sprouting' phenomenon: Woolf, Shortland and Coggeshall, 1992; Woolf et al., 1995; Lekan, Carlton and Coggeshall, 1996; Shortland, Kinman and Molander, 1997) (Figure 2.7). The original experiments describing sprouting used the cholera toxin B subunit as a selective tracer for A-fibres. The selectivity of this toxin is, however, altered after peripheral nerve injury (Tong et al., 1999; Shehab, Spike and Todd, 2003), which has probably led to an overestimation of the sprouting phenomenon (Hughes et al., 2003). Nevertheless, one of the

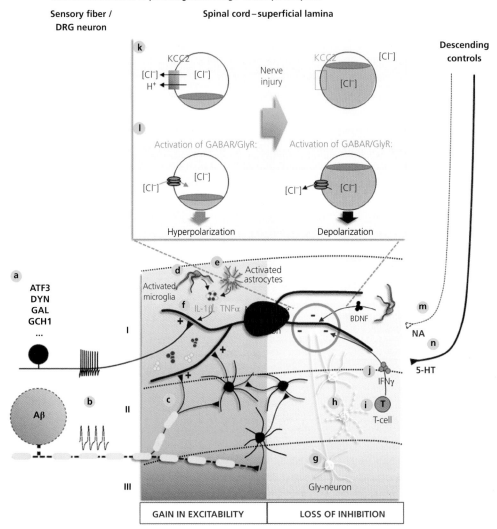

Central sensitisation in pathological settings – Neuropathic pain:

Sensory fiber /
DRG neuron

Spinal cord – superficial lamina

Descending
controls

k

KCC2

[Cl⁻] ← [Cl⁻]
H⁺ ←

Nerve
injury

KCC2   [Cl⁻]

[Cl⁻]

l

Activation of GABAR/GlyR:

Activation of GABAR/GlyR:

[Cl⁻]   [Cl⁻]

[Cl⁻] ← [Cl⁻]

Hyperpolarization

Depolarization

a

ATF3
DYN
GAL
GCH1
...

d

Activated
microglia

e Activated
astrocytes

f IL-1β, TNFα

m

BDNF

NA

n

5-HT

j
IFNγ

b

Aβ

c

h i T

T-cell

g

Gly-neuron

I

II

III

GAIN IN EXCITABILITY

LOSS OF INHIBITION

Key mechanisms:

GAIN IN EXCITABILITY

a- Changes in transcription profile

b- Ectopic activity of sensory fibers

c- Aβ fibers sprout into lamina II

d- Microglial activation

e- Astrocyte activation

f- IL-1β and TNF-α enhance AMPAR, NMDAR currents

LOSS OF INHIBITION

g- Reduction of Glycine inhibition

h- Apoptosis of GABAergic interneurons

i- Infiltration of T-cells that produce IFNγ

j- IFNγ reduces GABAergic currents

k- Loss of KCC2 causes a increase in [Cl⁻]ᵢ (see inset top line)

l- Loss of GABAAR/GlyR effects (see inset bottom line)

m- Decrease in descending NA controls

n- Shift of descending 5-HT controls into excitatory

key clinical and preclinical manifestations of neuropathic pain is mechanical allodynia; and both immunostaining for c-Fos in the nociceptive neurons (a marker of neuronal sustained activation: see Hunt, Pini and Evan, 1987) and electrophysiological recordings have clearly established that peripheral nerve injury causes large myelinated fibres to activate nociceptive neurons in the superficial laminae (Bester, Beggs and Woolf, 2000; Okamoto et al., 2001; Kohno et al., 2003). The lack of evidence for a massive direct rewiring of myelinated fibres onto nociceptive neurons suggests that a functional plasticity is more plausibly responsible for the development of allodynia – which points to a critical role of central sensitisation in this behavioural state.

Peripheral nerve injury triggers an extensive gliosis in the spinal cord segments contacted by the damaged nerve (Garrison, Dougherty and Carlton, 1994; Colburn et al., 1997), especially near the entry point of the dorsal roots (Tsuda et al., 2003; Beggs and Salter, 2007, 2010). The activation of astrocytes and microglial cells has been correlated with the apparition of pain hypersensitivity in various preclinical neuropathic pain models (Sweitzer, Schubert and DeLeo, 2001; Sweitzer and DeLeo, 2002; Raghavendra, Tanga and DeLeo, 2003; Milligan et al., 2004; Ledeboer et al., 2005; Latrémolière et al., 2008). Preventing the activation of glial cells, especially microglia, at the time of nerve injury attenuates the development of abnormal pain behaviours (Raghavendra et al., 2003; Ledeboer et al., 2005; Latrémolière et al., 2008; Beggs and Salter, 2010). The initial recruitment and activation of spinal glial cells is triggered by primary afferent fibres, and a total nerve blockade of the sciatic nerve significantly reduces microglial activation and proliferation after axotomy (Wen et al., 2007; Suter et al., 2009). The targeted blockade of TRPV1-positive C fibres is not, however, sufficient to prevent the microglial activation induced by nerve injury (Suter et al., 2009). This suggests that large A fibres release specific factors required for microglial activation (Suter et al., 2009), or that

---

Figure 2.7 Schematic representation of the changes contributing to central sensitisation in the superficial lamina in neuropathic pain. **Gain in excitability**: After peripheral nerve injury, both injured and non-injured sensory neurons in the dorsal root ganglion of the damaged nerve exhibit a massive change in transcription that alters their membrane properties, growth and transmitter function **(a)**, and also leads to the development of ectopic activity **(b)**. Some myelinated Aβ fibres sprout from deep to superficial laminae, to make contact with nociceptive neurons **(c)**. Recruitment and activation of microglial cells **(d)** and astrocytes **(e)** is an essential step in the development of pain after nerve injury and triggers central sensitisation by releasing pro-inflammatory cytokines (IL-1β and TNF-α **(f)**), which increase neuronal excitability. **Loss of inhibition**: Several mechanisms contribute to the phenomenon of disinhibition after nerve injury: there is a severe reduction in the inhibitory glycinergic currents onto the projection lamina I neuron **(g)**, as well as a loss of inhibitory GABAergic interneurons caused by excitotoxic apoptosis **(h)**. The infiltration of activated T-cells **(i)** produces and releases IFNγ, which also reduces GABAergic currents **(j)**. **(k)** KCC2 is critical to maintaining a low concentration of intracellular Cl⁻ (see inset). After peripheral nerve injury, KCC2 is down-regulated, which results in an increase in intracellular Cl⁻. **(l)** In normal conditions (left), the opening of Cl⁻-permeable channels (such as GABAA and glycine receptors) leads to an entry of Cl⁻ that hyperpolarizes the cell membrane. After nerve injury (right), the opening of Cl⁻ channels by activation of GABAR or GlyR now leads to an efflux of Cl⁻, which causes membrane depolarization. As a consequence, GABA and glycine excite the neurons upon binding to their receptors. **(m)** Descending inhibitory NAergic controls are reduced, and **(n)** descending 5-HTergic controls shift from inhibitory to excitatory. Projection lamina I neuron is shown in red. Excitatory interneurons are in black (the polysynaptic activation pathway from Aβ is highlighted in red) and the inhibitory interneurons are in white. Dashed neurites represent a loss of fibres. Lamina are marked I–III and separated by dotted lines.

the release of neurotransmitters, even from a limited number of sensory fibres, is sufficient to cause a spill-over of glutamate and ATP in the superficial dorsal horns (Nie and Weng, 2010) that could trigger microglial activation (Tsuda et al., 2003; Davalos et al., 2005; Ferrini et al., 2013). Another factor that contributes to the recruitment and activation of microglial cells is the chemokine fractalkine. Fractalkine is produced by sensory neurons. It is released upon nerve injury into the spinal cord, where it binds to its receptor, CX3CR1, which is expressed only by glial cells (Dorf et al., 2000; Verge et al., 2004); and it causes the activation of the p38 MAPK pathway in those cells (Zhuang et al., 2007). The peptide neuregulin-1 is also up-regulated and released by sensory neurons after nerve trauma. It acts as a chemo-attractant for nearby microglial cells by activating the PI3K pathway and causes the proliferation of these cells through the ERK pathway (Calvo et al., 2011). In addition, the binding of neuregulin-1 to the ErbB2 receptors triggers the activation of the microglial cells (Calvo et al., 2010). Together, these factors converge to cause a massive and sustained activation of glial cells in the dorsal horns of the spinal cord. Activated glial cells in turn produce and release a broad range of molecules, from tropic factors to neurotransmitters, cytokines and reactive oxygen species that contribute to the development and the maintenance of spinal neurons excitability (Watkins and Maier, 2002; Scholz and Woolf, 2007; Milligan and Watkins, 2009; Austin and Moalem-Taylor, 2010, Gao and Ji, 2010). Direct activation of spinal microglial cells through intrathecal injection of the human immunodeficiency virus-1 (HIV-1) envelope glycoprotein gp120 is sufficient to trigger a severe mechanical allodynia (Milligan et al., 2001), a behavioural manifestation strongly associated with central sensitisation. This is mostly mediated by the action of IL-1β and of TNF-α and by the activation of the neuronal isoform of NOS (Holguin et al., 2004; Milligan et al., 2005). IL-1β and TNF-α directly increase neuronal excitability by enhancing AMPAR and NMDAR currents (Kawasaki et al., 2008; Gruber-Schoffnegger et al., 2013), whereas NO increases pain sensitivity through an action on NMDARs (Kitto, Haley and Wilcox, 1992; Meller, Dykstra and Gebhart, 1996; Schmidtko, Tegeder, and Geisslinger, 2009) (Figure 2.7).

After peripheral nerve injury, astrocytes also become activated, and for a more prolonged time course than microglia. They appear to play a critical role in the maintenance of neuro-pathic pain hypersensitivity (Zhuang et al., 2005; Zhang and De Koninck, 2006; Gao et al., 2009) (Figure 2.7). Activated astrocytes can release several pro-inflammatory cytokines (Aloisi et al., 1992; Sawada et al., 1992; Schwaninger et al., 1999) that enhance excitatory synapses at the postsynaptic level (Milligan et al., 2001), or the chemokine CXCL1 (Gao et al., 2009; Zhang et al., 2013) that promotes presynaptic release of neurotransmitter (Chen, Park, et al., 2014). Interestingly, CXCL1 release from astrocyte is mediated by hemichannels (GAP junctions) containing the Cx43 connexin; and the administration of a Cx43 mimetic peptide that blocks connexin channel-facilitated intercellular communication reduces mechanical allodynia (Chen, Park, et al., 2014). Cx43 is the major connexin isoform; it forms gap junctions in astrocytes (Dermietzel et al., 1989) and plays a major role in the interastrocyte communication through calcium waves (Naus et al., 1997). Calcium waves propagate the information in a more widespread fashion than synaptic connections (Charles et al., 1991), which could promote a state of generalised hyperexcitability of the nociceptive neurons (central sensitisation). The administration of compounds that can decouple gap junctions reduces the symptoms of 'mirror-image pain' when the non-injured side develops hypersensitivity after injury (Spataro et al., 2004; for a detailed overview of neuroimmune interaction in neuropathic pain, see Chapter 4 in this volume).

Disinhibition

In neuropathic pain states, there is substantial loss of the segmental inhibitory controls normally carried by GABAergic and glycinergic currents (Moore et al., 2002). Several mechanisms converge to produce this nerve injury-induced disinhibition of the spinal nociceptive neurons.

In the weeks following the axotomy, there is a gradual loss of inhibitory interneurons in the dorsal horn of the spinal cord (Sugimoto, Bennett and Kajander, 1990; Scholz et al., 2005) (Figure 2.7). This neuronal death is possibly the result of an NMDAR-induced excitotoxicity that develops over time, rather than the result of the large amount of glutamate released at the time of nerve injury (Scholz et al., 2005). This loss of inhibitory interneurons appears to require a permanent denervation. Neuropathic pain models where regeneration of the injured fibres can occur do not exhibit signs of neuronal apoptosis (Polgár et al., 2003; Polgár et al., 2004). Interestingly, the reduction in glycinergic neurotransmission after nerve injury is not mediated by the activation of EP2 receptors, which further indicates that inflammatory and neuropathic pain mechanisms differ (Hosl et al., 2006). The molecular mechanisms responsible for the loss of glycinergic tone in neuropathic pain states have not yet been identified, but they contribute, together with the loss of GABAergic tone, to the opoids' lack of efficacy in relieving neuropathic pain symptoms (Chen, Chen and Pan, 2005). The restoration of an adequate inhibitory tone through transplantation of immature telencephalic GABAergic interneurons progressively abolishes nerve injury-induced mechanical allodynia (Braz et al., 2012), but it does not reduce inflammatory-mediated pain behaviours caused by intra-plantar formalin injection (Braz et al., 2012).

Another mechanism that contributes greatly to the reduction in segmental inhibition in lamina I neurons after nerve injury is caused by a BDNF-dependent change in the homeostatic chloride concentration gradient in the spinal cord neurons (Figure 2.7). Under normal conditions the intracellular concentrations of Cl^- are maintained by the opposed effects of Cl^- co-transporter K^+-Cl^- exporter 2 channels (KCC2) and Na^+-K^+-Cl^- exporter 1 channels (NKCC1) (Price, Cervero and De Koninck, 2005). KCC2 cotransports Cl^- and K^+ ions out of the cells, whereas NKCC1 is responsible for an influx of K^+, Na^+ and Cl^- into the cells. The net effect of these two co-transporters is a steady state with low intracellular Cl^- concentration and high extracellular Cl^- concentration. Because both GABAAR and glyR are chloride-permeable channels that open upon binding of their ligand (Zeilhofer et al., 2012), their activation triggers a massive entry of Cl^- into the cell that causes a membrane hyperpolarisation (Figure 2.7). After peripheral nerve injury, however, activated microglial cells produce and release large amounts of BDNF (Coull et al., 2003). The binding of BDNF to the TrkB receptors increases NMDAR transmission (Kerr et al., 1999), which leads to a massive calcium entry and to the Ca^{2+}-dependent activation of calpain, which in turn cleaves KCC2 (Zhou et al., 2012). Reduced levels of KCC2 in the spinal cord significantly diminish the efflux of Cl^- from the cell, so that intracellular concentrations are abnormally high, which leads to a loss of the

Cl⁻ gradient between intra and extracellular milieus (Coull et al., 2003; Miletic and Miletic, 2008) (Figure 2.7). As a result, the binding of GABA or glycine on their receptors causes a minimal entry of Cl⁻ into the cell; this entry is insufficient to hyperpolarise the nociceptive neurons or can even cause a depolarisation (Coull et al., 2003; Coull et al., 2005). Altogether, this loss of both GABA and glycine inhibitory effects leaves the nociceptive neurons in a state of disinhibition, allowing the development and maintenance of central sensitisation. Inhibition of KCC2 in uninjured animals is sufficient to cause disinhibition and increases pain sensitivity (Austin and Delpire, 2011), whereas rescuing K-Cl co-transport in neuropathic pain states restores spinal nociceptive neurons to a normal sensitivity (Lavertu, Cote and De Koninck, 2014). More importantly, this allows the neurons of the superficial lamina to return to a nociceptive-specific state, which suggests that the restoration of proper inhibitory controls could help resetting the neurons from their central sensitisation state (Lavertu et al., 2014).

After peripheral nerve injury there is, over time, a progressive infiltration of immune-competent cells into the dorsal horns of the spinal cord, notably T-cells (Watkins et al., 2007; Cao and DeLeo, 2008; Costigan, Moss, et al., 2009). These T-cells produce specific cytokines such as IFN-γ, which cause a disinhibition of the nociceptive neurons (Vikman et al., 2003) by reducing GABAergic currents in the dorsal horn (Vikman, Dugan and Siddall, 2007) through the activation of IFN-γ receptors expressed by nociceptive neurons (Vikman et al., 1998) (Figure 2.7).

In addition to the reduction of segmental inhibitory controls, there is an increase in descending excitatory controls from the rostral ventromedial medulla (RVM) along with a reduction in descending inhibitory controls (Gardell et al., 2003; Vogel et al., 2003; Vera-Portocarrero et al., 2006; Sikandar, Bannister and Dickenson, 2012; Kim et al., 2014), further enhancing the hyperexcitability of spinal nociceptive neurons (Figure 2.7). The effects of serotonin released from the raphe magnus neurons in particular appear to display a strong shift towards a gain in excitability. The specific disruption of the serotonin production through injection of shRNA (RNAi) of TPH2 into the RVM reveals a pronociceptive effect of 5-HT in the second phase of the formalin test without alteration of baseline nociception; this suggests a specific role in spinal plasticity (Wei et al., 2010). In support of this view, TPH2 blockade in the RVM strongly reduces thermal hyperalgesia, but also mechanical allodynia after peripheral nerve injury (Wei et al., 2010; Kim et al., 2014). Lesion of the dorsolateral funiculus, which contains in part descending fibres from serotoninergic neurons (Basbaum and Fields, 1979), strongly attenuates spontaneous signs of pain and mechanical allodynia after peripheral nerve injury (Wang et al., 2013). One potential mechanism through which serotonin increases nociceptive neurons excitability is the sensitisation of TRPV1 channels expressed by primary afferent fibres through 5-HT3R, which enhance synaptic transmission (Kim et al., 2014). Because nerve injury causes large-diameter neurons to express nociceptive-specific factors, including TRPV1, this could contribute to the maintenance of central sensitisation by innocuous stimuli (Kim et al., 2014). In addition, TRPV1 channels appear to be expressed by some GABAergic inhibitory interneurons (Kim et al., 2012). TRPV1 activation triggers the disinhibition phenomenon by causing a long-term depression (LTD) between the interneurons and the projection nociceptive neurons of lamina I; preventing this reduces the development of mechanical allodynia caused by nerve injury (Kim et al., 2012). Because 5-HT3R are also expressed by inhibitory interneurons, it is possible then that this mechanism participates to the pro-nociceptive action of 5-HT in neuropathic states.

Other diseases with chronic pain syndromes

There is an increasing body of evidence to suggest a critical role of central nociceptive plasticity in the development and maintenance of pain hypersensitivity syndromes observed in diseases such as CRPS, pancreatitis, whiplash injury, fibromyalgia, low back pain or migraine (Woolf, 2010). For most of these pathologies there are no adequate preclinical animal models that could identify the structures or mechanisms causing the pain symptoms. Several clinical signs and behavioural manifestations point, however, to the existence of central sensitisation and suggest that approaches otherwise efficient at altering this state in other conditions might be effective in the treatment of pain symptoms as well. The next section will briefly describe some recent key results and observations that support the hypothesis of the role of central sensitisation in these painful conditions.

Complex regional pain syndrome may develop after limb trauma – even after apparently minor injuries – and is characterised by chronic pain and sensory–motor and autonomic disturbances (Birklein, 2005; Marinus et al., 2011). Patients with CRPS can develop mechanical allodynia (Maihofner, Handwerker and Birklein, 2006) and contralateral pain in the chronic stages of the disease (Huge et al., 2008); this strongly suggests a spinal involvement. In support of this view, the intradermal injection of capsaicin in patients with CRPS causes a higher hypersensitivity than in healthy subjects, but also bilateral pain (Terkelsen et al., 2014), whereas the capsaicin-induced neurogenic inflammation is the same (Terkelsen et al., 2014). This defect in central pain controls was also found in another cohort of patients, where quantitative sensory testing of the affected region – but also of non-affected areas – showed a strong widespread muscle hyperalgesia (Van Rooijen, Marinus and Van Hilten, 2013). Two recent clinical trials tested the effect of intravenous ketamine infusions and found that these treatments produced significant pain relief (Schwartzman et al., 2009; Goldberg et al., 2010).

Pancreatitis is associated with chronic abdominal pain in the majority of patients (Lankisch, 2001), and the pain symptoms appear to be quite refractory to several current lines of medication (Gress et al., 1999). There is a growing body of evidence indicating that most, if not all the pain symptoms in chronic pancreatitis are central in nature (Drewes et al., 2008). For example, quantitative sensory testing through electric pain detection and pain tolerance paradigms revealed that patients with chronic pancreatitis had lower pain thresholds, including in cutaneous territories non-referred to the pancreas, and a significant loss of descending inhibitory controls (Bouwense, Ahmed Ali, et al., 2013; Bouwense, Olesen, et al., 2013). Temporal summation is strongly facilitated in patients with pancreatitis, as well as the expansion of their receptive fields (Dimcevski et al., 2007). A recent clinical trial revealed that a chronic pregabalin treatment could attenuate pain hypersensitivity, especially in dermatomes outside of the pancreas territory. This suggests an effect on descending rather than segmental controls (Bouwense et al., 2012).

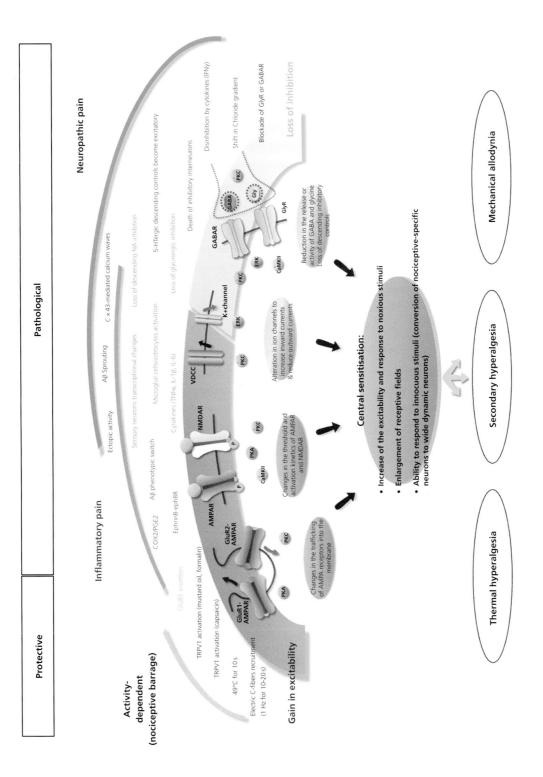

Protective

Pathological

Activity-dependent (nociceptive barrage)

Inflammatory pain

Neuropathic pain

Electric C-fibers recruitment (1 Hz for 10–20s)

49°C for 10s

TRPV1 activation (capsaicin)

TRPV1 activation (mustard oil, formalin)

GluR1 insertion

EphrinB-ephBR

COX2/PGE2

Aβ phenotypic switch

Cytokines (TNFα, IL-1β, IL-6)

Microglial cells/astrocytes activation

Sensory neurons transcriptional changes

Aβ Sprouting

Loss of glycinergic inhibition

5-HTergic descending controls become excitatory

Loss of descending NA inhibition

C × 43-mediated calcium waves

Ectopic activity

Death of inhibitory interneurons

Disinhibition by cytokines (IFNγ)

Shift in Chloride gradient

Blockade of GlyR or GABAR

Gain in excitability

Loss of inhibition

GluR1-AMPAR

GluR2-AMPAR

AMPAR

NMDAR

VDCC

K+channel

GABAR

GlyR

PKA

PKC

CaMKII

PKA

PKC

ERK

PKC

ERK

CaMKII

PKC

GABA

Gly

PKC

Changes in the trafficking of AMPA receptors into the membrane

Changes in the threshold and activation kinetics of AMPAR and NMDAR

Alteration in ion channels to increase inward currents & reduce outward currents

Reduction in the release or activity of GABA and glycine Loss of descending inhibitory controls

Central sensitisation:

- Increase of the excitability and response to noxious stimuli
- Enlargement of receptive fields
- Ability to respond to innocuous stimuli (conversion of nociceptive-specific neurons to wide dynamic neurons)

Thermal hyperalgesia

Secondary hyperalgesia

Mechanical allodynia

Patients suffering from either whiplash injury, which typically happens after car accidents, or fibromyalgia present a defect in centrally mediated inhibitory controls (Banic et al., 2004). Chronic whiplash-associated disorders appear to be accompanied by an almost complete loss of several types of descending controls (Daenen, Nijs, Cras, et al., 2013; Daenen, Nijs, Roussel, et al., 2013), which could explain the occurrence of pain symptoms in apparently healthy body areas. In patients with intractable neck pain, the anaesthetic infiltration of trigger points reduces the hypersensitivity observed in non-injured areas (Freeman, Nystrom and Centeno, 2009) – a phenomenon that confirms the central origin of this pain (Curatolo et al., 2001). Similarly, several symptoms in fibromyalgia point to a loss of descending inhibitory controls and to a generalised defect in spinal inhibition. Fibromyalgia patients have lower pressure thresholds and increased temporal summation to muscle stimulation (Graven-Nielsen et al., 2000; Staud et al., 2003), which can be reduced through systemic administration of ketamine (Graven-Nielsen et al., 2000). Experimental pain triggered by temporal summation protocol causes more pain in patients suffering from fibromyalgia and produces a greater extension of the nociceptive receptive fields (Staud et al., 2004). In addition, in this experimental setting, the hypersensitivity can be maintained using lower frequency than in healthy subjects, which again suggests that inhibitory controls are defective (Staud et al., 2004). Blocking sensory fibres through local intramuscular injections of an anaesthetic attenuates secondary hyperalgesia, a phenomenon that indicates that sensory input from muscles participates in the maintenance of the central hyperexcitability (Staud et al., 2009).

Concluding remarks

A summary of the factors causing central sensitisation through either a gain in excitability or a loss of inhibition in normal states and in chronic pain states is described in Figure 2.8.

Functional plasticity of the spinal nociceptive system is a critical feature of physiological pain. There are relatively few nociceptive neurons that project from the spinal cord to the brain and determine how much pain we feel and for how long. The nature of the signal they send is

Figure 2.8 Schematic representation of the factors that can trigger central sensitisation under normal or pathological conditions; the cellular processes involved; the three main features of central sensitisation; and the associated behavioural manifestations. Factors causing a gain in the excitability of the projection spinal nociceptive neuron are listed along the red arc, whereas those causing a loss of inhibition are listed along the blue arc. Activity-dependent inducers of central sensitisation are represented in black, inflammatory pain-specific mechanisms are represented in pink, neuropathic pain-specific mechanisms are represented in purple and mechanisms shared by inflammatory and neuropathic pains are represented in magenta. Central sensitisation can be triggered by a strong and sustained nociceptive activation, something referred to as activity-dependent central sensitisation. This represents a protective mechanism and is mostly mediated by a gain in the excitability of the spinal nociceptive neuron. Both chronic inflammatory pain and neuropathic pain exhibit many changes that can trigger or maintain central sensitisation. Some mechanisms overlap between the two types of diseases, whereas others are specific. Chronic inflammatory pain has a majority of changes that favour a gain in excitability, and neuropathic pain is associated with a severe loss of inhibition. **At the bottom**: Thermal hyperalgesia, secondary hyperalgesia and mechanical allodynia are symptoms that can be explained by the development of central sensitisation. Secondary hyperalgesia and mechanical allodynia are symptoms that strongly suggest functional changes at the spinal level.

therefore crucial to our reacting adequately to potential threats. The spinal cord is the first centre for the integration of the nociceptive input; the information is not just passively relayed to the brain areas but modified to convey the most relevant message possible. In normal states the nociceptive projection neurons are under strong local and distant (supraspinal) inhibitory and excitatory controls, whose balance affects the overall excitability of the cell. Sustained nociceptive activity triggers a state of hyperexcitability called 'activity-dependent central sensitisation', where nociceptive neurons display three key features: an increase in excitability and response to noxious stimuli; an enlargement of their receptor field; and an ability to respond to innocuous stimuli. In this state of central sensitisation the nociceptive system is no longer coupled to its environment. This can represent an adaptive advantage designed to promote behaviours and strategies that allow for a better protection of an injured body part, as the degree of nociceptive input required to trigger central sensitisation is almost always sufficient to cause tissue damage.

If central sensitisation occurs, or is maintained in the absence of tissue damage, this becomes a pathological and maladaptive plasticity of the nociceptive system. Nociception is not a sense any more, as it does not transmit relevant – or even real – information from the environment. Because nociception gives rise to pain and pain is an extremely complex and integrated cognitive state (Loeser and Melzack, 1999), an 'out-of-tune' nociceptive system will have dramatic chronic consequences that will eventually affect brain structures (Apkarian et al., 2004; Pelled et al., 2007; Wei and Zhuo, 2008; Valet et al., 2009).

Central sensitisation can be triggered by many different mechanisms, which can be schematically classed into two broad categories: a gain in excitability and a loss of inhibitory controls (Latrémolière and Woolf, 2009). Central sensitisation is not caused by a single defining molecular mechanism, but rather by one of many possible combinations that can affect the excitatory–inhibitory balance of the nociceptive spinal neuron. It seems logical therefore to infer that any pathological state associated with spinal alterations could affect the modulatory controls of spinal nociceptive neurons and could cause or facilitate central sensitisation and its associated behavioural manifestations. These alterations could originate directly at the spinal level or could affect supraspinal structures, which will in turn affect the excitatory–inhibitory balance of the nociceptive system.

Acknowledgements

I would like to address my sincere thanks to Dr Chloe Alexandre for her support and excellent advice.

References

Abbadie, C., Brown, J. L., Mantyh, P. W. and Basbaum, A. I. 1996. Spinal cord substance P receptor immunoreactivity increases in both inflammatory and nerve injury models of persistent pain. *Neuroscience* 70: 201–9.

Abbadie, C., Trafton, J., Liu, H., Mantyh, P. W. and Basbaum, A. I. 1997. Inflammation increases the distribution of dorsal horn neurons that internalize the neurokinin-1 receptor in response to noxious and non-noxious stimulation. *Journal of Neuroscience* 17: 8049–60.

Acosta, C., Djouhri, L., Watkins, R., Berry, C., Bromage, K. and Lawson, S. N. 2014. TREK2 expressed selectively in IB4-binding C-fiber nociceptors hyperpolarizes their membrane potentials and limits spontaneous pain. *Journal of Neuroscience* 34: 1494–509.

Adwanikar, H., Karim, F. and Gereau, R. W. T. 2004. Inflammation persistently enhances nocifensive behaviors mediated by spinal group I mGluRs through sustained ERK activation. *Pain* 111: 125–35.

Afrah, A. W., Fiska, A., Gjerstad, J., Gustafsson, H., Tjolsen, A., Olgart, L., Stiller, C. O., Hole, K. and Brodin, E. 2002. Spinal substance P release in vivo during the induction of long-term potentiation in dorsal horn neurons. *Pain* 96: 49–55.

Ahmadi, S., Lippross, S., Neuhuber, W. L. and Zeilhofer, H. U. 2002. PGE(2) selectively blocks inhibitory glycinergic neurotransmission onto rat superficial dorsal horn neurons. *Nature Neuroscience* 5: 34–40.

Aloisi, F., Care, A., Borsellino, G., Gallo, P., Rosa, S., Bassani, A., Cabibbo, A., Testa, U., Levi, G. and Peschle, C. 1992. Production of hemolymphopoietic cytokines (IL-6, IL-8, colony-stimulating factors) by normal human astrocytes in response to IL-1 beta and tumor necrosis factor-alpha. *Journal of Immunology* 149: 2358–66.

Alvarez, F. J., Kavookjian, A. M. and Light, A. R. 1992. Synaptic interactions between GABA-immunoreactive profiles and the terminals of functionally defined myelinated nociceptors in the monkey and cat spinal cord. *Journal of Neuroscience* 12: 2901–17.

Alvarez, F. J., Villalba, R. M., Carr, P. A., Grandes, P. and Somohano, P. M. 2000. Differential distribution of metabotropic glutamate receptors 1a, 1b, and 5 in the rat spinal cord. *Journal of Comparative Neurology* 422: 464–87.

Antal, M., Fukazawa, Y., Eordogh, M., Muszil, D., Molnar, E., Itakura, M., Takahashi, M. and Shigemoto, R. 2008. Numbers, densities, and colocalization of AMPA- and NMDA-type glutamate receptors at individual synapses in the superficial spinal dorsal horn of rats. *Journal of Neuroscience* 28: 9692–701.

Apkarian, A. V., Sosa, Y., Sonty, S., Levy, R. M., Harden, R. N., Parrish, T. B. and Gitelman, D. R. 2004. Chronic back pain is associated with decreased prefrontal and thalamic gray matter density. *Journal of Neuroscience* 24: 10410–5.

Austin, P. J. and Moalem-Taylor, G. 2010. The neuro-immune balance in neuropathic pain: Involvement of inflammatory immune cells, immune-like glial cells and cytokines. *Journal of Neuroimmunology* 229: 26–50.

Austin, T. M. and Delpire, E. 2011. Inhibition of KCC2 in mouse spinal cord neurons leads to hypersensitivity to thermal stimulation. *Anesthesia & Analgesia* 113: 1509–15.

Baba, H., Doubell, T. P. and Woolf, C. J. 1999. Peripheral inflammation facilitates Abeta fiber-mediated synaptic input to the substantia gelatinosa of the adult rat spinal cord. *Journal of Neuroscience* 19: 859–67.

Baba, H., Kohno, T., Moore, K. A. and Woolf, C. J. 2001. Direct activation of rat spinal dorsal horn neurons by prostaglandin E2. *Journal of Neuroscience* 21: 1750–6.

Baba, H., Ji, R. R., Kohno, T., Moore, K. A., Ataka, T., Wakai, A., Okamoto, M. and Woolf, C. J. 2003. Removal of GABAergic inhibition facilitates polysynaptic A fiber-mediated excitatory transmission to the superficial spinal dorsal horn. *Molecular and Cellular Neuroscience* 24: 818–30.

Balkowiec, A. and Katz, D. M. 2000. Activity-dependent release of endogenous brain-derived neurotrophic factor from primary sensory neurons detected by ELISA in situ. *Journal of Neuroscience* 20: 7417–23.

Banic, B., Petersen-Felix, S., Andersen, O. K., Radanov, B. P., Villiger, P. M., Arendt-Nielsen, L. and Curatolo, M. 2004. Evidence for spinal cord hypersensitivity in chronic pain after whiplash injury and in fibromyalgia. *Pain* 107: 7–15.

Banke, T. G., Bowie, D., Lee, H., Huganir, R. L., Schousboe, A. and Traynelis, S. F. 2000. Control of GluR1 AMPA receptor function by cAMP-dependent protein kinase. *Journal of Neuroscience* 20: 89–102.

Bardin, L., Lavarenne, J. and Eschalier, A. 2000a. Serotonin receptor subtypes involved in the spinal antinociceptive effect of 5-HT in rats. *Pain* 86: 11–8.

Bardin, L., Schmidt, J., Alloui, A. and Eschalier, A. 2000b. Effect of intrathecal administration of serotonin in chronic pain models in rats. *European Journal of Pharmacology* 409: 37–43.

Baron, R. 2006. Mechanisms of disease: Neuropathic pain: A clinical perspective. *Nature Clinical Practice Neurology* 2: 95–106.

Baron, R., Binder, A. and Wasner, G. 2010. Neuropathic pain: Diagnosis, pathophysiological mechanisms, and treatment. *Lancet Neurology* 9: 807–19.

Basbaum, A. I. and Fields, H. L. 1979. The origin of descending pathways in the dorsolateral funiculus of the spinal cord of the cat and rat: Further studies on the anatomy of pain modulation. *Journal of Comparative Neurology* 187: 513–31.

Basbaum, A. I., Bautista, D. M., Scherrer, G. and Julius, D. 2009. Cellular and molecular mechanisms of pain. *Cell* 139: 267–84.

Battaglia, A. A., Sehayek, K., Grist, J., Mcmahon, S. B. and Gavazzi, I. 2003. EphB receptors and ephrin-B ligands regulate spinal sensory connectivity and modulate pain processing. *Nature Neuroscience* 6: 339–40.

Beggs, S. and Salter, M. W. 2007. Stereological and somatotopic analysis of the spinal microglial response to peripheral nerve injury. *Brain, Behavior, and Immunity* 21: 624–33.

Beggs, S. and Salter, M. W. 2010. Microglia-neuronal signalling in neuropathic pain hypersensitivity 2.0. *Current Opinion in Neurobiology* 20: 474–80.

Bennett, D. L. and Woods, C. G. 2014. Painful and painless channelopathies. *Lancet Neurology* 13: 587–99.

Berthele, A., Boxall, S. J., Urban, A., Anneser, J. M., Zieglgansberger, W., Urban, L. and Tolle, T. R. 1999. Distribution and developmental changes in metabotropic glutamate receptor messenger RNA expression in the rat lumbar spinal cord. *Brain Research: Developmental Brain Research* 112: 39–53.

Bester, H., Beggs, S. and Woolf, C. J. 2000. Changes in tactile stimuli-induced behavior and c-Fos expression in the superficial dorsal horn and in parabrachial nuclei after sciatic nerve crush. *Journal of Comparative Neurology* 428: 45–61.

Beyer, C., Roberts, L. A. and Komisaruk, B. R. 1985. Hyperalgesia induced by altered glycinergic activity at the spinal cord. *Life Sciences* 37: 875–82.

Birklein, F. 2005. Complex regional pain syndrome. *Journal of Neurology* 252: 131–8.

Bliss, T. V. and Collingridge, G. L. 1993. A synaptic model of memory: Long-term potentiation in the hippocampus. *Nature* 361: 31–9.

Bohlhalter, S., Weinmann, O., Mohler, H. and Fritschy, J. M. 1996. Laminar compartmentalization of GABAA-receptor subtypes in the spinal cord: An immunohistochemical study. *Journal of Neuroscience* 16: 283–97.

Bormann, J. 2000. The 'ABC' of GABA receptors. *Trends in Pharmacological Sciences* 21: 16–9.

Bouwense, S. A., Olesen, S. S., Drewes, A. M., Frokjaer, J. B., Van Goor, H. and Wilder-Smith, O. H. 2013a. Is altered central pain processing related to disease stage in chronic pancreatitis patients with pain? An exploratory study. PLoS One 8: e55460.

Bouwense, S. A., Olesen, S. S., Drewes, A. M., Poley, J. W., Van Goor, H. and Wilder-Smith, O. H. 2012. Effects of pregabalin on central sensitization in patients with chronic pancreatitis in a randomized, controlled trial. PLoS One 7: e42096.

Bouwense, S. A., Ahmed Ali, U., Ten Broek, R. P., Issa, Y., Van Eijck, C. H., Wilder-Smith, O. H. and Van Goor, H. 2013b. Altered central pain processing after pancreatic surgery for chronic pancreatitis. *British Journal of Surgery* 100: 1797–804.

Bowery, N. G., Hudson, A. L. and Price, G. W. 1987. GABAA and GABAB receptor site distribution in the rat central nervous system. *Neuroscience* 20: 365–83.

Braz, J. M., Sharif-Naeini, R., Vogt, D., Kriegstein, A., Alvarez-Buylla, A., Rubenstein, J. L. and Basbaum, A. I. 2012. Forebrain GABAergic neuron precursors integrate into adult spinal cord and reduce injury-induced neuropathic pain. *Neuron* 74: 663–75.

Breivik, H., Collett, B., Ventafridda, V., Cohen, R. and Gallacher, D. 2006. Survey of chronic pain in Europe: Prevalence, impact on daily life, and treatment. *European Journal of Pain* 10: 287–333.

Brenner, G. J., Ji, R. R., Shaffer, S. and Woolf, C. J. 2004. Peripheral noxious stimulation induces phosphorylation of the NMDA receptor NR1 subunit at the PKC-dependent site, serine-896, in spinal cord dorsal horn neurons. *European Journal of Neuroscience* 20: 375–84.

Burstein, R. and Jakubowski, M. 2004. Analgesic triptan action in an animal model of intracranial pain: A race against the development of central sensitization. *Annals of Neurology* 55: 27–36.

Calvo, M., Zhu, N., Grist, J., Ma, Z., Loeb, J. A. and Bennett, D. L. 2011. Following nerve injury neuregulin-1 drives microglial proliferation and neuropathic pain via the MEK/ERK pathway. *Glia* 59: 554–68.

Calvo, M., Zhu, N., Tsantoulas, C., Ma, Z., Grist, J., Loeb, J. A. and Bennett, D. L. 2010. Neuregulin-ErbB signaling promotes microglial proliferation and chemotaxis contributing to microgliosis and pain after peripheral nerve injury. *Journal of Neuroscience* 30: 5437–50.

Cao, L. and Deleo, J. A. 2008. CNS-infiltrating CD4+ T lymphocytes contribute to murine spinal nerve transection-induced neuropathic pain. *European Journal of Immunology* 38: 448–58.

Carmichael, N. M., Charlton, M. P. and Dostrovsky, J. O. 2008. Activation of the 5-HT1B/D receptor reduces hindlimb neurogenic inflammation caused by sensory nerve stimulation and capsaicin. *Pain* 134: 97–105.

Carvalho, A. L., Duarte, C. B. and Carvalho, A. P. 2000. Regulation of AMPA receptors by phosphorylation. *Neurochemical Research* 25: 1245–55.

Caterina, M. J., Schumacher, M. A., Tominaga, M., Rosen, T. A., Levine, J. D. and Julius, D. 1997. The capsaicin receptor: A heat-activated ion channel in the pain pathway. *Nature* 389: 816–24.

Cavanaugh, D. J., Lee, H., Lo, L., Shields, S. D., Zylka, M. J., Basbaum, A. I. and Anderson, D. J. 2009. Distinct subsets of unmyelinated primary sensory fibers mediate behavioral responses to noxious thermal and mechanical stimuli. *Proceedings of the National Academy of Sciences of the United States of America* 106: 9075–80.

Charles, A. C., Merrill, J. E., Dirksen, E. R. and Sanderson, M. J. 1991. Intercellular signaling in glial cells: Calcium waves and oscillations in response to mechanical stimulation and glutamate. *Neuron* 6: 983–92.

Chen, B. S. and Roche, K. W. 2007. Regulation of NMDA receptors by phosphorylation. *Neuropharmacology* 53: 362–8.

Chen, G., Park, C. K., Xie, R. G., Berta, T., Nedergaard, M. and Ji, R. R. 2014a. Connexin-43 induces chemokine release from spinal cord astrocytes to maintain late-phase neuropathic pain in mice. *Brain* 137: 2193–209.

Chen, L. and Huang, L. Y. 1992. Protein kinase C reduces Mg2+ block of NMDA-receptor channels as a mechanism of modulation. *Nature* 356: 521–3.

Chen, S. R. and Pan, H. L. 2005. Distinct roles of group III metabotropic glutamate receptors in control of nociception and dorsal horn neurons in normal and nerve-injured Rats. *Journal of Pharmacology and Experimental Therapeutics* 312: 120–6.

Chen, W., Walwyn, W., Ennes, H. S., Kim, H., McRoberts, J. A. and Marvizon, J. C. 2014b. BDNF released during neuropathic pain potentiates NMDA receptors in primary afferent terminals. *Eur Journal of Neuroscience* 39: 1439–54.

Chen, Y. P., Chen, S. R. and Pan, H. L. 2005. Effect of morphine on deep dorsal horn projection neurons depends on spinal GABAergic and glycinergic tone: Implications for reduced opioid effect in neuropathic pain. *Journal of Pharmacology and Experimental Therapeutics* 315: 696–703.

Cheng, H. T., Suzuki, M., Hegarty, D. M., Xu, Q., Weyerbacher, A. R., South, S. M., Ohata, M. and Inturrisi, C. E. 2008. Inflammatory pain-induced signaling events following a conditional deletion of the N-methyl-D-aspartate receptor in spinal cord dorsal horn. *Neuroscience* 155: 948–58.

Chery, N. and De Koninck, Y. 1999. Junctional versus extrajunctional glycine and GABA(A) receptor-mediated IPSCs in identified lamina I neurons of the adult rat spinal cord. *Journal of Neuroscience* 19: 7342–55.

Choi, J. I., Svensson, C. I., Koehrn, F. J., Bhuskute, A. and Sorkin, L. S. 2010. Peripheral inflammation induces tumor necrosis factor dependent AMPA receptor trafficking and Akt phosphorylation in spinal cord in addition to pain behavior. *Pain* 149: 243–53.

Christensen, D., Gautron, M., Guilbaud, G. and Kayser, V. 2001. Effect of gabapentin and lamotrigine on mechanical allodynia-like behaviour in a rat model of trigeminal neuropathic pain. *Pain* 93: 147–53.

Chung, H. J., Xia, J., Scannevin, R. H., Zhang, X. and Huganir, R. L. 2000. Phosphorylation of the AMPA receptor subunit GluR2 differentially regulates its interaction with PDZ domain-containing proteins. *Journal of Neuroscience* 20: 7258–67.

Coderre, T. J. and Melzack, R. 1992. The role of NMDA receptor-operated calcium channels in persistent nociception after formalin-induced tissue injury. *Journal of Neuroscience* 12: 3671–5.

Colburn, R. W., Deleo, J. A., Rickman, A. J., Yeager, M. P., Kwon, P. and Hickey, W. F. 1997. Dissociation of microglial activation and neuropathic pain behaviors following peripheral nerve injury in the rat. *Journal of Neuroimmunology* 79: 163–75.

Collingridge, G. L., Isaac, J. T. and Wang, Y. T. 2004. Receptor trafficking and synaptic plasticity. *Nature Reviews Neuroscience* 5: 952–62.

Cook, A. J., Woolf, C. J. and Wall, P. D. 1986. Prolonged C-fibre mediated facilitation of the flexion reflex in the rat is not due to changes in afferent terminal or motoneurone excitability. *Neuroscience Letters* 70: 91–6.

Costigan, M., Scholz, J. and Woolf, C. J. 2009a. Neuropathic pain: A maladaptive response of the nervous system to damage. *Annual Review of Neuroscience* 32: 1–32.

Costigan, M., Befort, K., Karchewski, L., Griffin, R. S., D'Urso, D., Allchorne, A., Sitarski, J., Mannion, J. W., Pratt, R. E. and Woolf, C. J. 2002. Replicate high-density rat genome oligonucleotide microarrays reveal hundreds of regulated genes in the dorsal root ganglion after peripheral nerve injury. *BMC Neuroscience* 3: 16.

Costigan, M., Moss, A., Latrémoliere, A., Johnston, C., Verma-Gandhu, M., Herbert, T. A., Barrett, L., Brenner, G. J., Vardeh, D., Woolf, C. J., et al. 2009b. T-cell infiltration and signaling in the adult dorsal spinal cord is a major contributor to neuropathic pain-like hypersensitivity. *Journal of Neuroscience* 29: 14415–22.

Coull, J. A., Beggs, S., Boudreau, D., Boivin, D., Tsuda, M., Inoue, K., Gravel, C., Salter, M. W. and De Koninck, Y. 2005. BDNF from smicroglia causes the shift in neuronal anion gradient underlying neuropathic pain. *Nature* 438: 1017–21.

Coull, J. A., Boudreau, D., Bachand, K., Prescott, S. A., Nault, F., Sik, A., De Koninck, P. and De Koninck, Y. 2003. Trans-synaptic shift in anion gradient in spinal lamina I neurons as a mechanism of neuropathic pain. *Nature* 424: 938–42.

Cox, J. J., Reimann, F., Nicholas, A. K., Thornton, G., Roberts, E., Springell, K., Karbani, G., Jafri, H., Mannan, J., Raashid, Y., et al. 2006. An SCN9A channelopathy causes congenital inability to experience pain. *Nature* 444: 894–8.

Cox, J. J., Sheynin, J., Shorer, Z., Reimann, F., Nicholas, A. K., Zubovic, L., Baralle, M., Wraige, E., Manor, E., Levy, J., et al. 2010. Congenital insensitivity to pain: Novel SCN9A missense and in-frame deletion mutations. *Human Mutation* 31, E1670–86.

Curatolo, M., Petersen-Felix, S., Arendt-Nielsen, L., Giani, C., Zbinden, A. M. and Radanov, B. P. 2001. Central hypersensitivity in chronic pain after whiplash injury. *Clinical Journal of Pain* 17: 306–15.

Daenen, L., Nijs, J., Cras, P., Wouters, K. and Roussel, N. 2013a. Changes in pain modulation occur soon after whiplash trauma but are not related to altered perception of distorted visual feedback. *Pain Practice* 14: 588–98.

Daenen, L., Nijs, J., Roussel, N., Wouters, K., Van Loo, M. and Cras, P. 2013b. Dysfunctional pain inhibition in patients with chronic whiplash-associated disorders: An experimental study. *Clinical Rheumatology* 32: 23–31.

Dahlhaus, A., Ruscheweyh, R. and Sandkuhler, J. 2005. Synaptic input of rat spinal lamina I projection and unidentified neurones in vitro. *Journal of Physiology* 566: 355–68.

Dalva, M. B., Takasu, M. A., Lin, M. Z., Shamah, S. M., Hu, L., Gale, N. W. and Greenberg, M. E. 2000. EphB receptors interact with NMDA receptors and regulate excitatory synapse formation. *Cell* 103: 945–56.

Davalos, D., Grutzendler, J., Yang, G., Kim, J. V., Zuo, Y., Jung, S., Littman, D. R., Dustin, M. L. and Gan, W. B. 2005. ATP mediates rapid microglial response to local brain injury in vivo. *Nature Neuroscience* 8: 752–8.

Davies, S. N. and Lodge, D. 1987. Evidence for involvement of N-methylaspartate receptors in 'wind-up' of class 2 neurones in the dorsal horn of the rat. *Brain Research* 424: 402–6.

De Novellis, V., Siniscalco, D., Galderisi, U., Fuccio, C., Nolano, M., Santoro, L., Cascino, A., Roth, K. A., Rossi, F. and Maione, S. 2004. Blockade of glutamate mGlu5 receptors in a rat model of neuropathic pain prevents early over-expression of pro-apoptotic genes and morphological changes in dorsal horn lamina II. *Neuropharmacology* 46: 468–79.

Deleo, J. A. and Yezierski, R. P. 2001. The role of neuroinflammation and neuroimmune activation in persistent pain. *Pain* 90: 1–6.

Dermietzel, R., Traub, O., Hwang, T. K., Beyer, E., Bennett, M. V., Spray, D. C. and Willecke, K. 1989. Differential expression of three gap junction proteins in developing and mature brain tissues. *Proceedings of the National Academy of Sciences of the United States of America* 86: 10148–52.

Devor, M. 2009. Ectopic discharge in Abeta afferents as a source of neuropathic pain. *Experimental Brain Research* 196: 115–28.

Dickenson, A. H. and Sullivan, A. F. 1987. Evidence for a role of the NMDA receptor in the frequency dependent potentiation of deep rat dorsal horn nociceptive neurones following C fibre stimulation. *Neuropharmacology* 26: 1235–8.

Dickenson, A. H., Chapman, V. and Green, G. M. 1997. The pharmacology of excitatory and inhibitory amino acid-mediated events in the transmission and modulation of pain in the spinal cord. *General Pharmacology* 28: 633–8.

Dimcevski, G., Staahl, C., Andersen, S. D., Thorsgaard, N., Funch-Jensen, P., Arendt-Nielsen, L. and Drewes, A. M. 2007. Assessment of experimental pain from skin, muscle, and esophagus in patients with chronic pancreatitis. *Pancreas* 35: 22–9.

Dingledine, R. 1983. N-methyl aspartate activates voltage-dependent calcium conductance in rat hippocampal pyramidal cells. *Journal of Physiology* 343: 385–405.

Djouhri, L., Fang, X., Koutsikou, S. and Lawson, S. N. 2012. Partial nerve injury induces electrophysiological changes in conducting (uninjured) nociceptive and nonnociceptive DRG neurons: Possible relationships to aspects of peripheral neuropathic pain and paresthesias. *Pain* 153: 1824–36.

Djouhri, L., Koutsikou, S., Fang, X., Mcmullan, S. and Lawson, S. N. 2006. Spontaneous pain, both neuropathic and inflammatory, is related to frequency of spontaneous firing in intact C-fiber nociceptors. *Journal of Neuroscience* 26: 1281–92.

Dobler, T., Springauf, A., Tovornik, S., Weber, M., Schmitt, A., Sedlmeier, R., Wischmeyer, E. and Doring, F. 2007. TRESK two-pore-domain K+ channels constitute a significant component of background potassium currents in murine dorsal root ganglion neurones. *Journal of Physiology* 585: 867–79.

Dogrul, A., Ossipov, M. H., Lai, J., Malan, T. P., Jr. and Porreca, F. 2000. Peripheral and spinal antihyperalgesic activity of SIB-1757, a metabotropic glutamate receptor (mGLUR(5)) antagonist, in experimental neuropathic pain in rats. *Neuroscience Letters* 292: 115–8.

Dong, H., O'brien, R. J., Fung, E. T., Lanahan, A. A., Worley, P. F. and Huganir, R. L. 1997. GRIP: A synaptic PDZ domain-containing protein that interacts with AMPA receptors. *Nature* 386: 279–84.

Dorf, M. E., Berman, M. A., Tanabe, S., Heesen, M. and Luo, Y. 2000. Astrocytes express functional chemokine receptors. *Journal of Neuroimmunology* 111: 109–21.

Dougherty, P. M. and Willis, W. D. 1991. Enhancement of spinothalamic neuron responses to chemical and mechanical stimuli following combined micro-iontophoretic application of N-methyl-D-aspartic acid and substance P. *Pain* 47: 85–93.

Dougherty, P. M. and Willis, W. D. 1992. Enhanced responses of spinothalamic tract neurons to excitatory amino acids accompany capsaicin-induced sensitization in the monkey. *Journal of Neuroscience* 12: 883–94.

Drdla, R. and Sandkuhler, J. 2008. Long-term potentiation at C-fibre synapses by low-level presynaptic activity in vivo. *Molecular Pain* 4: 18.

Drewes, A. M., Krarup, A. L., Detlefsen, S., Malmstrom, M. L., Dimcevski, G. and Funch-Jensen, P. 2008. Pain in chronic pancreatitis: The role of neuropathic pain mechanisms. *Gut* 57: 1616–27.

Dyck, P. J., Peroutka, S., Rask, C., Burton, E., Baker, M. K., Lehman, K. A., Gillen, D. A., Hokanson, J. L. and O'brien, P. C. 1997. Intradermal recombinant human nerve growth factor induces pressure allodynia and lowered heat-pain threshold in humans. *Neurology* 48: 501–5.

Ehlers, M. D., Heine, M., Groc, L., Lee, M. C. and Choquet, D. 2007. Diffusional trapping of GluR1 AMPA receptors by input-specific synaptic activity. *Neuron* 54: 447–60.

Esteban, J. A., Shi, S. H., Wilson, C., Nuriya, M., Huganir, R. L. and Malinow, R. 2003. PKA phosphorylation of AMPA receptor subunits controls synaptic trafficking underlying plasticity. *Nature Neuroscience* 6: 136–43.

Everill, B. and Kocsis, J. D. 1999. Reduction in potassium currents in identified cutaneous afferent dorsal root ganglion neurons after axotomy. *Journal of Neurophysiology* 82: 700–8.

Fagni, L., Chavis, P., Ango, F. and Bockaert, J. 2000. Complex interactions between mGluRs, intracellular Ca2+ stores and ion channels in neurons. *Trends in Neurosciences* 23: 80–8.

Fang, L., Wu, J., Zhang, X., Lin, Q. and Willis, W. D. 2003a. Increased phosphorylation of the GluR1 subunit of spinal cord alpha-amino-3-hydroxy-5-methyl-4-isoxazole propionate receptor in rats following intradermal injection of capsaicin. *Neuroscience* 122: 237–45.

Fang, L., Wu, J., Lin, Q. and Willis, W. D. 2002. Calcium-calmodulin-dependent protein kinase II contributes to spinal cord central sensitization. *Journal of Neuroscience* 22: 4196–204.

Fang, L., Wu, J., Lin, Q. and Willis, W. D. 2003b. Protein kinases regulate the phosphorylation of the GluR1 subunit of AMPA receptors of spinal cord in rats following noxious stimulation. *Brain Research: Molecular Brain Research* 118: 160–5.

Ferrini, F., Trang, T., Mattioli, T. A., Laffray, S., Del'guidice, T., Lorenzo, L. E., Castonguay, A., Doyon, N., Zhang, W., Godin, A. G., et al. 2013. Morphine hyperalgesia gated through microglia-mediated disruption of neuronal Cl(-) homeostasis. *Nature Neuroscience* 16: 183–92.

Field, M. J., Cox, P. J., Stott, E., Melrose, H., Offord, J., Su, T. Z., Bramwell, S., Corradini, L., England, S., Winks, J., et al. 2006. Identification of the alpha2-delta-1 subunit of voltage-dependent calcium channels as a molecular target for pain mediating the analgesic actions of pregabalin. *Proceedings of the National Academy of Sciences of the United States of America* 103: 17537–42.

Field, M. J., Mccleary, S., Hughes, J. and Singh, L. 1999. Gabapentin and pregabalin, but not morphine and amitriptyline, block both static and dynamic components of mechanical allodynia induced by streptozocin in the rat. *Pain* 80: 391–8.

Fields, H. L., Basbaum, A. I., Clanton, C. H. and Anderson, S. D. 1977. Nucleus raphe magnus inhibition of spinal cord dorsal horn neurons. *Brain Research* 126: 441–53.

Foster, E., Wildner, H., Tudeau, L., Haueter, S., Ralvenius, W. T., Jegen, M., Johannssen, H., Hosli, L., Haenaets, K., Ghanem, A., et al. 2015. Targeted ablation, silencing, and activation establish glycinergic dorsal horn neurons as key components of a spinal gate for pain and itch. *Neuron*, 85: 1289–304.

Freeman, M. D., Nystrom, A. and Centeno, C. 2009. Chronic whiplash and central sensitization; an evaluation of the role of a myofascial trigger points in pain modulation. *Journal of Brachial Plexus and Peripheral Nerve Injury* 4: 2.

Fukuoka, T., Kondo, E., Dai, Y., Hashimoto, N. and Noguchi, K. 2001. Brain-derived neurotrophic factor increases in the uninjured dorsal root ganglion neurons in selective spinal nerve ligation model. *Journal of Neuroscience* 21: 4891–900.

Fundytus, M. E., Osborne, M. G., Henry, J. L., Coderre, T. J. and Dray, A. 2002. Antisense oligonucleotide knockdown of mGluR1 alleviates hyperalgesia and allodynia associated with chronic inflammation. *Pharmacology Biochemistry & Behavior* 73: 401–10.

Galan, A., Laird, J. M. and Cervero, F. 2004. In vivo recruitment by painful stimuli of AMPA receptor subunits to the plasma membrane of spinal cord neurons. *Pain* 112: 315–23.

Gao, Y. J. and Ji, R. R. 2010. Targeting astrocyte signaling for chronic pain. *Neurotherapeutics* 7: 482–93.

Gao, Y. J., Zhang, L., Samad, O. A., Suter, M. R., Yasuhiko, K., Xu, Z. Z., Park, J. Y., Lind, A. L., Ma, Q. and Ji, R. R. 2009. JNK-induced MCP-1 production in spinal cord astrocytes contributes to central sensitization and neuropathic pain. *Journal of Neuroscience* 29: 4096–108.

Gardell, L. R., Vanderah, T. W., Gardell, S. E., Wang, R., Ossipov, M. H., Lai, J. and Porreca, F. 2003. Enhanced evoked excitatory transmitter release in experimental neuropathy requires descending facilitation. *Journal of Neuroscience* 23: 8370–9.

Garrison, C. J., Dougherty, P. M. and Carlton, S. M. 1994. GFAP expression in lumbar spinal cord of naive and neuropathic rats treated with MK-801. *Experimental Neurology* 129: 237–43.

Gauriau, C. and Bernard, J. F. 2002. Pain pathways and parabrachial circuits in the rat. *Experimental Physiology* 87: 251–8.

Gebhart, G. F. 2004. Descending modulation of pain. *Neuroscience & Biobehavioral Reviews* 27: 729–37.

Geng, S. J., Liao, F. F., Dang, W. H., Ding, X., Liu, X. D., Cai, J., Han, J. S., Wan, Y. and Xing, G. G. 2010. Contribution of the spinal cord BDNF to the development of neuropathic pain by activation of the NR2B-containing NMDA receptors in rats with spinal nerve ligation. *Experimental Neurology* 222: 256–66.

Giles, P. A., Trezise, D. J. and King, A. E. 2007. Differential activation of protein kinases in the dorsal horn in vitro of normal and inflamed rats by group I metabotropic glutamate receptor subtypes. *Neuropharmacology* 53: 58–70.

Gold, M. S. and Gebhart, G. F. 2010. Nociceptor sensitization in pain pathogenesis. *Nature Medicine* 16: 1248–57.

Gold, M. S., Weinreich, D., Kim, C. S., Wang, R., Treanor, J., Porreca, F. and Lai, J. 2003. Redistribution of Na(V)1.8 in uninjured axons enables neuropathic pain. *Journal of Neuroscience* 23: 158–66.

Goldberg, M. E., Torjman, M. C., Schwartzman, R. J., Mager, D. E. and Wainer, I. W. 2010. Pharmacodynamic profiles of ketamine (R)- and (S)- with 5-day inpatient infusion for the treatment of complex regional pain syndrome. *Pain Physician* 13: 379–87.

Goosens, K. A. and Maren, S. 2002. Long-term potentiation as a substrate for memory: Evidence from studies of amygdaloid plasticity and Pavlovian fear conditioning. *Hippocampus* 12: 592–9.

Goudet, C., Chapuy, E., Alloui, A., Acher, F., Pin, J. P. and Eschalier, A. 2008. Group III metabotropic glutamate receptors inhibit hyperalgesia in animal models of inflammation and neuropathic pain. *Pain* 137: 112–24.

Graven-Nielsen, T., Aspegren Kendall, S., Henriksson, K. G., Bengtsson, M., Sorensen, J., Johnson, A., Gerdle, B. and Arendt-Nielsen, L. 2000. Ketamine reduces muscle pain, temporal summation, and referred pain in fibromyalgia patients. *Pain* 85: 483–91.

Gress, F., Schmitt, C., Sherman, S., Ikenberry, S. and Lehman, G. 1999. A prospective randomized comparison of endoscopic ultrasound- and computed tomography-guided celiac plexus block for managing chronic pancreatitis pain. *American Journal of Gastroenterology* 94: 900–5.

Gruber-Schoffnegger, D., Drdla-Schutting, R., Honigsperger, C., Wunderbaldinger, G., Gassner, M. and Sandkuhler, J. 2013. Induction of thermal hyperalgesia and synaptic long-term potentiation in the spinal cord lamina I by TNF-alpha and IL-1beta is mediated by glial cells. *Journal of Neuroscience* 33: 6540–51.

Guo, W., Robbins, M. T., Wei, F., Zou, S., Dubner, R. and Ren, K. 2006. Supraspinal brain-derived neurotrophic factor signaling: A novel mechanism for descending pain facilitation. *Journal of Neuroscience* 26: 126–37.

Guo, W., Wei, F., Zou, S., Robbins, M. T., Sugiyo, S., Ikeda, T., Tu, J. C., Worley, P. F., Dubner, R. and Ren, K. 2004. Group I metabotropic glutamate receptor NMDA receptor coupling and signaling cascade mediate spinal dorsal horn NMDA receptor 2B tyrosine phosphorylation associated with inflammatory hyperalgesia. *Journal of Neuroscience* 24: 9161–73.

Guo, W., Zou, S., Guan, Y., Ikeda, T., Tal, M., Dubner, R. and Ren, K. 2002. Tyrosine phosphorylation of the NR2B subunit of the NMDA receptor in the spinal cord during the development and maintenance of inflammatory hyperalgesia. *Journal of Neuroscience* 22: 6208–17.

Gustorff, B., Anzenhofer, S., Sycha, T., Lehr, S. and Kress, H. G. 2004. The sunburn pain model: The stability of primary and secondary hyperalgesia over 10 hours in a crossover setting. *Anesthesia & Analgesia* 98: 173–7, table of contents.

Gustorff, B., Sycha, T., Lieba-Samal, D., Rolke, R., Treede, R. D. and Magerl, W. 2013. The pattern and time course of somatosensory changes in the human UVB sunburn model reveal the presence of peripheral and central sensitization. *Pain* 154: 586–97.

Hamon, M., Gallissot, M. C., Menard, F., Gozlan, H., Bourgoin, S. and Verge, D. 1989. 5-HT3 receptor binding sites are on capsaicin-sensitive fibres in the rat spinal cord. *European Journal of Pharmacology* 164: 315–22.

Hanley, J. G. and Henley, J. M. 2005. PICK1 is a calcium-sensor for NMDA-induced AMPA receptor trafficking. *Embo Journal* 24: 3266–78.

Haroutounian, S., Nikolajsen, L., Bendtsen, T. F., Finnerup, N. B., Kristensen, A. D., Hasselstrom, J. B. and Jensen, T. S. 2014. Primary afferent input critical for maintaining spontaneous pain in peripheral neuropathy. *Pain* 155: 1272–9.

Hartmann, B., Ahmadi, S., Heppenstall, P. A., Lewin, G. R., Schott, C., Borchardt, T., Seeburg, P. H., Zeilhofer, H. U., Sprengel, R. and Kuner, R. 2004. The AMPA receptor subunits GluR-A and GluR-B reciprocally modulate spinal synaptic plasticity and inflammatory pain. *Neuron* 44: 637–50.

Harvey, R. J., Depner, U. B., Wassle, H., Ahmadi, S., Heindl, C., Reinold, H., Smart, T. G., Harvey, K., Schutz, B., Abo-Salem, O. M., et al. 2004. GlyR alpha3: An essential target for spinal PGE2-mediated inflammatory pain sensitization. *Science* 304: 884–7.

Heinricher, M. M., Tavares, I., Leith, J. L. and Lumb, B. M. 2009. Descending control of nociception: Specificity, recruitment and plasticity. *Brain Research Reviews* 60: 214–25.

Henry, J. L. 1976. Effects of substance P on functionally identified units in cat spinal cord. *Brain Res* 114: 439–51.

Heppenstall, P. A. and Lewin, G. R. 2001. BDNF but not NT-4 is required for normal flexion reflex plasticity and function. *Proceedings of the National Academy of Sciences of the United States of America* 98: 8107–12.

Hildebrand, M. E., Pitcher, G. M., Harding, E. K., Li, H., Beggs, S. and Salter, M. W. 2014. GluN2B and GluN2D NMDARs dominate synaptic responses in the adult spinal cord. *Scientific Reports* 4: 4094.

Hirai, H. 2001. Modification of AMPA receptor clustering regulates cerebellar synaptic plasticity. *Neuroscience Research* 39: 261–7.

Hitomi, S., Shinoda, M., Suzuki, I. and Iwata, K. 2012. Involvement of transient receptor potential vanilloid 1 in ectopic pain following inferior alveolar nerve transection in rats. *Neuroscience Letters* 513: 95–9.

Holguin, A., O'connor, K. A., Biedenkapp, J., Campisi, J., Wieseler-Frank, J., Milligan, E. D., Hansen, M. K., Spataro, L., Maksimova, E., Bravmann, C., et al. 2004. HIV-1 gp120 stimulates proinflammatory cytokine-mediated pain facilitation via activation of nitric oxide synthase-I (nNOS). *Pain* 110: 517–30.

Holzer, P. 1988. Local effector functions of capsaicin-sensitive sensory nerve endings: Involvement of tachykinins, calcitonin gene-related peptide and other neuropeptides. *Neuroscience* 24: 739–68.

Honore, P., Rogers, S. D., Schwei, M. J., Salak-Johnson, J. L., Luger, N. M., Sabino, M. C., Clohisy, D. R. and Mantyh, P. W. 2000. Murine models of inflammatory, neuropathic and cancer pain each generates a unique set of neurochemical changes in the spinal cord and sensory neurons. *Neuroscience* 98: 585–98.

Honore, P., Wade, C. L., Zhong, C., Harris, R. R., Wu, C., Ghayur, T., Iwakura, Y., Decker, M. W., Faltynek, C., Sullivan, J., et al. 2006. Interleukin-1alphabeta gene-deficient mice show reduced nociceptive sensitivity in models of inflammatory and neuropathic pain but not post-operative pain. *Behavioural Brain Research* 167: 355–64.

Hosl, K., Reinold, H., Harvey, R. J., Muller, U., Narumiya, S. and Zeilhofer, H. U. 2006. Spinal prostaglandin E receptors of the EP2 subtype and the glycine receptor alpha3 subunit, which mediate central inflammatory hyperalgesia, do not contribute to pain after peripheral nerve injury or formalin injection. *Pain* 126: 46–53.

Hu, H. J. and Gereau, R. W. T. 2003. ERK integrates PKA and PKC signaling in superficial dorsal horn neurons. II. Modulation of neuronal excitability. *Journal of Neurophysiology* 90: 1680–8.

Hua, X. Y., Svensson, C. I., Matsui, T., Fitzsimmons, B., Yaksh, T. L. and Webb, M. 2005. Intrathecal minocycline attenuates peripheral inflammation-induced hyperalgesia by inhibiting p38 MAPK in spinal microglia. *European Journal of Neuroscience* 22: 2431–40.

Huang, Y., Man, H. Y., Sekine-Aizawa, Y., Han, Y., Juluri, K., Luo, H., Cheah, J., Lowenstein, C., Huganir, R. L. and Snyder, S. H. 2005. S-nitrosylation of N-ethylmaleimide sensitive factor mediates surface expression of AMPA receptors. *Neuron* 46: 533–40.

Huge, V., Lauchart, M., Forderreuther, S., Kaufhold, W., Valet, M., Azad, S. C., Beyer, A. and Magerl, W. 2008. Interaction of hyperalgesia and sensory loss in complex regional pain syndrome type I (CRPS I). *PLoS One* 3: e2742.

Hughes, D. I., Scott, D. T., Todd, A. J. and Riddell, J. S. 2003. Lack of evidence for sprouting of Abeta afferents into the superficial laminas of the spinal cord dorsal horn after nerve section. *Journal of Neuroscience* 23: 9491–9.

Hunt, S. P., Pini, A. and Evan, G. 1987. Induction of c-fos-like protein in spinal cord neurons following sensory stimulation. *Nature* 328: 632–4.

Iggo, A. 1960. Cutaneous mechanoreceptors with afferent C fibres. *Journal of Physiology* 152: 337–53.

Ikeda, H., Heinke, B., Ruscheweyh, R. and Sandkuhler, J. 2003. Synaptic plasticity in spinal lamina I projection neurons that mediate hyperalgesia. *Science* 299: 1237–40.

Ikeda, H., Stark, J., Fischer, H., Wagner, M., Drdla, R., Jager, T. and Sandkuhler, J. 2006. Synaptic amplifier of inflammatory pain in the spinal dorsal horn. *Science* 312: 1659–62.

Indo, Y., Tsuruta, M., Hayashida, Y., Karim, M. A., Ohta, K., Kawano, T., Mitsubuchi, H., Tonoki, H., Awaya, Y. and Matsuda, I. 1996. Mutations in the TRKA/NGF receptor gene in patients with congenital insensitivity to pain with anhidrosis. *Nature Genetics* 13: 485–8.

Inquimbert, P., Bartels, K., Babaniyi, O. B., Barrett, L. B., Tegeder, I. and Scholz, J. 2012. Peripheral nerve injury produces a sustained shift in the balance between glutamate release and uptake in the dorsal horn of the spinal cord. *Pain* 153: 2422–31.

Ji, R. R., Baba, H., Brenner, G. J. and Woolf, C. J. 1999. Nociceptive-specific activation of ERK in spinal neurons contributes to pain hypersensitivity. *Nature Neuroscience* 2: 1114–9.

Ji, R. R., Befort, K., Brenner, G. J. and Woolf, C. J. 2002. ERK MAP kinase activation in superficial spinal cord neurons induces prodynorphin and NK-1 upregulation and contributes to persistent inflammatory pain hypersensitivity. *Journal of Neuroscience* 22: 478–85.

Ji, R. R., Kohno, T., Moore, K. A. and Woolf, C. J. 2003. Central sensitization and LTP: Do pain and memory share similar mechanisms? *Trends in Neurosciences* 26: 696–705.

Jones, T. L. and Sorkin, L. S. 2005. Activated PKA and PKC, but not CaMKIIalpha, are required for AMPA/ Kainate-mediated pain behavior in the thermal stimulus model. *Pain* 117: 259–70.

Jordt, S. E., Bautista, D. M., Chuang, H. H., Mckemy, D. D., Zygmunt, P. M., Hogestatt, E. D., Meng, I. D. and Julius, D. 2004. Mustard oils and cannabinoids excite sensory nerve fibres through the TRP channel ANKTM1. *Nature* 427: 260–5.

Julius, D. and Basbaum, A. I. 2001. Molecular mechanisms of nociception. *Nature* 413: 203–10.

Kang, D. and Kim, D. 2006. TREK-2 (K2P10.1) and TRESK (K2P18.1) are major background K+ channels in dorsal root ganglion neurons. *American Journal of Physiology: Cell Physiology* 291: C138–46.

Karim, F., Wang, C. C. and Gereau, R. W. T. 2001. Metabotropic glutamate receptor subtypes 1 and 5 are activators of extracellular signal-regulated kinase signaling required for inflammatory pain in mice. *Journal of Neuroscience* 21: 3771–9.

Katano, T., Furue, H., Okuda-Ashitaka, E., Tagaya, M., Watanabe, M., Yoshimura, M. and Ito, S. 2008. N-ethylmaleimide-sensitive fusion protein (NSF) is involved in central sensitization in the spinal cord through GluR2 subunit composition switch after inflammation. *European Journal of Neuroscience* 27: 3161–70.

Kawasaki, Y., Zhang, L., Cheng, J. K. and Ji, R. R. 2008. Cytokine mechanisms of central sensitization: Distinct and overlapping role of interleukin-1beta, interleukin-6, and tumor necrosis factor-alpha in regulating synaptic and neuronal activity in the superficial spinal cord. *Journal of Neuroscience* 28: 5189–94.

Kawasaki, Y., Kohno, T., Zhuang, Z. Y., Brenner, G. J., Wang, H., Van Der Meer, C., Befort, K., Woolf, C. J. and Ji, R. R. 2004. Ionotropic and metabotropic receptors, protein kinase A, protein kinase C, and Src contribute to C-fiber-induced ERK activation and cAMP response element-binding protein phosphorylation in dorsal horn neurons, leading to central sensitization. *Journal of Neuroscience* 24: 8310–21.

Kerr, B. J., Bradbury, E. J., Bennett, D. L., Trivedi, P. M., Dassan, P., French, J., Shelton, D. B., Mcmahon, S. B. and Thompson, S. W. 1999. Brain-derived neurotrophic factor modulates nociceptive sensory inputs and NMDA-evoked responses in the rat spinal cord. *Journal of Neuroscience* 19: 5138–48.

Kerr, R. C., Maxwell, D. J. and Todd, A. J. 1998. GluR1 and GluR2/3 subunits of the AMPA-type glutamate receptor are associated with particular types of neurone in laminae I-III of the spinal dorsal horn of the rat. *European Journal of Neuroscience* 10: 324–33.

Kessler, W., Kirchhoff, C., Reeh, P. W. and Handwerker, H. O. 1992. Excitation of cutaneous afferent nerve endings in vitro by a combination of inflammatory mediators and conditioning effect of substance P. *Experimental Brain Research* 91: 467–76.

Khasabov, S. G., Rogers, S. D., Ghilardi, J. R., Peters, C. M., Mantyh, P. W. and Simone, D. A. 2002. Spinal neurons that possess the substance P receptor are required for the development of central sensitization. *Journal of Neuroscience* 22: 9086–98.

Kim, Y. H., Back, S. K., Davies, A. J., Jeong, H., Jo, H. J., Chung, G., Na, H. S., Bae, Y. C., Kim, S. J., Kim, J. S., et al. 2012. TRPV1 in GABAergic interneurons mediates neuropathic mechanical allodynia and disinhibition of the nociceptive circuitry in the spinal cord. *Neuron* 74: 640–7.

Kim, Y. I., Na, H. S., Yoon, Y. W., Han, H. C., Ko, K. H. and Hong, S. K. 1997. NMDA receptors are important for both mechanical and thermal allodynia from peripheral nerve injury in rats. *Neuroreport* 8: 2149–53.

Kim, Y. S., Chu, Y., Han, L., Li, M., Li, Z., Lavinka, P. C., Sun, S., Tang, Z., Park, K., Caterina, M. J., et al. 2014. Central terminal sensitization of TRPV1 by descending serotonergic facilitation modulates chronic pain. *Neuron* 81: 873–87.

Kitto, K. F., Haley, J. E. and Wilcox, G. L. 1992. Involvement of nitric oxide in spinally mediated hyperalgesia in the mouse. *Neuroscience Letters* 148: 1–5.

Knabl, J., Witschi, R., Hosl, K., Reinold, H., Zeilhofer, U. B., Ahmadi, S., Brockhaus, J., Sergejeva, M., Hess, A., Brune, K., et al. 2008. Reversal of pathological pain through specific spinal GABAA receptor subtypes. *Nature* 451: 330–4.

Knabl, J., Zeilhofer, U. B., Crestani, F., Rudolph, U. and Zeilhofer, H. U. 2009. Genuine antihyperalgesia by systemic diazepam revealed by experiments in GABAA receptor point-mutated mice. *Pain* 141: 233–8.

Kohno, T., Moore, K. A., Baba, H. and Woolf, C. J. 2003. Peripheral nerve injury alters excitatory synaptic transmission in lamina II of the rat dorsal horn. *Journal of Physiology* 548: 131–8.

Kohno, T., Wang, H., Amaya, F. Brenner, G. J., Cheng, J. K., Ji, R. R. and Woolf, C. J. 2008. Bradykinin enhances AMPA and NMDA receptor activity in spinal cord dorsal horn neurons by activating multiple kinases to produce pain hypersensitivity. *Journal of Neuroscience* 28: 4533–40.

Koltzenburg, M., Lundberg, L. E. and Torebjork, H. E. 1992. Dynamic and static components of mechanical hyperalgesia in human hairy skin. *Pain* 51: 207–19.

Koltzenburg, M., Torebjork, H. E. and Wahren, L. K. 1994. Nociceptor modulated central sensitization causes mechanical hyperalgesia in acute chemogenic and chronic neuropathic pain. *Brain* 117 (Pt 3), 579–91.

Kopach, O., Kao, S. C., Petralia, R. S., Belan, P., Tao, Y. X. and Voitenko, N. 2011. Inflammation alters trafficking of extrasynaptic AMPA receptors in tonically firing lamina II neurons of the rat spinal dorsal horn. *Pain* 152: 912–23.

Kopach, O., Viatchenko-Karpinski, V., Atianjoh, F. E., Belan, P., Tao, Y. X. and Voitenko, N. 2013. PKCalpha is required for inflammation-induced trafficking of extrasynaptic AMPA receptors in tonically firing lamina II dorsal horn neurons during the maintenance of persistent inflammatory pain. *Journal of Pain* 14: 182–92.

Kress, M., Koltzenburg, M., Reeh, P. W. and Handwerker, H. O. 1992. Responsiveness and functional attributes of electrically localized terminals of cutaneous C-fibers in vivo and in vitro. *Journal of Neurophysiology* 68: 581–95.

Kullander, K. and Klein, R. 2002. Mechanisms and functions of Eph and ephrin signalling. *Nature Reviews Molecular Cell Biology* 3: 475–86.

Kwiat, G. C. and Basbaum, A. I. 1992. The origin of brainstem noradrenergic and serotonergic projections to the spinal cord dorsal horn in the rat. *Somatosensory and Motor Research* 9: 157–73.

Lamotte, R. H., Shain, C. N., Simone, D. A. and Tsai, E. F. 1991. Neurogenic hyperalgesia: Psychophysical studies of underlying mechanisms. *Journal of Neurophysiology* 66: 190–211.

Lankisch, P. G. 2001. Natural course of chronic pancreatitis. *Pancreatology* 1: 3–14.

Laporte, A. M., Doyen, C., Nevo, I. T., Chauveau, J., Hauw, J. J. and Hamon, M. 1996. Autoradiographic mapping of serotonin 5-HT1A, 5-HT1D, 5-HT2A and 5-HT3 receptors in the aged human spinal cord. *Journal of Chemical Neuroanatomy* 11: 67–75.

Larsson, M. and Broman, J. 2008. Translocation of GluR1-containing AMPA receptors to a spinal nociceptive synapse during acute noxious stimulation. *Journal of Neuroscience* 28: 7084–90.

Latrémoliere, A. and Costigan, M. 2011. GCH1, BH4 and pain. *Current Pharmaceutical Biotechnology* 12: 1728–41.

Latrémoliere, A., Latini, A., Andrews, S, N., Cronin, S. J., Fujita, M., Gorska, K., Hovius, R., Romero, C., Chuaiphichai, S., Painter, M., et al. 2015. Reduction of neuropathic and inflammatory pain through inhibition of the tetrahydrobiopterin pathway. *Neuron*, 86: 1393–406.

Latrémoliere, A. and Woolf, C. J. 2009. Central sensitization: A generator of pain hypersensitivity by central neural plasticity. *Journal of Pain* 10: 895–926.

Latrémoliere, A. and Woolf, C. J. 2010. Synaptic plasticity and central sensitization: Author reply. *Journal of Pain* 11: 801–3.

Latrémoliere, A., Mauborgne, A., Masson, J., Bourgoin, S., Kayser, V., Hamon, M. and Pohl, M. 2008. Differential implication of proinflammatory cytokine interleukin-6 in the development of cephalic versus extracephalic neuropathic pain in rats. *Journal of Neuroscience* 28: 8489–501.

Lavertu, G., Cote, S. L. and De Koninck, Y. 2014. Enhancing K-Cl co-transport restores normal spinothalamic sensory coding in a neuropathic pain model. *Brain* 137: 724–38.

Ledeboer, A., Sloane, E. M., Milligan, E. D., Frank, M. G., Mahony, J. H., Maier, S. F. and Watkins, L. R. 2005. Minocycline attenuates mechanical allodynia and proinflammatory cytokine expression in rat models of pain facilitation. *Pain* 115: 71–83.

Lee, S. E. and Kim, J. H. 2007. Involvement of substance P and calcitonin gene-related peptide in development and maintenance of neuropathic pain from spinal nerve injury model of rat. *Neuroscience Research* 58: 245–9.

Lekan, H. A., Carlton, S. M. and Coggeshall, R. E. 1996. Sprouting of A beta fibers into lamina II of the rat dorsal horn in peripheral neuropathy. *Neuroscience Letters* 208: 147–50.

Leonard, A. S. and Hell, J. W. 1997. Cyclic AMP-dependent protein kinase and protein kinase C phosphorylate N-methyl-D-aspartate receptors at different sites. *Journal of Biological Chemistry* 272: 12107–15.

Lever, I. J., Pezet, S., Mcmahon, S. B. and Malcangio, M. 2003. The signaling components of sensory fiber transmission involved in the activation of ERK MAP kinase in the mouse dorsal horn. *Molecular and Cellular Neuroscience* 24: 259–70.

Lin, D. T. and Huganir, R. L. 2007. PICK1 and phosphorylation of the glutamate receptor 2 (GluR2) AMPA receptor subunit regulates GluR2 recycling after NMDA receptor-induced internalization. *Journal of Neuroscience* 27: 13903–8.

Lin, Q., Peng, Y. B. and Willis, W. D. 1996. Inhibition of primate spinothalamic tract neurons by spinal glycine and GABA is reduced during central sensitization. *Journal of Neurophysiology* 76: 1005–14.

Lin, Q., Peng, Y. B., Wu, J. and Willis, W. D. 1997. Involvement of cGMP in nociceptive processing by and sensitization of spinothalamic neurons in primates. *Journal of Neuroscience* 17: 3293–302.

Lin, Q., Palecek, J., Paleckova, V., Peng, Y. B., Wu, J., Cui, M. and Willis, W. D. 1999. Nitric oxide mediates the central sensitization of primate spinothalamic tract neurons. *Journal of Neurophysiology* 81: 1075–85.

Liu, X. J., Gingrich, J. R., Vargas-Caballero, M., Dong, Y. N., Sengar, A., Beggs, S., Wang, S. H., Ding, H. K., Frankland, P. W. and Salter, M. W. 2008. Treatment of inflammatory and neuropathic pain by uncoupling Src from the NMDA receptor complex. *Nature Medicine* 14: 1325–32.

Loeser, J. D. and Melzack, R. 1999. Pain: An overview. *Lancet* 353: 1607–9.

Loomis, C. W., Khandwala, H., Osmond, G. and Hefferan, M. P. 2001. Coadministration of intrathecal strychnine and bicuculline effects synergistic allodynia in the rat: An isobolographic analysis. *Journal of Pharmacology and Experimental Therapeutics* 296: 756–61.

Lu, Y., Sun, Y. N., Wu, X., Sun, Q., Liu, F. Y., Xing, G. G. and Wan, Y. 2008. Role of alpha-amino-3-hydroxy-5-methyl-4-isoxazolepropionate (AMPA) receptor subunit GluR1 in spinal dorsal horn in inflammatory nociception and neuropathic nociception in rat. *Brain Research* 1200: 19–26.

Luo, C., Seeburg, P. H., Sprengel, R. and Kuner, R. 2008. Activity-dependent potentiation of calcium signals in spinal sensory networks in inflammatory pain states. *Pain* 140: 358–67.

Ma, Q. P. and Woolf, C. J. 1995. Noxious stimuli induce an N-methyl-D-aspartate receptor-dependent hypersensitivity of the flexion withdrawal reflex to touch: Implications for the treatment of mechanical allodynia. *Pain* 61: 383–90.

Macdermott, A. B., Mayer, M. L., Westbrook, G. L., Smith, S. J. and Barker, J. L. 1986. NMDA-receptor activation increases cytoplasmic calcium concentration in cultured spinal cord neurones. *Nature* 321: 519–22.

Maihofner, C., Handwerker, H. O. and Birklein, F. 2006. Functional imaging of allodynia in complex regional pain syndrome. *Neurology* 66: 711–7.

Malan, T. P., Mata, H. P. and Porreca, F. 2002. Spinal GABA(A) and GABA(B) receptor pharmacology in a rat model of neuropathic pain. *Anesthesiology* 96: 1161–7.

Malenka, R. C. and Bear, M. F. 2004. LTP and LTD: An embarrassment of riches. *Neuron* 44: 5–21.

Malmberg, A. B., Chen, C., Tonegawa, S. and Basbaum, A. I. 1997. Preserved acute pain and reduced neuropathic pain in mice lacking PKCgamma. *Science* 278: 279–83.

Man, H. Y., Sekine-Aizawa, Y. and Huganir, R. L. 2007. Regulation of {alpha}-amino-3-hydroxy-5-methyl-4-isoxazolepropionic acid receptor trafficking through PKA phosphorylation of the Glu receptor 1 subunit. *Proceedings of the National Academy of Sciences of the United States of America* 104: 3579–84.

Mannion, R. J., Costigan, M., Decosterd, I., Amaya, F., Ma, Q. P., Holstege, J. C., Ji, R. R., Acheson, A., Lindsay, R. M., Wilkinson, G. A., et al. 1999. Neurotrophins: Peripherally and centrally acting modulators of tactile stimulus-induced inflammatory pain hypersensitivity. *Proceedings of the National Academy of Sciences of the United States of America* 96: 9385–90.

Mantyh, P. W., Rogers, S. D., Honore, P., Allen, B. J., Ghilardi, J. R., Li, J., Daughters, R. S., Lappi, D. A., Wiley, R. G. and Simone, D. A. 1997. Inhibition of hyperalgesia by ablation of lamina I spinal neurons expressing the substance P receptor. *Science* 278: 275–9.

Mao, J., Price, D. D., Hayes, R. L., Lu, J., Mayer, D. J. and Frenk, H. 1993. Intrathecal treatment with dextrorphan or ketamine potently reduces pain-related behaviors in a rat model of peripheral mononeuropathy. *Brain Research* 605: 164–8.

Maricq, A. V., Peterson, A. S., Brake, A. J., Myers, R. M. and Julius, D. 1991. Primary structure and functional expression of the 5HT3 receptor, a serotonin-gated ion channel. *Science* 254: 432–7.

Marinus, J., Moseley, G. L., Birklein, F., Baron, R., Maihofner, C., Kingery, W. S. and Van Hilten, J. J. 2011. Clinical features and pathophysiology of complex regional pain syndrome. *Lancet Neurology* 10: 637–48.

Matsuda, S., Mikawa, S. and Hirai, H. 1999. Phosphorylation of serine-880 in GluR2 by protein kinase C prevents its C terminus from binding with glutamate receptor-interacting protein. *Journal of Neurochemistry* 73: 1765–8.

Mayer, M. L., Westbrook, G. L. and Guthrie, P. B. 1984. Voltage-dependent block by Mg2+ of NMDA responses in spinal cord neurones. *Nature* 309: 261–3.

Mcgaugh, J. L. 2000. Memory: A century of consolidation. *Science* 287: 248–51.

Mcnamara, C. R., Mandel-Brehm, J., Bautista, D. M., Siemens, J., Deranian, K. L., Zhao, M., Hayward, N. J., Chong, J. A., Julius, D., Moran, M. M., et al. 2007. TRPA1 mediates formalin-induced pain. *Proceedings of the National Academy of Sciences of the United States of America* 104: 13525–30.

Meller, S. T., Dykstra, C. and Gebhart, G. F. 1996. Acute thermal hyperalgesia in the rat is produced by activation of N-methyl-D-aspartate receptors and protein kinase C and production of nitric oxide. *Neuroscience* 71: 327–35.

Men, D. S. and Matsui, Y. 1994a. Activation of descending noradrenergic system by peripheral nerve stimulation. *Brain Research Bulletin* 34: 177–82.

Men, D. S. and Matsui, Y. 1994b. Peripheral nerve stimulation increases serotonin and dopamine metabolites in rat spinal cord. *Brain Research Bulletin* 33: 625–32.

Mendell, L. M. and Wall, P. D. 1965. Responses of Single Dorsal Cord Cells to Peripheral Cutaneous Unmyelinated Fibres. *Nature* 206: 97–9.

Miletic, G. and Miletic, V. 2008. Loose ligation of the sciatic nerve is associated with TrkB receptor-dependent decreases in KCC2 protein levels in the ipsilateral spinal dorsal horn. *Pain* 137: 532–9.

Millan, M. J. 1999. The induction of pain: An integrative review. *Progress in Neurobiology* 57: 1–164.

Millan, M. J. 2002. Descending control of pain. *Progress in Neurobiology* 66: 355–474.

Milligan, E. D. and Watkins, L. R. 2009. Pathological and protective roles of glia in chronic pain. *Nature Reviews Neuroscience* 10: 23–36.

Milligan, E. D., O'connor, K. A., Nguyen, K. T., Armstrong, C. B., Twining, C., Gaykema, R. P., Holguin, A., Martin, D., Maier, S. F. and Watkins, L. R. 2001. Intrathecal HIV-1 envelope glycoprotein gp120 induces enhanced pain states mediated by spinal cord proinflammatory cytokines. *Journal of Neuroscience* 21: 2808–19.

Milligan, E. D., Zapata, V., Chacur, M., Schoeniger, D., Biedenkapp, J., O'connor, K. A., Verge, G. M., Chapman, G., Green, P., Foster, A. C., et al. 2004. Evidence that exogenous and endogenous fractalkine can induce spinal nociceptive facilitation in rats. *European Journal of Neuroscience* 20: 2294–302.

Milligan, E., Zapata, V., Schoeniger, D., Chacur, M., Green, P., Poole, S., Martin, D., Maier, S. F. and Watkins, L. R. 2005. An initial investigation of spinal mechanisms underlying pain enhancement induced by fractalkine, a neuronally released chemokine. *European Journal of Neuroscience* 22: 2775–82.

Miraucourt, L. S., Dallel, R. and Voisin, D. L. 2007. Glycine inhibitory dysfunction turns touch into pain through PKCgamma interneurons. *PLoS One* 2: e1116.

Miyamoto, T., Dubin, A. E., Petrus, M. J. and Patapoutian, A. 2009. TRPV1 and TRPA1 mediate peripheral nitric oxide-induced nociception in mice. *PLoS One* 4: e7596.

Molliver, D. C., Wright, D. E., Leitner, M. L., Parsadanian, A. S., Doster, K., Wen, D., Yan, Q. and Snider, W. D. 1997. IB4-binding DRG neurons switch from NGF to GDNF dependence in early postnatal life. *Neuron* 19: 849–61.

Momiyama, A. 2000. Distinct synaptic and extrasynaptic NMDA receptors identified in dorsal horn neurones of the adult rat spinal cord. *Journal of Physiology* 523 (Pt 3): 621–8.

Moore, K. A., Kohno, T., Karchewski, L. A., Scholz, J., Baba, H. and Woolf, C. J. 2002. Partial peripheral nerve injury promotes a selective loss of GABAergic inhibition in the superficial dorsal horn of the spinal cord. *Journal of Neuroscience* 22: 6724–31.

Morales, M., Battenberg, E. and Bloom, F. E. 1998. Distribution of neurons expressing immunoreactivity for the 5HT3 receptor subtype in the rat brain and spinal cord. *Journal of Comparative Neurology* 402: 385–401.

Muller, F., Heinke, B. and Sandkuhler, J. 2003. Reduction of glycine receptor-mediated miniature inhibitory postsynaptic currents in rat spinal lamina I neurons after peripheral inflammation. *Neuroscience* 122: 799–805.

Munglani, R., Hudspith, M. J., Fleming, B., Harrisson, S., Smith, G., Bountra, C., Elliot, P. J., Birch, P. J. and Hunt, S. P. 1999. Effect of pre-emptive NMDA antagonist treatment on long-term Fos expression and hyperalgesia in a model of chronic neuropathic pain. *Brain Research* 822: 210–9.

Nagy, G. G., Al-Ayyan, M., Andrew, D., Fukaya, M., Watanabe, M. and Todd, A. J. 2004a. Widespread expression of the AMPA receptor GluR2 subunit at glutamatergic synapses in the rat spinal cord and phosphorylation of GluR1 in response to noxious stimulation revealed with an antigen-unmasking method. *Journal of Neuroscience* 24: 5766–77.

Nagy, G. G., Watanabe, M., Fukaya, M. and Todd, A. J. 2004b. Synaptic distribution of the NR1, NR2A and NR2B subunits of the N-methyl-d-aspartate receptor in the rat lumbar spinal cord revealed with an antigen-unmasking technique. *European Journal of Neuroscience* 20: 3301–12.

Naka, A., Gruber-Schoffnegger, D. and Sandkuhler, J. 2013. Non-Hebbian plasticity at C-fiber synapses in rat spinal cord lamina I neurons. *Pain* 154: 1333–42.

Naus, C. C., Bechberger, J. F., Zhang, Y., Venance, L., Yamasaki, H., Juneja, S. C., Kidder, G. M. and Giaume, C. 1997. Altered gap junctional communication, intercellular signaling, and growth in cultured astrocytes deficient in connexin43. *Journal of Neuroscience Res* 49: 528–40.

Neugebauer, V., Chen, P. S. and Willis, W. D. 1999. Role of metabotropic glutamate receptor subtype mGluR1 in brief nociception and central sensitization of primate STT cells. *Journal of Neurophysiology* 82: 272–82.

Neugebauer, V., Lucke, T. and Schaible, H. G. 1994. Requirement of metabotropic glutamate receptors for the generation of inflammation-evoked hyperexcitability in rat spinal cord neurons. *European Journal of Neuroscience* 6: 1179–86.

Neumann, S., Braz, J. M., Skinner, K., Llewellyn-Smith, I. J. and Basbaum, A. I. 2008. Innocuous, not noxious, input activates PKCgamma interneurons of the spinal dorsal horn via myelinated afferent fibers. *Journal of Neuroscience* 28: 7936–44.

Neumann, S., Doubell, T. P., Leslie, T. and Woolf, C. J. 1996. Inflammatory pain hypersensitivity mediated by phenotypic switch in myelinated primary sensory neurons. *Nature* 384: 360–4.

Nie, H. and Weng, H. R. 2010. Impaired glial glutamate uptake induces extrasynaptic glutamate spillover in the spinal sensory synapses of neuropathic rats. *Journal of Neurophysiology* 103: 2570–80.

Nimmerjahn, A., Kirchhoff, F. and Helmchen, F. 2005. Resting microglial cells are highly dynamic surveillants of brain parenchyma in vivo. *Science* 308: 1314–8.

Nishimune, A., Isaac, J. T., Molnar, E., Noel, J., Nash, S. R., Tagaya, M., Collingridge, G. L., Nakanishi, S. and Henley, J. M. 1998. NSF binding to GluR2 regulates synaptic transmission. *Neuron* 21: 87–97.

Noel, J., Ralph, G. S., Pickard, L., Williams, J., Molnar, E., Uney, J. B., Collingridge, G. L. and Henley, J. M. 1999. Surface expression of AMPA receptors in hippocampal neurons is regulated by an NSF-dependent mechanism. *Neuron* 23: 365–76.

Novakovic, S. D., Tzoumaka, E., Mcgivern, J. G., Haraguchi, M., Sangameswaran, L., Gogas, K. R., Eglen, R. M. and Hunter, J. C. 1998. Distribution of the tetrodotoxin-resistant sodium channel PN3 in rat sensory neurons in normal and neuropathic conditions. *Journal of Neuroscience* 18: 2174–87.

Obata, K., Yamanaka, H., Dai, Y., Mizushima, T., Fukuoka, T., Tokunaga, A. and Noguchi, K. 2004. Differential activation of MAPK in injured and uninjured DRG neurons following chronic constriction injury of the sciatic nerve in rats. *European Journal of Neuroscience* 20: 2881–95.

Obata, K., Yamanaka, H., Dai, Y., Tachibana, T., Fukuoka, T., Tokunaga, A., Yoshikawa, H. and Noguchi, K. 2003. Differential activation of extracellular signal-regulated protein kinase in primary afferent neurons regulates brain-derived neurotrophic factor expression after peripheral inflammation and nerve injury. *Journal of Neuroscience* 23: 4117–26.

Okamoto, M., Baba, H., Goldstein, P. A., Higashi, H., Shimoji, K. and Yoshimura, M. 2001. Functional reorganization of sensory pathways in the rat spinal dorsal horn following peripheral nerve injury. *Journal of Physiology* 532: 241–50.

Omana-Zapata, I., Khabbaz, M. A., Hunter, J. C., Clarke, D. E. and Bley, K. R. 1997. Tetrodotoxin inhibits neuropathic ectopic activity in neuromas, dorsal root ganglia and dorsal horn neurons. *Pain* 72: 41–9.

Pan, H. L., Eisenach, J. C. and Chen, S. R. 1999. Gabapentin suppresses ectopic nerve discharges and reverses allodynia in neuropathic rats. *Journal of Pharmacology and Experimental Therapeutics* 288: 1026–30.

Park, J. S., Voitenko, N., Petralia, R. S., Guan, X., Xu, J. T., Steinberg, J. P., Takamiya, K., Sotnik, A., Kopach, O., Huganir, R. L., et al. 2009. Persistent inflammation induces GluR2 internalization via NMDA receptor-triggered PKC activation in dorsal horn neurons. *Journal of Neuroscience* 29: 3206–19.

Park, J. S., Yaster, M., Guan, X., Xu, J. T., Shih, M. H., Guan, Y., Raja, S. N. and Tao, Y. X. 2008. Role of spinal cord alpha-amino-3-hydroxy-5-methyl-4-isoxazolepropionic acid receptors in complete Freund's adjuvant-induced inflammatory pain. *Molecular Pain* 4: 67.

Pelled, G., Chuang, K. H., Dodd, S. J. and Koretsky, A. P. 2007. Functional MRI detection of bilateral cortical reorganization in the rodent brain following peripheral nerve deafferentation. *Neuroimage* 37: 262–73.

Petralia, R. S., Wang, Y. X., Mayat, E. and Wenthold, R. J. 1997. Glutamate receptor subunit 2-selective antibody shows a differential distribution of calcium-impermeable AMPA receptors among populations of neurons. *Journal of Comparative Neurology* 385: 456–76.

Pezet, S., Malcangio, M., Lever, I. J., Perkinton, M. S., Thompson, S. W., Williams, R. J. and Mcmahon, S. B. 2002. Noxious stimulation induces Trk receptor and downstream ERK phosphorylation in spinal dorsal horn. *Molecular and Cellular Neuroscience* 21: 684–95.

Pezet, S., Marchand, F., D'mello, R., Grist, J., Clark, A. K., Malcangio, M., Dickenson, A. H., Williams, R. J. and Mcmahon, S. B. 2008. Phosphatidylinositol 3-kinase is a key mediator of central sensitization in painful inflammatory conditions. *Journal of Neuroscience* 28: 4261–70.

Pitcher, M. H., Ribeiro-Da-Silva, A. and Coderre, T. J. 2007. Effects of inflammation on the ultrastructural localization of spinal cord dorsal horn group I metabotropic glutamate receptors. *Journal of Comparative Neurology* 505: 412–23.

Polgar, E., Gray, S., Riddell, J. S. and Todd, A. J. 2004. Lack of evidence for significant neuronal loss in laminae I–III of the spinal dorsal horn of the rat in the chronic constriction injury model. *Pain* 111: 144–50.

Polgar, E., Hughes, D. I., Riddell, J. S., Maxwell, D. J., Puskar, Z. and Todd, A. J. 2003. Selective loss of spinal GABAergic or glycinergic neurons is not necessary for development of thermal hyperalgesia in the chronic constriction injury model of neuropathic pain. *Pain* 104: 229–39.

Potrebic, S., Ahn, A. H., Skinner, K., Fields, H. L. and Basbaum, A. I. 2003. Peptidergic nociceptors of both trigeminal and dorsal root ganglia express serotonin 1D receptors: Implications for the selective antimigraine action of triptans. *Journal of Neuroscience* 23: 10988–97.

Price, T. J., Cervero, F. and De Koninck, Y. 2005. Role of cation-chloride-cotransporters (CCC) in pain and hyperalgesia. *Current Topics in Medicinal Chemistry* 5: 547–55.

Price, T. J., Rashid, M. H., Millecamps, M., Sanoja, R., Entrena, J. M. and Cervero, F. 2007. Decreased nociceptive sensitization in mice lacking the fragile X mental retardation protein: Role of mGluR1/5 and mTOR. *Journal of Neuroscience* 27: 13958–67.

Puig, S., Rivot, J. P. and Besson, J. M. 1992. Effect of subcutaneous administration of the chemical algogen formalin, on 5-HT metabolism in the nucleus raphe magnus and the medullary dorsal horn: A voltammetric study in freely moving rats. *Brain Research* 590: 250–4.

Qu, X. X., Cai, J., Li, M. J., Chi, Y. N., Liao, F. F., Liu, F. Y., Wan, Y., Han, J. S. and Xing, G. G. 2009. Role of the spinal cord NR2B-containing NMDA receptors in the development of neuropathic pain. *Experimental Neurology* 215: 298–307.

Raghavendra, V., Tanga, F. and Deleo, J. A. 2003. Inhibition of microglial activation attenuates the development but not existing hypersensitivity in a rat model of neuropathy. *Journal of Pharmacology and Experimental Therapeutics* 306: 624–30.

Raghavendra, V., Tanga, F. Y. and Deleo, J. A. 2004. Complete Freunds adjuvant-induced peripheral inflammation evokes glial activation and proinflammatory cytokine expression in the CNS. *European Journal of Neuroscience* 20: 467–73.

Raman, I. M., Tong, G. and Jahr, C. E. 1996. Beta-adrenergic regulation of synaptic NMDA receptors by cAMP-dependent protein kinase. *Neuron* 16: 415–21.

Reeve, A. J., Patel, S., Fox, A., Walker, K. and Urban, L. 2000. Intrathecally administered endotoxin or cytokines produce allodynia, hyperalgesia and changes in spinal cord neuronal responses to nociceptive stimuli in the rat. *European Journal of Pain* 4: 247–57.

Rivot, J. P., Calvino, B. and Besson, J. M. 1987. Is there a serotonergic tonic descending inhibition on the responses of dorsal horn convergent neurons to C-fibre inputs? *Brain Research* 403: 142–6.

Roberts, L. A., Beyer, C. and Komisaruk, B. R. 1986. Nociceptive responses to altered GABAergic activity at the spinal cord. *Life Sciences* 39: 1667–74.

Rodriguez-Gaztelumendi, A., Rojo, M. L., Pazos, A. and Diaz, A. 2014. An altered spinal serotonergic system contributes to increased thermal nociception in an animal model of depression. *Experimental Brain Research* 232: 1793–803.

Rodriguez Parkitna, J., Korostynski, M., Kaminska-Chowaniec, D., Obara, I., Mika, J., Przewlocka, B. and Przewlocki, R. 2006. Comparison of gene expression profiles in neuropathic and inflammatory pain. *Journal of Physiology and Pharmacology* 57: 401–14.

Rogawski, M. A. and Loscher, W. 2004. The neurobiology of antiepileptic drugs for the treatment of nonepileptic conditions. *Nature Medicine* 10: 685–92.

Roh, D. H., Kim, H. W., Yoon, S. Y., Seo, H. S., Kwon, Y. B., Han, H. J., Beitz, A. J. and Lee, J. H. 2008. Depletion of capsaicin-sensitive afferents prevents lamina-dependent increases in spinal N-methyl-D-aspartate receptor subunit 1 expression and phosphorylation associated with thermal hyperalgesia in neuropathic rats. *European Journal of Pain* 12: 552–63.

Salter, M. W. and Kalia, L. V. 2004. Src kinases: A hub for NMDA receptor regulation. *Nature Reviews Neuroscience* 5: 317–28.

Samad, T. A., Moore, K. A., Sapirstein, A., Billet, S., Allchorne, A., Poole, S., Bonventre, J. V. and Woolf, C. J. 2001. Interleukin-1beta-mediated induction of Cox-2 in the CNS contributes to inflammatory pain hypersensitivity. *Nature* 410: 471–5.

Sandkuhler, J. 2010. Central sensitization versus synaptic long-term potentiation (LTP): A critical comment. *Journal of Pain* 11: 798–800.

Sawada, M., Suzumura, A. and Marunouchi, T. 1992. TNF alpha induces IL-6 production by astrocytes but not by microglia. *Brain Research* 583: 296–9.

Schmidtko, A., Tegeder, I. and Geisslinger, G. 2009. No NO, no pain? The role of nitric oxide and cGMP in spinal pain processing. *Trends in Neurosciences* 32: 339–46.

Scholz, J. and Woolf, C. J. 2007. The neuropathic pain triad: Neurons, immune cells and glia. *Nature Neuroscience* 10: 1361–8.

Scholz, J., Broom, D. C., Youn, D. H., Mills, C. D., Kohno, T., Suter, M. R., Moore, K. A., Decosterd, I., Coggeshall, R. E. and Woolf, C. J. 2005. Blocking caspase activity prevents transsynaptic neuronal apoptosis and the loss of inhibition in lamina II of the dorsal horn after peripheral nerve injury. *Journal of Neuroscience* 25: 7317–23.

Schreiber, K. L., Beitz, A. J. and Wilcox, G. L. 2008. Activation of spinal microglia in a murine model of peripheral inflammation-induced, long-lasting contralateral allodynia. *Neuroscience Letters* 440: 63–7.

Schwaninger, M., Sallmann, S., Petersen, N., Schneider, A., Prinz, S., Libermann, T. A. and Spranger, M. 1999. Bradykinin induces interleukin-6 expression in astrocytes through activation of nuclear factor-kappaB. *Journal of Neurochemistry* 73: 1461–6.

Schwartzman, R. J., Alexander, G. M., Grothusen, J. R., Paylor, T., Reichenberger, E. and Perreault, M. 2009. Outpatient intravenous ketamine for the treatment of complex regional pain syndrome: A double-blind placebo controlled study. *Pain* 147: 107–15.

Seltzer, Z. and Devor, M. 1979. Ephaptic transmission in chronically damaged peripheral nerves. *Neurology* 29: 1061–4.

Seltzer, Z., Cohn, S., Ginzburg, R. and Beilin, B. 1991. Modulation of neuropathic pain behavior in rats by spinal disinhibition and NMDA receptor blockade of injury discharge. *Pain* 45: 69–75.

Shaywitz, A. J. and Greenberg, M. E. 1999. CREB: A stimulus-induced transcription factor activated by a diverse array of extracellular signals. *Annual Review of Biochemistry* 68: 821–61.

Shehab, S. A., Spike, R. C. and Todd, A. J. 2003. Evidence against cholera toxin B subunit as a reliable tracer for sprouting of primary afferents following peripheral nerve injury. *Brain Research* 964: 218–27.

Sherman, S. E. and Loomis, C. W. 1994. Morphine insensitive allodynia is produced by intrathecal strychnine in the lightly anesthetized rat. *Pain* 56: 17–29.

Shields, S. D., Eckert, W. A., 3rd and Basbaum, A. I. 2003. Spared nerve injury model of neuropathic pain in the mouse: A behavioral and anatomic analysis. *Journal of Pain* 4: 465–70.

Shortland, P., Kinman, E. and Molander, C. 1997. Sprouting of A-fibre primary afferents into lamina II in two rat models of neuropathic pain. *European Journal of Pain* 1: 215–27.

Sikandar, S., Bannister, K. and Dickenson, A. H. 2012. Brainstem facilitations and descending serotonergic controls contribute to visceral nociception but not pregabalin analgesia in rats. *Neuroscience Letters* 519: 31–6.

Simmons, R. M., Webster, A. A., Kalra, A. B. and Iyengar, S. 2002. Group II mGluR receptor agonists are effective in persistent and neuropathic pain models in rats. *Pharmacology Biochemistry & Behavior* 73: 419–27.

Simone, D. A., Baumann, T. K. and Lamotte, R. H. 1989. Dose-dependent pain and mechanical hyperalgesia in humans after intradermal injection of capsaicin. *Pain* 38: 99–107.

Sivilotti, L. and Nistri, A. 1991. GABA receptor mechanisms in the central nervous system. *Progress in Neurobiology* 36: 35–92.

Sivilotti, L. and Woolf, C. J. 1994. The contribution of GABAA and glycine receptors to central sensitization: Disinhibition and touch-evoked allodynia in the spinal cord. *Journal of Neurophysiology* 72: 169–79.

Slack, S., Battaglia, A., Cibert-Goton, V. and Gavazzi, I. 2008. EphrinB2 induces tyrosine phosphorylation of NR2B via Src-family kinases during inflammatory hyperalgesia. *Neuroscience* 156: 175–83.

Slack, S. E., Grist, J., Mac, Q., Mcmahon, S. B. and Pezet, S. 2005. TrkB expression and phospho-ERK activation by brain-derived neurotrophic factor in rat spinothalamic tract neurons. *Journal of Comparative Neurology* 489: 59–68.

Smith, G. D., Wiseman, J., Harrison, S. M., Elliott, P. J. and Birch, P. J. 1994. Pre treatment with MK-801, a non-competitive NMDA antagonist, prevents development of mechanical hyperalgesia in a rat model of chronic neuropathy, but not in a model of chronic inflammation. *Neuroscience Letters* 165: 79–83.

Song, I., Kamboj, S., Xia, J., Dong, H., Liao, D. and Huganir, R. L. 1998. Interaction of the N-ethylmaleimide-sensitive factor with AMPA receptors. *Neuron* 21: 393–400.

South, S. M., Kohno, T., Kaspar, B. K., Hegarty, D., Vissel, B., Drake, C. T., Ohata, M., Jenab, S., Sailer, A. W., Malkmus, S., et al. 2003. A conditional deletion of the NR1 subunit of the NMDA receptor in adult spinal cord dorsal horn reduces NMDA currents and injury-induced pain. *Journal of Neuroscience* 23: 5031–40.

Spataro, L. E., Sloane, E. M., Milligan, E. D., Wieseler-Frank, J., Schoeniger, D., Jekich, B. M., Barrientos, R. M., Maier, S. F. and Watkins, L. R. 2004. Spinal gap junctions: Potential involvement in pain facilitation. *Journal of Pain* 5: 392–405.

Staud, R., Cannon, R. C., Mauderli, A. P., Robinson, M. E., Price, D. D. and Vierck, C. J., Jr. 2003. Temporal summation of pain from mechanical stimulation of muscle tissue in normal controls and subjects with fibromyalgia syndrome. *Pain* 102: 87–95.

Staud, R., Nagel, S., Robinson, M. E. and Price, D. D. 2009. Enhanced central pain processing of fibromyalgia patients is maintained by muscle afferent input: A randomized, double-blind, placebo-controlled study. *Pain* 145: 96–104.

Staud, R., Price, D. D., Robinson, M. E., Mauderli, A. P. and Vierck, C. J. 2004. Maintenance of windup of second pain requires less frequent stimulation in fibromyalgia patients compared to normal controls. *Pain* 110: 689–96.

Stephenson, F. A., Cousins, S. L. and Kenny, A. V. 2008. Assembly and forward trafficking of NMDA receptors. (Review.) *Molecular Membrane Biology* 25: 311–20.

Sugimoto, T., Bennett, G. J. and Kajander, K. C. 1990. Transsynaptic degeneration in the superficial dorsal horn after sciatic nerve injury: Effects of a chronic constriction injury, transection, and strychnine. *Pain* 42: 205–13.

Sullivan, A. F., Dashwood, M. R. and Dickenson, A. H. 1987. Alpha 2-adrenoceptor modulation of nociception in rat spinal cord: Location, effects and interactions with morphine. *European Journal of Pharmacology* 138: 169–77.

Sun, R. Q., Lawand, N. B. and Willis, W. D. 2003. The role of calcitonin gene-related peptide (CGRP) in the generation and maintenance of mechanical allodynia and hyperalgesia in rats after intradermal injection of capsaicin. *Pain* 104: 201–8.

Sun, R. Q., Tu, Y. J., Lawand, N. B., Yan, J. Y., Lin, Q. and Willis, W. D. 2004. Calcitonin gene-related peptide receptor activation produces PKA- and PKC-dependent mechanical hyperalgesia and central sensitization. *Journal of Neurophysiology* 92: 2859–66.

Suter, M. R., Berta, T., Gao, Y. J., Decosterd, I. and Ji, R. R. 2009. Large A-fiber activity is required for microglial proliferation and p38 MAPK activation in the spinal cord: Different effects of resiniferatoxin and bupivacaine on spinal microglial changes after spared nerve injury. *Molecular Pain* 5: 53.

Suter, M. R., Papaloizos, M., Berde, C. B., Woolf, C. J., Gilliard, N., Spahn, D. R. and Decosterd, I. 2003. Development of neuropathic pain in the rat spared nerve injury model is not prevented by a peripheral nerve block. *Anesthesiology* 99: 1402–8.

Svensson, C. I., Hua, X. Y., Protter, A. A., Powell, H. C. and Yaksh, T. L. 2003a. Spinal p38 MAP kinase is necessary for NMDA-induced spinal PGE(2) release and thermal hyperalgesia. *Neuroreport* 14: 1153–7.

Svensson, C. I., Marsala, M., Westerlund, A., Calcutt, N. A., Campana, W. M., Freshwater, J. D., Catalano, R., Feng, Y., Protter, A. A., Scott, B., et al. 2003b. Activation of p38 mitogen-activated protein kinase in spinal microglia is a critical link in inflammation-induced spinal pain processing. *Journal of Neurochemistry* 86: 1534–44.

Sweitzer, S. M. and Deleo, J. A. 2002. The active metabolite of leflunomide, an immunosuppressive agent, reduces mechanical sensitivity in a rat mononeuropathy model. *Journal of Pain* 3: 360–8.e

Sweitzer, S. M., Schubert, P. and Deleo, J. A. 2001. Propentofylline, a glial modulating agent, exhibits antiallodynic properties in a rat model of neuropathic pain. *Journal of Pharmacology and Experimental Therapeutics* 297: 1210–7.

Taguchi, K. and Suzuki, Y. 1992. The response of the 5-hydroxyindole oxidation current to noxious stimuli in the spinal cord of anesthetized rats: Modification by morphine. *Brain Research* 583: 150–4.

Takasu, M. A., Dalva, M. B., Zigmond, R. E. and Greenberg, M. E. 2002. Modulation of NMDA receptor-dependent calcium influx and gene expression through EphB receptors. *Science* 295: 491–5.

Tamamaki, N., Yanagawa, Y., Tomioka, R., Miyazaki, J., Obata, K. and Kaneko, T. 2003. Green fluorescent protein expression and colocalization with calretinin, parvalbumin, and somatostatin in the GAD67-GFP knock-in mouse. *Journal of Comparative Neurology* 467: 60–79.

Tanelian, D. L. and Maciver, M. B. 1991. Analgesic concentrations of lidocaine suppress tonic A-delta and C fiber discharges produced by acute injury. *Anesthesiology* 74: 934–6.

Tayeh, M. A. and Marletta, M. A. 1989. Macrophage oxidation of L-arginine to nitric oxide, nitrite, and nitrate: Tetrahydrobiopterin is required as a cofactor. *Journal of Biological Chemistry* 264: 19654–8.

Tegeder, I., Costigan, M., Griffin, R. S., Abele, A., Belfer, I., Schmidt, H., Ehnert, C., Nejim, J., Marian, C., Scholz, J., et al. 2006. GTP cyclohydrolase and tetrahydrobiopterin regulate pain sensitivity and persistence. *Nature Medicine* 12: 1269–77.

Terashima, A., Pelkey, K. A., Rah, J. C., Suh, Y. H., Roche, K. W., Collingridge, G. L., Mcbain, C. J. and Isaac, J. T. 2008. An essential role for PICK1 in NMDA receptor-dependent bidirectional synaptic plasticity. *Neuron* 57: 872–82.

Terkelsen, A. J., Gierthmuhlen, J., Finnerup, N. B., Hojlund, A. P. and Jensen, T. S. 2014. Bilateral hypersensitivity to capsaicin, thermal, and mechanical stimuli in unilateral complex regional pain syndrome. *Anesthesiology* 120: 1225–36.

Thompson, S. W., Woolf, C. J. and Sivilotti, L. G. 1993. Small-caliber afferent inputs produce a heterosynaptic facilitation of the synaptic responses evoked by primary afferent A-fibers in the neonatal rat spinal cord in vitro. *Journal of Neurophysiology* 69: 2116–28.

Thompson, S. W., Bennett, D. L., Kerr, B. J., Bradbury, E. J. and Mcmahon, S. B. 1999. Brain-derived neurotrophic factor is an endogenous modulator of nociceptive responses in the spinal cord. *Proceedings of the National Academy of Sciences of the United States of America* 96: 7714–8.

Tingley, W. G., Ehlers, M. D., Kameyama, K., Doherty, C., Ptak, J. B., Riley, C. T. and Huganir, R. L. 1997. Characterization of protein kinase A and protein kinase C phosphorylation of the N-methyl-D-aspartate receptor NR1 subunit using phosphorylation site-specific antibodies. *Journal of Biological Chemistry* 272: 5157–66.

Todd, A. J. 1990. An electron microscope study of glycine-like immunoreactivity in laminae I–III of the spinal dorsal horn of the rat. *Neuroscience* 39: 387–94.

Todd, A. J. 2010. Neuronal circuitry for pain processing in the dorsal horn. *Nature Reviews Neuroscience* 11: 823–36.

Tong, C. K. and Macdermott, A. B. 2006. Both Ca2+-permeable and -impermeable AMPA receptors contribute to primary synaptic drive onto rat dorsal horn neurons. *Journal of Physiology* 575: 133–44.

Tong, Y. G., Wang, H. F., Ju, G., Grant, G., Hokfelt, T. and Zhang, X. 1999. Increased uptake and transport of cholera toxin B-subunit in dorsal root ganglion neurons after peripheral axotomy: Possible implications for sensory sprouting. *Journal of Comparative Neurology* 404: 143–58.

Torebjork, H. E., Lundberg, L. E. and Lamotte, R. H. 1992. Central changes in processing of mechanoreceptive input in capsaicin-induced secondary hyperalgesia in humans. *Journal of Physiology* 448: 765–80.

Torrance, N., Smith, B. H., Bennett, M. I. and Lee, A. J. 2006. The epidemiology of chronic pain of predominantly neuropathic origin: Results from a general population survey. *Journal of Pain* 7: 281–9.

Torsney, C. and Macdermott, A. B. 2006. Disinhibition opens the gate to pathological pain signaling in superficial neurokinin 1 receptor-expressing neurons in rat spinal cord. *Journal of Neuroscience* 26: 1833–43.

Treede, R. D., Jensen, T. S., Campbell, J. N., Cruccu, G., Dostrovsky, J. O., Griffin, J. W., Hansson, P., Hughes, R., Nurmikko, T. and Serra, J. 2008. Neuropathic pain: Redefinition and a grading system for clinical and research purposes. *Neurology* 70: 1630–5.

Tsuchiya, M., Yamazaki, H. and Hori, Y. 1999. Enkephalinergic neurons express 5-HT3 receptors in the spinal cord dorsal horn: Single cell RT-PCR analysis. *Neuroreport* 10: 2749–53.

Tsuda, M., Shigemoto-Mogami, Y., Koizumi, S., Mizokoshi, A., Kohsaka, S., Salter, M. W. and Inoue, K. 2003. P2X4 receptors induced in spinal microglia gate tactile allodynia after nerve injury. *Nature* 424: 778–83.

Tulleuda, A., Cokic, B., Callejo, G., Saiani, B., Serra, J. and Gasull, X. 2011. TRESK channel contribution to nociceptive sensory neurons excitability: Modulation by nerve injury. *Molecular Pain* 7: 30.

Ultenius, C., Linderoth, B., Meyerson, B. A. and Wallin, J. 2006. Spinal NMDA receptor phosphorylation correlates with the presence of neuropathic signs following peripheral nerve injury in the rat. *Neuroscience Letters* 399: 85–90.

Urban, M. O. and Gebhart, G. F. 1999. Supraspinal contributions to hyperalgesia. *Proceedings of the National Academy of Sciences of the United States of America* 96: 7687–92.

Usoskin, D., Furlan, A., Islam, S., Abdo, H., Lonnerberg, P., Lou, D., Hjerling-Leffler, J., Haeggstrom, J., Kharchenko, O., Kharchenko, P. V., et al. 2014. Unbiased classification of sensory neuron types by large-scale single-cell RNA sequencing. *Nature Neuroscience*.

Valerio, A., Rizzonelli, P., Paterlini, M., Moretto, G., Knopfel, T., Kuhn, R., Memo, M. and Spano, P. 1997. mGluR5 metabotropic glutamate receptor distribution in rat and human spinal cord: A developmental study. *Neuroscience Research* 28: 49–57.

Valet, M., Gundel, H., Sprenger, T., Sorg, C., Muhlau, M., Zimmer, C., Henningsen, P. and Tolle, T. R. 2009. Patients with pain disorder show gray-matter loss in pain-processing structures: A voxel-based morphometric study. *Psychosomatic Medicine* 71: 49–56.

Van Hecke, O., Austin, S. K., Khan, R. A., Smith, B. H. and Torrance, N. 2014. Neuropathic pain in the general population: A systematic review of epidemiological studies. *Pain* 155: 654–62.

Van Rooijen, D. E., Marinus, J. and Van Hilten, J. J. 2013. Muscle hyperalgesia is widespread in patients with complex regional pain syndrome. *Pain* 154: 2745–9.

Vardeh, D., Wang, D., Costigan, M., Lazarus, M., Saper, C. B., Woolf, C. J., Fitzgerald, G. A. and Samad, T. A. 2009. COX2 in CNS neural cells mediates mechanical inflammatory pain hypersensitivity in mice. *Journal of Clinical Investigation* 119: 287–94.

Vasko, M. R., Campbell, W. B. and Waite, K. J. 1994. Prostaglandin E2 enhances bradykinin-stimulated release of neuropeptides from rat sensory neurons in culture. *Journal of Neuroscience* 14: 4987–97.

Vera-Portocarrero, L. P., Zhang, E. T., Ossipov, M. H., Xie, J. Y., King, T., Lai, J. and Porreca, F. 2006. Descending facilitation from the rostral ventromedial medulla maintains nerve injury-induced central sensitization. *Neuroscience* 140: 1311–20.

Verge, G. M., Milligan, E. D., Maier, S. F., Watkins, L. R., Naeve, G. S. and Foster, A. C. 2004. Fractalkine (CX3CL1) and fractalkine receptor (CX3CR1) distribution in spinal cord and dorsal root ganglia under basal and neuropathic pain conditions. *European Journal of Neuroscience* 20: 1150–60.

Vikman, K. S., Duggan, A. W. and Siddall, P. J. 2007. Interferon-gamma induced disruption of GABAergic inhibition in the spinal dorsal horn in vivo. *Pain* 133: 18–28.

Vikman, K. S., Rycroft, B. K. and Christie, M. J. 2008. Switch to Ca2+-permeable AMPA and reduced NR2B NMDA receptor-mediated neurotransmission at dorsal horn nociceptive synapses during inflammatory pain in the rat. *Journal of Physiology* 586: 515–27.

Vikman, K. S., Hill, R. H., Backstrom, E. and Kristensson, K. 2003. Interferon-gamma induces characteristics of central sensitization in spinal dorsal horn neurons in vitro. *Pain* 106: 241–51.

Vikman, K., Robertson, B., Grant, G., Liljeborg, A. and Kristensson, K. 1998. Interferon-gamma receptors are expressed at synapses in the rat superficial dorsal horn and lateral spinal nucleus. *Journal of Neurocytology* 27: 749–59.

Vogel, C., Mossner, R., Gerlach, M., Heinemann, T., Murphy, D. L., Riederer, P., Lesch, K. P. and Sommer, C. 2003. Absence of thermal hyperalgesia in serotonin transporter-deficient mice. *Journal of Neuroscience* 23: 708–15.

Von Hehn, C. A., Baron, R. and Woolf, C. J. 2012. Deconstructing the neuropathic pain phenotype to reveal neural mechanisms. *Neuron* 73: 638–52.

Walker, S. M., Meredith-Middleton, J., Lickiss, T., Moss, A. and Fitzgerald, M. 2007. Primary and secondary hyperalgesia can be differentiated by postnatal age and ERK activation in the spinal dorsal horn of the rat pup. *Pain* 128: 157–68.

Wang, R., King, T., De Felice, M., Guo, W., Ossipov, M. H. and Porreca, F. 2013. Descending facilitation maintains long-term spontaneous neuropathic pain. *Journal of Pain* 14: 845–53.

Watkins, L. R. and Maier, S. F. 2002. Beyond neurons: Evidence that immune and glial cells contribute to pathological pain states. *Physiological Reviews* 82: 981–1011.

Watkins, L. R., Milligan, E. D. and Maier, S. F. 2001. Glial activation: A driving force for pathological pain. *Trends in Neurosciences* 24: 450–5.

Watkins, L. R., Hutchinson, M. R., Milligan, E. D. and Maier, S. F. 2007. "Listening" and "talking" to neurons: Implications of immune activation for pain control and increasing the efficacy of opioids. *Brain Research Reviews* 56: 148–69.

Watson, A. H. 2004. Synaptic interactions between the terminals of slow-adapting type II mechanoreceptor afferents and neurones expressing gamma-aminobutyric acid- and glycine-like immunoreactivity in the rat spinal cord. *Journal of Comparative Neurology* 471: 168–79.

Waxman, S. G., Dib-Hajj, S., Cummins, T. R. and Black, J. A. 1999. Sodium channels and pain. *Proceedings of the National Academy of Sciences of the United States of America* 96: 7635–9.

Wei, F. and Zhuo, M. 2008. Activation of Erk in the anterior cingulate cortex during the induction and expression of chronic pain. *Molecular Pain* 4: 28.

Wei, F., Dubner, R. and Ren, K. 1999. Nucleus reticularis gigantocellularis and nucleus raphe magnus in the brain stem exert opposite effects on behavioral hyperalgesia and spinal Fos protein expression after peripheral inflammation. *Pain* 80: 127–41.

Wei, F., Dubner, R., Zou, S., Ren, K., Bai, G., Wei, D. and Guo, W. 2010. Molecular depletion of descending serotonin unmasks its novel facilitatory role in the development of persistent pain. *Journal of Neuroscience* 30: 8624–36.

Wei, F., Vadakkan, K. I., Toyoda, H., Wu, L. J., Zhao, M. G., Xu, H., Shum, F. W., Jia, Y. H. and Zhuo, M. 2006. Calcium calmodulin-stimulated adenylyl cyclases contribute to activation of extracellular signal-regulated kinase in spinal dorsal horn neurons in adult rats and mice. *Journal of Neuroscience* 26: 851–61.

Wen, Y. R., Suter, M. R., Kawasaki, Y., Huang, J., Pertin, M., Kohno, T., Berde, C. B., Decosterd, I. and Ji, R. R. 2007. Nerve conduction blockade in the sciatic nerve prevents but does not reverse the activation of p38 mitogen-activated protein kinase in spinal microglia in the rat spared nerve injury model. *Anesthesiology* 107: 312–21.

Willis, W. D. 2002. Long-term potentiation in spinothalamic neurons. *Brain Research: Brain Research Reviews* 40: 202–14.

Willis, W. D., Gerhart, K. D., Willcockson, W. S., Yezierski, R. P., Wilcox, T. K. and Cargill, C. L. 1984. Primate raphe- and reticulospinal neurons: Effects of stimulation in periaqueductal gray or VPLc thalamic nucleus. *Journal of Neurophysiology* 51: 467–80.

Woolf, C. J. 1983. Evidence for a central component of post-injury pain hypersensitivity. *Nature* 306: 686–8.

Woolf, C. J. 2010. Central sensitization: Implications for the diagnosis and treatment of pain. *Pain* 152: S2–15.

Woolf, C. J. and Doubell, T. P. 1994. The pathophysiology of chronic pain: Increased sensitivity to low threshold A beta-fibre inputs. *Current Opinion in Neurobiology* 4: 525–34.

Woolf, C. J. and King, A. E. 1990. Dynamic alterations in the cutaneous mechanoreceptive fields of dorsal horn neurons in the rat spinal cord. *Journal of Neuroscience* 10: 2717–26.

Woolf, C. J. and Ma, Q. 2007. Nociceptors: Noxious stimulus detectors. *Neuron* 55: 353–64.

Woolf, C. J., Shortland, P. and Coggeshall, R. E. 1992. Peripheral nerve injury triggers central sprouting of myelinated afferents. *Nature* 355: 75–8.

Woolf, C. J., Shortland, P., Reynolds, M., Ridings, J., Doubell, T. and Coggeshall, R. E. 1995. Reorganization of central terminals of myelinated primary afferents in the rat dorsal horn following peripheral axotomy. *Journal of Comparative Neurology* 360: 121–34.

Woolf, C. J. and Thompson, S. W. 1991. The induction and maintenance of central sensitization is dependent on N-methyl-D-aspartic acid receptor activation: Implications for the treatment of post-injury pain hypersensitivity states. *Pain* 44: 293–9.

Wu, G., Ringkamp, M., Murinson, B. B., Pogatzki, E. M., Hartke, T. V., Weerahandi, H. M., Campbell, J. N., Griffin, J. W. and Meyer, R. A. 2002. Degeneration of myelinated efferent fibers induces spontaneous activity in uninjured C-fiber afferents. *Journal of Neuroscience* 22: 7746–53.

Xiao, H. S., Huang, Q. H., Zhang, F. X., Bao, L., Lu, Y. J., Guo, C., Yang, L., Huang, W. J., Fu, G., Xu, S. H., et al. 2002. Identification of gene expression profile of dorsal root ganglion in the rat peripheral axotomy model of neuropathic pain. *Proceedings of the National Academy of Sciences of the United States of America* 99: 8360–5.

Xu, X. J., Dalsgaard, C. J. and Wiesenfeld-Hallin, Z. 1992a. Intrathecal CP-96,345 blocks reflex facilitation induced in rats by substance P and C-fiber-conditioning stimulation. *European Journal of Pharmacology* 216: 337–44.

Xu, X. J., Dalsgaard, C. J. and Wiesenfeld-Hallin, Z. 1992b. Spinal substance P and N-methyl-D-aspartate receptors are coactivated in the induction of central sensitization of the nociceptive flexor reflex. *Neuroscience* 51: 641–8.

Yaksh, T. L. 1985. Pharmacology of spinal adrenergic systems which modulate spinal nociceptive processing. *Pharmacology Biochemistry & Behavior* 22: 845–58.

Yaksh, T. L. 1989. Behavioral and autonomic correlates of the tactile evoked allodynia produced by spinal glycine inhibition: Effects of modulatory receptor systems and excitatory amino acid antagonists. *Pain* 37: 111–23.

Yang, F., Whang, J., Derry, W. T., Vardeh, D. and Scholz, J. 2014. Analgesic treatment with pregabalin does not prevent persistent pain after peripheral nerve injury in the rat. *Pain* 155: 356–66.

Yang, H. B., Yang, X., Cao, J., Li, S., Liu, Y. N., Suo, Z. W., Cui, H. B., Guo, Z. and Hu, X. D. 2011. cAMP-dependent protein kinase activated Fyn in spinal dorsal horn to regulate NMDA receptor function during inflammatory pain. *Journal of Neurochemistry* 116: 93–104.

Young, M. R., Fleetwood-Walker, S. M., Dickinson, T., Blackburn-Munro, G., Sparrow, H., Birch, P. J. and Bountra, C. 1997. Behavioural and electrophysiological evidence supporting a role for group I metabotropic glutamate receptors in the mediation of nociceptive inputs to the rat spinal cord. *Brain Research* 777: 161–9.

Yu, X. M. and Salter, M. W. 1999. Src, a molecular switch governing gain control of synaptic transmission mediated by N-methyl-D-aspartate receptors. *Proceedings of the National Academy of Sciences of the United States of America* 96: 7697–704.

Zeilhofer, H. U. and Zeilhofer, U. B. 2008. Spinal dis-inhibition in inflammatory pain. *Neuroscience Letters* 437: 170–4.

Zeilhofer, H. U., Wildner, H. and Yevenes, G. E. 2012. Fast synaptic inhibition in spinal sensory processing and pain control. *Physiological Reviews* 92: 193–235.

Zhang, H. M., Chen, S. R. and Pan, H. L. 2009. Effects of activation of group III metabotropic glutamate receptors on spinal synaptic transmission in a rat model of neuropathic pain. *Neuroscience* 158: 875–84.

Zhang, J. and De Koninck, Y. 2006. Spatial and temporal relationship between monocyte chemoattractant protein-1 expression and spinal glial activation following peripheral nerve injury. *Journal of Neurochemistry* 97: 772–83.

Zhang, R. X., Li, A., Liu, B., Wang, L., Ren, K., Zhang, H., Berman, B. M. and Lao, L. 2008. IL-1ra alleviates inflammatory hyperalgesia through preventing phosphorylation of NMDA receptor NR-1 subunit in rats. *Pain* 135: 232–9.

Zhang, Z. J., Cao, D. L., Zhang, X., Ji, R. R. and Gao, Y. J. 2013. Chemokine contribution to neuropathic pain: Respective induction of CXCL1 and CXCR2 in spinal cord astrocytes and neurons. *Pain* 154: 2185–97.

Zhou, H. Y., Chen, S. R., Byun, H. S., Chen, H., Li, L., Han, H. D., Lopez-Berestein, G., Sood, A. K. and Pan, H. L. 2012. N-methyl-D-aspartate receptor- and calpain-mediated proteolytic cleavage of K+-Cl- cotransporter-2 impairs spinal chloride homeostasis in neuropathic pain. *Journal of Biological Chemistry* 287: 33853–64.

Zhou, L. J., Zhong, Y., Ren, W. J., Li, Y. Y., Zhang, T. and Liu, X. G. 2008. BDNF induces late-phase LTP of C-fiber evoked field potentials in rat spinal dorsal horn. *Experimental Neurology* 212: 507–14.

Zhou, X. F. and Rush, R. A. 1996. Endogenous brain-derived neurotrophic factor is anterogradely transported in primary sensory neurons. *Neuroscience* 74: 945–53.

Zhu, C. Z., Wilson, S. G., Mikusa, J. P., Wismer, C. T., Gauvin, D. M., Lynch, J. J., 3rd, Wade, C. L., Decker, M. W. and Honore, P. 2004. Assessing the role of metabotropic glutamate receptor 5 in multiple nociceptive modalities. *European Journal of Pharmacology* 506: 107–18.

Zhuang, Z. Y., Gerner, P., Woolf, C. J. and Ji, R. R. 2005. ERK is sequentially activated in neurons, microglia, and astrocytes by spinal nerve ligation and contributes to mechanical allodynia in this neuropathic pain model. *Pain* 114: 149–59.

Zhuang, Z. Y., Kawasaki, Y., Tan, P. H., Wen, Y. R., Huang, J. and Ji, R. R. 2007. Role of the CX3CR1/p38 MAPK pathway in spinal microglia for the development of neuropathic pain following nerve injury-induced cleavage of fractalkine. *Brain, Behavior, and Immunity* 21: 642–51.

Ziegler, E. A., Magerl, W., Meyer, R. A. and Treede, R. D. 1999. Secondary hyperalgesia to punctate mechanical stimuli. Central sensitization to A-fibre nociceptor input. *Brain* 122 (Pt 12): 2245–57.

Zou, X., Lin, Q. and Willis, W. D. 2002. Role of protein kinase A in phosphorylation of NMDA receptor 1 subunits in dorsal horn and spinothalamic tract neurons after intradermal injection of capsaicin in rats. *Neuroscience* 115: 775–86.

Zou, X., Lin, Q. and Willis, W. D. 2004. Effect of protein kinase C blockade on phosphorylation of NR1 in dorsal horn and spinothalamic tract cells caused by intradermal capsaicin injection in rats. *Brain Research* 1020: 95–105.

CHAPTER 3

Symptoms and pathology in neuropathic pain

Matthew Thakur and Stephen B. McMahon

Introduction

Mr AP is a 55-year-old copy-editor who has had type 2 diabetes for ten years. At presentation, he is morbidly obese (BMI 40) and reports undertaking little exercise. He has OA in his left knee and takes occasional paracetamol and regular diclofenac for knee pain. Mr AP has come to the doctor because of pain in both feet that is causing him difficulty in walking, poor-quality sleep and feelings of despair. He finds it increasingly difficult to get out and about and is becoming more and more dependent on his family. Clinical examination reveals diminished touch and pin-prick perception in a 'sock' distribution in both feet, though blood tests show nothing abnormal. A full pain history indicates that his pain was gradual in onset but is now intense and continuous. Like the loss of fine sensation, it occurs primarily in a sock distribution and is characterised by spontaneous stabbing pain and by burning and tingling sensations. It is not relieved by hot packs or non-steroidal anti-inflammatory drugs (NSAIDs). Mr AP's depression score is high, and he admits to thoughts of suicide. His family is very concerned. Skin biopsy taken from both heels shows loss of intra-epidermal nerve fibre density. Questionnaire assessment tools confirm that Mr AP has neuropathic pain.

As his pain is currently not controlled by paracetamol or NSAIDs, his physician starts him on gabapentin. Mr AP says that the drug does take the edge off his pain, but he also finds that lethargy and difficulty in concentrating make it almost impossible for him to continue at work. He moves to a lower dose, which he says leaves him feeling 'less spaced out'. His physician advises that, as his pain is unlikely to go away and as no better treatment is currently available, adapting so as to cope with his pain in the long term is the only remaining option. Mr AP's experience is not uncommon – one in three people who visit the GP in the United Kingdom go because they are in pain (RCGPs, 2004, 2011).

Why is neuropathic pain so difficult to treat and manage, and why does it rarely resolve naturally?

This chapter will describe how neuropathic pain is defined, diagnosed and currently managed. A variety of quantitative sensory tests can be used to characterise the symptoms and signs found in neuropathic pain in greater detail. Animal and human experimental models allow exploration and manipulation of the pathology underlying the somatosensory changes present in neuropathic pain. Pathology can be studied in the peripheral and in the central nervous system (CNS), as well as in the immune system. Understanding the biology that produces the clinical features of neuropathic pain will help to ensure that patients receive the treatment most appropriate to their condition. The ultimate aim of research in neuropathic pain is to find new ways not just to manage pain, but to resolve it.

Defining and diagnosing neuropathic pain

To meet the diagnostic criteria for neuropathic pain, the pain described by a patient must arise as a direct consequence of a lesion or disease that affects the somatosensory system (Treede et al., 2008). This is a revision of the previous International Association of the Study of Pain (IASP) definition, which was 'pain initiated or caused by a primary lesion or dysfunction of the nervous system' (Merskey and Bogduk, 1994: 207). To diagnose whether a patient's pain meets these criteria, a clinician will evaluate a number of factors, including:

1 whether the pain has a plausible neuroanatomical distribution, such as a region corresponding to a peripheral innervation territory;
2 whether the patient's history suggests the relation of pain to a relevant lesion or disease, including a temporal relationship typical for the condition;
3 confirmatory demonstration of this lesion or disease, if appropriate;
4 whether tests performed as part of neurological examination confirm the presence of negative and/or positive neurological signs concordant with the distribution of pain.

These signs include mechanical/thermal allodynia and hyperalgesia, paroxysmal pain and dysaesthesia (positive signs) as well as numbness and loss of vibration/pinprick sensitivity (negative signs). Examination may be supplemented by laboratory and neurophysiological tests designed to uncover subclinical abnormalities such as reduced density of the intra-epidermal nerve fibres or changes in nerve conduction (see Figure 3.1).

The presence or absence of each of these features may then provide a diagnosis of definite, probable or possible neuropathic pain. An underlying logic of this system of classifying neuropathic pain is that it evaluates both the pathology associated with the pain and the signs or symptoms exhibited by the patient. One implication of the recent updating of the neuropathic pain definition is that, by evaluating both pathology and function, the new definition allows clinicians to distinguish, for example, true sciatic pain from referred pain arising from a degenerative facet joint.

The redefinition also clarifies the reason for the exclusion of conditions such as fibromyalgia, whiplash and irritable bowel syndrome (IBS), which could have met the previous definition of 'pain initiated or caused by a primary lesion or dysfunction of the nervous system'. As neurological pathology in these conditions is not (yet) well documented, they remain defined as functional or rheumatological disorders (Arendt-Nielsen and Graven-Nielsen, 2003; Kwan et al., 2005).

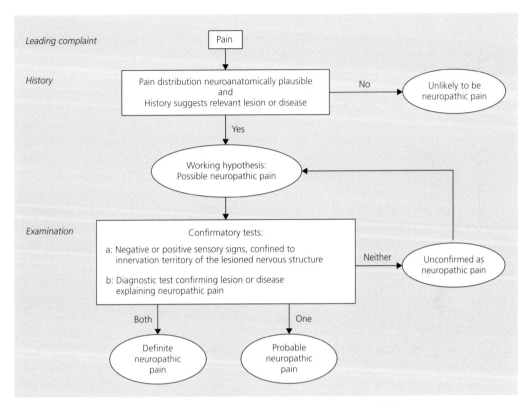

Leading complaint

Pain

History

Pain distribution neuroanatomically plausible
and
History suggests relevant lesion or disease

No → Unlikely to be neuropathic pain

Yes

Working hypothesis:
Possible neuropathic pain

Examination

Confirmatory tests:

a: Negative or positive sensory signs, confined to
   innervation territory of the lesioned nervous structure

b: Diagnostic test confirming lesion or disease
   explaining neuropathic pain

Neither → Unconfirmed as neuropathic pain

Both → Definite neuropathic pain

One → Probable neuropathic pain

Figure 3.1 Flow chart-grading schematic for neuropathic pain. Source: Treede 2008.

As an adjunct to full neurological examination and histopathological/neurophysiological tests, a number of screening tools and questionnaires are now available that use patients' descriptors of pain location, intensity, frequency and quality to indicate how likely their pain is to be of neuropathic origin. Two of the best validated are PainDETECT and DN4 (Freynhagen et al., 2006; Bouhassira et al., 2008). While these questionnaires are useful as a diagnostic aid, they are not yet fully validated as a way of assessing the effectiveness of therapeutic manipulations in reducing neuropathic pain (Steigerwald et al., 2012).

Epidemiology and management

Surveys suggest that one in five people in Europe suffer from chronic pain. This would correspond to 75 million people with chronic pain in the European Union as a whole. Two thirds of these patients describe the effectiveness of their medication as inadequate. In the United States, annual costs associated with persistent pain are estimated to amount to $600 billion (Gaskin and Richard, 2012). One in three patients visiting primary-care physicians in the United Kingdom are there because of chronic pain. Physicians describe it as one of the most challenging conditions to treat (RCGPs, 2004).

Why is it so difficult to relieve neuropathic pain? In some cases, such as peripheral nerve impingement or poorly managed diabetes, this is because the cause of nerve damage needs to be addressed before adequate pain relief can be achieved. Patients' expectations from their treatment regimen must be managed carefully – the clinical benchmark of success is a 30 per cent reduction in pain, which many patients may find disappointing. Other outcome measures, such as improvements in sleep quality or social and emotional function, must also be considered. Coexisting depression and anxiety may hinder the effectiveness of analgesia and should be addressed with specific treatment. This array of factors necessitates an interdisciplinary approach in specialist pain management, one designed to include pharmacological, cognitive behavioural, physical and occupational interventions (Baron, Hans and Dickenson, 2013).

Prospectively it remains impossible to predict treatment response in patients. This probably reflects poor understanding of the relevant mechanisms of current effective treatments. For this reason, the most common approach is a graded treatment plan, which trials different combinations of drugs in different doses in order to achieve the greatest possible pain relief with the least debilitating side effects.

The Monthly Index of Medical Specialities in the United Kingdom lists 85 simple, compound, neuropathic and opiate analgesic drugs. The majority of available drugs are opioids, non-steroidal anti-inflammatory drugs (NSAIDs) or selective cyclooxygenase (COX) inhibitors; the remainder act through calcium channel modulation, noradrenaline and serotonin re-uptake inhibition (SNRI) and topical mechanisms. In many cases the real mechanism of action of the drugs used to treat neuropathic pain is not known. The main aim of confirming a diagnosis of neuropathic pain is to guide the selection of therapies that are better suited to neuropathic pain than they are to nociceptive pain (i.e., pain in the absence of nerve damage).

Systematic meta-analysis indicates that the drugs with the most clearly documented and lowest numbers needed to treat (NNT) for neuropathic pain are tricyclic antidepressants (TCAs), followed by strong opioids and tramadol (shown in Figure 3.2). Certain other treatments – such as the gabapentinoids gabapentin and pregabalin, the combined SNRI/mu opioid agonist tapentadol (Cymbalta), local botulinum toxin or topical lidocaine – have also reported low NNTs in very small clinical trials, though this must be confirmed in larger cohorts (Finnerup et al., 2015).

Sensory signs and symptoms

The German Research Network on Neuropathic Pain (DFNS) has established a quantitative sensory testing (QST) protocol that allows the standardised quantification all of the measurable evoked sensory signs and symptoms of neuropathic pain. This quantification is accomplished by performing seven tests that measure 13 parameters of pain (see Figure 3.3a). Comparison of a patient's QST scores with appropriately matched healthy control scores allows his or her clinician to define precisely the nature of the sensory loss and/or gain (see Figure 3.3b). Note that these measures all assess pain that is evoked by stimuli; pain that occurs spontaneously must also be assessed through other means in order for us to obtain a full picture of the patient's pain experience. A large subgroup of patients have QST profiles

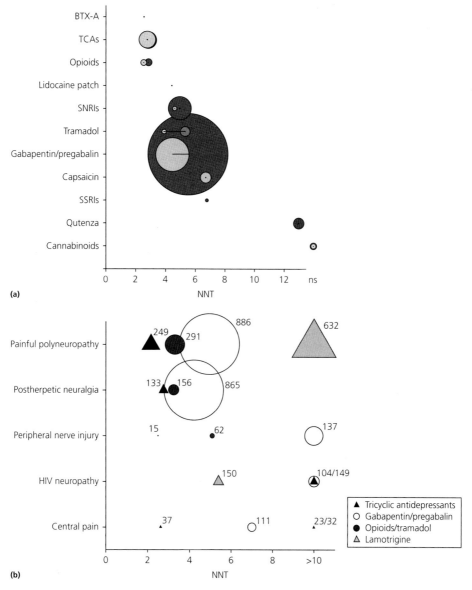

Figure 3.2 (a) The combined numbers needed to treat (NNT) values for various drug classes in all central and peripheral neuropathic pain conditions where data are available (not including trigeminal neuralgia). **(b)** The combined numbers needed to treat (NNT) values for different drug classes against specific disease aetiologies. **(a)** The figure illustrates the change from 2005 values in light grey to 2010 values in dark grey. The circle sizes indicate the relative number of patients who received active treatment drugs in trials for which dichotomous data were available. The relative NNTs of these treatments have subsequently been reassessed in larger meta-analyses. **(b)** The symbol sizes indicate the relative number of patients who received active treatment drugs in the trials for which dichotomous data were available. BTX-A = botulinum toxin type A; TCAs = tricyclic antidepressants; SNRIs = serotonin noradrenaline reuptake inhibitors; SSRIs = selective serotonin reuptake inhibitor. Sources: Finnerup 2010.

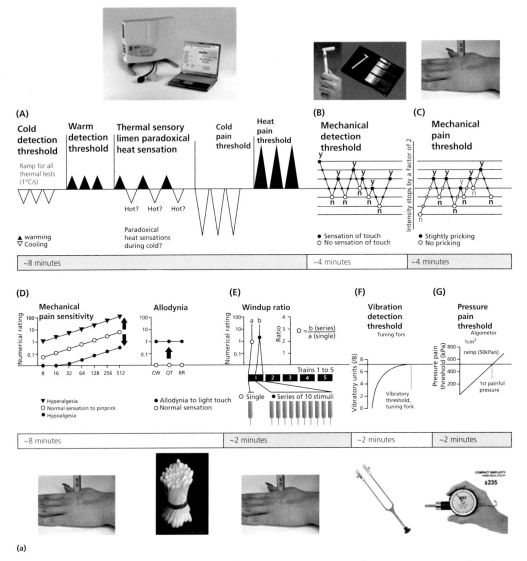

(a)

Figure 3.3 **(a)** Quantitative sensory testing (QST): A battery of sensory tests. **(b)** Z-score sensory profiles of two patients suffering from postherpetic neuralgia (PHN). **(a)** The standardised DFNS QST protocol assesses 13 parameters in seven test procedures **(A–G)**. The time taken to apply each test is indicated, as is the graphed output of the test. In total, testing takes around thirty minutes. Stimuli are applied to a consistent area of skin. **(A)** Thermal testing comprises detection and pain thresholds for cold, warm, or hot stimuli (C- and A-delta fibre mediated): cold detection threshold (CDT); warm detection threshold (WDT); number of paradoxical heat sensations (PHS) during the thermal sensory limen procedure (TSL) for alternating warm and cold stimuli; cold pain threshold (CPT); heat pain threshold (HPT). **(B)** Mechanical detection threshold (MDT) tests for A-beta fibre function using von Frey-filaments. **(C)** Mechanical pain threshold (MPT) tests for A-delta fibre-mediated hyper- or hypoalgesia to pinprick stimuli. **(D)** Stimulus–response functions: mechanical pain sensitivity (MPS) for pinprick stimuli and dynamic mechanical allodynia (ALL) assess A-delta mediated sensitivity to sharp stimuli (pinprick), and also A-beta fibre mediated pain sensitivity to stroking light touch (CW = cotton wisp; QT = cotton wool tip; BR = brush). **(E)** Wind-up ratio (WUR) compares the numerical ratings within five trains of a single pinprick stimulus (a) with a series (b) of 10 repetitive pinprick stimuli to calculate WUR as the ratio b/a. **(F)** Vibration detection threshold (VDT) tests for A-beta fibre function using a Rydel-Seiffer 64 Hz tuning fork. **(G)** Pressure pain threshold (PPT) is the only test for deep pain sensitivity, most probably mediated by muscle C- and A-delta fibres.

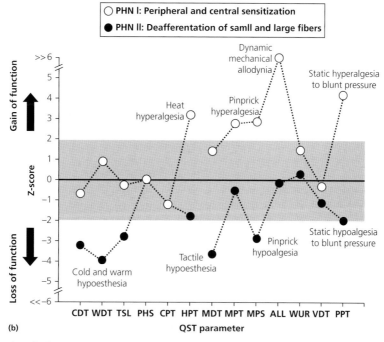

Figure 3.3 (*continued*) **(b)** Patient PHN I (open circles) is the QST profile of a 70-year-old woman suffering from PHN for eight years. Ongoing pain was 80 on a 0–100 numerical rating scale. The profile shows a predominant gain of sensory function in terms of heat pain hyperalgesia (HPT), pinprick mechanical hyperalgesia (MPS), dynamic mechanical allodynia (ALL), and static hyperalgesia to blunt pressure (PPT), which all exceed the 95% confidence interval of the distribution in healthy subjects (represented by the grey zone on the graph). This profile is consistent with a combination of peripheral and central sensitisation. Patient PHN II (filled circles) shows the QST profile of a 71-year-old woman with pain for eight months. Ongoing pain was 70 on a 0–100 numerical rating scale. The QST profile shows predominant loss of sensory function. Note that cold detection threshold (CDT), warm detection thresholds (WDT), thermal sensory limen (TSL), tactile detection thresholds (MDT), mechanical pain thresholds to pinprick stimuli (MPT), and pressure pain thresholds (PPT) are all outside the normal range as presented by the grey zone. This profile is consistent with a combined small and large fibre sensory deafferentation. (Note: a z-score is the number of standard deviations between patient data and a group-specific mean value.) Source: Rolke 2006.

that are indistinguishable from those of healthy controls – but these patients nonetheless experience significant spontaneous pain (Freeman et al., 2014).

What has this QST approach told us so far? One of the largest studies published to date was a study of 1,236 patients with polyneuropathy, postherpetic neuralgia, peripheral nerve injury, complex regional pain syndrome, trigeminal neuralgia and other conditions (Maier et al., 2010; see Figure 3.4).

This study confirms that, in almost all neuropathic pain states, the majority of patients experience both somatosensory loss and somatosensory gain of function. Loss most often occurs in fine perception and gain most often occurs in nociception. Mechanical hyperalgesia is the most common sign, more so than thermal hyperalgesia. Interestingly, enhanced wind-up of repetitive mechanical stimulation was seen in only one in ten patients (Maier et al., 2010). Wind-up is thought to be an indicator of central sensitisation. This implies either that

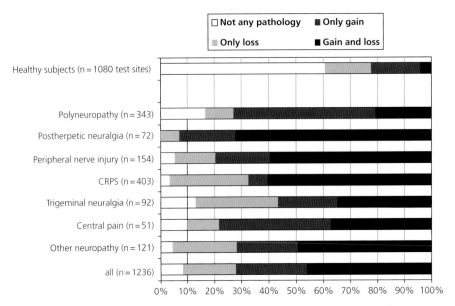

Figure 3.4 Sensory findings (gain or loss) according to the neurological syndrome. For each patient (n = 1236), QST data of the painful area were scored; for each healthy subject (n = 180), all six test areas were scored; 'without any pathology': none of the QST parameters was outside the 95% CI and there was no relative abnormality in evoked responses; 'only loss': at least one abnormally increased thermal or mechanical detection threshold, but neither thermal nor mechanical hyperalgesia; 'only gain': at least one abnormally decreased thermal or mechanical pain threshold, increased mechanical pain sensitivity, decreased pressure pain threshold or DMA, but neither thermal nor tactile hypoesthesia; 'gain and loss': at least one positive sign combined with at least one negative sign. Source: Maier 2010.

wind-up is not always an indicator of central sensitisation or that central sensitisation is less relevant than peripheral pathology in neuropathic pain.

Subsequent, more advanced analysis, which clustered patient QST results according to the patients' global sensory dysfunction profile instead of by disease type, provided striking results. There are four distinct profiles of sensory changes experienced by groups of patients with neuropathic pain, but these profiles do not correspond to an underlying disease. In other words, these results imply that patients could be defined according to the type of sensory dysfunction that is produced by their disease, rather than according to the cause. If these sensory dysfunction groups show distinctive responses to analgesia, this would imply that the existing way of clinically trialling analgesics (in large groups of patients with the same disease) is invalid; patient groups should be defined by their sensory dysfunction profile, not by their initiating disease (Freeman et al., 2014).

The versatility of the DFNS QST methodology is demonstrated by a study that combined QST with psychometric tests in order to investigate the association between depression and central sensitisation (Uhl et al., 2011). In healthy controls, it can be used to investigate the effect of various manipulations of somatosensory function, for example the 5 per cent lidocaine patch, or transcranial direct current stimulation of the primary sensory cortex (Grundmann et al., 2011; Krumova et al., 2011).

The proliferation of QST has already significantly advanced our understanding of the sensory phenotyping of neuropathic pain conditions. Pharmacological studies using QST will demonstrate which neuropathic symptoms are effectively relieved by current treatments and which ones persist. Clinical trials.gov already indexes more than sixty trials that used QST as an outcome measure. Soon the first data on QST-guided treatment will become available, and the validity of this approach will be put to the test.

Physiology and pathophysiology in neuropathic pain

Different types of painful and innocuous stimuli are sensed by a diverse range of peripheral nerve terminals found in the skin, the muscles and the viscera. These sensors are known as primary afferent fibres. A primary afferent nerve fibre that is only usually activated by potentially damaging environmental stimuli is known as a nociceptor. Every time a noxious stimulus depolarises a nociceptor, a series of action potentials is generated. This series travels from the peripheral terminals of the pain fibre, through its cell body in the dorsal root ganglion (DRG) and into the spinal cord, where it crosses synapses into projection neurones of the spinal cord (also called 'second-order neurones'). This first stage, of transmission into the spinal cord, is a key point at which the transformation and modulation of pain signals arising in the periphery can occur, before they proceed on to the brain. This process has been extensively covered by A. Todd in Chapter 1 of this volume.

Pain consists of both a somatosensory component (nociception) and a psychological, affective component. While nociception refers to the neural activity in the peripheral nervous system and in the CNS that is associated with a potentially painful stimulus, pain encompasses both this activity and the CNS-generated emotional and autonomic response to nociception. The relationship between nociceptive inputs arising from the periphery and the pain experienced by an individual is often complex. While nociception is usually the cause of pain, it is neither necessary nor sufficient and is very often not linearly related to the resulting subjective experience of pain. This is partially due to the presence of a feedback loop (descending noradrenergic and serotonergic controls) between the midbrain and the spinal cord that can alter the extent to which pain signals are allowed through the spinal cord.

The cardinal signs used to diagnose neuropathic pain include mechanical and thermal allodynia and hyperalgesia, paroxysmal pain and dysaesthesias (positive signs), as well as numbness and loss of vibration or pinprick sensitivity (negative signs). How can pathology produce both loss and gain of somatosensory function?

The answer is: through a mixture of structural and functional changes at both peripheral sites (nerve terminals, axons, DRG) and central sites (spinal cord, brainstem descending controls, higher centres in the brain). A. Latremolière covers phenomena of central sensitisation extensively in Chapter 2.

Analgesics can work at the level of the peripheral nerve, spinal cord or brain, or on a combination of these sites. The most effective analgesics, the opioids, are able to act on all of these sites. This also accounts for their broad array of undesirable side effects.

As the pathological changes that accompany persistent pain after nerve damage tend to occur in tissues that are not easily accessible for biopsy (i.e., peripheral nerve axon or cell bodies, spinal cord, brain), these changes are often investigated and manipulated by using animal models. Most of the information on mechanisms of pathology therefore derives from these model systems, which are described in Table 3.1, alongside the human pain conditions they are intended to model.

Types of models of neuropathic pain

Neuropathic pain models fit into three broad classes according to how the pain is induced: by traumatic nerve injury, by systemic neurotoxicity or by other initiating injuries/states. Typically, baseline behavioural responses, tissue status and transcription are assessed prior to induction of the model, then animals are assessed after injury, as they develop behavioural hypersensitivity to noxious and innocuous stimuli. Pharmacological, genetic, cell therapeutic or surgical interventions are applied to prevent or reverse the behavioural hypersensitivity. The effects of these interventions on tissue pathology, electrophysiology and gene transcription can also be assessed. In some cases recovery from the initiating injury will also be assessed over time.

Traumatic nerve injuries typically surgically constrict, ligate, crush or transect a nerve, producing a mixture of loss of sensation and gain of pain function in the nerve territory. They are reproducible, relatively simple to establish, and induce rapid changes in behaviour and pathology. They are the most popular type of animal model of neuropathic pain. Their limitation is the fact that traumatic nerve injuries are fairly rare clinically. Perhaps common mechanisms will link traumatic nerve injury and many other forms of neuropathic pain, but this may not be the case.

Systemic models use single or repeated treatments with chemotherapeutic, antiretroviral or other agents to induce a polyneuropathy that shares some of the features of the clinical condition being modelled. These models typically run over a longer time period than models of traumatic nerve injury. Pain symptoms develop more gradually than in traumatic models. A strength of these models is that the chemical inducing agent in clinical and preclinical cases is identical, even if the dosing regimen is different. A disadvantage is that, in the case of chemotherapeutic or antiretroviral toxicities, models lack the underlying condition (cancer or HIV), and thus may not have the same mechanisms of pathology as patients who have both the condition and treatment-associated harms.

The remaining models either use tissue injury to indirectly damage nerves (burn injury, facet joint distraction, tibial fracture, ischaemia/reperfusion) or induce neuropathy indirectly, via metabolic states (as in genetically diabetes-prone animals) or immunological states (as in cases of experimental autoimmune encephalitis). While these models may not automatically be viewed primarily as models of neuropathic pain, they all share a component of damage to neurones.

Limitations of neuropathic pain model systems

One significant difficulty in interpreting human studies of pathology in neuropathic pain is distinguishing between cause and effect. Which changes are the results of chronic painful inputs after nerve damage? Which changes are permissive, allowing some patients with acute nerve injuries to develop the distinctive non-resolving hallmarks of neuropathic pain – while others do not?

Table 3.1 Mononeuropathies, polyneuropathies and central neuropathic pain states and their corresponding animal models.

Type of Nerve Lesion	Disease	Model	Reference
Peripheral mononeuropathy	nerve transection pain (partial or complete)	Spinal nerve ligation (SNL, Chung); spared nerve injury	(Yoon et al., 1994; Decosterd and Woolf, 2000)
	plexus neuritis	mSNL	(Djouhri et al., 2006)
	neuritis	sciatic inflammatory neuritis	(Richards, Batty and Dilley, 2011)
	post-traumatic neuralgia	Partial sciatic nerve ligation (pSNL, Seltzer); sciatic nerve crush	(Malmberg and Basbaum, 1998; Decosterd and Woolf, 2000)
	entrapment syndromes	Chronic constriction injury (CCI, Bennett)	(Bennett and Xie, 1988)
		mCCI	(Nicotra et al., 2014)
	post surgery pain	Incision model	(Brennan, Vandermeulen and Gebhart, 1996; Field et al., 1999)
	painful scars	Full thickness burn injury	(Shields et al., 2012)
	postherpetic neuralgia	herpes simplex virus inoculation	(Dalziel et al., 2004)
	cancer pain	Bone cancer pain	(Urch, Donovan-Rodriguez and Dickenson, 2003)
	trigeminal neuralgia	infraorbital nerve chronic constriction injury (ION CCI); trigeminal ganglion compression	(Ahn et al., 2009, Alvarez et al., 2009)
	neuroma	Nerve transection	(Wall et al., 1979)
Peripheral polyneuropathy	diabetic neuropathy	streptozotocin model; Zucker fatty rat; Leptin deifcient mouse;	(Courteix, Eschalier and Lavarenne, 1993; André et al., 1996)
	ischaemic neuropathy	ischaemia/reperfusion injury	(Coderre et al., 2004; Laferrière et al., 2008; Seo et al., 2008)
	anti-retroviral-induced neuropathy	stavudine-induced neuropathy	(Joseph et al., 2004; Huang et al., 2013)
	chemotherapy-induced neuropathy	vincristine; oxaliplatin; cisplatin and paclitaxel-induced neuropathy	(Polomano et al., 2001; Thibault et al., 2008; Joseph and Levine, 2009)
	Charcot-Marie-Tooth disease	peripheral-myelin-protein 22 transgenic rat; C22 mouse	(Huxley et al., 1996; Sereda et al., 1996)
	Hereditary sensory and autonomic neuropathies		
	HIV	GP120-induced neuropathy	(Wallace et al., 2007)
	Guillain-Barré syndrome	experimental autoimmune neuritis	(Moalem-Taylor et al., 2007)
	Erythromelalgia		
Central	vascular lesions		
	multiple sclerosis	experimental autoimmune encephalitis	(Lu et al., 2012)
	traumatic spinal cord injury	spinal cord contusion / compression / transection	(Yezierski, 2000; Erichsen et al., 2005; Hao et al., 2006)
	traumatic brain injury		

Continued

Table 3.1 Continued

Type of Nerve Lesion	Disease	Model	Reference
	other CNS inflammatory diseases epilepsy parkinson's disease		
Mixed	complex regional pain syndromes	tibial fracture; hindlimb casting	(Guo et al., 2004)
	chronic low back pain with radiculopathy	facet joint distraction	(Lee and Winkelstein, 2009)
	phantom pain		

Source: Sandkühler, 2009; Baron, Binder and Wasner, 2010.

In humans, this question requires longitudinal study designs that track a large population of individuals at risk of neuropathic pain to see who develops it. These designs are difficult to implement with sufficient power. In animals, study of a subject before and after initiating injury is easy. But the question of risk factors is still hard to address because, unlike in humans, in models of neuropathic pain the initiating injury produces persistent pain in the vast majority of individual animals. Hence the pathology that accompanies injury-induced pain can be investigated and manipulated easily in model systems, but risk factors for neuropathic pain are not reflected in these models. Does this imply a difference in the initiating injury between patients and models? A species difference? A role for the immune system, genetics or epigenetics? A contribution of psychosocial factors and individual history? Because it is difficult to tackle preclinically or clinically, the mechanistic question of why individuals with a given injury do not all develop neuropathic pain remains largely unexplained.

The success rate for translating the analgesic mechanisms described in preclinical models into clinically useful therapeutics is strikingly low; this is especially true for neuropathic pain (Katz, Finnerup and Dworkin, 2008). Targets that appeared to be extremely well characterised and promising in preclinical models included the NK1 receptor found on dorsal horn projection neurones, the capsaicin receptor TRPV1, drugs that act on cannabinoid targets and blockers of voltage-gated sodium channels (Burgess and Williams, 2010). In all of these cases, clinical development was unsuccessful. In the case of blockers of TRPV1, this was due to unforeseen adverse effects on body temperature regulation; but in most cases the problem was insufficient efficacy (Burgess and Williams, 2010; Kort and Kym, 2012). What accounts for the poor translation record of neuropathic pain research?

To some extent, this record reflects a mismatch between the models of traumatic nerve injury that putative analgesics are most often tested on, and the actual patient populations recruited for clinical trials. Traumatic nerve injuries are relatively rare clinically – most neuropathic pain trials use cohorts of patients with painful postherpetic neuralgia or diabetic polyneuropathy (Dworkin et al., 2010). If, as is almost certainly the case, the mechanisms of pathology differ significantly in these states, then the widespread use of models of traumatic nerve injury as a default model of neuropathic pain should be rethought. Conversely, clinical trial failures could be explained by the finding that the initiating disease is a less suitable criterion for grouping patients than the profile of sensory dysfunction (Freeman et al., 2014).

Disease history also surely accounts for the mismatch between the preclinical and the clinical pictures. Clinically, patients are generally individuals more than 40 years old, who present chronic conditions and comorbidities that are rarely unmedicated. In contrast, model systems are, typically, unmedicated young animals whose response to monotherapy is studied over a two-week time frame after a single initiating injury. There are many challenges attendant on pursuing truly chronic models of pain in aged animals with comorbidities: animal husbandry costs and the cumulative severity of the model rise over time and the likelihood of adverse events increases, as do sources of biological variability (Soleman et al., 2012).

Another factor not represented in model systems is the unusually high influence of the placebo effect in clinical trials of analgesics by comparison to trials of most other drugs (Dworkin, Katz and Gitlin, 2005). The extremely high placebo effects observed in analgesic trials reduce the power of trials, potentially leading to widespread underestimation of drug efficacy (for a comprehensive overview of placebo effects, see Chapter 9 in this volume); thus, according to this school of thought, the apparently poor translation record in pain research may not be due to problems in the translational relevance of the models used, but to unique difficulties in the design of the pain clinical trials supposed to assess preclinical targets (Katz et al., 2008). Alternative study designs may address these placebo confounds (Dworkin et al., 2009; Dworkin et al., 2010).

Perhaps the most significant limitations of preclinical model systems of neuropathic pain are the available outcome measures. Clinically, pain assessment encompasses patient self-reports on pain quality, pain frequency, interference in day-to-day life and emotional impact (Turk et al., 2003). In model systems, the most commonly used and pharmacologically manipulated outcome measures assess mechanical and thermal withdrawal thresholds (reflexes). Implementing outcome measures that may reflect the impact of pain on quality of life in model systems is a greater challenge. Assays of the ability of analgesics to condition place preference or to restore natural behaviours such as burrowing and nest building in injured animals are the most promising method of bridging this outcome-measure gap (Andrews et al., 2011; King et al., 2011).

Peripheral pathology in neuropathic pain

Peripheral nerve fibre loss

Negative symptoms, such as loss of fine tactile sensation, seen in neuropathic pain are likely to be linked to the loss of peripheral nerve fibre innervation. In some types of injury (mainly traumatic injuries) this will entail Wallerian degeneration, whereby Schwann cells coordinate the phagocytosis and clearance of remnants of injured axons by macrophages, removing the entire nerve apparatus between the injury site and the innervated tissue – a process described in greater detail in what follows (see also Gaudet, Popovich and Ramer, 2011).

One way of assessing nerve fibre loss clinically is to quantify the density of intra-epidermal nerve fibres (IENF) in punch biopsy from skin at distal sites (usually the lower leg). There is a strong correlation between reduced IENF density and other clinical measures of neuropathy, such as heat-elicited or electrically evoked potentials and axon density on nerve biopsy, though this correlation does not necessarily extend to pain (Sorensen, Molyneaux and Yue, 2006; Casanova-Molla et al., 2011).

In certain other cases, withdrawal of peripheral terminals will occur alone, in the absence of any pathology in axons distal to DRG nerve cell bodies (Bennett et al., 2011; Latronico et al., 2013). This axonal loss can be modelled in a clinical experimental setting by using topical capsaicin to trigger reduction in IENF density in healthy volunteers (O'Neill et al., 2012).

Loss of target-derived factors

The peripheral terminals of both motor and sensory nerves receive a variety of growth factors from the cutaneous and visceral tissues they innervate. These include nerve growth factor (NGF), glial cell line-derived factor (GDNF), brain-derved growth factor (BDNF) and neurotrophin 3 (NT-3). Fibres damaged through nerve injury are no longer able to access these growth factors (Bennett et al., 1998).

Growth factors clearly have a role in the regulation of the pain sensation. Transgenic manipulations that increase the expression of NGF in skin result in the hypertrophy of DRG neurons and in an increase in the number and proportion of small fibres exhibiting nociceptive functional properties (Davis et al., 1993; Goodness et al., 1997; Stucky et al., 1999; Molliver et al., 2005). Acute injection of NGF appears to induce a peripheral sensitisation of nerve fibres (Hirth et al., 2013). Antibody therapies that bind and sequester NGF are one of the most promising novel therapeutics, and have shown great efficacy in reducing osteoarthritis pain (Lane et al., 2010).

Damage to peripheral axons interrupts the supply of target-derived growth factors, leading (among other things) to the up-regulation of a number of regeneration-associated genes (RAGs). RAGs promote plasticity in the injured neuron, so that it can begin to reinnervate the denervated tissue. Examples of RAGs are activating transcription factor-3 (ATF-3), GAP-43, C-JUN, STAT3 and SMAD1 (MacGillavry et al., 2011). ATF-3 is an especially useful marker for detecting neuronal damage (Bráz and Basbaum, 2010). Around 80 per cent of the neurons in the DRGs at the relevant spinal levels will up-regulate ATF-3 following transection of the sciatic nerve. Introducing exogenous NGF or GDNF reduces this expression to 35 per cent or to 25 per cent respectively –a result that demonstrates the key role that loss of access to these growth factors plays in the induction of RAGs (Averill et al., 2004). Many of the genes regulated by RAGs also have key roles in the regulation of somatosensory function (MacGillavry et al., 2011).

Spontaneous electrophysiological activity

One of the cardinal symptoms of neuropathic pain is paroxysmal pain: bursts of pain, often with a 'tingling' or 'electric-shock' quality, that are not associated with movement or stimulation but occur spontaneously.

What changes must occur in the somatosensory system in order for it to allow spontaneous activity?

One proposal is that, following nerve injury, postganglionic sympathetic efferent fibres form aberrant connections with DRG cells. Sympathetic fibres sprout into the DRG and enwrap cell bodies, forming basket structures. Sympathetic efferent noradrenaline release then acts as a trigger for DRG ectopic firing, producing 'sympathetically maintained pain'.

Pathological changes in sympathetic fibre innervation do certainly occur in rats after nerve injury – and experimental manipulations that aim to reduce sympathetic–DRG interaction do relieve some measures of neuropathic pain. However, attempts to translate this research into clinical practice have been inconsistent, producing a consensus view that, while these mechanisms may sometimes account for the spontaneous attacks experienced by individuals with neuropathic pain, they are unlikely to be the most influential mechanisms in the majority of patients and therefore are not a good therapeutic target (discussed in Max and Gilron, 1999).

One case where autonomic interaction with the somatosensory system is likely to be more clinically relevant is complex regional pain syndrome. Here patients often report autonomic dysfunction (Wei et al., 2009).

Another mechanism of spontaneous activity is illustrated by the phenomenon of the neuroma. A neuroma is a swelling that sometimes forms on the proximal stump of transected peripheral nerves. The mechanisms that determine whether a neuroma occurs after transection are not at all clear. The relative expression of depolarising and hyperpolarising ion channels in neuromas is very different from that found in healthy peripheral nerves, producing unusual membrane biophysics. This may generate *de novo* sensitivity to mechanical pressure in the neuroma, as well as spontaneous activity (England et al., 1998; Kretschmer et al., 2002; Huang et al., 2008). This mechanism will only account for spontaneous pain in patients with nerve transection-induced neuromas (which are relatively rare). But the principle of changes in ion channel expression that alter the membrane's capacity to produce spontaneous action potentials or *de novo* sensitivity to innocuous stimuli is likely to apply widely in neuropathic pain (Markman and Dworkin, 2006).

Spontaneous action potentials in peripheral nerve fibres can be assessed using microneurography. This technique uses sharp electrode recordings from either DRG cell bodies in animals (Djouhri et al., 2006) or bundles of C fibre axons in animals or humans, in combination with a repetitive stimulation protocol designed to identify spontaneous action potentials in single C fibres (Serra, 2009). Spontaneous firing can be detected both in animal models and in patients, both in polyneuropathy and in traumatic nerve injury states (Serra et al., 2011). One caveat is that, in similar whole-animal anaesthetised preparations where electrophysiological recordings are made from spinal cord projection neurons, very little spontaneous activity is seen (Chapman, Suzuki and Dickenson, 1998). This does raise the question of whether peripheral spontaneous activity contributes significantly to pain percepts or has some other role (such as triggering the efferent release of neuropeptides).

Interaction between injured fibres and their uninjured neighbours

With the exception of spinal nerve transections, most nerve injuries will produce a situation in the DRG where cell bodies of uninjured neurons are adjacent to cell bodies of neurons that have been damaged. At the level of central and peripheral primary afferent terminals, this will also be the case – the arbors of injured and uninjured fibres will often sit adjacently to each other.

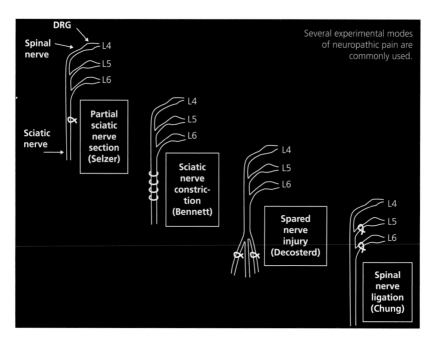

Figure 3.5 Schematic overview of models of surgically induced nerve injury.

Do injured fibres have any cross-sensitisation effects on their uninjured neighbours? This question can be examined by using various models of traumatic nerve injury (see Figure 3.5).

The spinal nerve ligation (SNL) model, also referred to as 'the Chung model', surgically ligates the L5 spinal nerve distal to the DRG. Long-lasting mechanical and thermal hypersensitivity develops, accompanied by ongoing pain in the ipsilteral hindlimb, where the receptive fields of the ligated neurons will intermix with those of uninjured neighbours originating from DRGs L4 and L6.

What is the main drive for this pain? The distal terminals of the ligated L5 DRG are now disconnected from the CNS and will undergo Wallerian degeneration. Therefore stimulus-evoked pain cannot arise via these neurons. The neuronal damage marker, ATF-3, is up-regulated in the cell bodies of neurons at L5. They develop spontaneous activity, which suggests that the ongoing firing from damaged neurons at L5 could maintain the pain. However, after another ligation is applied to the central root of L5, hypersensitivity persists. This implies that ongoing inputs from the damaged neurons are not essential for maintaining stimulus-evoked pain. But applying a ligation to the central root of the DRGs at adjacent levels will reduce pain behaviours (see Figure 3.6). This implies that the main drivers of hypersensitivity after traumatic nerve injury in the SNL model are the uninjured fibres in adjacent ganglia, which share peripheral and central receptive fields with the injured neurons (Li et al., 2000).

One caveat is that these observations can be explained by the hypothesis of intra-operative damage to neurons at the L4 level during L5 spinal nerve ligation: in contrast to the completely axotomosed L5 DRG neurons, the neurons at L4 would be under

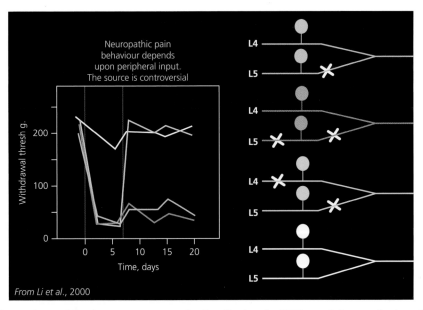

Figure 3.6 Experimental ligation nerve roots proximal or distal to the DRG reveal the contribution of injured and uninjured fibres. Source: Adapted from Li 2000.

partial injury. This interpretation is partially supported by experiments that used a refined surgical approach to avoid intra-operative damage: the results demonstrated that the previously described changes in neuropeptide expression at L4 following L5 spinal nerve ligation were in fact a product of intra-operative nerve damage (Fukuoka et al., 2011). But, even in the absence of these changes, a mechanism whereby changes induced by injured afferents can alter uninjured afferent peripheral activation or CNS transmission of inputs is still both plausible and likely.

Another traumatic nerve injury model, the spared nerve injury, ligates two of the three branches of the sciatic nerve to produce a tightly spatially defined hypersensitivity in the hindpaw in the peripheral receptive field of the spared branch. In this model, injured cell bodies will sit in the same DRG as uninjured cell bodies, but peripheral terminals will not mix. Here hypersensitivity is found only in the spared receptive field, which underlies the importance of uninjured fibres and suggests that crosstalk between injured and uninjured primary afferent peripheral terminals is not a driving mechanism in pain after nerve injury (Duraku et al., 2012).

Central pathology in neuropathic pain

This section describes pathology in the CNS in neuropathic pain. As always, distinguishing cause and effect in these changes is challenging. Do comorbidities such as major depression class as pathological changes produced by pain or as a risk factor? In reality, depression can probably fit into both of these categories.

As changes in the brain are discussed in greater detail elsewhere, this section mainly focuses on the spinal cord. Broadly, pathological changes in CNS pathology are a consequence of changes at the molecular level (synaptic strength, receptor or neuropeptide expression, membrane biophysics) rather than of gross anatomical alterations or cell death, although neuroimmune cell infiltration or proliferation in the cord may also occur (Polgár, 2005). By comparison to peripheral mechanisms, central pathophysiology is much harder to assess clinically, necessitating a greater reliance on mechanisms inferred from preclinical models.

A human experimental model of central sensitisation pain

The chemical capsaicin, which is responsible for the burning sensation produced by chilli peppers, has become widely used in pain research. Capsaicin activates the ion channel TRPV1, which is highly expressed on heat-sensitive primary afferent sensory neurons. Analgesics that attempt to block TRPV1 have not been successful in clinical trials thus far (Kort and Kym, 2012). Nonetheless, capsaicin stimulation of nociceptors produces a number of interesting phenomena, which can be used therapeutically or as basic and clinical research tools.

One of these interesting phenomena is secondary hyperalgesia, which shows how short-term functional plasticity in the CNS can change the way painful and non-painful inputs are processed. The subdermal injection of capsaicin results (unsurprisingly) in heat and mechanical hyperalgesia and spontaneous pain at the site of injection, which persists for a number of hours. This can be explained purely in terms of hypersensitivity induced by TRPV1 receptors on the fibres activated by the capsaicin injection. Interestingly, a large area of the skin, centimetres away from the injection site, also becomes hypersensitive to light mechanical stimulation; this phenomenon is termed 'secondary mechanical hyperalgesia' (also known as 'mechanical allodynia' – see Figure 3.7).

This is not simply due to diffusion of the capsaicin laterally in the skin adjacent to the injection site. A fibres in the secondary mechanical hyperalgesia zone show no electrophysiological abnormalities when assessed through microneurography (Baumann et al., 1991). Local anaesthesia of the cutaneous innervation territory in which mechanical hyperalgesia has occurred prior to the intraneural stimulation of A fibres does not prevent the perception of secondary mechanical hyperalgesia (Torebjörk, Lundberg and LaMotte, 1992). Additionally, A fibre selective nerve block can reverse secondary mechanical hyperalgesia (Magerl et al., 2001). Thus secondary mechanical hyperalgesia illustrates how, in the absence of electrophysiological changes in the primary afferent A fibres that innervate the secondary zone around the injection, a change in the central handling of those fibres' inputs can produce pain from light touch.

This change in the central handling of inputs from the secondary zone is a form of temporary plasticity brought about by the intense burst of inputs generated at the primary site. In the context of the capsaicin model, and also in chronic pain conditions, this phenomenon is termed 'central sensitisation' and is often observed in clinical research carried out on neuropathic pain patients. It is the result of persistent functional changes in any of the elements of the dorsal horn of the spinal cord. (These changes will be described below.)

Pain from light touch (allodynia) is often seen in neuropathic pain, as is an exaggerated response to repetitive stimulation (wind-up). Are the mechanisms producing allodynia in

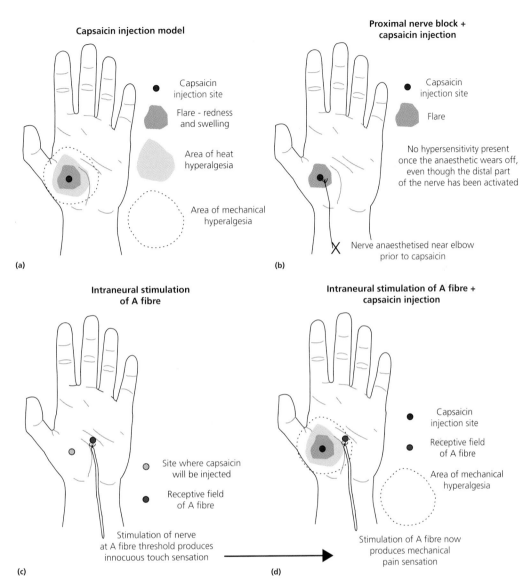

Figure 3.7 Four variations of the capsaicin-chilli extract injection model. **(a)** The standard model. After injection, an area of flare develops close to the injection site (redness, swelling); an area of heat hyperalgesia and a larger area of mechanical hyperalgesia accompany this. **(b)** Local anaesthetic block of the nerve at elbow level prevents the development of hypersensitivity, though flare still occurs. Hence flare requires only excitation of the peripheral part of the nerve, while hyperalgesia requires inputs to reach the spinal cord. **(c)** Insertion of a microelectrode into the nerve allows the selective stimulation of peripheral nerve fibres with A fibre intensity. This causes a sensation of innocuous touch. **(d)** Following capsacin injection, the previously innocuous A fibre stimulation is perceived as noxious. This suggests that the capsaicin injection triggers changes in the way normally non-painful inputs are processed at the spinal level. Source: Thakur, 2012; O'Neill 2012.

patients similar to those elucidated in the experimental studies of capsaicin pain? Both types of pain can be reduced through treatment with the analgesic gabapentin (Iannetti et al., 2005). The fact that gabapentin is selectively effective in neuropathic pain suggests that similar mechanisms produce mechanical allodynia in both states. One effect of gabapentinoids is the modulation of the membrane activity of the voltage-gated calcium channel accessory subunit alpha 2 delta (Bauer et al., 2009). This effect suggests that a change in synaptic connectivity in the spinal cord may account for gabapentin's ability to reduce allodynia.

The dorsal horn: Transformation of pain

The gate-control theory of pain, published by Wall and Melzack in 1965, postulated that the extent to which painful inputs arriving from primary afferent fibres proceeded through the spinal cord to be projected up to the brain could be modulated by lateral inhibitions produced by non-noxious inputs from nearby receptive fields, which acted via inhibitory interneurons in the substantia gelatinosa (Mendell, 2014). This process is conceptually similar to the processes that occur in the retina, where a similar diversity of excitatory and inhibitory cell types transduce and transform visual inputs before they are sent on to the visual cortex (Wässle, 2004). Certain details of the originally proposed gate-control model have not been supported by subsequent empirical evidence. However, thinking of the dorsal horn as a point along the pain pathway where inputs can be transformed so that the extent to which they are passed on to the brain is either amplified or reduced has proven to have great explanatory power and is born out by molecular dissections of spinal cord circuitry (Duan et al., 2014).

The dorsal horn can be viewed simply as a system comprising five key contributors: primary afferent neurons (i.e., inputs); interneurons; descending controls from the brain; non-neuronal cells (astrocytes, microglia, immune cells); and projection neurons (i.e., outputs). The diversity of sensory phenomena and mechanisms of analgesia observed in neuropathic pain can be produced through combinations of changes in the way these elements handle inputs from the periphery, transform them and pass them on to the brain. (These mechanisms are described in detail in Chapter 1.)

Peripheral and central pathology in neuropathic pain: Which is more important?

A hotly debated question is whether the crucial pathological changes responsible for neuropathic pain occur in the peripheral nervous system or in the CNS. In addition to the traumatic nerve injuries described above, the existence of central neuropathic pain (neuropathic pain following stroke or spinal cord injury) indicates that a purely central lesion is sufficient to induce pain (Klit, Finnerup and Jensen, 2009). Would a purely peripheral lesion likewise be sufficient for the full expression of neuropathic pain? Unfortunately, as nerve injury is always associated with secondary changes in the CNS, this question is impossible to answer, although it is striking that not all patients with a given peripheral nerve lesion will develop neuropathic pain.

An illuminating approach to unpicking peripheral and central contributions is to use outbred strains of rat with high and low pain behaviours after an injury. One distinctive feature of the high-pain lines generated through this approach is the higher expression of the neuropeptide CGRP in DRG following nerve injury (Nitzan-Luques, Devor and Tal, 2011). A similar approach applied to mice confirms variation in peripheral CGRP (rather than in any central factor) as a determinant of sensitivity to heat pain (Mogil et al., 2005). This would suggest that variation in peripheral mechanisms is sufficient to account for a large portion of the neuropathic pain phenotype. The relative rarity of wind-up in QST profiles of neuropathic pain patients (Maier et al., 2010) and the ineffectiveness of NK1 antagonists as analgesics (Hill, 2000), both of which would rely on alteration in spinal cord projection neurons, also suggest that these specific CNS pathological changes do not underpin neuropathic pain in most cases.

However, another key difference between high- and low-pain outbred rat lines is the extent to which they recruit brainstem descending inhibitory controls after nerve injury. This suggests that identical peripheral lesions do not produce the same extent of neuropathic pain in these animals, chiefly because of differing central mechanisms (De Felice et al., 2011). A similar differentiation of high- and low-pain groups by descending modulatory capacity can be observed in clinical research (Denk, McMahon and Tracey, 2014). Consider also that gabapentinoids and tricyclic antidepressants, two of the most effective therapeutics in neuro-pathic pain, most likely act on CNS targets (Micó et al., 2006; Bauer et al., 2009). Lastly, most of the genes linked through genomic studies to variability in chronic pain risk are expressed in the CNS, not in peripheral tissues (Thakur, Dawes and McMahon, 2013).

Roles of the immune system in neuropathic pain

It is becoming increasingly clear that the immune system interacts with the nervous system at various points along the pain pathway. This has been made clear by transcriptomic studies of tissue from nerve injuries, in which genes from the immune system are overrepresented. This almost certainly reflects the phenomenon of immune cells infiltrating the nervous tissues that are being assayed (Lacroix-Fralish et al., 2011). These immune interactions modify how neu-rons respond to nerve injury, and they may determine the extent to which other nerve injury-induced pathologies persist. Neuroimmune interactions occur in the axon, in the DRG, in the spinal cord areas associated with a damaged nerve, and even in the brain (Marchand, Perretti and McMahon, 2005). At the systemic level, circulating auto-antibodies may also contribute to the pathogenesis of various painful polyneuropathies (Klein et al., 2012).

Leukocytes in the periphery following nerve injury

Following nerve injury, circulating monocytes and neutrophils extravasate into the injured nerve after a loosening of the blood–nerve barrier. One of their roles is to participate in the clearing of damaged nerve fibre debris through Wallerian degeneration – a process in which Schwann cells play an important role (Gaudet et al., 2011). Macrophages may even help to initiate this fragmentation process, as they are present at the degenerating nerve long before fragmentation begins (Rosenberg et al., 2012). The number of monocytes that infiltrate the nerve immediately after injury can be reduced by disrupting the TRPM2, TLR2 or CCR2 genes; behavioural hypersensitivity is also reduced in these mice (Abbadie et al., 2003; Haraguchi

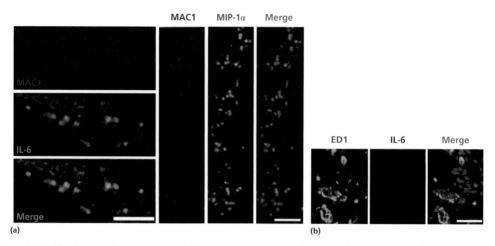

Figure 3.8 Following sciatic nerve injury, MAC1+ macrophages in the injured nerve express the pro-inflammatory cytokines IL-6 and MIP-1α, which are both pro-nociceptive **(a)**. ED1+ macrophages in the injured nerve do not express these molecules. Source: **(b)** Echeverry, Wu and Zhang, 2013.

et al., 2012). The macrophages in the injured nerve are a diverse population and will include cells expressing pro-inflammatory cytokines and cells that do not express these cytokines but express CGRP (Lee and Zhang, 2012; Echeverry, Wu and Zhang, 2013). Whether these populations correspond to the putative classical/cytotoxic M1 and alternative/proresolving M2 macrophage subclasses is currently unknown. Reducing the infiltration of chemokine-expressing macrophages reduces pain (Echeverry et al., 2013; see Figure 3.8).

Immune cells infiltrate the DRG in a variety of situations. Traumatic nerve injury or systemic paclitaxel neuropathy trigger the immediate infiltration of macrophages and lymphocytes (Vega-Avelaira, Géranton and Fitzgerald, 2009; Liu et al., 2010; Kim and Moalem-Taylor, 2011). In a primate model of HIV-neuropathy, the extent of infiltration of macrophages into DRG correlates strongly with reduced conduction velocity in the nerve. Interestingly, in this case macrophage infiltration precedes reduced conduction velocity and intra-epidermal nerve fibre density loss, showing that macrophages in these contexts are not simply acting as phagocytes that clear injured tissue but are recruited early on, and may even initiate the injury (Laast et al., 2011).

Central roles of leukocytes following nerve injury

In the spinal cord, one of the most striking features of pathology after nerve injury is microgliosis. A detailed overview of the pivotal role of microglia in neuropathic pain is described and discussed in Chapter 4 of this volume. In brief, microglia are resident monocytes of the CNS – a population of non-haematopoetic monocytes of yolk-sac lineage. In the steady state they contribute to the control of synapses, they regulate CNS blood–brain barrier integrity, and they may even respond to neurotransmitters (Calvo and Bennett, 2012). They react to injury by changing their morphology, proliferating and migrating towards the central terminals or cell bodies of injured neurons (McMahon and Malcangio, 2009). In parallel, some claim that monocytes from the circulation extravasate into the spinal cord and differentiate into microglia (Yamasaki et al., 2014). However, these two populations are hard to distinguish, and more recent experiments using refined labelling techniques suggest that infiltration of circulating monocytes does not actually occur following nerve injury (Zhang, Hoffman and Sheikh, 2014).

Treatments that disrupt gliosis – such as spinal minocycline, fluorocitrate or inhibitors of cathepsin/fractalkine signalling – are able to prevent or reverse the hypersensitivity observed in many models of chronic pain (Clark and Malcangio, 2012). However, drug treatments targeted to microglia have shown little clinical promise in neuropathic pain (Sweitzer and De Leo, 2011). One reason for this could be that the human microgliosis response is not similar to that seen in animal models. Currently very little information is available on the existence or nature of microgliosis that accompanies pain in humans – beyond preliminary imaging studies and one analysis of post-mortem tissue (Banati et al., 2001; Shi et al., 2012; Loggia et al., 2015). Heterogeneity within the microglial population in the spinal cord may also be a reason why broadly targeted microglial-inhibiting therapy was unsuccessful. For example, it could be that resident microglia differ in function from microglia derived from infiltrating cells, and that suppressing all these cells pharmacologically confounds the resolution of chronic pain after nerve injury.

Endothelial cells lining blood vessels in the CNS form a physical and biochemical barrier, as they oppose the free passage of cells and soluble factors from the circulation into the nervous system. This is true of the brain, of the spinal cord, and even of nerves. These blood-neural barriers are produced by astrocyte endfeet – tight junctions between endothelial cells and an array of multi-solute transporters that return to the circulation any solutes that do diffuse through the barrier (Kanda, 2012). In the mouse, after nerve injury, the blood–nerve barrier is disrupted distally to the injury (i.e., where Wallerian degeneration occurs). This allows the passage of various pain-increasing substances such as fibrinogen or pro-inflammatory cytokines into the nerve, potentially providing a mechanism whereby systemic immune activation can exacerbate pain following nerve injury (Lim et al., 2014). In contrast, the blood–spinal cord barrier opens for around two days but is rapidly reinstated, and the extent of this opening does not appear to relate to the degree of pain (Cahill et al., 2014).

A role for lymphocytes in neuropathic pain

Lymphocytes have only been investigated in a small number of studies, but appear to have a strong influence on the extent of neuropathic pain sensitivity after injury in rodent models. This is despite the fact that only relatively small numbers of CD3+ T cells infiltrate the cord after injury. There are roughly twenty cells per section of mouse L4 dorsal horn; by comparison there are at least ten times more microglia in an injured cord (Costigan et al., 2009 – see Figure 3.9). Splenectomised and transgenic mice lacking lymphocytes or the interferon-gamma receptor develop much milder pain after nerve injury; preliminary evidence suggests that this effect is more relevant in females than in males, and in adults than in juveniles (Costigan et al., 2009; Sorge et al., 2011). Another outbred rat study, similar to those already described above, identified paracrine DRG-to-T cell serine protease inhibitor signalling as a key determinant of higher neuropathic pain hypersensitivity (Vicuña et al., 2015).

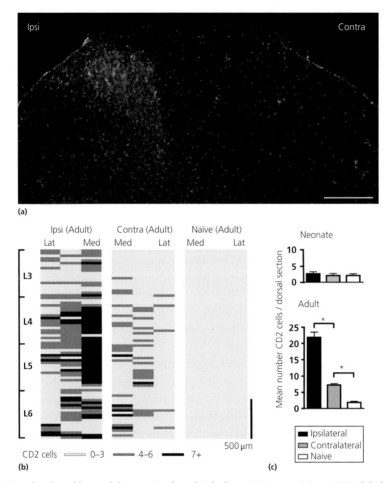

(a)

(b)

(c)

Figure 3.9 (a) Lumbar dorsal horn of the rat spinal cord 7 d after sciatic nerve injury (SNI), labeled with the microglial marker Iba1 (green) and the T cell marker CD2 (red). **(b)** Distribution of CD2-immunoreactive T cells in the dorsal horn of the spinal cord. **(c)** While CD2-immunoreactive T cells do not accumulate in the neonatal dorsal horn ipsilateral to SNI, T cells do accumulate both ipsilaterally and contralaterally in adults 7d after injury. **(a)** Both immune cells types are predominantly present in the ipsilateral (ipsi) dorsal horn. Contra denotes 'contralateral'. Scale bar = 250 μm. **(b)** Location of T cells is shown as an intensity map, darker shading representing greater number of profiles per region counted. The T cell distribution is shown for adult ipsilateral (ipsi) and contralateral (contra) 7d SNI and naïve cord. Scale bar = 500 μm. Source: Costigan 2009.

Lymphocytes are also recruited from the circulation into the damaged nerve (Echeverry et al., 2013). A subset of these cells, namely regulatory T cells (Tregs), are well characterised endogenous immune suppressor cells. Increasing Treg numbers after traumatic nerve injury reduces subsequent pain behaviours, while Treg depletion exacerbates pain (Austin et al., 2012). Although they form only a small body of evidence, these studies suggest that, while lymphocytes are relatively few by comparison to other neuroimmune interactions, modulation of neuronal function by lymphocytes could well contribute to neuropathic pain.

Humoural factors in painful neuropathies

The definition of neuropathic pain deliberately excludes conditions such as fibromyalgia, which are defined as functional, as they occur in the absence of pathological changes in nerve tissue. However, novel research indicates that humoural, rather than neurological, pathology may define a subset of painful neurological conditions.

Guillain-Barré syndrome causes acute muscular paralysis, as well as pain and fatigue. Its pathogenesis involves molecular mimicry and a cross-reactive immune response to gangliosides present on peripheral nerve axons. Some patients' neuromuscular symptoms respond well to intravenous immunoglobulins and plasma exchange; is it not currently known whether these treatments also reduce pain (van Doorn, Ruts and Jacobs, 2008).

Guillain-Barré syndrome could be an example of humourally maintained neuropathic pain. A similar case is Sjögren syndrome, an autoimmune condition that affects the exocrine glands. These patients also experience burning pain and length-dependent small-fibre polyneuropathy (Birnbaum, 2010). Positive response to immunoglobulin therapy has been reported in patients with Sjögren's, complex regional pain disorder and fibromyalgia (Caro, Winter and Dumas, 2008; Birnbaum, 2010; Goebel et al., 2013). In all of these groups, mild demyelinating polyneuropathy has been reported in a subset of patients (Oaklander et al., 2006; Caro et al., 2008; Birnbaum, 2010; Üçeyler et al., 2013), as well as circulating antibodies to receptors found on small autonomic efferent fibres in CRPS (Kohr et al., 2011). The majority of patients positive for serum IgG to voltage-gated potassium channels also show pain and neurological manifestations not attributable to any other cause (Klein et al., 2012).

The frequent coexistence of chronic pain in primary rheumatological conditions, together with evidence of neurological abnormalities, may force a reassessment of whether the confirmatory demonstration of this lesion or disease required for a diagnosis of neuropathic pain (Treede et al., 2008) could include seropositivity for auto-antibodies to neuronal antigens. It may be that a new category of humourally maintained neuropathic pain is required to encompass these conditions, which are likely to demand distinctive management.

Conclusion

Convergent approaches in clinical and preclinical research have considerably broadened our understanding of the symptoms and pathophysiology of neuropathic pain. However, greater knowledge has proven difficult to translate into better management of these conditions. Appreciation of the roles of the immune system and other non-neuronal cells in maintaining neuropathic pain is likely to develop considerably and may offer various novel treatment avenues. The proliferation of clinical research using quantitative sensory testing is also a promising means of bridging the translational gap, in order to better treat this costly and debilitating disease.

References

Abbadie, C., Lindia, J. A., Cumiskey, A. M., Peterson, L. B., Mudgett, J. S., Bayne, E. K., DeMartino, J. A., MacIntyre, D. E. and Forrest, M. J. 2003. Impaired neuropathic pain responses in mice lacking the chemokine receptor CCR2. *Proceedings of the National Academy of Sciences* 100 (13): 7947–52.

Ahn, D. K., Lim, E. J., Kim, B. C., Yang, G. Y., Lee, M. K., Ju, J. S., Han, S. R. and Bae, Y. C. 2009. Compression of the trigeminal ganglion produces prolonged nociceptive behavior in rats. *European Journal of Pain* 13 (6): 568–75.

Alvarez, P., Dieb, W., Hafidi, A., Voisin, D. L. and Dallel, R. 2009. Insular cortex representation of dynamic mechanical allodynia in trigeminal neuropathic rats. *Neurobiology of Disease* 33 (1): 89–95.

André, I., Gonzalez, A., Wang, B., Katz, J., Benoist, C. and Mathis, D. 1996. Checkpoints in the progression of auto-immune disease: Lessons from diabetes models. *Proceedings of the National Academy of Sciences* 93 (6): 2260–3.

Andrews, N., Legg, E., Lisak, D., Issop, Y., Richardson, D., Harper, S., Huang, W., Burgess, G., Machin, I. and Rice, A. S. C. 2011. Spontaneous burrowing behaviour in the rat is reduced by peripheral nerve injury or inflammation associated pain. *European Journal of Pain* 16: 485–95.

Arendt-Nielsen, L. and Graven-Nielsen, T. 2003. Central sensitization in fibromyalgia and other musculoskel-etal disorders. *Current Pain and Headache Reports* 7 (5): 355–61.

Austin, P. J., Kim, C. F., Perera, C. J. and Moalem-Taylor, G. (2012). Regulatory T cells attenuate neuropathic pain following peripheral nerve injury and experimental autoimmune neuritis. *Pain* 153 (9): 1916–31.

Averill, S., Michael, G., Shortland, P., Leavesley, R., King, V., Bradbury, E., McMahon, S. and Priestley, J. 2004. NGF and GDNF ameliorate the increase in ATF3 expression which occurs in dorsal root ganglion cells in response to peripheral nerve injury. *European Journal of Neuroscience* 19 (6): 1437–45.

Banati, R. B., Cagnin, A., Brooks, D. J., Gunn, R. N., Myers, R., Jones, T., Birch, R. and Anand, P. 2001. Long-term trans-synaptic glial responses in the human thalamus after peripheral nerve injury. *Neuroreport* 12 (16): 3439–42.

Baron, R., Binder, A. and Wasner, G. 2010. Neuropathic pain: Diagnosis, pathophysiological mechanisms, and treatment. *Lancet Neurology* 9 (8): 807–19.

Baron, R., Hans, G. and Dickenson, A. 2013. Peripheral input and its importance for central sensitization. *Annals of Neurology* 75 (5): 630–6.

Bauer, C. S., Nieto, M.-Rostro, Rahman, W., Tran-Van-Minh, A., Ferron, L., Kadurin, D. I., Sri Ranjan, Y., Fernandez-Alacid, L., Millar, N. S., Dickenson, A. H., R. Lujan and Dolphin, A. C. 2009. The increased traf-ficking of the calcium channel subunit alpha2delta-1 to presynaptic terminals in neuropathic pain is inhib-ited by the alpha2delta ligand pregabalin. *Journal of Neuroscience* 29 (13): 4076–88.

Baumann, T. K., Simone, D. A., Shain, C. N. and LaMotte, R. H. 1991. Neurogenic hyperalgesia: The search for the primary cutaneous afferent fibers that contribute to capsaicin-induced pain and hyperalgesia. *Journal of Neurophysiology* 66 (1): 212–227.

Bennett, D., Michael, G., Ramachandran, N., Munson, J., Averill, S., Yan, Q., McMahon, S. and Priestley, J. 1998. A distinct subgroup of small DRG cells express GDNF receptor components and GDNF is protective for these neurons after nerve injury. *Journal of Neuroscience* 18 (8): 3059–72.

Bennett, G. J. and Xie, Y.-K. 1988. A peripheral mononeuropathy in rat that produces disorders of pain sensa-tion like those seen in man. *Pain* 33 (1): 87–107.

Bennett, G. J., Liu, G. K., Xiao, W. H., Jin, H. W. and Siau, C. 2011. Terminal arbor degeneration: A novel lesion produced by the antineoplastic agent paclitaxel. *European Journal of Neuroscience* 33 (9): 1667–76.

Birnbaum, J. 2010. Peripheral nervous system manifestations of Sjögren syndrome: Clinical patterns, diagnos-tic paradigms, etiopathogenesis, and therapeutic strategies. *The Neurologist* 16 (5): 287–97.

Bouhassira, D., Lanteri-Minet, M., Attal, N., Laurent, B. and Touboul, C. 2008. Prevalence of chronic pain with neuropathic characteristics in the general population. *Pain* 136 (3): 380–7.

Bráz, J. M. and Basbaum, A. I. 2010. Differential ATF3 expression in dorsal root ganglion neurons reveals the profile of primary afferents engaged by diverse noxious chemical stimuli. *Pain* 150 (2): 290–301.

Brennan, T. J., Vandermeulen, E. P. and Gebhart, G. 1996. Characterization of a rat model of incisional pain. *Pain* 64 (3): 493–502.

Burgess, G. and Williams, D. 2010. The discovery and development of analgesics: New mechanisms, new modalities. *The Journal of Clinical Investigation* 120 (11): 3753–9.

Cahill, L. S., Laliberté, C. L., Liu, X. J., Bishop, J., Nieman, B. J., Mogil, J. S., Sorge, R. E., Jones, C. D., Salter, M. W. and Henkelman, R. M. 2014. Quantifying blood-spinal cord barrier permeability after peripheral nerve injury in the living mouse. *Molecular Pain* 10 (60). doi: 10.1186/1744-8069-10-60.

Calvo, M. and Bennett, D. L. 2012. The mechanisms of microgliosis and pain following peripheral nerve injury. *Experimental Neurology* 234 (2): 271–82.

Caro, X., Winter, E. and Dumas, A. 2008. A subset of fibromyalgia patients have findings suggestive of chronic inflammatory demyelinating polyneuropathy and appear to respond to IVIg. *Rheumatology* 47 (2): 208–11.

Casanova-Molla, J., Grau-Junyent, J. M., Morales, M. and Valls-Solé, J. 2011. On the relationship between nociceptive evoked potentials and intraepidermal nerve fiber density in painful sensory polyneuropathies. *Pain* 152 (2): 410–18.

Chapman, V., Suzuki, R. and Dickenson, A. H. 1998. Electrophysiological characterization of spinal neuronal response properties in anaesthetized rats after ligation of spinal nerves L5–L6. *Journal of Physiology* 507 (Pt 3): 881–94.

Clark, A. K. and Malcangio, M. 2012. Microglial signalling mechanisms: Cathepsin S and fractalkine. *Experimental Neurology* 234 (2): 283–92.

Coderre, T. J., Xanthos, D. N., Francis, L. and Bennett, G. J. 2004. Chronic post-ischemia pain (CPIP): A novel animal model of complex regional pain syndrome-type I (CRPS-I; reflex sympathetic dystrophy) produced by prolonged hindpaw ischemia and reperfusion in the rat. *Pain* 112 (1): 94–105.

Costigan, M., Moss, A., Latremoliere, A., Johnston, C., Verma-Gandhu, M., Herbert, T. A., Barrett, L., Brenner, G. J., D. Vardeh, Woolf, C. J. and Fitzgerald, M. 2009. T-cell infiltration and signaling in the adult dorsal spinal cord is a major contributor to neuropathic pain-like hypersensitivity. *Journal of Neuroscience* 29 (46): 14415–22.

Courteix, C., Eschalier, A. and Lavarenne, J. 1993. Streptozocin-induced diabetic rats: Behavioural evidence for a model of chronic pain. *Pain* 53 (1): 81–8.

Dalziel, R. G., Bingham, S., Sutton, D., Grant, D., Champion, J. M., Dennis, S. A., Quinn, J. P., Bountra, C. and Mark, M. A. 2004. Allodynia in rats infected with varicella zoster virus: A small animal model for post-herpetic neuralgia. *Brain Research Reviews* 46 (2): 234–42.

Davis, B., Lewin, G., Mendell, L., Jones, M. and Albers, K. 1993. Altered expression of nerve growth factor in the skin of transgenic mice leads to changes in response to mechanical stimuli. *Neuroscience* 56 (4): 789–92.

De Felice, M., Sanoja, R., Wang, R., Vera-Portocarrero, L., Oyarzo, J., King, T., Ossipov, M. H., Vanderah, T. W., Lai, J. and Dussor, G. O. 2011. Engagement of descending inhibition from the rostral ventromedial medulla protects against chronic neuropathic pain. *Pain* 152 (12): 2701–9.

Decosterd, I. and Woolf, C. J. 2000. Spared nerve injury: An animal model of persistent peripheral neuropathic pain. *Pain* 87 (2): 149–58.

Denk, F., McMahon, S. B. and Tracey, I. 2014. Pain vulnerability: A neurobiological perspective. *Nature neuroscience* 17 (2): 192–200.

Djouhri, L., Koutsikou, S., Fang, X., McMullan, S. and Lawson, S. N. 2006. Spontaneous pain, both neuropathic and inflammatory, is related to frequency of spontaneous firing in intact C-fiber nociceptors. *Journal of Neuroscience* 26 (4): 1281–92.

Duan, B., Cheng, L., Bourane, S., Britz, O., Padilla, C., Garcia-Campmany, L., Krashes, M., Knowlton, W., Velasquez, T. and Ren, X. 2014. Identification of spinal circuits transmitting and gating mechanical pain. *Cell* 159 (6): 1417–32.

Duraku, L. S., Hossaini, M., Hoendervangers, S., Falke, L. L., Kambiz, S., Mudera, V. C., Holstege, J. C., Walbeehm, E. T. and Ruigrok, T. J. H. 2012. Spatiotemporal dynamics of re-innervation and hyperinnervation patterns by uninjured CGRP fibers in the rat foot sole epidermis after nerve injury. *Molecular Pain* 8 (61). doi: 10.1186/1744-8069-8-61.

Dworkin, R. H., Katz, J. and Gitlin, M. J. 2005. Placebo response in clinical trials of depression and its implications for research on chronic neuropathic pain. *Neurology* 65 (12 Suppl. 4): S7–S19.

Dworkin, R. H., Turk, D. C., McDermott, M. P., Peirce-Sandner, S., Burke, L. B., Cowan, P., Farrar, J. T., Hertz, S., Raja, S. N. and Rappaport, B. A. 2009. Interpreting the clinical importance of group differences in chronic pain clinical trials: IMMPACT recommendations. *Pain* 146 (3): 238–44.

Dworkin, R. H., Turk, D. C., Peirce-Sandner, S., Baron, R., Bellamy, N., Burke, L. B., Chappell, A., Chartier, K., Cleeland, C. S., Costello, A., Cowan, P., Dimitrova, R., Ellenberg, S., Farrar, J. T., French, J. A., Gilron, I., Hertz, S., Jadad, A. R., Jay, G. W., Kalliomäki, J., Katz, N. P., Kerns, R. D., Manning, D. C., McDermott, M. P., McGrath, P. J., Narayana, A., Porter, L., Quessy, S., Rappaport, B. A., Rauschkolb, C., Reeve, B. B., Rhodes, T., Sampaio, C., Simpson, D. M., Stauffer, J. W., Stucki, G., Tobias, J., White, R. E. and Witter, J. 2010. Research design considerations for confirmatory chronic pain clinical trials: IMMPACT recommendations. *Pain* 149 (2): 177–93.

Echeverry, S., Wu, Y. and Zhang, J. 2013. Selectively reducing cytokine/chemokine expressing macrophages in injured nerves impairs the development of neuropathic pain. *Experimental neurology* 240: 205–218.

England, J. D., Happel, L. T., Liu, Z., Thouron, C. L. and Kline, D. G. 1998. Abnormal distributions of potassium channels in human neuromas. *Neuroscience letters* 255 (1): 37–40.

Erichsen, H. K., Hao, J.-X., Xu, X.-J. and Blackburn-Munro, G. 2005. Comparative actions of the opioid analgesics morphine, methadone and codeine in rat models of peripheral and central neuropathic pain. *Pain* 116 (3): 347–58.

Field, M. J., Carnell, A. J., Gonzalez, M. I., McCleary, S., Oles, R. J., Smith, R., Hughes, J. and Singh, L. 1999. Enadoline, a selective κ-opioid receptor agonist shows potent antihyperalgesic and antiallodynic actions in a rat model of surgical pain. *Pain* 80 (1): 383–9.

Finnerup, N. B., Sindrup, S. H. and Jensen, T. S. 2010. The evidence for pharmacological treatment of neuropathic pain. *Pain* 150 (3): 573–81.

Finnerup, N. B., Attal, N., Haroutounian, S., McNicol, E., Baron, R., Dworkin, R. H., Gilron, I., Haanpää, M., Hansson, P., Jensen, T. S., Kamerman, P. R., Lund, K., Moore, A., Raja, S. N., Rice, A. S. C., Rowbotham, M., Sena, E., Siddall, P., Smith, B. H. and Wallace, M. 2015. Pharmacotherapy for neuropathic pain in adults: A systematic review and meta-analysis. *The Lancet Neurology* 14 (2): 162–73.

Freeman, R., Baron, R., Bouhassira, D., Cabrera, J. and Emir, B. 2014. Sensory profiles of patients with neuropathic pain based on the neuropathic pain symptoms and signs. *Pain* 155 (2): 367–76.

Freynhagen, R., Baron, R., Gockel, U. and Tolle, T. R. 2006. painDETECT: A new screening questionnaire to identify neuropathic components in patients with back pain. *Current Medical Research and Opinion* 22 (10): 1911–20.

Fukuoka, T., Yamanaka, H., Kobayashi, K., Okubo, M., Miyoshi, K., Dai, Y. and Noguchi, K. 2011. Re-evaluation of the phenotypic changes in L4 dorsal root ganglion neurons after L5 spinal nerve ligation. *Pain* 153 (1): 68–79.

Gaskin, D. J. and Richard, P. 2012. The economic costs of pain in the United States. *The Journal of Pain* 13 (8): 715–24.

Gaudet, A. D., Popovich, P. G. and Ramer, M. S. 2011. Wallerian degeneration: Gaining perspective on inflammatory events after peripheral nerve injury. *Journal of Neuroinflammation* 8 (110). doi:10.1186/1742-2094-8-110.

Goebel, A., Misbah, S., MacIver, K., Haynes, L., Burton, J., Philips, C., Frank, B. and Poole, H. 2013. Immunoglobulin maintenance therapy in long-standing complex regional pain syndrome, an open study. *Rheumatology* 52 (11): 2091–3.

Goodness, T. P., Albers, K. M., Davis, F. E. and Davis, B. M. 1997. Overexpression of nerve growth factor in skin increases sensory neuron size and modulates Trk receptor expression. *European Journal of Neuroscience* 9 (8): 1574–85.

Grundmann, L., Rolke, R., Nitsche, M. A., Pavlakovic, G., Happe, S., Treede, R.-D., Paulus, W. and Bachmann, C. G. 2011. Effects of transcranial direct current stimulation of the primary sensory cortex on somatosensory perception. *Brain Stimulation* 4 (4): 253–60.

Guo, T.-Z., Offley, S. C., Boyd, E. A., Jacobs, C. R. and Kingery, W. S. 2004. Substance P signaling contributes to the vascular and nociceptive abnormalities observed in a tibial fracture rat model of complex regional pain syndrome type I. *Pain* 108 (1): 95–107.

Hao, J.-X., Stöhr, T., Selve, N., Wiesenfeld-Hallin, Z. and Xu, X.-J. 2006. Lacosamide, a new anti-epileptic, alleviates neuropathic pain-like behaviors in rat models of spinal cord or trigeminal nerve injury. *European Journal of Pharmacology* 553 (1): 135–40.

Haraguchi, K., Kawamoto, A., Isami, K., Maeda, S., Kusano, A., Asakura, K., Shirakawa, H., Mori, Y., Nakagawa, T. and Kaneko, S. 2012. TRPM2 contributes to inflammatory and neuropathic pain through the aggravation of pronociceptive inflammatory responses in mice. *The Journal of Neuroscience* 32 (11): 3931–41.

Hill, R. 2000. NK1 (substance P) receptor antagonists: Why are they not analgesic in humans? *Trends in Pharmacological Sciences* 21 (7): 244–6.

Hirth, M., Rukwied, R., Gromann, A., Turnquist, B., Weinkauf, B., Francke, K., Albrecht, P., Rice, F., Hägglöf, B. and Ringkamp, M. 2013. Nerve growth factor induces sensitization of nociceptors without evidence for increased intraepidermal nerve fiber density. *Pain* 154 (11): 2500–11.

Huang, H.-L., Cendan, C.-M., Roza, C., Okuse, K., Cramer, R., Timms, J. F. and Wood, J. N. 2008. Proteomic profiling of neuromas reveals alterations in protein composition and local protein synthesis in hyper-excitable nerves. *Molecular Pain* 4 (33). Doi:10.1186/1744-8069-4-33.

Huang, W., Calvo, M., Karu, K., Olausen, H. R., Bathgate, G., Okuse, K., Bennett, D. L. and Rice, A. S. 2013. A clinically relevant rodent model of the HIV antiretroviral drug stavudine induced painful peripheral neuropathy. *Pain* 154 (4): 560–75.

Huxley, C., Passage, E., Manson, A., Putzu, G., Figarella-Branger, D., Pellissier, J. and Fontes, M. 1996. Construction of a mouse model of Charcot–Marie–Tooth disease type 1A by pronuclear injection of human YAC DNA. *Human molecular genetics* 5 (5): 563–9.

Iannetti, G. D., Zambreanu, L., Wise, R. G., Buchanan, T. J., Huggins, J. P., Smart, T. S., Vennart, W. and Tracey, I. 2005. Pharmacological modulation of pain-related brain activity during normal and central sensitization states in humans. *Proceedings of the National Academy of Sciences of the United States of America* 102 (50): 18195–200.

Joseph, E. K. and Levine, J. D. 2009. Comparison of oxaliplatin-and cisplatin-induced painful peripheral neuropathy in the rat. *The Journal of Pain* 10 (5): 534–41.

Joseph, E. K., Chen, X., Khasar, S. G. and Levine, J. D. 2004. Novel mechanism of enhanced nociception in a model of AIDS therapy-induced painful peripheral neuropathy in the rat. *Pain* 107 (1): 147–58.

Kanda, T. 2012. Biology of the blood–nerve barrier and its alteration in immune mediated neuropathies. *Journal of Neurology, Neurosurgery & Psychiatry*. doi: 10.1136/jnnp-2012–302312.

Katz, J., Finnerup, N. B. and Dworkin, R. H. 2008. Clinical trial outcome in neuropathic pain: Relationship to study characteristics. *Neurology* 70 (4): 263–72.

Kim, C. F. and Moalem-Taylor, G. 2011. Interleukin-17 contributes to neuroinflammation and neuropathic pain following peripheral nerve injury in mice. *The Journal of Pain* 12 (3): 370–83.

King, T., Qu, C., Okun, A., Mercado, R., Ren, J., Brion, T., Lai, J. and Porreca, F. 2011. Contribution of afferent pathways to nerve injury-induced spontaneous pain and evoked hypersensitivity. *Pain* 152 (9): 1997–2005.

Klein, C. J., Lennon, V. A., Aston, P. A., McKeon, A. and Pittock, S. J. 2012. Chronic pain as a manifestation of potassium channel-complex autoimmunity. *Neurology* 79 (11): 1136–44.

Klit, H., Finnerup, N. B. and Jensen, T. S. 2009. Central post-stroke pain: Clinical characteristics, pathophysiology, and management. *The Lancet Neurology* 8 (9): 857–68.

Kohr, D., Singh, P., Tschernatsch, M., Kaps, M., Pouokam, E., Diener, M., Kummer, W., Birklein, F., Vincent, A. and Goebel, A. 2011. Autoimmunity against the β 2 adrenergic receptor and muscarinic-2 receptor in complex regional pain syndrome. *Pain* 152 (12): 2690–700.

Kort, M. E. and Kym, P. R. 2012. 2 TRPV1 Antagonists: Clinical setbacks and prospects for future development. *Progress in medicinal chemistry* 51: 57–70.

Kretschmer, T., England, J. D., Happel, L. T., Liu, Z., Thouron, C. L., Nguyen, D. H., Beuerman, R. W. and Kline, D. G. 2002. Ankyrin G and voltage gated sodium channels colocalize in human neuroma: Key proteins of membrane remodeling after axonal injury. *Neuroscience Letters* 323 (2): 151–5.

Krumova, E. K., Zeller, M., Westermann, A. and Maier, C. 2011. Lidocaine patch (5%) induces only partial blockade of Aδ- and C-fibers to variable extent. Pain 153 (2): 273–80.

Kwan, C. L., Diamant, N. E., Pope, G., Mikula, K., Mikulis, D. J. and Davis, K. D. 2005. Abnormal forebrain activity in functional bowel disorder patients with chronic pain. *Neurology* 65 (8): 1268–77.

Laast, V. A., Shim, B., Johanek, L. M., Dorsey, J. L., Hauer, P. E., Tarwater, P. M., Adams, R. J., Pardo, C. A., McArthur, J. C. and Ringkamp, M. 2011. Macrophage-mediated dorsal root ganglion damage precedes altered nerve conduction in SIV-infected macaques. *The American Journal of Pathology* 179 (5): 2337–45.

Lacroix-Fralish, M. L., Austin, J.-S., Zheng, F. Y., Levitin, D. J. and Mogil, J. S. 2011. Patterns of pain: Meta-analysis of microarray studies of pain. *Pain* 152 (8): 1888–98.

Laferrière, A., Millecamps, M., Xanthos, D. N., Xiao, W. H., Siau, C., de Mos, M., Sachot, C., Ragavendran, J. V., Huygen, F. and Bennett, G. J. 2008. Cutaneous tactile allodynia associated with microvascular dysfunction in muscle. *Molecular Pain* 4 (49). doi: 10.1186/1744-8069-4-49.

Lane, N. E., Schnitzer, T. J., Birbara, C. A., Mokhtarani, M., Shelton, D. L., Smith, M. D. and Brown, M. T. 2010. Tanezumab for the treatment of pain from osteoarthritis of the knee. *New England Journal of Medicine* 363 (16): 1521–31.

Latronico, N., Filosto, M., Fagoni, N., Gheza, L., Guarneri, B., Todeschini, A., Lombardi, R., Padovani, A. and Lauria, G. 2013. Small nerve fiber pathology in critical illness. *PloS One* 8 (9): e75696.

Lee, K. E. and Winkelstein, B. A. 2009. Joint distraction magnitude is associated with different behavioral outcomes and substance P levels for cervical facet joint loading in the rat. *The Journal of Pain* 10 (4): 436–45.

Lee, S. and Zhang, J. 2012. Heterogeneity of macrophages in injured trigeminal nerves: Cytokine/chemokine expressing vs phagocytic macrophages. *Brain, Behavior, and Immunity* 26 (6): 891–903.

Li, Y., Dorsi, M. J., Meyer, R. A. and Belzberg, A. J. 2000. Mechanical hyperalgesia after an L5 spinal nerve lesion in the rat is not dependent on input from injured nerve fibers. *Pain* 85 (3): 493–502.

Lim, T. K., Shi, X. Q., Martin, H. C., Huang, H., Luheshi, G., Rivest, S. and Zhang, J. 2014. Blood-nerve barrier dysfunction contributes to the generation of neuropathic pain and allows targeting of injured nerves for pain relief. *Pain* 155 (5): 954–67.

Liu, C.-C., Lu, N., Cui, Y., Yang, T., Zhao, Z.-Q., Xin, W.-J. and Liu, X.-G. 2010. Prevention of Paclitaxel-induced allodynia by minocycline: Effect on loss of peripheral nerve fibers and infiltration of macrophages in rats. *Molecular Pain* 6 (76). doi: 10.1186/1744-8069-6-76.

Loggia, M. L., Chonde, D. B., Akeju, O., Arabasz, G., Catana, C., Edwards, R. R., Hill, E., Hsu, S., Izquierdo-Garcia, D. and Ji, R.-R. 2015. Evidence for brain glial activation in chronic pain patients. *Brain* 138: 604–15.

Lu, J., Kurejova, M., Wirotanseng, L. N., Linker, R. A., Kuner, R. and Tappe-Theodor, A. 2012. Pain in experimental autoimmune encephalitis: A comparative study between different mouse models. *Journal of Neuroinflammation* 9 (233). doi: 10.1186/1742-2094-9-233.

MacGillavry, H. D., Cornelis, J., L. R. van der Kallen, Sassen, M. M., Verhaagen, J., Smit, A. B. and van Kesteren, R. E. 2011. Genome-wide gene expression and promoter binding analysis identifies NFIL3 as a repressor of C/EBP target genes in neuronal outgrowth. *Molecular and Cellular Neurosciences* 46: 460–8.

Magerl, W., Fuchs, P. N., Meyer, R. A. and Treede, R. D. 2001. Roles of capsaicin-insensitive nociceptors in cutaneous pain and secondary hyperalgesia. *Brain* 124 (9): 1754–64.

Maier, C., Baron, R., Tölle, T. R., Binder, A., Birbaumer, N., Birklein, F., Gierthmühlen, J., Flor, H., Geber, C., Huge, V., Krumova, E. K., Landwehrmeyer, G. B., Magerl, W., Maihöfner, C., Richter, H., Rolke, R., Scherens, A., Schwarz, A., Sommer, C., Tronnier, V., Uçeyler, N., Valet, M., Wasner, G. and Treede, R.-D. 2010. Quantitative sensory testing in the German Research Network on Neuropathic Pain (DFNS): Somatosensory abnormalities in 1236 patients with different neuropathic pain syndromes. *Pain* 150 (3): 439–50.

Malmberg, A. B. and Basbaum, A. I. 1998. Partial sciatic nerve injury in the mouse as a model of neuropathic pain: Behavioral and neuroanatomical correlates. *Pain* 76 (1): 215–22.

Marchand, F., Perretti, M. and McMahon, S. B. 2005. Role of the immune system in chronic pain. *Nature Reviews Neuroscience* 6 (7): 521–32.

Markman, J. D. and Dworkin, R. H. 2006. Ion channel targets and treatment efficacy in neuropathic pain. *The Journal of Pain* 7 (1): S38–S47.

Max, M. B. and Gilron, I. 1999. Sympathetically maintained pain: Has the emperor no clothes? *Neurology* 52 (5): 905–7.

McMahon, S. B. and Malcangio, M. 2009. Current challenges in glia-pain biology. *Neuron* 64 (1): 46–54.

Mendell, L. M. 2014. Constructing and deconstructing the gate theory of pain. *Pain* 155 (2): 210–16.

Merskey, H. and Bogduk, N. 1994. *Classification of Chronic Pain*. Seattle, WA: IASP Press.

Micó, J. A., Ardid, D., Berrocoso, E. and Eschalier, A. 2006. Antidepressants and pain. *Trends in Pharmacological Sciences* 27 (7): 348–54.

Moalem-Taylor, G., Allbutt, H. N., Iordanova, M. D. and Tracey, D. J. 2007. Pain hypersensitivity in rats with experimental autoimmune neuritis, an animal model of human inflammatory demyelinating neuropathy. *Brain, Behavior, and Immunity* 21 (5): 699–710.

Mogil, J. S., Miermeister, F., Seifert, F., Strasburg, K., Zimmermann, K., Reinold, H., Austin, J.-S., Bernardini, N., Chesler, E. J., Hofmann, H. A., C. Hordo, K. Messlinger, Nemmani, K. V. S., Rankin, A. L., J. Ritchie, A. Siegling, Smith, S. B., S. Sotocinal, Vater, A., Lehto, S. G., Klussmann, S., Quirion, R., Michaelis, M., Devor, M. and Reeh, P. W. 2005. Variable sensitivity to noxious heat is mediated by differential expression of the CGRP gene. *Proceedings of the National Academy of Sciences of the United States of America* 102 (36): 12938–43.

Molliver, D. C., Lindsay, J., Albers, K. M. and Davis, B. M. 2005. Overexpression of NGF or GDNF alters transcriptional plasticity evoked by inflammation. *Pain* 113 (3): 277–84.

Nicotra, L., Tuke, J., Grace, P. M., Rolan, P. E. and Hutchinson, M. R. 2014. Sex differences in mechanical allodynia: How can it be preclinically quantified and analyzed? *Frontiers in Behavioral Neuroscience* 8. doi: 10.3389/fnbeh.2014.00040 DOI:10.3389%2Ffnbeh.2014.00040 .

Nitzan-Luques, A., Devor, M. and Tal, M. 2011. Genotype-selective phenotypic switch in primary afferent neurons contributes to neuropathic pain. *Pain* 152 (10): 2413–26.

O'Neill, J., Brock, C., Olesen, A. E., Andresen, T., Nilsson, M. and Dickenson, A. H. 2012. Unravelling the mystery of capsaicin: A tool to understand and treat pain. *Pharmacological Reviews* 64 (4): 939–71.

Oaklander, A. L., Rissmiller, J. G., Gelman, L. B., Zheng, L., Chang, Y. and Gott, R. 2006. Evidence of focal small-fiber axonal degeneration in complex regional pain syndrome-I (reflex sympathetic dystrophy). *Pain* 120 (3): 235–43.

Polgár, E. 2005. Loss of neurons from laminas I–III of the spinal dorsal horn is not required for development of tactile allodynia in the spared nerve injury model of neuropathic pain. *Journal of Neuroscience* 25 (28): 6658–66.

Polomano, R. C., Mannes, A. J., Clark, U. S. and Bennett, G. J. 2001. A painful peripheral neuropathy in the rat produced by the chemotherapeutic drug, paclitaxel. *Pain* 94 (3): 293–304.

RCGPs (2004). *A Practical Guide to the provision of Chronic Pain Services for Adults in Primary Care*. London: The British Pain Society and the Royal College of General Practitioners.

RCGPs (2011). *A Case of Neuropathic Pain*. London: British Pain Society and the Royal College of General Practitioners.

Richards, N., Batty, T. and Dilley, A. 2011. CCL2 has similar excitatory effects to TNF- in a subgroup of inflamed C-fiber axons. *Journal of Neurophysiology* 106 (6): 2838–48.

Rolke, R., Baron, R., Maier, C., Tolle, T. R., Treede, R. D., Beyer, A., Binder, A., Birbaumer, N., Birklein, F., Botefur, I. C., Braune, S., Flor, H., Huge, V., Klug, R., Landwehrmeyer, G. B., Magerl, W., Maihofner, C., Rolko, C., Schaub, C., Scherens, A., Sprenger, T., Valet, M. and Wasserka, B. 2006. Quantitative sensory testing in the German Research Network on Neuropathic Pain (DFNS): Standardized protocol and reference values. *Pain* 123 (3): 231–43.

Rosenberg, A. F., Wolman, M. A., Franzini-Armstrong, C. and Granato, M. 2012. In vivo nerve: Macrophage interactions following peripheral nerve injury. *The Journal of Neuroscience* 32 (11): 3898–909.

Sandkühler, J. 2009. Models and mechanisms of hyperalgesia and allodynia. *Physiological Reviews* 89 (2): 707–58.

Seo, H.-S., Kim, H.-W., Roh, D.-H., Yoon, S.-Y., Kwon, Y.-B., Han, H.-J., Chung, J. M., Beitz, A. J. and Lee, J.-H. 2008. A new rat model for thrombus-induced ischemic pain (TIIP): Development of bilateral mechanical allodynia. *Pain* 139 (3): 520–32.

Sereda, M., Griffiths, I., Pühlhofer, A., Stewart, H., Rossner, M. J., Zimmermann, F., Magyar, J. P., Schneider, A., Hund, E. and Meinck, H.-M. 1996. A transgenic rat model of Charcot-Marie-Tooth disease. *Neuron* 16 (5): 1049–60.

Serra, J. 2009. Re-emerging microneurography. *The Journal of Physiology* 587 (2): 295–6.

Serra, J., Bostock, H., Solà, R., Aleu, J., García, E., Cokic, B., Navarro, X. and Quiles, C. 2011. Microneurographic identification of spontaneous activity in C-nociceptors in neuropathic pain states in humans and rats. *Pain* 153 (1): 42–55.

Shi, Y., Gelman, B. B., Lisinicchia, J. G. and Tang, S. J. 2012. Chronic-pain-associated astrocytic reaction in the spinal cord dorsal horn of human immunodeficiency virus-infected patients. *Journal of Neuroscience* 32 (32): 10833–40.

Shields, S. D., Cheng, X., Uceyler, N., Sommer, C., Dib-Hajj, S. D. and Waxman, S. G. 2012. Sodium channel Na(v)1.7 is essential for lowering heat pain threshold after burn injury. *Journal of Neuroscience* 32 (32): 10819–32.

Soleman, S., Yip, P. K., Duricki, D. A. and Moon, L. D. 2012. Delayed treatment with chondroitinase ABC promotes sensorimotor recovery and plasticity after stroke in aged rats. *Brain* 135 (4): 1210–23.

Sorensen, L., Molyneaux, L. and Yue, D. K. 2006. The relationship among pain, sensory loss, and small nerve fibers in diabetes. *Diabetes Care* 29 (4): 883–7.

Sorge, R. E., LaCroix-Fralish, M. L., Tuttle, A. H., Sotocinal, S. G., J.-S. Austin, Ritchie, J., Chanda, M. L., Graham, A. C., Topham, L. and Beggs, S. 2011. Spinal cord Toll-like receptor 4 mediates inflammatory and neuropathic hypersensitivity in male but not female mice. *The Journal of Neuroscience* 31 (43): 15450–4.

Steigerwald, I., Muller, M., Kujawa, J., Balblanc, J. C. and Calvo-Alen, J. 2012. Effectiveness and safety of tapentadol prolonged release with tapentadol immediate release on-demand for the management of severe, chronic osteoarthritis-related knee pain: Results of an open-label, phase 3b study. *Journal of Pain Research* 5: 121–38.

Stucky, C., Koltzenburg, M., Schneider, M., Engle, M., Albers, K. and Davis, B. 1999. Overexpression of nerve growth factor in skin selectively affects the survival and functional properties of nociceptors. *The Journal of Neuroscience* 19 (19): 8509–16.

Sweitzer, S. and De Leo, J. 2011. Propentofylline: Glial modulation, neuroprotection, and alleviation of chronic pain. *Handbook of Experimental Pharmacology* 200: 235–50.

Thakur, M. 2015. Neurobiology of pain. Fastbleep: Biology notes. At http://www.fastbleep.com/biology-notes/39/142/888 (accessed 27 September 2015).

Thakur, M., Dawes, J. M. and McMahon, S. B. 2013. Genomics of pain in osteoarthritis. *Osteoarthritis Cartilage* 21 (9): 1374–82.

Thibault, K., Elisabeth, B., Sophie, D., Claude, F.-Z. M., Bernard, R. and Bernard, C. 2008. Antinociceptive and anti-allodynic effects of oral PL37, a complete inhibitor of enkephalin-catabolizing enzymes, in a rat model of peripheral neuropathic pain induced by vincristine. *European Journal of Pharmacology* 600 (1): 71–7.

Torebjörk, H., Lundberg, L. and LaMotte, R. 1992. Central changes in processing of mechanoreceptive input in capsaicin-induced secondary hyperalgesia in humans. *The Journal of Physiology* 448 (1): 765–80.

Treede, R.-D., Jensen, T. S., Campbell, J. N., Cruccu, G., Dostrovsky, J. O., Griffin, J. W., Hansson, P., Hughes, R., Nurmikko, T. and Serra, J. 2008. Neuropathic pain: Redefinition and a grading system for clinical and research purposes. *Neurology* 70 (18): 1630–5.

Turk, D. C., Dworkin, R. H., Allen, R. R., Bellamy, N., Brandenburg, N., Carr, D. B., Cleeland, C., Dionne, R., Farrar, J. T. and Galer, B. S. 2003. Core outcome domains for chronic pain clinical trials: IMMPACT recommendations. *Pain* 106 (3): 337–45.

Üçeyler, N., Zeller, D., Kahn, A.-K., Kewenig, S., Kittel-Schneider, S., Schmid, A., Casanova-Molla, J., Reiners, K. and Sommer, C. 2013. Small fibre pathology in patients with fibromyalgia syndrome. *Brain* 136 (6): 1857–67.

Uhl, I., Krumova, E. K., Regeniter, S., Bar, K. J., Norra, C., Richter, H., Assion, H. J., Westermann, A., Juckel, G. and Maier, C. 2011. Association between wind-up ratio and central serotonergic function in healthy subjects and depressed patients. *Neuroscience Letters* 504 (2): 176–80.

Urch, C., Donovan-Rodriguez, T. and Dickenson, A. 2003. Alterations in dorsal horn neurones in a rat model of cancer-induced bone pain. *Pain* 106 (3): 347–56.

van Doorn, P. A., Ruts, L. and Jacobs, B. C. 2008. Clinical features, pathogenesis, and treatment of Guillain-Barré syndrome. *The Lancet Neurology* 7 (10): 939–50.

Vega-Avelaira, D., Géranton, S. M. and Fitzgerald, M. 2009. Differential regulation of immune responses and macrophage/neuron interactions in the dorsal root ganglion in young and adult rats following nerve injury. *Molecular Pain* 5 (70). doi: 10.1186/1744-8069-5-70.

Vicuña, L., Strochlic, D. E., Latremolière, A., Bali, K. K., Simonetti, M., Husainie, D., Prokosch, S., Riva, P., Griffin, R. S., Njoo, C., Gehrig, S., Mall, M. A., Arnold, B., Devor, M., Woolf, C. J., Liberles, S. D., Costigan, M. and Kuner, R. 2015. The serine protease inhibitor SerpinA3N attenuates neuropathic pain by inhibiting T cell-derived leukocyte elastase. *Nature Medicine* 21 (5): 518–23.

Wässle, H. 2004. Parallel processing in the mammalian retina. *Nature Reviews Neuroscience* 5 (10): 747–57.

Wall, P., Devor, M., Inbal, R., Scadding, J., Schonfeld, D., Seltzer, Z. and Tomkiewicz, M. 1979. Autotomy following peripheral nerve lesions: Experimental anesthesia dolorosa. *Pain* 7 (2): 103–13.

Wallace, V. C., Blackbeard, J., Pheby, T., Segerdahl, A. R., Davies, M., Hasnie, F., Hall, S., McMahon, S. B. and Rice, A. S. 2007. Pharmacological, behavioural and mechanistic analysis of HIV-1 gp120 induced painful neuropathy. *Pain* 133 (1): 47–63.

Wei, T., Li, W.-W., Guo, T.-Z., Zhao, R., Wang, L., Clark, D. J., Oaklander, A. L., Schmelz, M. and Kingery, W. S. 2009. Post-junctional facilitation of substance P signaling in a tibia fracture rat model of complex regional pain syndrome type I. *Pain* 144 (3): 278–86.

Yamasaki, R., Lu, H., Butovsky, O., Ohno, N., Rietsch, A. M., Cialic, R., Wu, P. M., Doykan, C. E., Lin, J. and Cotleur, A. C. 2014. Differential roles of microglia and monocytes in the inflamed central nervous system. *The Journal of Experimental Medicine* 211 (8): 1533–49.

Yezierski, R. P. 2000. Pain following spinal cord injury: Pathophysiology and central mechanisms. *Progress in Brain Research* 129: 429–49.

Yoon, C., Wook, Y. Y., Sik, N. H., Ho, K. S. and Mo, C. J. 1994. Behavioral signs of ongoing pain and cold allodynia in a rat model of neuropathic pain. *Pain* 59 (3): 369–76.

Zhang, G., Hoffman, P. N. and Sheikh, K. A. 2014. Axonal degeneration in dorsal columns of spinal cord does not induce recruitment of hematogenous macrophages. *Experimental Neurology* 252: 57–62.

CHAPTER 4

Recent advances in neuroimmune interactions in neuropathic pain: The role of microglia

Elizabeth A. Old, Louise S. C. Nicol and Marzia Malcangio

Introduction

The International Association for the Study of Pain (IASP) defines pain as 'an unpleasant sensory and emotional experience associated with actual or potential tissue damage' (www. isap.org). Under physiological conditions, pain is a vital protective mechanism and persists only for the duration of the noxious stimulus. The importance of this protective mechanism becomes apparent in individuals with the rare phenotype of insensitivity to pain, which results in frequent inadvertent self-mutilation and injury throughout life (Cox et al., 2006; Verpoorten et al., 2006).

The experience of acute pain is mediated by specialised primary afferent fibres, namely the nociceptors, whose cell bodies lay in the dorsal root ganglia (DRG) and which detect noxious stimuli in the periphery and transmit electrical impulses centrally to the spinal cord on their way to the brain, where pain is perceived. Descending pathways from the brain to the spinal cord exert modulatory control of pain processing alongside dorsal horn excitatory and inhibitory interneurons (Todd, 2010; for a complete overview of the spinal cord microcircuitry involved in pain pathways, see Todd's Chapter 1 in this volume). At the level of the spinal cord, nociceptive transmission is mainly mediated by glutamate release and activation of postsynaptic ionotropic receptors. However, the co-release of peptides such as substance P (SP) and calcitonin gene-related peptide (CGRP) and the activation of NK1 and CGRP receptors result in postsynaptic modulation of glutamatergic transmission (Go and Yaksh, 1987; Cheunsuang and Morris, 2000; Malcangio and Bowery, 1994; Lever, Grant, et al., 2003; Oku et al., 1987; Seybold et al., 2003).

More commonly than pain insensitivity, individuals may experience chronic pain, which results from a malfunction of the nociceptive system and outlives the causative stimuli and the physiological role of acute pain. Chronic pain states fall into two broad categories: neuropathic and inflammatory. Here neuropathic pain states will be described in brief and specific attention will be paid to the contribution of neuroimmune interactions to spinal cord mechanisms of chronic pain. (For a comprehensive overview of neuropathic pain symptoms and pathology, see Chapter 3 in this volume.)

An Introduction to Pain and its Relation to Nervous System Disorders, First Edition. Edited by Anna A. Battaglia.
© 2016 John Wiley & Sons, Ltd. Published 2016 by John Wiley & Sons, Ltd.

Chronic neuropathic pain

There are profound differences between acute and chronic pain conditions, as dramatic changes occur in peripheral nociceptors and in central nociceptive pathways. In chronic pain conditions nociceptors become sensitised, thereby generating exaggerated responses to noxious stimuli (hyperalgesia) and inappropriate responses to non-noxious stimuli (allodynia) (Woolf and Mannion, 1999; Baron, 2006; for more details see Latremolière's Chapter 2 on spinal plasticity in this volume).

Neuropathic pain occurs as a consequence of a lesion to the nervous system, be that the result of physical trauma or of disease. Typically, painful neuropathy may arise from the crushing, stretching or severing of a peripheral nerve. Neuropathic pain can also result from the pathogenesis of a disease such as diabetes mellitus and viral infections. Furthermore, painful neuropathy can occur after toxic insult or as a side effect of antiretrovirals or certain chemotherapeutics (Baron, Binder and Wasner, 2010).

The treatment of neuropathic pain is a considerable clinical problem, as many patients do not experience sufficient pain relief from the currently available analgesics; less than 50 per cent of patients experience effective pain relief and as many as 64 per cent report that their treatment is inadequate for controlling their pain (Breivik et al., 2006). Several classes of drugs, including selective serotonin re-uptake inhibitors (SSRIs) and opiates, have shown efficacy in clinical trials and in meta-analyses (Baron et al., 2010; Dworkin et al., 2007; O'Connor and Dworkin, 2009). However, many drugs require long-term use, which is accompanied by adverse and intolerable side effects. Furthermore, no clear predictors of treatment response have been identified in chronic pain patients (Baron et al., 2010). The difficulty in treating this condition is likely a result of the heterogeneity of the mechanisms underlying the development and maintenance of chronic neuropathic pain.

Animal models have provided a good understanding of the pathogenesis of neuropathic pain. Models in rodents typically involve the surgical injury of a peripheral nerve, usually the sciatic or a branch thereof. Such injuries induce a sustained hypersensitivity to the application of noxious stimuli in the ipsilateral hind paw (the contralateral paw acting as an internal control) and involve the measurement of nociceptive or flexor withdrawal reflex. Thus the application of mechanical or thermal stimuli to the glabrous skin of the hind paw evokes reflex withdrawals; these are spinally mediated and initiated through the activation of primary nociceptors under descending control from supraspinal structures in the brain stem and in the midbrain (Fields and Heinricher, 1989). Symptoms and pathology of neuropathic pain have been considered in detail by Thakur and McMahin in Chapter 3.

Under neuropathic pain conditions, damaged and undamaged afferent fibres undergo changes in ion channel expression and distribution, which may initiate ectopic charges that spread via the cross-excitation of nerves (ephaptic transmission). The increased nociceptive input from the periphery triggers a physiological plasticity defined as central sensitization: dorsal horn neurons show decreased thresholds to peripheral noxious stimuli, *de novo* excitation by innocuous stimuli, increased spontaneous activity and the expansion of receptive fields (Kuner, 2010; Latremolière and Woolf, 2009). The increased release of glutamate from the central terminals of sensitized nociceptive fibres leads to phosphorylation of N-methyl-D-aspartate (NMDA) receptors, an increase in intracellular calcium concentration and subsequent long-lasting transcriptional and post-translational changes. Glial cells in the spinal cord, such as microglia and astrocytes, respond to peripheral tissue and nerve injury by proliferating, presenting antigens (microglia) and releasing mediators that contribute to dorsal horn (spinal) mechanisms of chronic pain (Clark and Malcangio, 2012; Old and Malcangio, 2012; Clark, Old and Malcangio, 2013). Cortical and subcortical structures facilitate spinal cord excitation and signalling (D'Mello and Dickenson, 2008). Here we will describe and define the modality of microglial and astrocyte responses in the dorsal horn after a remote injury in the peripheral nerve. We will then discuss the role played by microglial and astrocyte responses in mechanisms of chronic neuropathic pain, with a specific focus on chemokine and purinergic signalling.

Microglia
Origin and physiological role
Microglia are defined as the resident macrophages of the central nervous system (CNS) (Lawson, Perry and Gordon, 1992; Perry, Hume and Gordon, 1985). They were first described by del Río-Hortega in the 1920s, as cells different from macroglia, during a series of studies in which silver carbon labelling and light microscopy were used to visualise these cells. These cells were considered to arise either from invasion of the CNS by mesodermal pial elements or from blood mononuclear cells (Kettenmann et al., 2011; Prinz and Mildner, 2011). Similarly, pathologist W. Ford Robertson described a population of cells that he considered to be of mesodermal origin, which did not appear to contact blood vessels and withdrew their processes in response to injury (Kettenmann et al., 2011). Río-Hortega's and Robertson's views were widely accepted by most but not by all, and several alternative theories were hypothesised, including differentiation from glioblasts of the neuroectoderm (Paterson et al., 1973; Kitamura, Miyake and Fujita, 1984). The view that microglial cells were of monocytic origin remained popular and was later verified by Leblond and colleagues through autoradiography; here it was demonstrated that amoeboid microglia exhibited monocytic characteristics and later matured into ramified microglia (Imamoto and Leblond, 1978; Ling, Penney and Leblond, 1980). While the precise origin of microglia has long been the subject of debate in the glia field, it remains widely accepted that these cells are of a monocytic/myeloid lineage and infiltrate during postnatal development. Microglia can be positively identified with antibodies against the macrophage antigens F4/80 (Perry et al., 1985) and CR3 (detected with an OX42 antibody; Graeber, Streit and Kreutzberg, 1988) and do not share immunohistochemical markers with astrocytes or oligodendrocytes. These observations indicate that microglia are unlikely to have derived from neuroectodermal glioblasts. The lineage of microglia has been confirmed conclusively by their absence from the CNS in mice deficient of PU.1, a key transcription factor in the control of myeloid differentiation (Beers et al., 2006; McKercher et al., 1996). Current evidence indicates that microglia appear early during development and originate from primitive macrophages in the foetal yolk sac, which migrate into the CNS during early embryogenesis (Saijo and Glass, 2011). There is also evidence from postnatal rodent studies using fluorescent-labelled cells; this evidence indicates that, at least in the early stages of postnatal development, microglia can differentiate themselves from monocytes that enter the CNS from the circulation (Perry et al., 1985). In adult rodents, circulating monocytes

are able to enter the CNS under conditions where there is disruption to the blood–brain barrier (BBB) (King, Dickendesher and Segal, 2009). However, evidence obtained from immune irradiated mice indicates that CNS colonisation by monocytes is transient and does not contribute to the resident microglial cell pool (Ajami et al., 2011). This means that microglial populations are maintained locally by proliferation of existing microglia and independently of circulating monocytes (Ginhoux et al., 2010).

Microglia perform immune surveillance of the CNS parenchyma. They are characterised by a small soma, expression of receptors for complement components (Fcγ receptor of IgG), low expression of cell surface antigens and the possession of highly motile ramified processes (Nimmerjahn, Kirchhoff and Helmchen, 2005). Once activated by a CNS or peripheral nervous system (PNS) injury, their phagocytic capacity is enhanced, they engage in antigen presentation via major histocompatibility complex (MHC) class I and II molecules and they release cytokines, chemokines, proteases and reactive oxygen species (Hoftberger et al., 2004; Gehrmann, Matsumoto and Kreutzberg, 1995; Perry, 1998; Tambuyzer et al., 2009).

Activation of microglia under neuropathic pain states

As the resident immune cells of the CNS, microglia respond rapidly to tissue trauma, pathogen products (e.g., bacterial lipopolysaccharides) and mediators released from damaged neurons and other immune cells. In models of neuropathic pain induced by peripheral nerve injury, spinal microglia respond promptly to enhance primary afferent fibre input at the dorsal horn. Indeed, peripheral nerve conduction blockade via the application of a local anaesthetic prevents nerve injury-induced microglial activation in the spinal cord (Wen et al., 2007; Hathway et al., 2009; Suter et al., 2009). Additionally, electrical stimulation of the peripheral nerve, at a strength that recruits high-threshold nociceptive fibres, induces a significant microglial response in the ipsilateral dorsal horn of the spinal cord in naïve mice observed at both 24 and 48 hours after stimulation (Hathway et al., 2009).

Microglia express a number of receptors for the neurotransmitters that are released from the central terminals of nociceptive fibres, namely amino-3-hydroxy-5-methyl-4-isoxazole propionate (AMPA) and NMDA receptors for glutamate, neurokinin-1 (NK1) receptors and the calcitonin gene-related peptide receptor for SP and CGRP respectively, tropomyosin kinase B (TrkB) receptors for brain-derived neurotrophic factor (BDNF) (Rasley et al., 2002; Pezet et al., 2002; McMahon and Malcangio, 2009; Ransohoff and Perry, 2009), and purinergic receptors – including P2X4 and P2X7, which are substantially up-regulated after peripheral nerve injury (Tsuda et al., 2003; Chessell et al., 2005; Kobayashi et al., 2011). Microglial activation in the spinal cord has been documented in models of neuropathic pain, and changes in cell morphology are visible immunohisto-chemically as early as three days after peripheral nerve injury (Echeverry, Shi and Zhang, 2008) and maintained for at least fifty days (Clark, Gentry, et al., 2007); however, real-time polymerase chain reaction (PCR) analysis of spinal cord lysates is able to detect changes in the levels of microglial specific mRNA transcripts (e.g., Raghavendra and DeLeo, 2004). Microglial proliferation and activation are most abundant in the superficial dorsal horn ipsilateral to the injury in the periphery, where the central terminals of the injured primary afferent fibres end; but microgliosis is also observed in the ventral horn, due to the combination of sensory and motor fibre in the damaged nerve. The temporal profile of microglial activation in the spinal cord following peripheral damage is concomitant with the presence of behavioural hypersensitivity (Zhang et al., 2007; Peters et al., 2007; Wodarski et al., 2009) and the administration of glial inhibitors attenuates pain in neuropathic models (Clark, Gentry, et al., 2007; Tawfik et al., 2007), which indicates that microglial activation does indeed contribute to the development of pain.

The activation of microglia in neuropathic pain states is associated with the recruitment of downstream signalling cascades or pathways – such as the mitogen-activated protein kinase family of signalling molecules, which consist of extracellular signal-regulated kinase (ERK), p38 and c-Jun-N-terminal kinase (JNK). Activation of both ERK and p38 is evident in microglia after peripheral nerve injury; the expression of phosphorylated ERK (pERK) peaks approximately three days after injury and the inhibition of its signalling pathway attenuates injury-induced hypersensitivity (Zhuang et al., 2005; Tsuda et al., 2008). Similarly, the expression of phosphorylated p38 (p-p38) peaks approximately three days after injury and remains elevated for several weeks. However, while an increase in the expression of pERK is observed in dorsal horn neurons and astrocytes as well as in microglia, p-p38 is found mainly in microglia (Jin et al., 2003; Tsuda et al., 2004; Clark, Yip, et al., 2007). Additionally, current evidence indicates that, unlike the activation of many other protein kinases, the activation of mitogen-activated protein kinases (MAPKs) is predominantly a pathophysiological response, as opposed to contributing to normal sensory function (Crown, 2012). Several studies have now examined the role of MAPK-dependent signalling pathways within the CNS in the development and maintenance of chronic pain states. Constriction of the peripheral nerve increases the detectable level of p-p38 in spinal microglia (Kim et al., 2002); this activation can be observed within 24 hours and persists for several weeks after injury (Jin et al., 2003; Zhuang et al., 2006). Furthermore, the intrathecal administration of a p38 MAPK inhibitor can both prevent the development of mechanical hypersensitivity following injury and reverse established pain-like behaviour (Wen et al., 2007). In accordance with this study, the oral administration, twice daily for two weeks, of a centrally penetrant and selective p38 MAPK inhibitor, dilmapimod, to traumatic nerve injury patients reduced the pain score when compared to a placebo treated group (Anand et al., 2011). Following injury, a persistent increase in pERK is also observed in spinal microglia, and this lasts for up to ten days – in contrast to increases in neuronal pERK, which is more transient and of shorter duration (Zhuang et al., 2005). Relevantly, the pharmacological inhibition of signalling proteins upstream of ERK, such as MEK, significantly reduces mechanical allodynia in neuropathic animals (Obata et al., 2003; Seino et al., 2006; Zhuang et al., 2005). The intracellular consequences of ERK and p38 phosphorylation include the activation of transcription factors that mediate an up-regulation of several pronociceptive mediators such as IL-1β, TNF, IL-18, CX_3CL_1, CCL_2, prostaglandins and BDNF (McMahon and Malcangio, 2009) (Figure 4.1).

In summary, spinal cord microglia promptly respond to a remote injury in the peripheral nerve and contribute to pain signalling facilitation through the release of mediators some of which are discussed below in terms of their role in glial cell communication and in the development and maintenance of neuropathic pain.

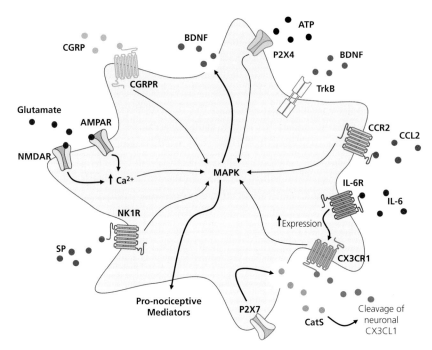

Figure 4.1 Examples of mechanisms of microglial activation in models of chronic pain. The activation of a number of receptors – including neuropeptides, neurotransmitters and chemokines – results in the activation of intracellular MAPK signalling pathways, which in turn leads to the release of pronociceptor mediators. Source: Created by Elizabeth Old for this volume.

Astrocytes
Origin and physiological role
Astrocytes are the most abundant glial cells in the CNS and have traditionally been considered to provide little more than structural and homeostatic support to neurons. However, as research in the field has evolved, it is now recognised that astrocytes are critical participants not only in intercellular communication, but in almost all the processes that occur in the CNS (Volterra and Meldolesi, 2005). Derived from the neuroectoderm, these star-shaped cells envelope many neuronal elements, including synapses, axons and dendrites; each astrocyte is able to make contact with several neurons (Halassa et al., 2007; Haydon, 2001). Due to their close contact with neurons, astrocytes are able to provide support and nourishment to these cells and assist in the regulation of the extracellular environment around synapses (Gao and Ji, 2010b). Astrocytes play a key role in the regulation of extracellular K^+ concentration ($[K^+]_o$). Fluctuations in $[K^+]_o$ occurs physiologically as a result of neuronal activity; however, this must be tightly regulated in order to prevent a pathological build-up of K^+ that could cause hyperexcitability in neurons (Scemes and Spray, 2012). $[K^+]_o$ is maintained by astrocytes via several mechanisms, such as uptake by ion channels and/or transporters (predominantly Na-K-ATPase pumps or Na-K-Cl co-transporters), or spatial buffering. Spatial buffering is the process whereby cells that are highly permeable to K^+ facilitate the movement of the ions from regions of high to regions of low extracellular concentrations (Kofuji and Newman, 2004).

One of the features of astrocytes that allow spatial buffering to occur is the presence of gap junctions between these cells; these intracellular connections comprise two hemi-channels formed of connexin subunits, which allow ions and small molecules to pass between cells. In the adult, connexins (Cx) 43 and 30 are the predominant forms found in astrocytes (Dermietzel et al., 1991; Giaume et al., 1991; Nagy and Rash, 2000); dye-coupling experiments have demonstrated that gap junction-mediated intracellular communication is abolished in the Cx43/Cx30 double knock-out mouse (Wallraff et al., 2006; Rouach et al., 2008).

Extracellular K^+ regulation is not the only homeostatic mechanism in which astrocytes play a major role. Due to the intimate relationship between these cells and the BBB (Abbott, Ronnback and Hansson, 2006), astrocytes are in a prime position to assist in the maintenance of the BBB (Cahoy et al., 2008). Additionally, in a similar fashion to $[K^+]_o$ regulation, astrocytes contribute to the regulation of extracellular neurotransmitter concentrations by actively uptaking excess glutamate. After depolarisation, most of the glutamate is released from the presynaptic terminal into the extracellular space, while a small quantity is released from the astrocytes themselves as a result of both Ca^{2+}-dependent and independent mechanisms (Araque et al., 2000; Hamilton and Attwell, 2010; Santello, Bezzi and Volterra, 2011). Glutamate is removed from the extracellular space by excitatory amino acid transporters (EAATs), of which two (GLT1 and GLAST) are preferentially expressed on astrocytes (Coulter and Eid, 2012). Failure of this system to regulate glutamate uptake can result in glutamate excitotoxicity. This phenomenon was first described in retinal neurons by Lucas and Newhouse in 1957 (Lucas and Newhouse, 1957), and in CNS neurons later, in 1969, by Olney and Sharpe (Olney and Sharpe, 1969); it consists of a process whereby prolonged activation of glutamate receptors leads to cell death.

Activation of astrocytes under neuropathic pain states

Under physiological conditions, in their 'active' state, astrocytes possess many thin processes that contact neurons. However, following activation these cells undergo hypertrophy and increase their expression of several proteins, including glial fibrillary acid protein (GFAP), which is typically used as a marker of astrocyte activation (Garrison, Dougherty and Carlton, 1994; Vallejo et al., 2010), vimentin, and S100B, a calcium-binding peptide (Ridet et al., 1997; Pekny and Nilsson, 2005). These astrocytes are referred to as 'reactive'.

As in the case for microglia, the activation of astrocytes, often referred to as astrogliosis, has been demonstrated in several models of chronic pain (Garrison et al., 1991; Colburn, Rickman and DeLeo, 1999; Sweitzer et al., 1999), and several pieces of evidence support the hypothesis of a contribution of these cells to the maintenance of neuropathic pain after peripheral nerve injury. Astrogliosis typically occurs several days after injury and lasts longer than microgliosis (Colburn et al., 1999; Tanga et al., 2004; Romero-Sandoval et al., 2008). Additionally, the administration of astrocytic inhibitors such as fluorocitrate attenuates pain behaviours in many of the models (Milligan et al., 2003; Watkins et al., 1997; Okada-Ogawa et al., 2009) and can prevent the development of mechanical allodynia when administered prior to nerve transection (Sweitzer, Schubert and DeLeo, 2001). Furthermore, mice deficient in the astrocytic protein GFAP develop a pain phenotype of shorter duration than in their wild-type counterparts, and the administration of a GFAP anti-sense mRNA reverses established neuropathic pain (Kim et al., 2009).

Astrocytes can be activated via a multitude of pathways, due to the expression of neurotransmitter receptors such as NMDA, mGluR, purinergic, SP and CGRP receptors (Porter and McCarthy, 1997; McMahon and Malcangio, 2009). They are therefore considered to be excitatory cells, albeit not in the classical sense that describes the activity of neurons, but in relation to their responsiveness to factors released from both neuronal and non-neuronal cells as a consequence of injury. For example, it has been demonstrated that increased GFAP expression associated with peripheral injury requires the NMDA receptor and neuronal activity (Guo et al., 2007; Garrison et al., 1994). Additionally, in a similar fashion to microglia, astrocytes express receptors for chemokines and cytokines that can be released from neuronal and non-neuronal cells of the CNS (including IL-1 and TNF) (Miyoshi et al., 2008; Gao et al., 2009; Gao and Ji, 2010b). In turn, as a result of the activation of intercellular signalling pathways, astrocytes release mediators such as glutamate and adenosine triphosphate ATP, which facilitate enhanced nociceptive transmission within the dorsal horn of the spinal cord (Hansen and Malcangio, 2013).

While astrocytes undergo many changes after activation, there are several key pathways that are likely to be responsible for the contribution of these cells to enhanced nociceptive transmission. Following an initial increase after nerve injury, astrocytic expression of the glutamate transporters GLT1 and GLAST decreases (Sung, Lim and Mao, 2003; Wang et al., 2008; Xin, Weng and Dougherty, 2009; Tawfik et al., 2008), causing elevation in spinal extracellular glutamate concentrations and increased nociceptive signalling (Liaw et al., 2005; Weng, Chen and Cata, 2006). Reactive astrocytes express pJNK and pERK (Zhuang et al., 2006; Gao and Ji, 2010a; Weyerbacher et al., 2010; Ma and Quirion, 2002), which, in a fashion similar to the pERK and p-p38 activation in microglia, stimulate the release of pro-inflammatory mediators via the activation of transcription factors. Indeed, after peripheral nerve injury, pJNK expression – specifically, the JNK1 isoform – is almost exclusively astrocytic within the spinal cord. Increased pJNK expression within the spinal cord is evident as early as three days after injury and, like p-p38 in microglia, persists for several weeks (Ma and Quirion, 2002; Zhuang et al., 2006). Furthermore, the intrathecal infusion of a JNK inhibitor can both prevent and reverse pain-related behaviours (but not the increase in GFAP associated with reactive astrocytes), while an astrocyte-specific toxin dose-dependently suppresses established neuropathic allodynia with a concurrent decrease in GFAP immunoreactivity (Zhuang et al., 2006).

In summary, reactive astrocytes in the dorsal horn of the spinal cord contribute to the pronociceptive state that develops following peripheral nerve injury.

Glial pronociceptive mediators
CX_3CL_1/CX_3CR_1

The chemokines are a family of proteins critically involved in the migration of leukocytes. CX_3CL_1, also known as fractalkine (FKN), is no exception to this. As determined by structure, it is the only member of the CX_3C family of chemokines and was first described as a potent chemoattractant of T cells and monocytes (Bazan et al., 1997). CX_3CL_1 can exist in two forms – as a membrane-bound tethered protein and as a soluble chemokine domain (Bazan et al., 1997) – each of which mediates distinct biological actions. In its membrane-bound form, CX_3CL_1

serves as an adhesion protein, promoting the firm adhesion of leukocytes to the endothelium without the activation of integrins (Fong et al., 1998). Soluble CX_3CL_1, on the other hand, is a potent chemoattractant for monocytes, NK cells, T cells and B cells (Imai et al., 1997; Corcione et al., 2009). The precise structure of CX_3CL_1 is well established and mutational analysis of the protein has identified which regions are most critical for function; for example, the affinity and efficacy of chemokine domain binding to the CX_3CR_1 receptor are independent of stalk length. However, cells possessing the entire mucin-like domain are more efficient at capturing leukocytes than the chemokine domain alone (Harrison et al., 2001; Fong et al., 2000). In addition, CX_3CL_1 possesses several basic residues in its chemokine domain that contribute to the binding affinity but not to the receptor activation capacity of this protein (Harrison et al., 2001; Nishiyori et al., 1998).

Soluble CX_3CL_1 is a product of the cleavage of the membrane-bound CX_3CL_1 by the metalloproteases ADAM10 and ADAM17 (Bazan et al., 1997; Hundhausen et al., 2003; Hundhausen et al., 2007) and the cysteine protease cathepsin S (Clark, Staniland and Malcangio, 2011). Cathepsin S is an endopeptidase expressed by antigen-presenting cells, including macrophages and microglia, where its secretion is enhanced by pro-inflammatory mediators, including cytokines (Liuzzo, Petanceska, and Devi, 1999; Liuzzo, Petanceska, Moscatelli, et al., 1999) and is dependent on signalling via the purinergic receptor P2X7 (Clark et al., 2010). On the other hand, ADAM10 and ADAM17 are responsible, respectively, for the constitutive and inducible shedding of soluble CX_3CL_1 from endothelial and epithelial cells, where they are themselves expressed.

CX_3CL_1 exerts its biological effects by binding to its only receptor, CX_3CR_1, which binds no other chemokines – as confirmed through radio-ligand binding analysis (Harrison et al., 1998). The expression of this receptor is highest in the spleen and peripheral blood leukocytes; however, high levels are also observed in neural tissues, including in several brain regions and in the spinal cord (Raport et al., 1995). The expression of CX_3CR_1 within the CNS is of microglial origin; the expression analysis of the rat homology (RBS-11) revealed CX_3CR_1 mRNA to be relatively abundant in the brain and ten times higher in cultured primary microglia than in whole brain samples (Harrison et al., 1998; Harrison, Barber and Lynch, 1994). Furthermore, microglia express functional CX_3CR_1 protein as well as the mRNA; this can be demonstrated by treating these cells with CX_3CL_1, which induced calcium mobilisation and chemotaxis – that can be blocked by an anti- CX_3CR_1 antibody (Harrison et al., 1998). Additionally, the pattern of expression of CX_3CR_1 in the mouse has been analysed in depth, thanks to the development of a transgenic mouse in which the CX_3CR_1 gene has had a green fluorescent protein (GFP) reporter inserted, so that the CX_3CR_1 protein is co-expressed with the GFP and is easily visible. In the periphery, murine CX_3CR_1 is predominantly monocytic; but it is also expressed on dendritic cells, as well as on Langerhans cells. In the rat and mouse CNS, CX_3CR_1 expression is found on microglia but not astrocytes or oligodendrocytes (Jung et al., 2000). There is evidence to suggest that pathophysiological changes may alter the expression profile of CX_3CR_1; however, this typically follows an inflammatory stimulus rather than the development of a neuropathic state (Wang et al., 2012).

The pattern of expression of CX_3CL_1 is less well defined than that of CX_3CR_1. While CX_3CL_1 mRNA is abundant in the brain and in the heart, it is present in most other tissues that were tested, with the exception of peripheral blood leukocytes. Importantly, while expression of CX_3CL_1 in endothelial cells is constitutive, it is enhanced by exposure to pro-inflammatory cytokines IL-1β and IFNγ (Pan et al., 1997). Within the CNS, CX_3CL_1 is constitutively expressed in neuronal cells, but not in endothelial cells (Nishiyori et al., 1998; Harrison et al., 1998; Ludwig and Mentlein, 2008; Verge et al., 2004). In the brain it is restricted to neurons within the hippocampus and the striatum and to discrete layers of the cortex; and, although CX_3CL_1 is found in dorsal horn neurons, it may be absent or expressed at very low levels in CGRP-positive neuronal cell bodies in the DRG (Kim et al., 2011). Current evidence suggests that CX_3CL_1 expression in the dorsal horn of the spinal cord is not altered following peripheral nerve injury (Verge et al., 2004; Clark, Yip and Malcangio, 2009), while the level of soluble CX_3CR_1 in the cerebrospinal fluid is increased (Clark et al., 2009; Dalrymple et al., 2009). These observations suggest the occurrence of increased shedding of the transmembrane protein under neuropathic conditions; in humans a similar increase in soluble CX_3CL_1 is observed in the cerebrospinal fluid (CSF) of patients with autoimmune disease of the CNS and PNS (Sporer et al., 2003; Ludwig and Mentlein, 2008).

CX_3CL_1 is pronociceptive; the intrathecal spinal administration of the soluble chemokine domain causes mechanical allodynia and thermal hypersensitivity in rodents (Sun et al., 2013; Clark, Yip, et al., 2007; Milligan et al., 2004; Milligan, Zapata, et al., 2005). Additionally, injection of the chemokine domain into the periaqueductal gray (PAG) area of the brain, which plays a crucial role in the ascending and descending modulation of pain, is also pronociceptive (Chen et al., 2007). Supportingly, the central administration of an anti-CX_3CL_1 antibody attenuates pain-related behaviours in neuropathic animals (Clark, Yip, et al., 2007). It is likely that neuronal CX_3CL_1 in the CNS is what contributes to pain in these animals, as it is a centrally administered anti-CX_3CL_1 antibody that is antinociceptive. Furthermore, current evidence indicates that it is soluble rather than membrane-bound CX_3CL_1 that is pronociceptive. Inhibiting the cleavage of CX_3CL_1 into its soluble form through the administration of an irreversible cathepsin S inhibitor attenuates pain to a similar degree to that seen with anti-CX_3CL_1 antibodies; and only the soluble, not the membrane-bound form of CX_3CL_1 is able to induce pain-related behaviours in rodents (Clark et al., 2012; Clark, Yip, et al., 2007). Accordingly, the elevated level of soluble CX_3CL_1 seen in the CSF of neuropathic animals is reduced as a consequence of treatment with cathepsin S inhibitors (Clark et al., 2009). Once cleaved from its membrane tether, CX_3CL_1 does not exert a direct effect on neurons (Zhuang et al., 2007); rather, following its liberation from neurons, CX_3CL_1 exerts its pronociceptive effects via the activation of CX_3CR_1 on microglia.

Several models of chronic pain have provided evidence for a role of microglia and of the CX_3CR_1 receptor in the development of these conditions. After peripheral nerve injury, pain behaviours are observed concomitantly with the development of microgliosis and an enhanced expression of CX_3CR_1 in the ipsilateral dorsal horn of the spinal cord (Zhuang et al., 2007; Staniland et al., 2010). The mechanism by which the expression of this receptor is up-regulated has not yet been fully elucidated; however, an interleukin-6 (IL-6)-dependent pathway has been proposed. Pre-treatment with an IL-6 neutralising antibody prevents the increase in CX_3CR_1 expression that typically follows a peripheral nerve injury, and the intrathecal administration of exogenous recombinant rat IL-6 (rrIL-6) induces a significant increase in

CX_3CR_1 expression in the spinal cord (Lee, Jeon and Cho, 2010). It is becoming increasingly apparent that the interaction between CX_3CL_1 and CX_3CR_1 is critical for the pathophysiological changes in both behaviour and glial cell activation within the spinal cord; mice deficient in CX_3CR_1 do not develop thermal hyperalgesia or mechanical allodynia following peripheral nerve injury and show reduced microgliosis and expression of p-p38 in comparison to wild-type littermate controls (Staniland et al., 2010). Likewise, both the behavioural and the microglial responses to peripheral nerve injury can be attenuated through the intrathecal administration of a CX_3CR_1 neutralising antibody (Zhuang et al., 2007; Milligan, Zapata, et al., 2005; Milligan et al., 2004).

Within the spinal cord, the activation of CX_3CR_1 results in an increase in intracellular calcium levels (Harrison et al., 1998), the phosphorylation of p38 MAPK (Clark et al., 2007; Zhuang et al., 2007) and a release of pronociceptive mediators – including IL-1b, IL-6 and NO (Milligan, Langer, et al., 2005), which are able to sensitise neurons and enhance nociceptive transmission (Owolabi and Saab, 2006). Altogether this evidence suggests that CX_3CR_1 and CX_3CL_1 contribute substantially to neuron–glia communication in the spinal cord during neuropathic states.

CCL_2/CCR_2

Monocyte chemoattractant protein 1 (MCP-1), more commonly referred to as CCL_2, belongs to the CC family of chemokines and is one of four other monocyte-attracting chemokines that bear highly homologous structures (Van, Van and Opdenakker, 1999). Unlike CX_3CL_1, under physiological conditions CCL_2 is almost absent from the CNS; negligible basal expression is observed within the DRG, spinal cord and brain of naïve animals (White et al., 2005; Jung et al., 2009; Flugel et al., 2001). Following peripheral nerve injury, however, CCL_2 expression is rapidly up-regulated within neurons in the superficial dorsal horn of the spinal cord (Zhang and De, 2006; Tanaka et al., 2004) and within small and medium neuronal cell bodies in the DRG (White et al., 2005; Jung et al., 2009), where it co-localises with CGRP in large and dense core vesicles and is released in an activity-dependent fashion (Jung et al., 2008). It should also be noted that astrocytes are able to express this protein in response to a number of neurological pathologies, for example following mechanical/traumatic brain injury (Glabinski et al., 1996) and in models of multiple sclerosis (Hulkower et al., 1993; Ransohoff et al., 1993). Here CCL_2 is released in a TNF-JNK-dependent fashion and intrathecally administered TNF has been shown to induced pain hypersensitivity and CCL_2 up-regulation that could be suppressed by a JNK inhibitor (Gao et al., 2009). Furthermore, animals that overexpress CCL_2 in astrocytes display enhanced nociceptive responses (Menetski et al., 2007) and impaired microglial function (Huang et al., 2005). The localisation of the CCL_2 principal receptor (CCR_2), while well established in an immunological setting, is more heavily debated in terms of expression within the nervous system. The expression of CCR_2 under physiological conditions is thought to be neuronal (Gosselin et al., 2005); however, CCR_2 mRNA is often undetectable in the spinal cord of rodents (Eltayeb et al., 2007). Most of the evidence concerns the expression of this protein after peripheral nerve injury, where CCR_2 mRNA has been reported to increase at the site of injury and in lumbar DRG, and increased immunoreactivity for the CCR_2 protein has been observed in microglia (Abbadie et al., 2003; Gosselin et al., 2005) and in astrocytes (Knerlich-Lukoschus et al., 2008) in the dorsal horn of the spinal cord.

Typical of the chemokines, CCR_2 is a G-protein coupled receptor and is associated with a G-protein of the $G\alpha i$ class; thus activation of this receptor results in the inhibition of adenylate cyclase and in a subsequent decrease in the production of intracellular cyclic AMP. Additionally there is activation of PLC-$\beta2$; and, via an IP3-dependent mechanism, there is Ca^{2+} release from intracellular stores, which triggers several signalling cascades that result in the activation and nuclear translocation of several transcription factors. This contributes to the release of pronociceptive mediators. CCL_2-CCR_2 signalling is certainly pronociceptive and plays a role in the development of a chronic pain state; the intrathecal administration of CCL_2 is associated with the development of mechanical and thermal hypersensitivity (Gao et al., 2009; Dansereau et al., 2008; Thacker et al., 2009; Abbadie et al., 2003). Furthermore, CCR_2-deficient mice demonstrate a significant reduction in the development of the pain behaviours associated with nerve injury (Zhang et al., 2007; Abbadie et al., 2003), and the central administration of a CCR_2 antagonist or CCL_2 neutralising antibody exhibits similar effects and is able to reverse established pain (Gao et al., 2009; Thacker et al., 2009; Pevida et al., 2013; Serrano et al., 2010; Bhangoo et al., 2007).

CCL_2-CCR_2 signalling is thought to contribute to the mechanisms of chronic pain via two mechanisms. First, the modulation of neuronal activity: CCL_2 released from glia rapidly induces the excitation and sensitisation of dorsal horn neurons, which is indicated by the phosphorylation of neuronally expressed ERK. Furthermore, the application of CCL_2 to spinal cord slices enhances spontaneous EPSCs and potentiates AMPA- and NMDA-induced currents, which results in enhanced glutamatergic transmission in lamina II neurons (Gao et al., 2009). Additionally, the *in vivo* application of a CCR_2 antagonist to the spinal cord attenuates the activity of wide dynamic range (WDR) neurons in neuropathic animals (Serrano et al., 2010) – an activity that would otherwise contribute substantially to central sensitisation. CCL_2-CCR_2 signalling may also modulate neuronal activity via the disinhibition of GABAergic synaptic transmission; co-administration of CCL_2 with GABA to neurons resulted in a concentration-dependent reduction of GABA-induced inward currents. The mechanism by which this occurs has not yet been fully elucidated, but it appears to be G-protein independent, as the effect is not prevented by the omission of guanosine triphosphate GTP (Gosselin et al., 2005). Secondly, current evidence indicates that CCL_2 released by primary afferents can activate glia. Intrathecally administered, CCL_2 induces microgliosis that is absent in CCR_2-deficient mice (Thacker et al., 2009; Pevida et al., 2013; Abbadie et al., 2003); and the same antibodies or antagonists that attenuate pain behaviours in neuropathic animals reduce microglia activation in these models (Thacker et al., 2009; Pevida et al., 2013).

Purinergic signalling: P2X4 and P2X7

There is a growing body of evidence that implicates ATP as a critical neuron–glia signalling molecule. ATP is the endogenous ligand of the P2 receptors, a family of receptor proteins encompassing the ionotropic P2X receptors and the metabotrophic P2Y receptors (Trang, Beggs and Salter, 2012); and ATP is known to activate microglia both *in vitro* and *in vivo* (Honda et al., 2001; Tsuda et al., 2003; Davalos et al., 2005). Here we have limited the discussion to the P2X4 and P2X7 receptors, which have, respectively, high and low affinity for ATP.

A critical role for microglial purinergic signalling under chronic pain states was first indicated by the demonstration that P2X4 in microglia (but not in neurons or in astrocytes) is up-regulated as early as one day after peripheral nerve injury (Tsuda et al., 2003). This finding has since been confirmed by using a P2X4 knock-out mouse, which exhibited reduced mechanical hypersensitivity after peripheral nerve injury by comparison to wild-type controls (Ulmann et al., 2008; Tsuda et al., 2009). Furthermore, pharmacological blockade of the P2X4 receptor attenuated mechanical allodynia in nerve-injured rats, and suppressing the expression of the P2X4 receptor with an anti-sense oligonucleotide prevented the development of neuropathic pain (Tsuda et al., 2003). Interestingly, the up-regulation of several markers of microglial activation such as ionized calcium binding adaptor molecule-1 (Iba1) does not appear to be P2X4-dependent, as the prevention of P2X4 up-regulation does not affect the expression of these proteins and is unable to suppress microgliosis (Tsuda et al., 2003; Ulmann et al., 2008). Current evidence indicates that stimulation of microglial P2X4 is sufficient to induce pain behaviours, as the intraspinal administration of microglia activated via P2X4 produced allodynia in naïve rats (Tsuda et al., 2003).

While it was understood that microglia–neuron communication was playing a key role in the development of chronic pain states, how the mechanism by which ATP and the P2X4 receptor were contributing to this process was not known until a series of elegant experiments were performed and published by Coull and colleagues in 2005. Here the team demonstrated that ATP stimulated microglia to modulate nociceptive transmission by evoking a depolarizing shift in the anion reversal potential of lamina I neurons, which resulted in an inversion of polarity in GABA-A-induced currents (outward to inward) and in a subsequent reduction in inhibition (Coull et al., 2005). This effect had previously been attributed to a decrease in the neuronal expression of a potassium chloride exporter (KCC2) (Coull et al., 2003); this phenomenon is known to occur after a peripheral nerve injury in a temporal profile – an injury that coincides with an increase in BDNF expression (Miletic and Miletic, 2008). Under neuropathic conditions, activity-dependent release of BDNF from sensory neurons is absent (Lever, Cunningham, et al., 2003) and microglia become the primary source for BDNF (Coull et al., 2005). In consequence, it was hypothesised that the P2X4-microglia effect on neurons was mediated by BDNF; and indeed the application of BDNF to spinal cord slices resulted in a depolarising shift in the anion potential of lamina I neurons that was similar to the one observed after ATP activation of microglia (Coull et al., 2005). Additionally, blocking BDNF-TrkB signalling prevents mechanical allodynia evoked by P2X4-activated microglia; and, when these cells were pre-treated with an siRNA directed against BDNF, the microglia were no longer able to induce behavioural hypersensitivity in rats (Coull et al., 2005). Accordingly, ATP stimulation of microglia from P2X4 knock-out mice does not induce BDNF release (whereas it does in the wild type) (Ulmann et al., 2008). Moreover, in P2X4 knock-out mice an accumulation of BDNF is observed within spinal microglia after peripheral nerve injury (Ulmann et al., 2008). Collectively these data indicate that BDNF is released from microglia under neuropathic pain states and significantly contributes to aberrant signalling and to the development of behavioural hypersensitivity. It is now known that the intracellular pathway mediating the release of BDNF subsequent to P2X4 activation is p38 MAPK-dependent. Influx of Ca^{2+} through the ionotrophic P2X4 receptor results in a Ca^{2+}-dependent activation of p38, which in turn results in the synthesis and release of BDNF; pharmacological blockade of p38 activity prevents the release of BDNF. The release of BDNF is biphasic: the initial exocytotic release peaks five minutes after ATP stimulation, and this is thought to be the release of intracellular stores of the protein; it is then followed by a second release, which peaks 60 minutes after ATP stimulation (Trang et al., 2009).

A second purinergic receptor, P2X7, has been heavily implicated in the development of neuropathic pain behaviours after peripheral nerve injury. Microglia express functional P2X7, the activation of which is associated with proliferation (Bianco et al., 2005; Monif et al., 2009) and release of pro-inflammatory mediators (Brough et al., 2002; Clark et al., 2010; Chakfe et al., 2002; Trang et al., 2012). In addition to this, P2X7 is implicated in pain due to the reduction in hypersensitivity observed in P2X7 knock-out mice after partial ligation of the sciatic nerve (Chessell et al., 2005) or pharmacological blockade of the receptors' function (Broom et al., 2008; Dell'Antonio et al., 2002; Honore et al., 2006; Honore et al., 2009). The influence of this protein on microglia in pain, as opposed to its influence on other cells on which it is expressed, stems in part from the fact (for which there is evidence) that activation of P2X7 on microglia causes the release of IL-1β and cathepsin S (via p38 activation); both of these proteins have been demonstrated to be pronociceptive (Clark, Yip, et al., 2007; Clark et al., 2010) and have been implicated in neuron–glia communication (Old and Malcangio, 2012; Clark et al., 2013). Furthermore, P2X7 has been demonstrated to be up-regulated in the DRG and peripheral nerve of chronic neuropathic pain patients (Chessell et al., 2005).

Concluding remarks

Both microglia and astrocytes contribute substantially to the development and maintenance of the pain-associated behaviours that arise as the result of peripheral nerve injury. Under neuropathic conditions, the microglial response within the spinal cord is associated with the development of pain, whereas inhibition of these cells attenuates pain behaviours. The activation of intracellular signalling pathways in microglia results in the release of a number of pronociceptive mediators, some of which have been discussed here. Similarly, the activation of astrocytes – a process termed astrogliosis – results in the release of pronociceptive mediators. Astrogliosis is, however, typically observed slightly later, in the course of the cellular changes that result from the injury. Yet the inhibition of these cells is able to impact on the behavioural hypersensitivity displayed in various animal models.

Clinical evidence for microglial and astrocyte activation as well as for their contribution to clinical neuropathic pain has started to grow steadily. Astrogliosis has been reported in the post-mortem spinal cord of HIV-positive patients with a history of chronic pain, but not in tissues obtained from HIV patients without a history of chronic pain or from healthy control patients (Shi et al., 2012). In this tissue an increase in astrocytic markers is associated with increased levels of pro-inflammatory cytokines and intracellular MAP kinases, which is consistent with the preclinical evidence. Microglial and astrocyte activation has also been observed in the spinal cord of a patient who had suffered with chronic pain for several years after a peripheral nerve injury (Del, Schwartzman and Alexander, 2009), which thus validates preclinical observations.

The number of clinical trials on the possible analgesic effect of microglial and astrocyte inhibitors in chronic pain is still small. However, beside the lack of efficacy of propentofylline in post-herpetic pain (Landry et al., 2012), an inhibitor of p38 MAPK has recently shown analgesic effects in neuropathic pain (Anand et al., 2011), and the microglia inhibitor minocycline has shown anti-hyperalgesic efficacy in neuropathic patients (Sumracki et al., 2012). In conclusion, these initial pilot results provide evidence suggesting that some aspects of the preclinical findings translate to the human condition and that microglial and astrocyte targets represent potential therapeutic avenues for the treatment of neuropathic pain.

References

Abbadie, C., Lindia, J. A., Cumiskey, A. M., Peterson, L. B., Mudgett, J. S., Bayne, E. K., DeMartino, J. A., MacIntyre, D. E. and Forrest, M. J. 2003. Impaired neuropathic pain responses in mice lacking the chemokine receptor CCR_2. *Proceedings of the National Academy of Sciences of the United States of America* 100: 7947–52.

Abbott, N. J., Ronnback, L. and Hansson, E. 2006. Astrocyte-endothelial interactions at the blood-brain barrier. *Nature Reviews Neuroscience* 7: 41–53.

Ajami, B., Bennett, J. L., Krieger, C., McNagny, K. M. and Rossi, F. M. 2011. Infiltrating monocytes trigger EAE progression, but do not contribute to the resident microglia pool. *Nature Neuroscience* 14: 1142–9.

Anand, P., Shenoy, R., Palmer, J. E., Baines, A. J., Lai, R. Y., Robertson, J., Bird, N., Ostenfeld, T. and Chizh, B. A. 2011. Clinical trial of the p38 MAP kinase inhibitor dilmapimod in neuropathic pain following nerve injury. *European Journal of Pain* 15: 1040–8.

Araque, A., Li, N., Doyle, R. T. and Haydon, P. G. 2000. SNARE protein-dependent glutamate release from astrocytes. *Journal of Neuroscience* 20: 666–73.

Baron, R. 2006. Mechanisms of disease: Neuropathic pain – A clinical perspective. *Nature Clinical Practice Neurology* 2: 95–106.

Baron, R., Binder, A. and Wasner, G. 2010. Neuropathic pain: Diagnosis, pathophysiological mechanisms, and treatment. *Lancet Neurology* 9: 807–19.

Bazan, J. F., Bacon, K. B., Hardiman, G., Wang, W., Soo, K., Rossi, D., Greaves, D. R., Zlotnik, A. and Schall, T. J. 1997. A new class of membrane-bound chemokine with a CX3C motif. *Nature* 385: 640–4.

Beers, D. R., Henkel, J. S., Xiao, Q., Zhao, W., Wang, J., Yen, A. A., Siklos, L., McKercher, S. R. and Appel, S. H. 2006. Wild-type microglia extend survival in PU1 knockout mice with familial amyotrophic lateral sclerosis. *Proceedings of the National Academy of Sciences of the United States of America* 103: 16021–6.

Bhangoo, S., Ren, D., Miller, R. J., Henry, K. J., Lineswala, J., Hamdouchi, C., Li, B., Monahan, P. E., Chan, D. M., Ripsch, M. S. and White, F. A. 2007. Delayed functional expression of neuronal chemokine receptors following focal nerve demyelination in the rat: A mechanism for the development of chronic sensitization of peripheral nociceptors. *Molecular Pain* 3: 38.

Bianco, F., Pravettoni, E., Colombo, A., Schenk, U., Moller, T., Matteoli, M. and Verderio, C. 2005. Astrocyte-derived ATP induces vesicle shedding and, I. L.-1 beta release from microglia. *Journal of Immunology* 174: 7268–77.

Breivik, H., Collett, B., Ventafridda, V., Cohen, R. and Gallacher, D. 2006. Survey of chronic pain in Europe: Prevalence, impact on daily life, and treatment. *European Journal of Pain* 10: 287–333.

Broom, D. C., Matson, D. J., Bradshaw, E., Buck, M. E., Meade, R., Coombs, S., Matchett, M., Ford, K. K., Yu, W., Yuan, J., Sun, S. H., Ochoa, R., Krause, J. E., Wustrow, D. J. and Cortright, D. N. 2008. Characterization of N-(adamantan-1-ylmethyl)-5-[(3R-amino-pyrrolidin-1-yl)methyl]-2-chloro-benzamide, a P2X7 antagonist in animal models of pain and inflammation. *Journal of Pharmacology and Experimental Therapeutics* 327: 620–33.

Brough, D., Le Feuvre, R. A., Iwakura, Y. and Rothwell, N. J. 2002. Purinergic (P2X7) receptor activation of microglia induces cell death via an interleukin-1-independent mechanism. *Molecular and Cellular Neuroscience* 19: 272–80.

Cahoy, J. D., Emery, B., Kaushal, A., Foo, L. C., Zamanian, J. L., Christopherson, K. S., Xing, Y., Lubischer, J. L., Krieg, P. A., Krupenko, S. A., Thompson, W. J. and Barres, B. A. 2008. A transcriptome database for astrocytes, neurons, and oligodendrocytes: A new resource for understanding brain development and function. *Journal of Neuroscience* 28: 264–78.

Chakfe, Y., Seguin, R., Antel, J. P., Morissette, C., Malo, D., Henderson, D. and Seguela, P. 2002. ADP and AMP induce interleukin-1beta release from microglial cells through activation of ATP-primed P2X7 receptor channels. *Journal of Neuroscience* 22: 3061–9.

Chen, X., Geller, E. B., Rogers, T. J. and Adler, M. W. 2007. The chemokine CX3CL1/fractalkine interferes with the antinociceptive effect induced by opioid agonists in the periaqueductal grey of rats. *Brain Research* 1153: 52–7.

Chessell, I. P., Hatcher, J. P., Bountra, C., Michel, A. D., Hughes, J. P., Green, P., Egerton, J., Murfin, M., Richardson, J., Peck, W. L., Grahames, C. B., Casula, M. A., Yiangou, Y., Birch, R., Anand, P. and Buell, G. N. 2005. Disruption of the P2X7 purinoceptor gene abolishes chronic inflammatory and neuropathic pain. *Pain* 114: 386–96.

Cheunsuang, O. and Morris, R. 2000. Spinal lamina I neurons that express neurokinin 1 receptors: Morphological analysis. *Neuroscience* 97: 335–45.

Clark, A. K. and Malcangio, M. 2012. Microglial signalling mechanisms: Cathepsin S and Fractalkine. *Experimental Neurology* 234: 283–92.

Clark, A. K., Old, E. A. and Malcangio, M. 2013. Neuropathic pain and cytokines: Current perspectives. *Journal of Pain Research* 6: 803–14.

Clark, A. K., Staniland, A. A. and Malcangio, M. 2011. Fractalkine/CX$_3$CR$_1$ signalling in chronic pain and inflammation. *Current Pharmaceutical Biotechnology* 12: 1707–14.

Clark, A. K., Yip, P. K. and Malcangio, M. 2009. The liberation of fractalkine in the dorsal horn requires microglial cathepsin S. *Journal of Neuroscience* 29: 6945–54.

Clark, A. K., Gentry, C., Bradbury, E. J., McMahon, S. B. and Malcangio, M. 2007a. Role of spinal microglia in rat models of peripheral nerve injury and inflammation. *European Journal of Pain* 11: 223–30.

Clark, A. K., Grist, J., Al-Kashi, A., Perretti, M. and Malcangio, M. 2012. Spinal cathepsin S and fractalkine contribute to chronic pain in the collagen-induced arthritis model. *Arthritis & Rheumatology* 64: 2038–47.

Clark, A. K., Staniland, A. A., Marchand, F., Kaan, T. K., McMahon, S. B. and Malcangio, M. 2010. P2X7-dependent release of interleukin-1beta and nociception in the spinal cord following lipopolysaccharide. *Journal of Neuroscience* 30: 573–82.

Clark, A. K., Yip, P. K., Grist, J., Gentry, C., Staniland, A. A., Marchand, F., Dehvari, M., Wotherspoon, G., Winter, J., Ullah, J., Bevan, S. and Malcangio, M. 2007b. Inhibition of spinal microglial cathepsin S for the reversal of neuropathic pain. *Proceedings of the National Academy of Sciences of the United States of America* 104: 10655–60.

Colburn, R. W., Rickman, A. J. and DeLeo, J. A. 1999. The effect of site and type of nerve injury on spinal glial activation and neuropathic pain behavior. *Experimental Neurology* 157: 289–304.

Corcione, A., Ferretti, E., Bertolotto, M., Fais, F., Raffaghello, L., Gregorio, A., Tenca, C., Ottonello, L., Gambini, C., Furtado, G., Lira, S. and Pistoia, V. 2009. CX3CR$_1$ is expressed by human B lymphocytes and mediates [corrected] CX3CL1 driven chemotaxis of tonsil centrocytes. *PLoS One* 4: e8485.

Coull, J. A., Boudreau, D., Bachand, K., Prescott, S. A., Nault, F., Sik, A., De, K. P., De, K. Y. 2003. Trans-synaptic shift in anion gradient in spinal lamina I neurons as a mechanism of neuropathic pain. *Nature* 424: 938–42.

Coull, J. A., Beggs, S., Boudreau, D., Boivin, D., Tsuda, M., Inoue, K., Gravel, C., Salter, M. W. and De, K. Y. 2005. BDNF from microglia causes the shift in neuronal anion gradient underlying neuropathic pain. *Nature* 438: 1017–21.

Coulter, D. A. and Eid, T. 2012. Astrocytic regulation of glutamate homeostasis in epilepsy. *Glia* 60: 1215–26.

Cox, J. J., Reimann, F., Nicholas, A. K., Thornton, G., Roberts, E., Springell, K., Karbani, G., Jafri, H., Mannan, J., Raashid, Y., Al-Gazali, L., Hamamy, H., Valente, E. M., Gorman, S., Williams, R., McHale, D. P., Wood, J. N., Gribble, F. M. and Woods, C. G. 2006. An SCN9A channelopathy causes congenital inability to experience pain. *Nature* 444: 894–8.

Crown, E. D. 2012. The role of mitogen activated protein kinase signaling in microglia and neurons in the initiation and maintenance of chronic pain. *Experimental Neurology* 234: 330–9.

D'Mello, R. and Dickenson, A. H. 2008. Spinal cord mechanisms of pain. *British Journal of Anaesthesia* 101: 8–16.

Dalrymple, S., Grist, J., Clark, A. K. and Malcangio, M. 2009. Systemic inhibition of cathepsin S attenuates vincristine-induced neuropathic hypersensitivity [abstract]. At http://www.virobayinc.com/docs/VBY285_Vincristine_Poster_SfN_Oct_09.pdf (accessed 30 September 2015).

Dansereau, M. A., Gosselin, R. D., Pohl, M., Pommier, B., Mechighel, P., Mauborgne, A., Rostene, W., Kitabgi, P., Beaudet, N., Sarret, P. and Melik-Parsadaniantz, S. 2008. Spinal CCL_2 pronociceptive action is no longer effective in CCR_2 receptor antagonist-treated rats. *Journal of Neurochemistry* 106: 757–69.

Davalos, D., Grutzendler, J., Yang, G., Kim, J. V., Zuo, Y., Jung, S., Littman, D. R., Dustin, M. L. and Gan, W. B. 2005. ATP mediates rapid microglial response to local brain injury in vivo. *Nature Neuroscience* 8: 752–8.

Del, V. L., Schwartzman, R. J. and Alexander, G. 2009. Spinal cord histopathological alterations in a patient with longstanding complex regional pain syndrome. *Brain, Behavior, and Immunity* 23: 85–91.

Dell'Antonio, G., Quattrini, A., Dal, C. E., Fulgenzi, A. and Ferrero, M. E. 2002. Antinociceptive effect of a new P(2Z)/P2X7 antagonist, oxidized ATP in arthritic rats. *Neuroscience Letters* 327: 87–90.

Dermietzel, R., Hertberg, E. L., Kessler, J. A. and Spray, D. C. 1991. Gap junctions between cultured astrocytes: Immunocytochemical, molecular, and electrophysiological analysis. *Journal of Neuroscience* 11: 1421–32.

Dworkin, R. H., O'Connor, A. B., Backonja, M., Farrar, J. T., Finnerup, N. B., Jensen, T. S., Kalso, E. A., Loeser, J. D., Miaskowski, C., Nurmikko, T. J., Portenoy, R. K., Rice, A. S., Stacey, B. R., Treede, R. D., Turk, D. C. and Wallace, M. S. 2007. Pharmacologic management of neuropathic pain: Evidence-based recommendations. *Pain* 132: 237–51.

Echeverry, S., Shi, X. Q. and Zhang, J. 2008. Characterization of cell proliferation in rat spinal cord following peripheral nerve injury and the relationship with neuropathic pain. *Pain* 135: 37–47.

Eltayeb, S., Berg, A. L., Lassmann, H., Wallstrom, E., Nilsson, M., Olsson, T., Ericsson-Dahlstrand, A. and Sunnemark, D. 2007. Temporal expression and cellular origin of CC chemokine receptors CCR1, CCR_2 and CCR_5 in the central nervous system: Insight into mechanisms of MOG-induced EAE. *Journal of Neuroinflammation* 4 (14). doi: 10.1186/1742-2094-4-14.

Fields, H. L. and Heinricher, M. M. 1989. Brainstem modulation of nociceptor-driven withdrawal reflexes. *Annals of the New York Academy of Sciences* 563: 34–44.

Flugel, A., Hager, G., Horvat, A., Spitzer, C., Singer, G. M., Graeber, M. B., Kreutzberg, G. W. and Schwaiger, F. W. 2001. Neuronal MCP-1 expression in response to remote nerve injury. *Journal of Cerebral Blood Flow and Metabolism* 21: 69–76.

Fong, A. M., Erickson, H. P., Zachariah, J. P., Poon, S., Schamberg, N. J., Imai, T. and Patel, D. D. 2000. Ultrastructure and function of the fractalkine mucin domain in CX(3)C chemokine domain presentation. *Journal of Biological Chemistry* 275: 3781–6.

Fong, A. M., Robinson, L. A., Steeber, D. A., Tedder, T. F., Yoshie, O., Imai, T. and Patel, D. D. 1998. Fractalkine and CX_3CR_1 mediate a novel mechanism of leukocyte capture, firm adhesion, and activation under physiologic flow. *Journal of Experimental Medicine* 188: 1413–19.

Gao, Y. J. and Ji, R. R. 2010a. Light touch induces ERK activation in superficial dorsal horn neurons after inflammation: Involvement of spinal astrocytes and JNK signaling in touch-evoked central sensitization and mechanical allodynia. *Journal of Neurochemistry* 115: 505–14.

Gao, Y. J. and Ji, R. R. 2010b. Targeting astrocyte signaling for chronic pain. *Neurotherapeutics* 7: 482–93.

Gao, Y. J., Zhang, L., Samad, O. A., Suter, M. R., Yasuhiko, K., Xu, Z. Z., Park, J. Y., Lind, A. L., Ma, Q. and Ji, R. R. 2009. JNK-induced MCP-1 production in spinal cord astrocytes contributes to central sensitization and neuropathic pain. *Journal of Neuroscience* 29: 4096–108.

Garrison, C. J., Dougherty, P. M. and Carlton, S. M. 1994. GFAP expression in lumbar spinal cord of naive and neuropathic rats treated with MK-801. *Experimental Neurology* 129: 237–43.

Garrison, C. J., Dougherty, P. M., Kajander, K. C. and Carlton, S. M. 1991. Staining of glial fibrillary acidic protein (GFAP) in lumbar spinal cord increases following a sciatic nerve constriction injury. *Brain Research* 565: 1–7.

Gehrmann, J., Matsumoto, Y. and Kreutzberg, G. W. 1995. Microglia: Intrinsic immuneffector cell of the brain. *Brain Research: Brain Research Reviews* 20: 269–87.

Giaume, C., Fromaget, C., el, A. A., Cordier, J., Glowinski, J. and Gros, D. 1991. Gap junctions in cultured astrocytes: Single-channel currents and characterization of channel-forming protein. *Neuron* 6: 133–43.

Ginhoux, F., Greter, M., Leboeuf, M., Nandi, S., See, P., Gokhan, S., Mehler, M. F., Conway, S. J., Ng, L. G., Stanley, E. R., Samokhvalov, I. M. and Merad, M. 2010. Fate mapping analysis reveals that adult microglia derive from primitive macrophages. *Science* 330: 841–5.

Glabinski, A. R., Balasingam, V., Tani, M., Kunkel, S. L., Strieter, R. M., Yong, V. W. and Ransohoff, R. M. 1996. Chemokine monocyte chemoattractant protein-1 is expressed by astrocytes after mechanical injury to the brain. *Journal of Immunology* 156: 4363–8.

Go, V. L. and Yaksh, T. L. 1987. Release of substance P from the cat spinal cord. *Journal of Physiology* 391: 141–67.

Gosselin, R. D., Varela, C., Banisadr, G., Mechighel, P., Rostene, W., Kitabgi, P. and Melik-Parsadaniantz, S. 2005. Constitutive expression of CCR_2 chemokine receptor and inhibition by MCP-1/CCL_2 of GABA-induced currents in spinal cord neurones. *Journal of Neurochemistry* 95: 1023–34.

Graeber, M. B., Streit, W. J. and Kreutzberg, G. W. 1988. Axotomy of the rat facial nerve leads to increased CR3 complement receptor expression by activated microglial cells. *Journal of Neuroscience Research* 21: 18–24.

Guo, W., Wang, H., Watanabe, M., Shimizu, K., Zou, S., LaGraize, S. C., Wei, F., Dubner, R. and Ren, K. 2007. Glial–cytokine–neuronal interactions underlying the mechanisms of persistent pain. *Journal of Neuroscience* 27: 6006–18.

Halassa, M. M., Fellin, T., Takano, H., Dong, J. H. and Haydon, P. G. 2007. Synaptic islands defined by the territory of a single astrocyte. *Journal of Neuroscience* 27: 6473–7.

Hamilton, N. B. and Attwell, D. 2010. Do astrocytes really exocytose neurotransmitters? *Nature Reviews Neuroscience* 11: 227–38.

Hansen, R. R. and Malcangio, M. 2013. Astrocytes: Multitaskers in chronic pain. *European Journal of Pharmacology* 716 (1–3): 120–8.

Harrison, J. K., Barber, C. M. and Lynch, K. R. 1994. cDNA cloning of a G-protein-coupled receptor expressed in rat spinal cord and brain related to chemokine receptors. *Neuroscience Letters* 169: 85–9.

Harrison, J. K., Fong, A. M., Swain, P. A., Chen, S., Yu, Y. R., Salafranca, M. N., Greenleaf, W. B., Imai, T. and Patel, D. D. 2001. Mutational analysis of the fractalkine chemokine domain: Basic amino acid residues differentially contribute to CX_3CR_1 binding, signaling, and cell adhesion. *Journal of Biological Chemistry* 276: 21632–41.

Harrison, J. K., Jiang, Y., Chen, S., Xia, Y., Maciejewski, D., McNamara, R. K., Streit, W. J., Salafranca, M. N., Adhikari, S., Thompson, D. A., Botti, P., Bacon, K. B. and Feng, L. 1998. Role for neuronally derived fractalkine in mediating interactions between neurons and CX3CR1-expressing microglia. *Proceedings of the National Academy of Sciences of the United States of America* 95: 10896–901.

Hathway, G. J., Vega-Avelaira, D., Moss, A., Ingram, R. and Fitzgerald, M. 2009. Brief, low frequency stimulation of rat peripheral C-fibres evokes prolonged microglial-induced central sensitization in adults but not in neonates. *Pain* 144: 110–18.

Haydon, P. G. 2001. GLIA: Listening and talking to the synapse. *Nature Reviews Neuroscience* 2: 185–93.

Hoftberger, R., Aboul-Enein, F., Brueck, W., Lucchinetti, C., Rodriguez, M., Schmidbauer, M., Jellinger, K. and Lassmann, H. 2004. Expression of major histocompatibility complex class I molecules on the different cell types in multiple sclerosis lesions. *Brain Pathology* 14: 43–50.

Honda, S., Sasaki, Y., Ohsawa, K., Imai, Y., Nakamura, Y., Inoue, K. and Kohsaka, S. 2001. Extracellular ATP or ADP induce chemotaxis of cultured microglia through Gi/o-coupled P2Y receptors. *Journal of Neuroscience* 21: 1975–82.

Honore, P., Donnelly-Roberts, D., Namovic, M., Zhong, C., Wade, C., Chandran, P., Zhu, C., Carroll, W., Perez-Medrano, A., Iwakura, Y. and Jarvis, M. F. 2009. The antihyperalgesic activity of a selective P2X7 receptor antagonist, A-839977, is lost in IL-1alphabeta knockout mice. *Behavioural Brain Research* 204: 77–81.

Honore, P., Donnelly-Roberts, D., Namovic, M. T., Hsieh, G., Zhu, C. Z., Mikusa, J. P., Hernandez, G., Zhong, C., Gauvin, D. M., Chandran, P., Harris, R., Medrano, A. P., Carroll, W., Marsh, K., Sullivan, J. P., Faltynek, C. R. and Jarvis, M. F. 2006. A-740003 [N-(1-{[(cyanoimino)(5-quinolinylamino) methyl]amino}-2,2-dimethylpropyl)-2-(3,4-dimethoxyphenyl)acetamide], a novel and selective P2X7 receptor antagonist, dose-dependently reduces neuropathic pain in the rat. *Journal of Pharmacology and Experimental Therapeutics* 319: 1376–85.

Huang, D., Wujek, J., Kidd, G., He, T. T., Cardona, A., Sasse, M. E., Stein, E. J., Kish, J., Tani, M., Charo, I. F., Proudfoot, A. E., Rollins, B. J., Handel, T. and Ransohoff, R. M. 2005. Chronic expression of monocyte chemoattractant protein-1 in the central nervous system causes delayed encephalopathy and impaired microglial function in mice. *FASEB Journal* 19: 761–72.

Hulkower, K., Brosnan, C. F., Aquino, D. A., Cammer, W., Kulshrestha, S., Guida, M. P., Rapoport, D. A. and Berman, J. W. 1993. Expression of CSF-1, c-fms, and MCP-1 in the central nervous system of rats with experimental allergic encephalomyelitis. *Journal of Immunology* 150: 2525–33.

Hundhausen, C., Schulte, A., Schulz, B., Andrzejewski, M. G., Schwarz, N., von Hundhausen, P., Winter, U., Paliga, K., Reiss, K., Saftig, P., Weber, C., Ludwig, A. 2007. Regulated shedding of transmembrane chemokines by the disintegrin and metalloproteinase 10 facilitates detachment of adherent leukocytes. *Journal of Immunology* 178: 8064–72.

Hundhausen, C., Misztela, D., Berkhout, T. A., Broadway, N., Saftig, P., Reiss, K., Hartmann, D., Fahrenholz, F., Postina, R., Matthews, V., Kallen, K. J., Rose-John, S. and Ludwig, A. 2003. The disintegrin-like metalloproteinase ADAM10 is involved in constitutive cleavage of CX3CL1 (fractalkine) and regulates CX3CL1-mediated cell–cell adhesion. *Blood* 102: 1186–95.

Imai, T., Hieshima, K., Haskell, C., Baba, M., Nagira, M., Nishimura, M., Kakizaki, M., Takagi, S., Nomiyama, H., Schall, T. J. and Yoshie, O. 1997. Identification and molecular characterization of fractalkine receptor CX_3CR_1, which mediates both leukocyte migration and adhesion. *Cell* 91: 521–30.

Imamoto, K. and Leblond, C. P. 1978. Radioautographic investigation of gliogenesis in the corpus callosum of young rats. II: Origin of microglial cells. *Journal of Comparative Neurology* 180: 139–63.

Jin, S. X., Zhuang, Z. Y., Woolf, C. J. and Ji, R. R. 2003. p38 mitogen-activated protein kinase is activated after a spinal nerve ligation in spinal cord microglia and dorsal root ganglion neurons and contributes to the generation of neuropathic pain. *Journal of Neuroscience* 23: 4017–22.

Jung, H., Toth, P. T., White, F. A. and Miller, R. J. 2008. Monocyte chemoattractant protein-1 functions as a neuromodulator in dorsal root ganglia neurons. *Journal of Neurochemistry* 104: 254–63.

Jung, H., Bhangoo, S., Banisadr, G., Freitag, C., Ren, D., White, F. A. and Miller, R. J. 2009. Visualization of chemokine receptor activation in transgenic mice reveals peripheral activation of CCR_2 receptors in states of neuropathic pain. *Journal of Neuroscience* 29: 8051–62.

Jung, S., Aliberti, J., Graemmel, P., Sunshine, M. J., Kreutzberg, G. W., Sher, A. and Littman, D. R. 2000. Analysis of fractalkine receptor CX(3)CR1 function by targeted deletion and green fluorescent protein reporter gene insertion. *Molecular and Cellular Biology* 20: 4106–14.

Kettenmann, H., Hanisch, U. K., Noda, M. and Verkhratsky, A. 2011. Physiology of microglia. *Physiological Reviews* 91: 461–553.

Kim, D. S., Figueroa, K. W., Li, K. W., Boroujerdi, A., Yolo, T. and Luo, Z. D. 2009. Profiling of dynamically changed gene expression in dorsal root ganglia post peripheral nerve injury and a critical role of injury-induced glial fibrillary acidic protein in maintenance of pain behaviors [corrected]. *Pain* 143: 114–22.

Kim, K. W., Vallon-Eberhard, A., Zigmond, E., Farache, J., Shezen, E., Shakhar, G., Ludwig, A., Lira, S. A. and Jung, S. 2011. In vivo structure/function and expression analysis of the CX3C chemokine fractalkine. *Blood* 118: e156–e167.

Kim, S. Y., Bae, J. C., Kim, J. Y., Lee, H. L., Lee, K. M., Kim, D. S. and Cho, H. J. 2002. Activation of p38 MAP kinase in the rat dorsal root ganglia and spinal cord following peripheral inflammation and nerve injury. *Neuroreport* 13: 2483–6.

King, I. L., Dickendesher, T. L. and Segal, B. M. 2009. Circulating Ly-6C+ myeloid precursors migrate to the CNS and play a pathogenic role during autoimmune demyelinating disease. *Blood* 113: 3190–7.

Kitamura, T., Miyake, T. and Fujita, S. 1984. Genesis of resting microglia in the gray matter of mouse hippocampus. *Journal of Comparative Neurology* 226: 421–33.

Knerlich-Lukoschus, F., Juraschek, M., Blomer, U., Lucius, R., Mehdorn, H. M. and Held-Feindt, J. 2008. Force-dependent development of neuropathic central pain and time-related CCL_2/CCR_2 expression after graded spinal cord contusion injuries of the rat. *Journal of Neurotrauma* 25: 427–48.

Kobayashi, K., Takahashi, E., Miyagawa, Y., Yamanaka, H. and Noguchi, K. 2011. Induction of the P2X7 receptor in spinal microglia in a neuropathic pain model. *Neuroscience Letters* 504: 57–61.

Kofuji, P. and Newman, E. A. 2004. Potassium buffering in the central nervous system. *Neuroscience* 129: 1045–56.

Kuner, R. 2010. Central mechanisms of pathological pain. *Nature Medicine* 16: 1258–66.

Landry, R. P., Jacobs, V. L., Romero-Sandoval, E. A. and DeLeo, J. A. 2012. Propentofylline, a CNS glial modulator does not decrease pain in post-herpetic neuralgia patients: In vitro evidence for differential responses in human and rodent microglia and macrophages. *Experimental Neurology* 234: 340–50.

Latremolière, A. and Woolf, C. J. 2009. Central sensitization: A generator of pain hypersensitivity by central neural plasticity. *Journal of Pain* 10: 895–926.

Lawson, L. J., Perry, V. H. and Gordon, S. 1992. Turnover of resident microglia in the normal adult mouse brain. *Neuroscience* 48: 405–15.

Lee, K. M., Jeon, S. M. and Cho, H. J. 2010. Interleukin-6 induces microglial CX_3CR_1 expression in the spinal cord after peripheral nerve injury through the activation of p38 MAPK. *European Journal of Pain* 14: 682–12.

Lever, I. J., Grant, A. D., Pezet, S., Gerard, N. P., Brain, S. D. and Malcangio, M. 2003a. Basal and activity-induced release of substance P from primary afferent fibres in NK1 receptor knockout mice: Evidence for negative feedback. *Neuropharmacology* 45: 1101–10.

Lever, I., Cunningham, J., Grist, J., Yip, P. K. and Malcangio, M. 2003b. Release of BDNF and GABA in the dorsal horn of neuropathic rats. *European Journal of Neuroscience* 18: 1169–74.

Liaw, W. J., Stephens, R. L., Jr., Binns, B. C., Chu, Y., Sepkuty, J. P., Johns, R. A., Rothstein, J. D. and Tao, Y. X. 2005. Spinal glutamate uptake is critical for maintaining normal sensory transmission in rat spinal cord. *Pain* 115: 60–70.

Ling, E. A., Penney, D. and Leblond, C. P. 1980. Use of carbon labeling to demonstrate the role of blood monocytes as precursors of the 'ameboid cells' present in the corpus callosum of postnatal rats. *Journal of Comparative Neurology* 193: 631–57.

Liuzzo, J. P., Petanceska, S. S. and Devi, L. A. (1999a) Neurotrophic factors regulate cathepsin S in macrophages and microglia: A role in the degradation of myelin basic protein and amyloid beta peptide. *Molecular Medicine* 5: 334–43.

Liuzzo, J. P., Petanceska, S. S., Moscatelli, D. and Devi, L. A. (1999b) Inflammatory mediators regulate cathepsin S in macrophages and microglia: A role in attenuating heparan sulfate interactions. *Molecular Medicine* 5: 320–33.

Lucas, D. R. and Newhouse, J. P. 1957. The toxic effect of sodium L-glutamate on the inner layers of the retina. *AMA Archives of Ophthalmology* 58: 193–201.

Ludwig, A. and Mentlein, R. 2008. Glial cross-talk by transmembrane chemokines CX3CL1 and CXCL16. *Journal of Neuroimmunology* 198: 92–7.

Ma, W. and Quirion, R. 2002. Partial sciatic nerve ligation induces increase in the phosphorylation of extracellular signal-regulated kinase (ERK) and c-Jun N-terminal kinase (JNK) in astrocytes in the lumbar spinal dorsal horn and the gracile nucleus. *Pain* 99: 175–84.

Malcangio, M. and Bowery, N. G. 1994. Spinal cord, S. P. release and hyperalgesia in monoarthritic rats: Involvement of the GABAB receptor system. *British Journal of Pharmacology* 113: 1561–6.

McKercher, S. R., Torbett, B. E., Anderson, K. L., Henkel, G. W., Vestal, D. J., Baribault, H., Klemsz, M., Feeney, A. J., Wu, G. E., Paige, C. J. and Maki, R. A. 1996. Targeted disruption of the, PU1 gene results in multiple hematopoietic abnormalities. *EMBO Journal* 15: 5647–58.

McMahon, S. B. and Malcangio, M. 2009. Current challenges in glia-pain biology. *Neuron* 64: 46–54.

Menetski, J., Mistry, S., Lu, M., Mudgett, J. S., Ransohoff, R. M., DeMartino, J. A., MacIntyre, D. E. and Abbadie, C. 2007. Mice overexpressing chemokine ligand 2 (CCL_2) in astrocytes display enhanced nociceptive responses. *Neuroscience* 149: 706–14.

Miletic, G. and Miletic, V. 2008. Loose ligation of the sciatic nerve is associated with TrkB receptor-dependent decreases in KCC2 protein levels in the ipsilateral spinal dorsal horn. *Pain* 137: 532–9.

Milligan, E., Zapata, V., Schoeniger, D., Chacur, M., Green, P., Poole, S., Martin, D., Maier, S. F. and Watkins, L. R. 2005a. An initial investigation of spinal mechanisms underlying pain enhancement induced by fractalkine, a neuronally released chemokine. *European Journal of Neuroscience* 22: 2775–82.

Milligan, E. D., Twining, C., Chacur, M., Biedenkapp, J., O'Connor, K., Poole, S., Tracey, K., Martin, D., Maier, S. F. and Watkins, L. R. 2003. Spinal glia and proinflammatory cytokines mediate mirror-image neuropathic pain in rats. *Journal of Neuroscience* 23: 1026–40.

Milligan, E. D., Langer, S. J., Sloane, E. M., He, L., Wieseler-Frank, J., O'Connor, K., Martin, D., Forsayeth, J. R., Maier, S. F., Johnson, K., Chavez, R. A., Leinwand, L. A. and Watkins, L. R. 2005b. Controlling pathological pain by adenovirally driven spinal production of the anti-inflammatory cytokine, interleukin-10. *European Journal of Neuroscience* 21: 2136–48.

Milligan, E. D., Zapata, V., Chacur, M., Schoeniger, D., Biedenkapp, J., O'Connor, K. A., Verge, G. M., Chapman, G., Green, P., Foster, A. C., Naeve, G. S., Maier, S. F., Watkins, L. R. 2004. Evidence that exogenous and endogenous fractalkine can induce spinal nociceptive facilitation in rats. *European Journal of Neuroscience* 20: 2294–302.

Miyoshi, K., Obata, K., Kondo, T., Okamura, H. and Noguchi, K. 2008. Interleukin-18-mediated microglia/astrocyte interaction in the spinal cord enhances neuropathic pain processing after nerve injury. *Journal of Neuroscience* 28: 12775–87.

Monif, M., Reid, C. A., Powell, K. L., Smart, M. L. and Williams, D. A. 2009. The P2X7 receptor drives microglial activation and proliferation: A trophic role for P2X7R pore. *Journal of Neuroscience* 29: 3781–91.

Nagy, J. I and Rash, J. E. 2000. Connexins and gap junctions of astrocytes and oligodendrocytes in the CNS. *Brain Research: Brain Research Reviews* 32: 29–44.

Nimmerjahn, A., Kirchhoff, F. and Helmchen, F. 2005. Resting microglial cells are highly dynamic surveillants of brain parenchyma in vivo. *Science* 308: 1314–18.

Nishiyori, A., Minami, M., Ohtani, Y., Takami, S., Yamamoto, J., Kawaguchi, N., Kume, T., Akaike, A. and Satoh, M. 1998. Localization of fractalkine and CX_3CR_1 mRNAs in rat brain: Does fractalkine play a role in signaling from neuron to microglia? *FEBS Letters* 429: 167–72.

O'Connor, A. B. and Dworkin, R. H. 2009. Treatment of neuropathic pain: An overview of recent guidelines. *American Journal of Medicine* 122: S22–S32.

Obata, K., Yamanaka, H., Dai, Y., Tachibana, T., Fukuoka, T., Tokunaga, A., Yoshikawa, H. and Noguchi, K. 2003. Differential activation of extracellular signal-regulated protein kinase in primary afferent neurons regulates brain-derived neurotrophic factor expression after peripheral inflammation and nerve injury. *Journal of Neuroscience* 23: 4117–26.

Okada-Ogawa, A., Suzuki, I., Sessle, B. J., Chiang, C. Y., Salter, M. W., Dostrovsky, J. O., Tsuboi, Y., Kondo, M., Kitagawa, J., Kobayashi, A., Noma, N., Imamura, Y. and Iwata, K. 2009. Astroglia in medullary dorsal horn (trigeminal spinal subnucleus caudalis) are involved in trigeminal neuropathic pain mechanisms. *Journal of Neuroscience* 29: 11161–71.

Oku, R., Satoh, M., Fujii, N., Otaka, A., Yajima, H. and Takagi, H. 1987. Calcitonin gene-related peptide promotes mechanical nociception by potentiating release of substance P from the spinal dorsal horn in rats. *Brain Research* 403: 350–4.

Old, E. A. and Malcangio, M. 2012. Chemokine mediated neuron-glia communication and aberrant signalling in neuropathic pain states. *Current Opinion in Pharmacology* 12: 67–73.

Olney, J. W. and Sharpe, L. G. 1969. Brain lesions in an infant rhesus monkey treated with monsodium glutamate. *Science* 166: 386–8.

Owolabi, S. A. and Saab, C. Y. 2006. Fractalkine and minocycline alter neuronal activity in the spinal cord dorsal horn. *FEBS Letters* 580: 4306–10.

Pan, Y., Lloyd, C., Zhou, H., Dolich, S., Deeds, J., Gonzalo, J. A., Vath, J., Gosselin, M., Ma, J., Dussault, B., Woolf, E., Alperin, G., Culpepper, J., Gutierrez-Ramos, J. C. and Gearing, D. 1997. Neurotactin, a membrane-anchored chemokine upregulated in brain inflammation. *Nature* 387: 611–17.

Paterson, J. A., Privat, A., Ling, E. A. and Leblond, C. P. 1973. Investigation of glial cells in semithin sections. 3. Transformation of subependymal cells into glial cells, as shown by radioautography after 3 H-thymidine injection into the lateral ventricle of the brain of young rats. *Journal of Comparative Neurology* 149: 83–102.

Pekny, M. and Nilsson, M. 2005. Astrocyte activation and reactive gliosis. *Glia* 50: 427–34.

Perry, V. H. 1998. A revised view of the central nervous system microenvironment and major histocompatibility complex class II antigen presentation. *Journal of Neuroimmunology* 90: 113–21.

Perry, V. H., Hume, D. A. and Gordon, S. 1985. Immunohistochemical localization of macrophages and microglia in the adult and developing mouse brain. *Neuroscience* 15: 313–26.

Peters, C. M., Jimenez-Andrade, J. M., Kuskowski, M. A., Ghilardi, J. R. and Mantyh, P. W. 2007. An evolving cellular pathology occurs in dorsal root ganglia, peripheral nerve and spinal cord following intravenous administration of paclitaxel in the rat. *Brain Research* 1168: 46–59.

Pevida, M., Lastra, A., Hidalgo, A., Baamonde, A. and Menendez, L. 2013. Spinal CCL_2 and microglial activation are involved in paclitaxel-evoked cold hyperalgesia. *Brain Research Bulletin* 95: 21–7.

Pezet, S., Malcangio, M., Lever, I. J., Perkinton, M. S., Thompson, S. W., Williams, R. J. and McMahon, S. B. 2002. Noxious stimulation induces Trk receptor and downstream ERK phosphorylation in spinal dorsal horn. *Molecular and Cellular Neuroscience* 21: 684–95.

Porter, J. T. and McCarthy, K. D. 1997. Astrocytic neurotransmitter receptors in situ and in vivo. *Progress in Neurobiology* 51: 439–55.

Prinz, M. and Mildner, A. 2011. Microglia in the CNS: Immigrants from another world. *Glia* 59: 177–87.

Ransohoff, R. M. and Perry, V. H. 2009. Microglial physiology: Unique stimuli, specialized responses. *Annual Review of Immunology* 27: 119–45.

Ransohoff, R. M., Hamilton, T. A., Tani, M., Stoler, M. H., Shick, H. E., Major, J. A., Estes, M. L., Thomas, D. M. and Tuohy, V. K. 1993. Astrocyte expression of mRNA encoding cytokines, I. P.-10 and, J. E./MCP-1 in experimental autoimmune encephalomyelitis. *FASEB Journal* 7: 592–600.

Raport, C. J., Schweickart, V. L., Eddy, R. L., Jr., Shows, T. B. and Gray, P. W. 1995. The orphan G-protein-coupled receptor-encoding gene V28 is closely related to genes for chemokine receptors and is expressed in lymphoid and neural tissues. *Gene* 163: 295–9.

Rasley, A., Bost, K. L., Olson, J. K., Miller, S. D. and Marriott, I. 2002. Expression of functional, NK-1 receptors in murine microglia. *Glia* 37: 258–67.

Ridet, J. L., Malhotra, S. K., Privat, A. and Gage, F. H. 1997. Reactive astrocytes: Cellular and molecular cues to biological function. *Trends in Neurosciences* 20: 570–7.

Romero-Sandoval, A., Chai, N., Nutile-McMenemy, N. and DeLeo, J. A. 2008. A comparison of spinal Iba1 and GFAP expression in rodent models of acute and chronic pain. *Brain Research* 1219: 116–26.

Rouach, N., Koulakoff, A., Abudara, V., Willecke, K. and Giaume, C. 2008. Astroglial metabolic networks sustain hippocampal synaptic transmission. *Science* 322: 1551–5.

Saijo, K. and Glass, C. K. 2011. Microglial cell origin and phenotypes in health and disease. *Nature Reviews Immunology* 11: 775–87.

Santello, M., Bezzi, P. and Volterra, A. 2011. TNFalpha controls glutamatergic gliotransmission in the hippocampal dentate gyrus. *Neuron* 69: 988–1001.

Scemes, E. and Spray, D. C. 2012. Extracellular K(+) and astrocyte signaling via connexin and pannexin channels. *Neurochemical Research* 37: 2310–16.

Seino, D., Tokunaga, A., Tachibana, T., Yoshiya, S., Dai, Y., Obata, K., Yamanaka, H., Kobayashi, K. and Noguchi, K. 2006. The role of ERK signaling and the P2X receptor on mechanical pain evoked by movement of inflamed knee joint. *Pain* 123: 193–203.

Serrano, A., Pare, M., McIntosh, F., Elmes, S. J., Martino, G., Jomphe, C., Lessard, E., Lembo, P. M., Vaillancourt, F., Perkins, M. N. and Cao, C. Q. 2010. Blocking spinal CCR_2 with AZ889 reversed hyperalgesia in a model of neuropathic pain. *Molecular Pain* 6: 90.

Seybold, V. S., McCarson, K. E., Mermelstein, P. G., Groth, R. D. and Abrahams, L. G. 2003. Calcitonin gene-related peptide regulates expression of neurokinin1 receptors by rat spinal neurons. *Journal of Neuroscience* 23: 1816–24.

Shi, Y., Gelman, B. B., Lisinicchia, J. G. and Tang, S. J. 2012. Chronic-pain-associated astrocytic reaction in the spinal cord dorsal horn of human immunodeficiency virus-infected patients. *Journal of Neuroscience* 32: 10833–40.

Sporer, B., Kastenbauer, S., Koedel, U., Arendt, G. and Pfister, H. W. 2003. Increased intrathecal release of soluble fractalkine in HIV-infected patients. *AIDS Research and Human Retroviruses* 19: 111–16.

Staniland, A. A., Clark, A. K., Wodarski, R., Sasso, O., Maione, F., D'Acquisto, F. and Malcangio, M. 2010. Reduced inflammatory and neuropathic pain and decreased spinal microglial response in fractalkine receptor (CX_3CR_1) knockout mice. *Journal of Neurochemistry* 114: 1143–57.

Sumracki, N. M., Hutchinson, M. R., Gentgall, M., Briggs, N., Williams, D. B. and Rolan, P. 2012. The effects of pregabalin and the glial attenuator minocycline on the response to intradermal capsaicin in patients with unilateral sciatica. *PLoS One* 7: e38525.

Sun, J. L., Xiao, C., Lu, B., Zhang, J., Yuan, X. Z., Chen, W., Yu, L. N., Zhang, F. J., Chen, G. and Yan, M. 2013. CX3CL1/CX_3CR_1 regulates nerve injury-induced pain hypersensitivity through the ERK5 signaling pathway. *Journal of Neuroscience Research* 91: 545–53.

Sung, B., Lim, G. and Mao, J. 2003. Altered expression and uptake activity of spinal glutamate transporters after nerve injury contribute to the pathogenesis of neuropathic pain in rats. *Journal of Neuroscience* 23: 2899–910.

Suter, M. R., Berta, T., Gao, Y. J., Decosterd, I. and Ji, R. R. 2009. Large A-fiber activity is required for microglial proliferation and p38 MAPK activation in the spinal cord: Different effects of resiniferatoxin and bupivacaine on spinal microglial changes after spared nerve injury. *Molecular Pain* 5: 53.

Sweitzer, S. M., Schubert, P. and DeLeo, J. A. 2001. Propentofylline, a glial modulating agent, exhibits antiallodynic properties in a rat model of neuropathic pain. *Journal of Pharmacology and Experimental Therapeutics* 297: 1210–17.

Sweitzer, S. M., Colburn, R. W., Rutkowski, M. and DeLeo, J. A. 1999. Acute peripheral inflammation induces moderate glial activation and spinal, IL-1beta expression that correlates with pain behavior in the rat. *Brain Research* 829: 209–21.

Tambuyzer, B. R., Bergwerf, I., De, V. N., Reekmans, K., Daans, J., Jorens, P. G., Goossens, H., Ysebaert, D. K., Chatterjee, S., Van, M. E., Berneman, Z. N. and Ponsaerts, P. 2009. Allogeneic stromal cell implantation in brain tissue leads to robust microglial activation. *Immunology & Cell Biology* 87: 267–73.

Tanaka, T., Minami, M., Nakagawa, T. and Satoh, M. 2004. Enhanced production of monocyte chemoattractant protein-1 in the dorsal root ganglia in a rat model of neuropathic pain: Possible involvement in the development of neuropathic pain. *Neuroscience Research* 48: 463–9.

Tanga, F. Y., Raghavendra, V. and DeLeo, J. A. 2004. Quantitative real-time RT-PCR assessment of spinal microglial and astrocytic activation markers in a rat model of neuropathic pain. *Neurochemistry International* 45: 397–407.

Tawfik, V. L., Nutile-McMenemy, N., Lacroix-Fralish, M. L. and DeLeo, J. A. 2007. Efficacy of propentofylline, a glial modulating agent, on existing mechanical allodynia following peripheral nerve injury. *Brain, Behavior, and Immunity* 21: 238–246.

Tawfik, V. L., Regan, M. R., Haenggeli, C., Lacroix-Fralish, M. L., Nutile-McMenemy, N., Perez, N., Rothstein, J. D. and DeLeo, J. A. 2008. Propentofylline-induced astrocyte modulation leads to alterations in glial glutamate promoter activation following spinal nerve transection. *Neuroscience* 152: 1086–92.

Thacker, M. A., Clark, A. K., Bishop, T., Grist, J., Yip, P. K., Moon, L. D., Thompson, S. W., Marchand, F. and McMahon, S. B. 2009. CCL_2 is a key mediator of microglia activation in neuropathic pain states. *European Journal of Pain* 13: 263–72.

Todd, A. J. 2010. Neuronal circuitry for pain processing in the dorsal horn. *Nature Reviews Neuroscience* 11: 823–36.

Trang, T., Beggs, S. and Salter, M. W. 2012. ATP receptors gate microglia signaling in neuropathic pain. *Experimental Neurology* 234: 354–61.

Trang, T., Beggs, S., Wan, X., Salter, M. W. 2009. P2X4-receptor-mediated synthesis and release of brain-derived neurotrophic factor in microglia is dependent on calcium and p38-mitogen-activated protein kinase activation. *Journal of Neuroscience* 29: 3518–28.

Tsuda, M., Mizokoshi, A., Shigemoto-Mogami, Y., Koizumi, S. and Inoue, K. 2004. Activation of p38 mitogen-activated protein kinase in spinal hyperactive microglia contributes to pain hypersensitivity following peripheral nerve injury. *Glia* 45: 89–95.

Tsuda, M., Ueno, H., Kataoka, A., Tozaki-Saitoh, H. and Inoue, K. 2008. Activation of dorsal horn microglia contributes to diabetes-induced tactile allodynia via extracellular signal-regulated protein kinase signaling. *Glia* 56: 378–86.

Tsuda, M., Kuboyama, K., Inoue, T., Nagata, K., Tozaki-Saitoh, H. and Inoue, K. 2009. Behavioral phenotypes of mice lacking purinergic P2X4 receptors in acute and chronic pain assays. *Molecular Pain* 5: 28.

Tsuda, M., Shigemoto-Mogami, Y., Koizumi, S., Mizokoshi, A., Kohsaka, S., Salter, M. W. and Inoue, K. 2003. P2X4 receptors induced in spinal microglia gate tactile allodynia after nerve injury. *Nature* 424: 778–83.

Ulmann, L., Hatcher, J. P., Hughes, J. P., Chaumont, S., Green, P. J., Conquet, F., Buell, G. N., Reeve, A. J., Chessell, I. P. and Rassendren, F. 2008. Up-regulation of P2X4 receptors in spinal microglia after peripheral nerve injury mediates BDNF release and neuropathic pain. *Journal of Neuroscience* 28: 11263–8.

Vallejo, R., Tilley, D., Vogel, L. and Benyamin, R. 2010. The role of glia and the immune system in the development and maintenance of neuropathic pain. *Pain Practice* 10: 167–84.

Van, C. E., Van, D. J. and Opdenakker, G. 1999. The MCP/eotaxin subfamily of CC chemokines. *Cytokine & Growth Factor Reviews* 10: 61–86.

Verge, G. M., Milligan, E. D., Maier, S. F., Watkins, L. R., Naeve, G. S. and Foster, A. C. 2004. Fractalkine (CX3CL1) and fractalkine receptor (CX$_3$CR$_1$) distribution in spinal cord and dorsal root ganglia under basal and neuropathic pain conditions. *European Journal of Neuroscience* 20: 1150–60.

Verpoorten, N., Claeys, K. G., Deprez, L., Jacobs, A., Van, G., V, Lagae, L., Arts, W. F., De, M. L., Keymolen, K., Ceuterick-de, G. C., De, J. P., Timmerman, V. and Nelis, E. 2006. Novel frameshift and splice site mutations in the neurotrophic tyrosine kinase receptor type 1 gene (NTRK1) associated with hereditary sensory neuropathy type IV. *Neuromuscular Disorders* 16: 19–25.

Volterra, A. and Meldolesi, J. 2005. Astrocytes, from brain glue to communication elements: The revolution continues. *Nature Reviews Neuroscience* 6: 626–40.

Wallraff, A., Kohling, R., Heinemann, U., Theis, M., Willecke, K. and Steinhauser, C. 2006. The impact of astrocytic gap junctional coupling on potassium buffering in the hippocampus. *Journal of Neuroscience* 26: 5438–47.

Wang, S., Song, L., Tan, Y., Ma, Y., Tian, Y., Jin, X., Lim, G., Zhang, S., Chen, L. and Mao, J. 2012. A functional relationship between trigeminal astroglial activation and NR1 expression in a rat model of temporomandibular joint inflammation. *Pain Medicine* 13: 1590–600.

Wang, W., Wang, W., Wang, Y., Huang, J., Wu, S. and Li, Y. Q. 2008. Temporal changes of astrocyte activation and glutamate transporter-1 expression in the spinal cord after spinal nerve ligation-induced neuropathic pain. *The Anatomical Record (Hoboken)* 291: 513–18.

Watkins, L. R., Martin, D., Ulrich, P., Tracey, K. J. and Maier, S. F. 1997. Evidence for the involvement of spinal cord glia in subcutaneous formalin induced hyperalgesia in the rat. *Pain* 71: 225–35.

Wen, Y. R., Suter, M. R., Kawasaki, Y., Huang, J., Pertin, M., Kohno, T., Berde, C. B., Decosterd, I. and Ji, R. R. 2007. Nerve conduction blockade in the sciatic nerve prevents but does not reverse the activation of p38 mitogen-activated protein kinase in spinal microglia in the rat spared nerve injury model. *Anesthesiology* 107: 312–21.

Weng, H. R., Chen, J. H. and Cata, J. P. 2006. Inhibition of glutamate uptake in the spinal cord induces hyperalgesia and increased responses of spinal dorsal horn neurons to peripheral afferent stimulation. *Neuroscience* 138: 1351–60.

Weyerbacher, A. R., Xu, Q., Tamasdan, C., Shin, S. J. and Inturrisi, C. E. 2010. N-Methyl-D-aspartate receptor (NMDAR) independent maintenance of inflammatory pain. *Pain* 148: 237–46.

White, F. A., Sun, J., Waters, S. M., Ma, C., Ren, D., Ripsch, M., Steflik, J., Cortright, D. N., Lamotte, R. H. and Miller, R. J. 2005. Excitatory monocyte chemoattractant protein-1 signaling is up-regulated in sensory neurons after chronic compression of the dorsal root ganglion. *Proceedings of the National Academy of Sciences of the United States of America* 102: 14092–7.

Wodarski, R., Clark, A. K., Grist, J., Marchand, F. and Malcangio, M. 2009. Gabapentin reverses microglial activation in the spinal cord of streptozotocin-induced diabetic rats. *European Journal of Pain* 13: 807–11.

Woolf, C. J. and Mannion, R. J. 1999. Neuropathic pain: Aetiology, symptoms, mechanisms, and management. *Lancet* 353: 1959–64.

Xin, W. J., Weng, H. R. and Dougherty, P. M. 2009. Plasticity in expression of the glutamate transporters GLT-1 and GLAST in spinal dorsal horn glial cells following partial sciatic nerve ligation. *Molecular Pain* 5: 15.

Zhang, J. and De, K. Y. 2006. Spatial and temporal relationship between monocyte chemoattractant protein-1 expression and spinal glial activation following peripheral nerve injury. *Journal of Neurochemistry* 97: 772–83.

Zhang, J., Shi, X. Q., Echeverry, S., Mogil, J. S., De, K. Y. and Rivest, S. 2007. Expression of CCR$_2$ in both resident and bone marrow-derived microglia plays a critical role in neuropathic pain. *Journal of Neuroscience* 27: 12396–12406.

Zhuang, Z. Y., Gerner, P., Woolf, C. J. and Ji, R. R. 2005. ERK is sequentially activated in neurons, microglia, and astrocytes by spinal nerve ligation and contributes to mechanical allodynia in this neuropathic pain model. *Pain* 114: 149–59.

Zhuang, Z. Y., Kawasaki, Y., Tan, P. H., Wen, Y. R., Huang, J. and Ji, R. R. 2007. Role of the CX$_3$CR$_1$/p38 MAPK pathway in spinal microglia for the development of neuropathic pain following nerve injury-induced cleavage of fractalkine. *Brain, Behavior, and Immunity* 21: 642–51.

Zhuang, Z. Y., Wen, Y. R., Zhang, D. R., Borsello, T., Bonny, C., Strichartz, G. R., Decosterd, I. and Ji, R. R. 2006. A peptide c-Jun N-terminal kinase (JNK) inhibitor blocks mechanical allodynia after spinal nerve ligation: Respective roles of JNK activation in primary sensory neurons and spinal astrocytes for neuropathic pain development and maintenance. *Journal of Neuroscience* 26: 3551–60.

CHAPTER 5

Genetics and epigenetics of pain

Franziska Denk and Stephen B. McMahon

Genetics: Introduction

With the advent of large-scale sequencing and the mapping of the human genome (Collins. Morgan and Patrinos, 2003), it has become possible to study genetic variation and its consequent effects on health and disease in great detail. The most common type of variation between individuals are single base pair (bp) changes, known as single nucleotide polymorphisms (SNPs), which occur roughly in one out of every 1,200 bp and the total number of which is estimated around 10 million in the human genome. SNPs are usually studied in so-called haplotypes – groups of adjacent SNPs that are inherited together as a result of genetic linkage. This allows researchers to capture variation in a given population by genotyping a smaller number of tagging SNPs (tSNPs) rather than by assessing every individual SNP within a haplotype block.

The identity and frequency of haplotypes varies greatly between different human populations (International HapMap, 2003), Africans showing the largest degree of heterogeneity due to the population bottlenecks associated with the emigration of the first humans into other continents around sixty thousand years ago (Reich et al., 2001; Marth et al., 2004). This is a very important fact to bear in mind when designing or assessing genetic studies, as careless cohort selection may lead to serious population-based confounds. Even when a homogeneous population is used, stratification may occur where unknown subpopulations account for some of the variation in the frequency of alleles.

Different approaches can be used to link genetic variation to a phenotype. Linkage studies identify regions of the genome that segregate with disease in one or more families (Dawn Teare and Barrett, 2005). They are best suited to the study of diseases with Mendelian inheritance patterns and require only a low density of genetic markers. However, the risk regions returned are, typically, quite large and require further fine mapping, done for example by using positional cloning or SNP genotyping. Moreover, suitably large, multigenerational families can be difficult to find. In contrast, association studies can be carried out using simpler case-control designs. Groups of affected and unaffected individuals are genotyped, and a statistical association is sought between individual SNPs or haplotypes and disease. This method can be

An Introduction to Pain and its Relation to Nervous System Disorders, First Edition. Edited by Anna A. Battaglia.
© 2016 John Wiley & Sons, Ltd. Published 2016 by John Wiley & Sons, Ltd.

applied both at candidate gene and at genome-wide level. A positive association between a disease and a given SNP requires careful interpretation. Very often the SNP in question is only in linkage disequilibrium – in other words it is correlated only with the actual disease-causing SNP. As mentioned above, confounds due to population stratification are very common and can be to some extent counteracted through careful cohort selection or use of family-based designs, such as sibling pairs or parent–offspring trios. Statistical noise and therefore type-1 errors are also major issues, as are real but spurious associations with non-causal markers or with multiple causal markers. For instance, if SNP A is associated with skin thickness and SNP B with both skin thickness and pain sensitivity, a genome-wide association study (GWAS) for pain sensitivity might erroneously identify SNP A as a causal SNP (Hirschhorn and Daly, 2005; Platt, Vilhjalmsson and Nordborg, 2010).

Other sources of genetic variation between people that have recently received more attention derive from rare single-nucleotide variants (SNVs) (Gibson, 2011; Nelson et al., 2012; Tennessen et al., 2012) and from structural variants, such as changes in the copy number of genes or insertions and deletions of large stretches of the DNA sequence (Feuk, Carson and Scherer, 2006). SNVs are simply SNPs that are less common in a given population. The cut-offs are arbitrarily defined and still quite fluid. Commonly, a single base pair change observed with a minor allelic frequency of less than 0.5 per cent is referred to as a rare variant, while any change with a frequency of above 5 per cent would be classed as a SNP. The study of SNVs is difficult, since it requires deep sequencing as well as family-based designs in order to distinguish between sequencing errors and genuine variation. Similarly, the detection of structural variants is very complex, traditionally relying on technically challenging and low-resolution hybridisation techniques – such as fluorescence in situ (FISH) and comparative genomic hybridisation (aCGH). The continually decreasing cost of next-generation sequencing, coupled with better statistical tools and databases, is beginning to greatly facilitate research in these areas (Figure 5.1).

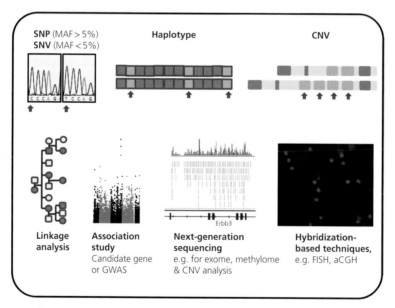

Figure 5.1 Sources of genetic variation and analysis methods.

Genetics of chronic pain

There is little doubt from twin and epidemiological studies that pain sensitivity per se is heritable (Norbury et al., 2007; Nielsen et al., 2008), as is the tendency to develop chronic pain of varying aetiologies (MacGregor et al., 2004; Hartvigsen et al., 2009; Williams, Spector and MacGregor, 2010; Altman et al., 2011; Livshits et al., 2011; Hocking et al., 2012). Estimates of heritability –that is, the proportion of genetic variance out of the total phenotypic variance – range from 9 per cent to 60 per cent, depending on the specific condition examined, and tend to increase with pain severity (Table 5.1).

With regard to these data, research has been trying to identify the genes that might be responsible for the genetic contribution to chronic pain conditions. The majority of studies in the field so far have fallen into two categories: linkage analyses in families with rare Mendelian pain-related disorders; and candidate-gene association studies in the general population (see Raouf, Quick and Wood, 2010; Mogil, 2012 for in-depth reviews). There are a number of familial syndromes that have pain as a defining symptom, such as familial migraines, primary erythermalgia and paroxysmal extreme pain disorder (PEPD). Moreover, there are several

Table 5.1 Heritability estimates for pain sensitivity and chronic pain conditions.

Authors	Subject details	Cohort	Heritability estimate	Phenotype examined
(Livshits et al., 2011)	Female twins (371 MZ, 698 DZ)	Twins UK	0.43–0.68	Lower back pain
(Battie et al., 2007)	Male twins (147 MZ, 153 DZ)	Finnish twin cohort	0.3–0.46	Lower back pain
(MacGregor et al., 2004)	Female twins (181 MZ, 351 DZ)	Twins UK	0.52–0.68 0.35–0.58	Lower back pain Neck pain
(Hartvigsen et al., 2009)	Mixed twins (6,700 MZ, 8,500 DZ)	Danish twin registry	0.33–0.39	Back pain (lumbar, thoracic and neck)
(Altman et al., 2011)	Female twins (1,867 MZ, 1,293 DZ)	Swedish twin registry	0.3	Bladder pain syndrome
(Williams et al., 2010)	Mixed twins (991 MZ, 1,074 DZ)	Twins UK	0.46	Pain at different body sites (neck, back, elbow, knee, thigh, hands, feet)
(Markkula et al., 2009)	Mixed twins (12,500)	Finnish twin cohort	0.51	fibromyalgia[1]
(Kato, Sullivan and Pedersen, 2010)	Mixed twins (28,531 pairs)	Swedish twin registry	0.09 (men) 0.13 (women)	Chronic widespread pain
(Hocking et al., 2012)	2,195 extended families	Scottish family health study	0.16 0.3	Any chronic pain Severe chronic pain
(Norbury et al., 2007)	Female twins (51 MZ, 47 DZ)	Twins UK	0.22–0.55	Quantitative sensory testing scores (i.a.: punctate hyperalgesia, heat pain threshold, itch, acid & ATP responses)
(Nielsen et al., 2008)	Mixed twins (53 MZ, 39 DZ)	Norwegian twin registry	0.6 0.26	Cold-pressor pain Heat contact pain

[1] Indirect phenotypic measure, using data from 49 fibromyalgia patients to infer fibromyalgia-like pain patterns in twins.

families across the world whose members display congenital analgesia, suffering as they do from channelopathy-associated indifference to pain (CIPA) or from various hereditary sensory and autonomic neuropathies (HSAN I-VI). The disease-causing polymorphisms have been identified in the majority of cases (for details, view relevant entries in the Online Mendelian Inheritance in Man database at http://www.ncbi.nlm.nih.gov/omim). They vary according to family and syndrome, but two biological groups have emerged as particularly relevant, namely ion channel subunits and nerve growth factor (NGF)-related genes. The calcium and sodium channel genes CACNL1A4 and SCN1A have both been linked to migraines (Ophoff et al., 1996; Dichgans et al., 2005), and mutations in the sodium channel subunit $Na_v1.7$ (SCN9a) can cause erythermalgia (Yang et al., 2004), PEPD (Fertleman et al., 2006), other severe pain (Meijer et al., 2013), and, conversely, congenital insensitivity to pain with anhidrosis (CIPA) (Cox et al., 2006; Nilsen et al., 2009). Equally, the β subunit of NGF (NGFB) and the NGF receptor TrkA (NTRK1) have been implicated in HSANs IV and V (Indo et al., 1996; Einarsdottir et al., 2004; Minde et al., 2009). SCN9A and NGF are now being targeted by the drug development industry, and one antibody against NGF looks particularly promising: tanezumab is in phase III clinical trials for osteoarthritis (Spierings et al., 2013), and several earlier stage trials for other types of chronic pain, such as back pain and interstitial cystitis (Evans et al., 2011; Kivitz et al., 2013).

While genetic evidence from monogenic diseases has thus helped to identify and substantiate the importance of some of the biological mechanisms involved in chronic pain, candidate-gene association studies present a slightly more confusing picture. Ion channels, again, feature high on the list of targets of interest, and significant associations have been reported, for example to alleles in SCN9a (Reimann et al., 2010; Valdes et al., 2011; Vargas-Alarcon et al., 2012), KCNS1 (Costigan et al., 2010; Hendry et al., 2013) and CACNG2 (Nissenbaum et al., 2010; Greenbaum et al., 2012). Similarly, neurotransmitter-related genes have been studied, most frequently catechol-O-methyl transferase, COMT (Belfer and Segall, 2011; Tammimaki and Mannisto, 2012; Belfer et al., 2015), the μ-opioid receptor, OPRM1 (Walter and Lotsch, 2009) and GTP cyclohydrolase 1, GCH1 (Tegeder et al., 2006; Campbell et al., 2009). Finally, positive associations have been detected to inflammatory genes, among others IL1A (Solovieva et al., 2004) and TNF (Reyes-Gibby et al., 2009).[1]

Yet the field has been plagued by inconsistent results, associations not replicating across populations or across pain conditions (Kim and Dionne, 2007, Hendry et al., 2013). To give a typical example, a SNP in SCN9A was found to associate with pain sensitivity in healthy individuals as well as with pain scores from patients suffering from a variety of pains, including osteoarthrits (Reimann et al., 2010). However, the link to this particular SNP (rs6746030) could not be replicated in a different population of osteoarthritis patients by Valdes et al. (2011), who instead reported an association with multiple regional pain. This last finding was in turn contested in a third publication, which reported no association between rs6746030 and widespread pain in four independent cohorts, totalling 1,071 case and 3,214 control subjects (Holliday et al., 2012). What can we make of these and many other similarly conflicting findings? Their potential origins and our options for future alternative work will be discussed in the following pages.

---

[1] References are selective. For a more comprehensive list of studies, including details on the risk allele identified and the phenotype studied refer to Jeff Mogil's supplementary table in Mogil, 2012.

A simple and common reason for inconsistencies between different association studies is lack of power. Early studies in particular tended to use smaller sample sizes, and any associations uncovered were therefore often either false or overestimated in terms of their effect size. Any power calculations based on initial effect sizes were then consequently distorted, and subsequent replication cohorts were too small. This latter phenomenon is referred to as the 'winner's curse' – although maybe it should be more appropriately called the 'follower's curse', since negative data, even if true, tend to be more difficult to publish. The problem of issues with population selection, a second common shortcoming of association studies, has already been introduced. Thus, in the example of SCN9A, the different publications used Finnish, North American and British cohorts, which might not always compare in terms of allele and haplotype frequencies. There may be genuine differences between these ethnic groups, preventing replication; or population stratification may have occurred, where a spurious difference between groups that is due to unknown subsets is observed. Use of more family-based designs might give protection against these shortcomings. Lastly, and perhaps most importantly in the case of chronic pain, populations may have differed in terms of their phenotypic composition. To reach sufficiently large sample sizes, genotyping studies often have to use cohorts that have not been specifically designed with chronic pain in mind. Measurements of pain differ widely across studies: they range from no direct assessment – where data are available only on the underlying painful condition: for example the extent of joint damage in osteoarthritis (Limer et al., 2009) – to fully standardised, quantitative sensory testing (QST) (Kremeyer et al., 2010). Inclusion and exclusion criteria are often based on symptoms, not on clinical signs, which can impact classification, for example in the case of neuropathy (Robinson-Papp et al., 2010). As a result, comparison between different datasets can be very challenging in the pain field, not least because one basic question remains unresolved – namely whether there will be many genes for pain or nociception per se, like SCN9a, or whether genes are more likely to be associated with a particular pain modality (say, thermal vs mechanical) or with the underlying painful condition. A comprehensive analysis of different mouse strains indicated that both heat-induced and mechanical hypersensitivity were dependent on genotype but were inversely correlated. The modality of pain therefore seemed to be more crucial in determining an animal's sensitivity than the original source of that sensitivity, which was either neuropathic or inflammatory (Mogil, 1999; Mogil et al., 1999). Studies have attempted to address this issue in humans, but the data remain contradictory (see, e.g., Neddermeyer et al., 2008 or Nielsen et al., 2008).

In addition to improving candidate gene association approaches, there are other avenues of genetic research that are beginning to be explored in the field of chronic pain. GWAS have been conducted, providing unbiased screens for common variants associated with disorders that feature pain as a major symptom, like osteoarthritis (arcOgen Consortium et al., 2012; Styrkarsdottir et al., 2014), lumbar disc degeneration (Williams et al., 2012) and endometriosis (Nyholt et al., 2012). There have also been several GWAS in which pain was directly measured and correlated with genotype: work on migraine (Freilinger et al., 2012; Esserlind et al., 2013; de Vries et al., 2015), opioid sensitivity (Nishizawa et al., 2012), chronic widespread pain (Peters et al., 2013) and pain associated with acute tooth extraction (Kim et al., 2009). With the exception of the migraine field, data are still sparse and it is not clear whether any of the risk alleles identified will stand the test of time. GWAS suffer from the same real and potential problems that are associated with candidate gene exploration and, given the extra expense and effort associated with them, particular care should be taken to ensure adequate phenotyping, cohort selection and sample size.

Also, GWAS are only able to identify common SNPs, to the neglect of rare and copy number variants. Yet it has been argued that this second group – rare and copy number variants – may account for a significant proportion of the variance between individuals (Gibson, 2011). In pain research, exome sequencing is still infrequent and mostly used to identify mutations in familial disease (e.g., Leipold et al., 2013). One exception is a study by Williams and colleagues (2012). They explored SNVs in a cohort of divergent twins and correlated them with heat pain sensitivity in an unbiased fashion. No single SNV survived multiple comparison correction, but directed pathway analysis revealed a network of genes, centred around angiotensin II, that was significantly correlated with heat pain thresholds.

Another promising approach has been to link results from association studies to preclinical data, in an attempt to better characterise the putative underlying biology (Sorge et al., 2012; Peters et al., 2013). For example, Sorge et al. employed a genome-wide linkage analysis in different mouse strains in order to identify a haplotype in the P2X7 receptor gene that was associated with hypersensitivity to mechanical stimuli. They then demonstrated a functional effect on pore formation and nociception, before ultimately testing for and confirming the presence of the haplotype in human patient populations.

Finally, researchers are starting to consider gene x gene interactions, also called epistasis, and gene x environment interactions. For instance, an epistatic effect has been reported by Reyes-Gibby and colleagues (2007); they found both COMT and OPRM1 alleles to act jointly to mediate morphine responsiveness. Similarly, an analysis of archival mouse behavioural data indicated that 18 per cent of the variance in phenotype was accounted for by gene x environment interaction (Chesler et al., 2002). In humans, interactions between sex and genotype were reported in case of the OPRM1 variant A118G (Fillingim et al., 2005) and between sex, stress and genotype in the case of a vasopressin receptor polymorphism (Mogil et al., 2011).

In summary, while it is clear that chronic pain conditions are heritable, the underlying genetic risk factors are still being elucidated. Approaches that lead to, or are based on, a deeper understanding of the mechanistic link between a particular genotype and its corresponding pain phenotype may prove to be the most promising.

Epigenetics: Introduction

In addition to genetics, an organism's phenotype is determined by environmental influences – sources of non-genetic variation after fertilisation. Epidemiological research tends to distinguish between shared and non-shared environment, where 'shared' is used to describe factors that are common to a specific study group, such as the influence of parents on siblings or of the laboratory environment on a cohort of rats. In contrast, 'non-shared' environment is an umbrella term for systematic sources of variation within groups (e.g., birth order), any non-systematic, stochastic sources of variation and, finally, measurement error that occurs as part of the research itself.

Box 5.1 Classic epigenetic modifications.

DNA methylation: DNA methylation refers to the addition of methyl groups to cytosines within the sequence to form 5-methylcytosine or 5-hydroxymethylcytosine. DNA methylation is tissue-specific, helps determine cell lineage during development and can be inherited via imprinting (Smith and Meissner, 2013). In the traditional literature, it was described as a stable postnatal modification that served a gene-silencing function by binding mainly to cytosine-guanine runs (CpG islands) found within promoter regions. More recently it has been shown that a significant proportion (~30%) of non-CG methylation occurs in embryonic stem cells as well as in adult mouse brain (Lister et al., 2009; Xie et al., 2012), that DNA can be actively demethylated in adult tissues (Wu and Zhang, 2011) and that methylation patterns can change rapidly (within four hours) even in postmitotic cells (Guo et al., 2011). Research is also beginning to appreciate the importance of 5-hydroxymethylcytosine, which may be of particular importance to neurons (Kriaucionis and Heintz, 2009). **Histone modifications**: The human genome is 3 billion basepairs (bp) in length, which makes a single DNA molecule about 0.9 meters long (at 0.3nM per bp). To fit inside cell nuclei of a size of 3–10 micrometers, the DNA is therefore highly condensed. The smallest packaging block is the nucleosome; it consists of basic histone proteins that assemble into octamers around which the DNA is wrapped in two 147 bp turns. Histones occur in different variants and have lysine residues that can be post-translationally modified, e.g. through phosphorylation, methylation or acetylation. The type of histone as well as its specific modification profile affects chromatin conformation and hence can impact gene function (Bannister and Kouzarides, 2011; Maze, Noh and Allis, 2013). Histone modifications are dynamically regulated with cellular activity and most likely act in concert to form a code or language that can be interpreted by reader proteins, which then recruit or dismantle protein complexes involved in transcription or chromatin remodelling (Strahl and Allis, 2000; Borrelli et al., 2008; Lee, Smith and Shilatifard, 2010).

Contrary to popular belief, the shared environment tends on average to explain very little of the variance within a population. Thus siblings do not resemble each other much more than they resemble members of the general population (Plomin and Daniels, 2011), twins raised apart are not very different from twins raised together (Bouchard et al., 1990), and reducing shared environmental variation in laboratory animals has very little impact on overall variation in phenotype (Gartner, 1990). One potential confound for this common observation is that interaction effects between the environment and genetics are very hard to model and therefore could cause an overestimation of the importance of non-shared environmental components. Another possibility is that the overall importance of shared environmental factors is underestimated as a result of differential behavioural outcomes. For instance one child might thrive under a particular treatment, while the other is negatively affected (see Box 5.1).

Regardless of the precise relative contribution of shared versus non-shared environment components, it is clear that non-genetic factors play a sizable role in determining whether a particular individual develops a particular disease or a condition like chronic pain. It is therefore important to understand the potential origins and molecular basis of such diseases or conditions. One large potential source of non-shared variation that has already been identified more than thirty years ago, through embryo transfer experiments, is the zygote environment (Gartner, 1990). Artificially created, genetically identical calves are more similar to each other when they are born from the same zygote than when they are born from two different zygotes. Post-natally, some commonly reported environmental risk factors for chronic pain are a prior history of pain (Thomas et al., 1999); initial pain severity (Kehlet, Jensen and Woolf, 2006; Deumens et al., 2013); psychosocial factors such as stress, for example due to unemployment or overwork; depression; and personality traits like pessimism and a tendency to catastrophise (Pavlin et al., 2005; Patten et al., 2008). Finally, the incidence of chronic pain tends mostly to increase with age.

What molecular mechanisms could account for these environmental risk factors? One possibility is that epigenetic processes are at play. Epigenetics is the study of heritable and stable changes in gene function that are not due to changes to the DNA sequence itself. Epigenetic mechanisms, such as DNA methylation and histone modifications, determine lineage specificity during development, can be heritable (e.g., in the case of imprinting) (Smith and Meissner, 2013) and, according to recent literature, may change as a result of environmental influences, coming to function as a kind of molecular memory (Dulac, 2010; Peleg et al., 2010; Guo et al., 2011; Telese et al., 2013). In theory, they could there-fore drive some of the risk associated with chronic pain, both during development and in adult organisms.

Epigenetics in chronic pain

What is the evidence that epigenetic processes are involved in chronic pain? Publications have reported links between pain and both histone acetylation and DNA methylation. Histone deacetylase inhibitors (HDACi), which interfere with the removal of acetyl groups from histones, have repeatedly been shown to have analgesic effects in both neuropathic and inflammatory pain models when they are delivered systemically (Chiechio et al., 2009), directly into the spinal cord (Bai et al., 2010; Denk et al., 2013) or into the raphe nucleus – an important descending pain control region (Zhang et al., 2011). The compounds may also reduce endometriosis and visceral hypersensitivity in the rat (Liu et al., 2012; Tran et al., 2012), though these latter findings remain to be replicated by independent groups. Histone acetyltransferase (HAT) inhibitors – that is, drugs that inhibit the enzymes that add acetyl groups – have also been employed. The HAT p300 has been claimed to show increased expression in the spinal cord after chronic constriction injury, and intrathecal administration of a p300 inhibitor was reported to reduce mechanical and thermal hypersensitivity (Zhu et al., 2012; Zhu et al., 2013). Similar results have been published about treating nerve injury with anarcadic acid – a compound that, among many other things, also inhibits HATs (Kiguchi et al., 2011). It seems slightly counterintuitive that preventing both the removal and the addition of acetyl groups has analgesic effects. More work will be required in this area, employing careful dosing and selective compounds, to confirm results and to start investigat-ing potential mechanisms.

In line with the behavioural changes seen after HATi and HDACi administration, histone acetylation has been observed to be altered globally in complete Freund's adjuvant (CFA)-induced inflammatory pain (Zhang et al., 2011) and morphine-induced conditioned place preference (Bie et al., 2012). Moreover, local changes in acetylation were found at pain rele-vant promoters, such as NGF, GAD2, CXCL2 and CXCR2, in various regions and models (Kiguchi et al., 2011; Zhang et al., 2011; Bie et al., 2012). Lastly, quite convincing evidence from the field of addiction research showed altered histone methylation at lysine residue 9 of histone 3 in the nucleus accumbens after morphine administration (Sun et al., 2012). It is conceivable that similar mechanisms are at play in pain and analgesia.

What about DNA methylation? Here, too, inhibitors exist that can be used as initial screening tools, for instance drugs that prevent the action of DNA methyltransferases (DNMTs) and hence the addition of methyl groups. However, so far only one article has been published using the DNMT inhibitor 5-azacytidine, and its results are inconclusive. The authors reported reduced hypersensitivity after intrathecal delivery in a chronic constriction injury model. Yet 5-azacytidine's mode of action relies on integration into desoxyribonucleic acid (DNA) or ribonucleic acid (RNA) sequence upon replication, which calls into question its effect on postmitotic cells. Moreover, the drug is highly cytotoxic and unstable. Confirmation of the findings with the more suitable direct DNMT inhibitor RG108 would therefore be of great interest.

While a causal connection between DNA methylation and pain is therefore still lacking, there are many hints from correlational data that methylation status may be important. Changes in DNA methylation have been observed at the glucocorticoid receptor gene in a model of visceral hypersensitivity (Tran et al., 2012). Moreover, the endothelin receptor B gene was found to be hypermethylated and transcriptionally silenced in painful human oral cancers, but not in their non-painful counterparts. Reexpression of the gene attenuated hypersensitivity in a cancer model in the mouse. Finally, global decreases in methylation were registered in the prefrontal cortex and amygdala after spared nerve injury in the mouse, which, in the case of the prefrontal cortex, correlated with the degree of hypersensitivity (Tajerian et al., 2013).

A third strand of evidence derives from the study of the methyl-CpG-binding protein 2 (MeCP2), a reader molecule that binds to methylated CpG sites and, among other functions, can act as a transcriptional repressor. MeCP2 is down-regulated after nerve injury in the dorsal root ganglion (Tochiki et al., 2012), while its targets are up-regulated in the spinal cord after CFA injection in the ankle joint (Geranton, Morenilla-Palao and Hunt, 2007).

So far, the experiments that have been discussed are largely based on animal models. Yet the evidence for a link between epigenetics and chronic pain is not limited to non-human species. Recently, several profiling experiments have been conducted that posit a connection. The most preliminary, tentative suggestion arises from a GWAS in hip osteoarthritis patients that found a SNP in the DOT1L gene to be associated with reduced disease risk. Dot1L is a histone lysine methyltransferase that methylates lysine residue 79 of H3. Yet results need to be interpreted with caution, as pain was not measured in the cohort and the association may be due to Dot1L involvement in joint degradation. In fact the authors show that Dot1L is involved in chondrogenesis *in vitro* (Castano Betancourt et al., 2012). Stronger evidence comes from the following studies. Increased genome-wide methylation was observed in the blood of patients who received chronic opioid treatment, which correlated with pain scores (Doehring et al., 2013). Altered methylation was observed at the PARK2 locus in a twin patient group with lumbar disc degeneration (Williams et al., 2012). Similarly, increased methylation at the promoter of the SPARC gene was reported, together with disc degeneration, in

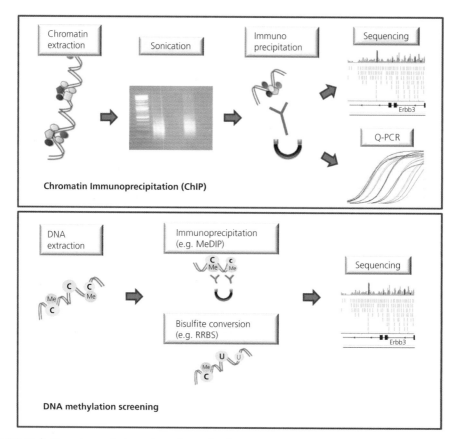

Figure 5.2 Techniques for exploring epigenetic mechanisms.

both mice and humans (Tajerian et al., 2011). Finally, the most extensive study to date is a genome-wide DNA methylation profile conducted by using blood from twin pairs whose members diverged in their heat pain sensitivity. Nine differentially methylated regions were identified, the strongest association being with the promoter region of the TRPA1 gene. Fittingly, TRPA1 expression was found to be increased in the skin of individuals with high pain thresholds – a result consistent with the relative decrease in methylation observed (Bell et al., 2014).

In summary, the study of epigenetic mechanisms in chronic pain is still in its infancy. So far, the evidence suggests that persistent pain states are correlated with histone modification and DNA methylation changes. This connection is likely to be functional, as its disruption through pharmacological intervention affects pain processing. Despite this promising-looking start, the field faces many challenges. Too frequently, papers still suffer from basic methodological flaws, such as inappropriate use of compounds (Wang et al., 2011) and lack of adequate controls, especially in the case of chromatin immunoprecipitation (Mashayekhi et al., 2012; Zhu et al., 2012; Imai et al., 2013; Liang, Li and Clark, 2013; Uchida, Matsushita and Ueda, 2013). While these problems could be easily resolved, others are more general and technically difficult to address.

First and foremost, there is the question of tissue specificity. In which cell type do these changes occur, and could some of the current results have been misinterpreted as a result of the mixture of cells examined? In preclinical research, cell-sorting techniques and the use of tissue culture models may be able to provide some answers, though cell number is a rate-limiting factor: most of the current chromatin immunoprecipitation protocols (ChIPs) have lower limits of around 10,000 cells. Very recently, a few promising technical reports have emerged that may alleviate this issue by allowing precipitation down to hundreds of cells (Lara-Astiaso et al., 2014; Brind'Amour et al., 2015).

Secondly, there are questions about timing, both the turnover time of the marks themselves and their relevance in a pain state persistent over time. Thus, histone modifications have been reported to have relatively fast turnover times: acetylation marks in HEK cells have a half-life of about 50–80 minutes, depending on the lysine residue (Evertts et al., 2013), while methylation marks turn over about once a day (Zee et al., 2010). What happens in postmitotic cells is unclear. Histones themselves have very long half-lives – 220 days in mouse brain (Commerford, Carsten and Cronkite, 1982) and one year at least in rat brain (Savas et al., 2012) – though there is evidence for a replication-independent exchange of histone variants (Maze, Noh and Allis, 2013). As to the modifications themselves, data from quiescent fibroblasts indicate that non-replicating cells might show slower turnover rates (Evertts et al., 2013), but neurons in particular remain to be studied in this regard. In the case of DNA methylation, timing is also a big unknown. Until very recently it was believed that DNA was not being actively demethylated in adulthood. Now a DNA demethylation mechanism has been uncovered (Wu and Zhang, 2010), and there are reports of rapid alterations in the methylation state, for instance removal of DNA methyl groups in hippocampal neurons within a period of only four hours (Guo et al., 2011). Finally, no one knows how this relates to persistent pain versus acute pain. Do the epigenetic changes observed in the field so far relate only to the emergence, or also to the maintenance of hypersensitivity? And are there epigenetic profiles from birth that predispose towards chronic pain or co-determine pain sensitivity? Initial exploration in this area indicates that monozygotic twins have very similar DNA methylation at birth but that differences emerge with time, the functionality of which is still undetermined (Fraga et al., 2005; Bell et al., 2012).

This brings us to a third important question faced by researchers the field of epigenetics. Many voices have been raised to argue that the functionality of marks cannot simply be assumed, especially since many of the data are correlational (Henikoff and Shilatifard, 2011; Graur et al., 2013). Few would probably argue that epigenetic modifications can never be of relevance. Histone modifications have been firmly linked to transcription (e.g. Ling et al., 1996; Ernst et al., 2011; Chen et al., 2013), and DNA methylation is an important mechanism of gene silencing (Jones, 2012). The question is how many and which of the marks are functional and how their function differs under different conditions, such as cell type, nucleosome position or developmental state. For example, non-promoter DNA methylation has been reported to increase transcription (Wu et al., 2010) and histone acetylation does not always spell enhanced gene expression (Wang et al., 2009).

Ultimately, though, while many challenges lie ahead, the study of epigenetics in chronic pain and in neuroscience in general promises to be a very worthwhile enterprise. It provides potentially compelling molecular mechanisms by which risk for chronic pain could be assessed and environmental influences could be made to have a persistent impact on nociceptive pathways and on the later development of pain conditions.

References

Altman, D., Lundholm, C., Milsom, I., Peeker, R., Fall, M., Iliadou, A. N. and Pedersen, N. L. 2011. The genetic and environmental contribution to the occurrence of bladder pain syndrome: An empirical approach in a nationwide population sample. *European Urology* 59: 280–5.

arcOgen Consortium, arcOgen Collaborators, Zeggini, E., Panoutsopoulou, K., Southam, L., Rayner, N. W., Day-Williams, A. G., Lopes, M. C., Boraska, V., Esko, T., et al. 2012. Identification of new susceptibility loci for osteoarthritis (arcOGEN): A genome-wide association study. *Lancet* 380: 815–23.

Bai, G., Wei, D., Zou, S., Ren, K. and Dubner, R. 2010. Inhibition of class II histone deacetylases in the spinal cord attenuates inflammatory hyperalgesia. *Molecular Pain* 6: 51.

Bannister, A. J. and Kouzarides, T. 2011. Regulation of chromatin by histone modifications. *Cell Research* 21: 381–95.

Battie, M. C., Videman, T., Levalahti, E., Gill, K. and Kaprio, J. 2007. Heritability of low back pain and the role of disc degeneration. *Pain* 131: 272–80.

Belfer, I. and Segall, S. 2011. COMT genetic variants and pain. Drugs of today 47: 457–67.

Belfer, I., Dai, F., Kehlet, H., Finelli, P., Qin, L., Bittner, R. and Aasvang, E. K. 2015. Association of functional variations in COMT and GCH1 genes with postherniotomy pain and related impairment. *Pain* 156: 273–9.

Bell, J. T., Tsai, P. C., Yang, T. P., Pidsley, R., Nisbet, J., Glass, D., Mangino, M., Zhai, G., Zhang, F., Valdes, A., et al. 2012. Epigenome-wide scans identify differentially methylated regions for age and age-related phenotypes in a healthy ageing population. *PLoS genetics* 8: e1002629.

Bell, J. T., Loomis, A. K., Butcher, L. M., Gao, F., Zhang, B., Hyde, C. L., Sun, J., Wu, H., Ward, K., Harris, J., et al. 2014. Differential methylation of the TRPA1 promoter in pain sensitivity. *Nature Communications* 5: 2978.

Bie, B., Wang, Y., Cai, Y. Q., Zhang, Z., Hou, Y. Y. and Pan, Z. Z. 2012. Upregulation of nerve growth factor in central amygdala increases sensitivity to opioid reward. *Neuropsychopharmacology: Official Publication of the American College of Neuropsychopharmacology* 37: 2780–8.

Borrelli, E., Nestler, E. J., Allis, C. D. and Sassone-Corsi, P. 2008. Decoding the epigenetic language of neuronal plasticity. *Neuron* 60: 961–74.

Bouchard, T. J., Jr., Lykken, D. T., McGue, M., Segal, N. L. and Tellegen, A. 1990. Sources of human psychological differences: The Minnesota Study of Twins Reared Apart. *Science* 250: 223–8.

Brind'Amour, J., Liu, S., Hudson, M., Chen, C., Karimi, M. M. and Lorincz, M. C. 2015. An ultra-low-input native ChIP-seq protocol for genome-wide profiling of rare cell populations. *Nature Communications* 6: 6033.

Campbell, C. M., Edwards, R. R., Carmona, C., Uhart, M., Wand, G., Carteret, A., Kim, Y. K., Frost, J. and Campbell, J. N. 2009. Polymorphisms in the GTP cyclohydrolase gene (GCH1) are associated with ratings of capsaicin pain. *Pain* 141: 114–18.

Castano Betancourt, M. C., Cailotto, F., Kerkhof, H. J., Cornelis, F. M., Doherty, S. A., Hart, D. J., Hofman, A., Luyten, F. P., Maciewicz, R. A., Mangino, M., et al. 2012. Genome-wide association and functional studies identify the DOT1L gene to be involved in cartilage thickness and hip osteoarthritis. *Proceedings of the National Academy of Sciences of the United States of America* 109: 8218–23.

Chen, M., Licon, K., Otsuka, R., Pillus, L. and Ideker, T. 2013. Decoupling epigenetic and genetic effects through systematic analysis of gene position. *Cell reports* 3: 128–37.

Chesler, E. J., Wilson, S. G., Lariviere, W. R., Rodriguez-Zas, S. L. and Mogil, J. S. 2002. Influences of laboratory environment on behavior. *Nature Neuroscience* 5: 1101–2.

Collins, F. S., Morgan, M. and Patrinos, A. 2003. The Human Genome Project: Lessons from large-scale biology. *Science* 300: 286–90.

Commerford, S. L., Carsten, A. L. and Cronkite, E. P. 1982. Histone turnover within nonproliferating cells. *Proceedings of the National Academy of Sciences of the United States of America* 79: 1163–5.

Costigan, M., Belfer, I., Griffin, R. S., Dai, F., Barrett, L. B., Coppola, G., Wu, T., Kiselycznyk, C., Poddar, M., Lu, Y., et al. 2010. Multiple chronic pain states are associated with a common amino acid-changing allele in KCNS1. *Brain* 133: 2519–27.

Cox, J. J., Reimann, F., Nicholas, A. K., Thornton, G., Roberts, E., Springell, K., Karbani, G., Jafri, H., Mannan, J., Raashid, Y., Al-Gazali, L., Hamamy, H., Valente, E. M., Gorman, S., Williams, R., McHale, D. P., Wood, J. N., Gribble, F. M. and Woods, C. G. 2006. An SCN9A channelopathy causes congenital inability to experience pain. *Nature* 444: 894–8.

Dawn Teare, M. and Barrett, J. H. 2005. Genetic linkage studies. *Lancet* 366, 1036–44.

de Vries, B., Anttila, V., Freilinger, T., Wessman, M., Kaunisto, M. A., Kallela, M., Artto, V., Vijfhuizen, L. S., Gobel, H., Dichgans, M., et al. 2015. Systematic re-evaluation of genes from candidate gene association studies in migraine using a large genome-wide association data set. *Cephalalgia: An International Journal of Headache* 84 (21): 2132–45.

Denk, F., Huang, W., Sidders, B., Bithell, A., Crow, M., Grist, J., Sharma, S., Ziemek, D., Rice, A. S., Buckley, N. J. and McMahon, S. B. 2013. HDAC inhibitors attenuate the development of hypersensitivity in models of neuropathic pain. *Pain* 154 (9): 1668–79.

Deumens, R., Steyaert, A., Forget, P., Schubert, M., Lavand'homme, P., Hermans, E. and De Kock, M. 2013. Prevention of chronic postoperative pain: Cellular, molecular, and clinical insights for mechanism-based treatment approaches. *Progress in Neurobiology* 104: 1–37.

Dichgans, M., Freilinger, T., Eckstein, G., Babini, E., Lorenz-Depiereux, B., Biskup, S., Ferrari, M. D., Herzog, J., van den Maagdenberg, A. M., Pusch, M. and Strom, T. M. 2005. Mutation in the neuronal voltage-gated sodium channel SCN1A in familial hemiplegic migraine. *Lancet* 366: 371–77.

Doehring, A., Oertel, B. G., Sittl, R. and Lotsch, J. 2013. Chronic opioid use is associated with increased DNA methylation correlating with increased clinical pain. *Pain* 154: 15–23.

Dulac, C. 2010. Brain function and chromatin plasticity. *Nature* 465: 728–35.

Einarsdottir, E., Carlsson, A., Minde, J., Toolanen, G., Svensson, O., Solders, G., Holmgren, G., Holmberg, D. and Holmberg, M. 2004. A mutation in the nerve growth factor beta gene (NGFB) causes loss of pain perception. *Human Molecular Genetics* 13: 799–805.

Ernst, J., Kheradpour, P., Mikkelsen, T. S., Shoresh, N., Ward, L. D., Epstein, C. B., Zhang, X., Wang, L., Issner, R., Coyne, M., Ku, M., Durham, T., Kellis, M. and Bernstein, B. E. 2011. Mapping and analysis of chromatin state dynamics in nine human cell types. *Nature* 473: 43–9.

Esserlind, A. L., Christensen, A. F., Le, H., Kirchmann, M., Hauge, A. W., Toyserkani, N. M., Hansen, T., Grarup, N., Werge, T., Steinberg, S., Bettella, F., Stefansson, H. and Olesen, J. 2013. Replication and meta-analysis of common variants identifies a genome-wide significant locus in migraine. *European Journal of Neurology: The Official Journal of the European Federation of Neurological Societies* 20: 765–72.

Evans, R. J., Moldwin, R. M., Cossons, N., Darekar, A., Mills, I. W. and Scholfield, D. 2011. Proof of concept trial of tanezumab for the treatment of symptoms associated with interstitial cystitis. *The Journal of Urology* 185: 1716–21.

Evertts, A. G., Zee, B. M., Dimaggio, P. A., Gonzales-Cope, M., Coller, H. A. and Garcia, B. A. 2013. Quantitative dynamics of the link between cellular metabolism and histone acetylation. *The Journal of Biological Chemistry* 288: 12142–51.

Fertleman, C. R., Baker, M. D., Parker, K. A., Moffatt, S., Elmslie, F. V., Abrahamsen, B., Ostman, J., Klugbauer, N., Wood, J. N., Gardiner, R. M. and Rees, M. 2006. SCN9A mutations in paroxysmal extreme pain disorder: Allelic variants underlie distinct channel defects and phenotypes. *Neuron* 52: 767–74.

Feuk, L., Carson, A. R. and Scherer, S. W. 2006. Structural variation in the human genome. *Nature Reviews Genetics* 7: 85–97.

Fillingim, R. B., Kaplan, L., Staud, R., Ness, T. J., Glover, T. L., Campbell, C. M., Mogil, J. S. and Wallace, M. R. 2005. The A118G single nucleotide polymorphism of the mu-opioid receptor gene (OPRM1) is associated with pressure pain sensitivity in humans. *Journal of Pain* 6: 159–67.

Fraga, M. F., Ballestar, E., Paz, M. F., Ropero, S., Setien, F., Ballestar, M. L., Heine-Suner, D., Cigudosa, J. C., Urioste, M., Benitez, J., et al. 2005. Epigenetic differences arise during the lifetime of monozygotic twins. *Proceedings of the National Academy of Sciences of the United States of America* 102: 10604–9.

Freilinger, T., Anttila, V., de Vries, B., Malik, R., Kallela, M., Terwindt, G. M., Pozo-Rosich, P., Winsvold, B., Nyholt, D. R., van Oosterhout, W. P., et al. 2012. Genome-wide association analysis identifies susceptibility loci for migraine without aura. *Nature Genetics* 44: 777–82.

Gartner, K. 1990. A third component causing random variability beside environment and genotype. A reason for the limited success of a 30 year long effort to standardize laboratory animals? *Laboratory Animals* 24: 71–77.

Geranton, S. M., Morenilla-Palao, C. and Hunt, S. P. 2007. A role for transcriptional repressor methyl-CpG-binding protein 2 and plasticity-related gene serum- and glucocorticoid-inducible kinase 1 in the induction of inflammatory pain states. *Journal of Neuroscience* 27: 6163–73.

Gibson, G. 2011. Rare and common variants: Twenty arguments. *Nature Reviews Genetics* 13: 135–45.

Graur, D., Zheng, Y., Price, N., Azevedo, R. B., Zufall, R. A. and Elhaik, E. 2013. On the immortality of television sets: 'Function' in the human genome according to the evolution-free gospel of ENCODE. *Genome Biology and Evolution* 5: 578–90.

Greenbaum, L., Tegeder, I., Barhum, Y., Melamed, E., Roditi, Y. and Djaldetti, R. 2012. Contribution of genetic variants to pain susceptibility in Parkinson disease. *European Journal of Pain* 16: 1243–50.

Guo, J. U., Ma, D. K., Mo, H., Ball, M. P., Jang, M. H., Bonaguidi, M. A., Balazer, J. A., Eaves, H. L., Xie, B., Ford, E., Zhang, K., Ming, G. L., Gao, Y. and Song, H. 2011. Neuronal activity modifies the DNA methylation landscape in the adult brain. *Nature Neuroscience* 14: 1345–51.

Hartvigsen, J., Nielsen, J., Kyvik, K. O., Fejer, R., Vach, W., Iachine, I. and Leboeuf-Yde, C. 2009. Heritability of spinal pain and consequences of spinal pain: A comprehensive genetic epidemiologic analysis using a population-based sample of 15,328 twins ages 20–71 years. *Arthritis & Rheumatology* 61: 1343–51.

Hendry, L., Lombard, Z., Wadley, A. and Kamerman, P. 2013. KCNS1, but not GCH1, is associated with pain intensity in a black southern African population with HIV-associated sensory neuropathy: A genetic association study. *Journal of Acquired Immune Deficiency Syndromes* 63: 27–30.

Henikoff, S. and Shilatifard, A. 2011. Histone modification: Cause or cog? *Trends in Genetics* 27: 389–96.

Hirschhorn, J. N. and Daly, M. J. 2005. Genome-wide association studies for common diseases and complex traits. *Nature Reviews Genetics* 6: 95–108.

Hocking, L. J., Morris, A. D., Dominiczak, A. F., Porteous, D. J. and Smith, B. H. 2012. Heritability of chronic pain in 2195 extended families. *European Journal of Pain* 16 (7): 1053–63.

Holliday, K. L., Thomson, W., Neogi, T., Felson, D. T., Wang, K., Wu, F. C., Huhtaniemi, I. T., Bartfai, G., Casanueva, F., Forti, G., Kula, K., Punab, M., Vanderschueren, D., Macfarlane, G. J., Horan, M. A., Ollier, W., Payton, A., Pendleton, N. and McBeth, J. 2012. The non-synonymous SNP, R1150W, in SCN9A is not associated with chronic widespread pain susceptibility. *Molecular Pain* 8: 72.

Imai, S., Ikegami, D., Yamashita, A., Shimizu, T., Narita, M., Niikura, K., Furuya, M., Kobayashi, Y., Miyashita, K., Okutsu, D., et al. 2013. Epigenetic transcriptional activation of monocyte chemotactic protein 3 contributes to long-lasting neuropathic pain. *Brain* 136: 828–43.

Indo, Y., Tsuruta, M., Hayashida, Y., Karim, M. A., Ohta, K., Kawano, T., Mitsubuchi, H., Tonoki, H., Awaya, Y. and Matsuda, I. 1996. Mutations in the TRKA/NGF receptor gene in patients with congenital insensitivity to pain with anhidrosis. *Nature Genetics* 13: 485–8.

International HapMap. 2003. The International HapMap Project. *Nature* 426: 789–96.

Jones, P. A. 2012. Functions of DNA methylation: Islands, start sites, gene bodies and beyond. *Nature Reviews Genetics* 13: 484–92.

Kato, K., Sullivan, P. F. and Pedersen, N. L. 2010. Latent class analysis of functional somatic symptoms in a population-based sample of twins. *Journal of Psychosomatic Research* 68: 447–53.

Kehlet, H., Jensen, T. S. and Woolf, C. J. 2006. Persistent postsurgical pain: Risk factors and prevention. *Lancet* 367: 1618–25.

Kiguchi, N., Kobayashi, Y., Maeda, T., Fukazawa, Y., Tohya, K., Kimura, M. and Kishioka, S. 2011. Epigenetic augmentation of the MIP-2/CXCR2 axis through histone H3 acetylation in injured peripheral nerves elicits neuropathic pain. *The Journal of Pharmacology and Experimental Therapeutics* 340 (3): 577–87.

Kim, H. and Dionne, R. A. 2007. Lack of influence of GTP cyclohydrolase gene (GCH1) variations on pain sensitivity in humans. *Molecular Pain* 3: 6.

Kim, H., Ramsay, E., Lee, H., Wahl, S. and Dionne, R. A. 2009. Genome-wide association study of acute post-surgical pain in humans. *Pharmacogenomics* 10: 171–9.

Kivitz, A. J., Gimbel, J. S., Bramson, C., Nemeth, M. A., Keller, D. S., Brown, M. T., West, C. R. and Verburg, K. M. 2013. Efficacy and safety of tanezumab versus naproxen in the treatment of chronic low back pain. *Pain* 154: 1009–21.

Kremeyer, B., Lopera, F., Cox, J. J., Momin, A., Rugiero, F., Marsh, S., Woods, C. G., Jones, N. G., Paterson, K. J., Fricker, F. R., Villegas, A., Acosta, N., Pineda-Trujillo, N. G., Ramirez, J. D., Zea, J., Burley, M. W., Bedoya, G., Bennett, D. L., Wood, J. N. and Ruiz-Linares, A. 2010. A gain-of-function mutation in TRPA1 causes familial episodic pain syndrome. *Neuron* 66: 671–80.

Kriaucionis, S. and Heintz, N. 2009. The nuclear DNA base 5-hydroxymethylcytosine is present in Purkinje neurons and the brain. *Science* 324: 929–30.

Lara-Astiaso, D., Weiner, A., Lorenzo-Vivas, E., Zaretsky, I., Jaitin, D. A., David, E., Keren-Shaul, H., Mildner, A., Winter, D., Jung, S., Friedman, N. and Amit, I. 2014. Immunogenetics. Chromatin state dynamics during blood formation. *Science* 345: 943–9.

Lee, J. S., Smith, E. and Shilatifard, A. 2010. The language of histone crosstalk. *Cell* 142: 682–5.

Leipold, E., Liebmann, L., Korenke, G. C., Heinrich, T., Giesselmann, S., Baets, J., Ebbinghaus, M., Goral, R. O., Stodberg, T., Hennings, J. C., et al. 2013. A de novo gain-of-function mutation in SCN11A causes loss of pain perception. *Nature Genetics* 45: 1399–404.

Liang, D. Y., Li, X. and Clark, J. D. 2013. Epigenetic regulation of opioid-induced hyperalgesia, dependence, and tolerance in mice. *Journal of Pain* 14: 36–47.

Limer, K. L., Tosh, K., Bujac, S. R., McConnell, R., Doherty, S., Nyberg, F., Zhang, W., Doherty, M., Muir, K. R. and Maciewicz, R. A. 2009. Attempt to replicate published genetic associations in a large, well-defined osteoarthritis case-control population (the GOAL study). *Osteoarthritis and Cartilage* 17: 782–9.

Ling, X., Harkness, T. A., Schultz, M. C., Fisher-Adams, G. and Grunstein, M. 1996. Yeast histone H3 and H4 amino termini are important for nucleosome assembly in vivo and in vitro: Redundant and position-independent functions in assembly but not in gene regulation. *Genes & Development* 10: 686–99.

Lister, R., Pelizzola, M., Dowen, R. H., Hawkins, R. D., Hon, G., Tonti-Filippini, J., Nery, J. R., Lee, L., Ye, Z., Ngo Q-M, Edsall, L., Antosiewicz-Bourget, J., Stewart, R., Ruotti, V., Millar aH, Thomson Ja, Ren, B. and Ecker, J. R. 2009. Human DNA methylomes at base resolution show widespread epigenomic differences. *Nature* 462: 315–22.

Liu, M., Liu, X., Zhang, Y. and Guo, S. W. 2012. Valproic acid and progestin inhibit lesion growth and reduce hyperalgesia in experimentally induced endometriosis in rats. *Reproductive Sciences* 19: 360–73.

Livshits, G., Popham, M., Malkin, I., Sambrook, P. N., Macgregor, A. J., Spector, T. and Williams, F. M. 2011. Lumbar disc degeneration and genetic factors are the main risk factors for low back pain in women: The UK Twin Spine Study. *Annals of the Rheumatic Diseases* 70: 1740–5.

MacGregor, A. J., Andrew, T., Sambrook, P. N. and Spector, T. D. 2004. Structural, psychological, and genetic influences on low back and neck pain: A study of adult female twins. *Arthritis and Rheumatology* 51: 160–7.

Markkula, R., Jarvinen, P., Leino-Arjas, P., Koskenvuo, M., Kalso, E. and Kaprio, J. 2009. Clustering of symptoms associated with fibromyalgia in a Finnish Twin Cohort. Eur *Journal of Pain* 13: 744–50.

Marth, G. T., Czabarka, E., Murvai, J. and Sherry, S. T. 2004. The allele frequency spectrum in genome-wide human variation data reveals signals of differential demographic history in three large world populations. *Genetics* 166: 351–72.

Mashayekhi, F. J., Rasti, M., Rahvar, M., Mokarram, P., Namavar, M. R. and Owji, A. A. 2012. Expression levels of the BDNF gene and histone modifications around its promoters in the ventral tegmental area and locus ceruleus of rats during forced abstinence from morphine. *Neurochemical Research* 37: 1517–23.

Maze, I., Noh, K. M. and Allis, C. D. 2013. Histone regulation in the CNS: Basic principles of epigenetic plasticity. *Neuropsychopharmacology: Official Publication of the American College of Neuropsychopharmacology* 38: 3–22.

Meijer, I. A., Vanasse, M., Nizard, S., Robitaille, Y. and Rossignol, E. 2013. An atypical case of SCN9A mutation presenting with global motor delay and a severe pain disorder. *Muscle & Nerve* 49 (1):134–8.

Minde, J., Andersson, T., Fulford, M., Aguirre, M., Nennesmo, I., Remahl, I. N., Svensson, O., Holmberg, M., Toolanen, G. and Solders, G. 2009. A novel NGFB point mutation: A phenotype study of heterozygous patients. *Journal of Neurology, Neurosurgery, and Psychiatry* 80: 188–95.

Mogil, J. S. 1999. The genetic mediation of individual differences in sensitivity to pain and its inhibition. Proceedings of the National Academy of Sciences of the United States of America 96: 7744–7751.

Mogil, J. S. 2012. Pain genetics: Past, present and future. *Trends in Genetics* 28: 258–66.

Mogil, J. S., Wilson, S. G., Bon, K., Lee, S. E., Chung, K., Raber, P., Pieper, J. O., Hain, H. S., Belknap, J. K., Hubert, L., Elmer, G. I., Chung, J. M. and Devor, M. 1999. Heritability of nociception. II: 'Types' of nociception revealed by genetic correlation analysis. *Pain* 80: 83–93.

Mogil, J. S., Sorge, R. E., LaCroix-Fralish, M. L., Smith, S. B., Fortin, A., Sotocinal, S. G., Ritchie, J., Austin, J. S., Schorscher-Petcu, A., Melmed, K., et al. 2011. Pain sensitivity and vasopressin analgesia are mediated by a gene-sex-environment interaction. *Nature Neuroscience* 14: 1569–73.

Neddermeyer, T. J., Fluhr, K. and Lotsch, J. 2008. Principle components analysis of pain thresholds to thermal, electrical, and mechanical stimuli suggests a predominant common source of variance. *Pain* 138: 286–91.

Nelson, M. R., Wegmann, D., Ehm, M. G., Kessner, D., St Jean, P., Verzilli, C., Shen, J., Tang, Z., Bacanu, S. A., Fraser, D., et al. 2012. An abundance of rare functional variants in 202 drug target genes sequenced in 14,002 people. *Science* 337: 100–4.

Nielsen, C. S., Stubhaug, A., Price, D. D., Vassend, O., Czajkowski, N. and Harris, J. R. 2008. Individual differences in pain sensitivity: Genetic and environmental contributions. *Pain* 136: 21–9.

Nilsen, K. B., Nicholas, A. K., Woods, C. G., Mellgren, S. I., Nebuchennykh, M. and Aasly, J. 2009. Two novel SCN9A mutations causing insensitivity to pain. *Pain* 143: 155–8.

Nishizawa, D., Fukuda, K., Kasai, S., Hasegawa, J., Aoki, Y., Nishi, A., Saita, N., Koukita, Y., Nagashima, M., Katoh, R., et al. 2012. Genome-wide association study identifies a potent locus associated with human opioid sensitivity. *Molecular Psychiatry* 19 (1): 55–62.

Nissenbaum, J., Devor, M., Seltzer, Z., Gebauer, M., Michaelis, M., Tal, M., Dorfman, R., Abitbul-Yarkoni, M., Lu, Y., Elahipanah, T., delCanho, S., Minert, A., Fried, K., Persson, A. K., Shpigler, H., Shabo, E., Yakir, B., Pisante, A. and Darvasi, A. 2010. Susceptibility to chronic pain following nerve injury is genetically affected by CACNG2. *Genome Research* 20: 1180–90.

Norbury, T. A., MacGregor, A. J., Urwin, J., Spector, T. D. and McMahon, S. B. 2007. Heritability of responses to painful stimuli in women: A classical twin study. *Brain* 130: 3041–9.

Nyholt, D. R., Low, S. K., Anderson, C. A., Painter, J. N., Uno, S., Morris, A. P., MacGregor, S., Gordon, S. D., Henders, A. K., Martin, N. G., et al. 2012. Genome-wide association meta-analysis identifies new endometriosis risk loci. *Nature Genetics* 44: 1355–9.

Ophoff, R. A., Terwindt, G. M., Vergouwe, M. N., van Eijk, R., Oefner, P. J., Hoffman, S. M., Lamerdin, J. E., Mohrenweiser, H. W., Bulman, D. E., Ferrari, M., Haan, J., Lindhout, D., van Ommen, G. J., Hofker, M. H., Ferrari, M. D. and Frants, R. R. 1996. Familial hemiplegic migraine and episodic ataxia type-2 are caused by mutations in the Ca2+ channel gene CACNL1A4. *Cell* 87: 543–52.

Patten, S. B., Williams, J. V., Lavorato, D. H., Modgill, G., Jette, N. and Eliasziw, M. 2008. Major depression as a risk factor for chronic disease incidence: Longitudinal analyses in a general population cohort. *General Hospital Psychiatry* 30: 407–13.

Pavlin, D. J., Sullivan, M. J., Freund, P. R. and Roesen, K. 2005. Catastrophizing: A risk factor for postsurgical pain. *The Clinical Journal of Pain* 21: 83–90.

Peleg, S., Sananbenesi, F., Zovoilis, A., Burkhardt, S., Bahari-Javan, S., Agis-Balboa, R. C., Cota, P., Wittnam, J. L., Gogol-Doering, A., Opitz, L., Salinas-Riester, G., Dettenhofer, M., Kang, H., Farinelli, L., Chen, W. and Fischer, A. 2010. Altered histone acetylation is associated with age-dependent memory impairment in mice. *Science* 328: 753–6.

Peters, M. J., Broer, L., Willemen, H. L., Eiriksdottir, G., Hocking, L. J., Holliday, K. L., Horan, M. A., Meulenbelt, I., Neogi, T., Popham, M., et al. 2013. Genome-wide association study meta-analysis of chronic widespread pain: Evidence for involvement of the 5p15.2 region. *Annals of the Rheumatic Diseases* 72: 427–36.

Platt, A., Vilhjalmsson, B. J. and Nordborg, M. 2010. Conditions under which genome-wide association studies will be positively misleading. *Genetics* 186: 1045–52.

Plomin, R. and Daniels, D. 2011. Why are children in the same family so different from one another? *International Journal of Epidemiology* 40: 563–82.

Raouf, R., Quick, K. and Wood, J. N. 2010. Pain as a channelopathy. *The Journal of Clinical Investigation* 120: 3745–52.

Reich, D. E., Cargill, M., Bolk, S., Ireland, J., Sabeti, P. C., Richter, D. J., Lavery, T., Kouyoumjian, R., Farhadian, S. F., Ward, R. and Lander, E. S. 2001. Linkage disequilibrium in the human genome. *Nature* 411: 199–204.

Reimann, F., Cox, J. J., Belfer, I., Diatchenko, L., Zaykin, D. V., McHale, D. P., Drenth, J. P., Dai, F., Wheeler, J., Sanders, F., et al. 2010. Pain perception is altered by a nucleotide polymorphism in SCN9A. *Proceedings of the National Academy of Sciences of the United States of America* 107: 5148–53.

Reyes-Gibby, C. C., Shete, S., Rakvag, T., Bhat, S. V., Skorpen, F., Bruera, E., Kaasa, S. and Klepstad, P. 2007. Exploring joint effects of genes and the clinical efficacy of morphine for cancer pain: OPRM1 and COMT gene. *Pain* 130: 25–30.

Reyes-Gibby, C. C., Spitz, M. R., Yennurajalingam, S., Swartz, M., Gu, J., Wu, X., Bruera, E. and Shete, S. 2009. Role of inflammation gene polymorphisms on pain severity in lung cancer patients. *Cancer Epidemiology, Biomarkers & Prevention* 18: 2636–42.

Robinson-Papp, J., Morgello, S., Vaida, F., Fitzsimons, C., Simpson, D. M., Elliott, K. J., Al-Lozi, M., Gelman, B. B., Clifford, D., Marra, C. M., McCutchan, J. A., Atkinson, J. H., Dworkin, R. H., Grant, I. and Ellis, R. 2010. Association of self-reported painful symptoms with clinical and neurophysiologic signs in HIV-associated sensory neuropathy. *Pain* 151: 732–6.

Savas, J. N., Toyama, B. H., Xu, T., Yates, J. R., 3rd and Hetzer, M. W. 2012. Extremely long-lived nuclear pore proteins in the rat brain. *Science* 335 (942). doi: 10.1126/science.

Smith, Z. D. and Meissner, A. 2013. DNA methylation: Roles in mammalian development. *Nature Reviews Genetics* 14: 204–20.

Solovieva, S., Leino-Arjas, P., Saarela, J., Luoma, K., Raininko, R. and Riihimaki, H. 2004. Possible association of interleukin 1 gene locus polymorphisms with low back pain. *Pain* 109: 8–19.

Sorge, R. E., Trang, T., Dorfman, R., Smith, S. B., Beggs, S., Ritchie, J., Austin, J. S., Zaykin, D. V., Vander Meulen, H., Costigan, M., et al. (2012) Genetically determined P2X7 receptor pore formation regulates variability in chronic pain sensitivity. *Nature Medicine* 18: 595–9.

Spierings, E. L., Fidelholtz, J., Wolfram, G., Smith, M. D., Brown, M. T. and West, C. R. 2013. A phase III placebo- and oxycodone-controlled study of tanezumab in adults with osteoarthritis pain of the hip or knee. *Pain* 154 (9): 1603–12.

Styrkarsdottir, U., Thorleifsson, G., Helgadottir, H. T., Bomer, N., Metrustry, S., Bierma-Zeinstra, S., Strijbosch, A. M., Evangelou, E., Hart, D., Beekman, M., et al. 2014. Severe osteoarthritis of the hand associates with common variants within the ALDH1A2 gene and with rare variants at 1p31. *Nature Genetics* 46: 498–502.

Sun, H., Maze, I., Dietz, D. M., Scobie, K. N., Kennedy, P. J., Damez-Werno, D., Neve, R. L., Zachariou, V., Shen, L. and Nestler, E. J. 2012. Morphine epigenomically regulates behavior through alterations in histone H3 lysine 9 dimethylation in the nucleus accumbens. *The Journal of Neuroscience:* 32: 17454–64.

Tajerian, M., Alvarado, S., Millecamps, M., Vachon, P., Crosby, C., Bushnell, M. C., Szyf, M. and Stone, L. S. 2013. Peripheral nerve injury is associated with chronic, reversible changes in global DNA methylation in the mouse prefrontal cortex. *PloS One* 8: e55259.

Tajerian, M., Alvarado, S., Millecamps, M., Dashwood, T., Anderson, K. M., Haglund, L., Ouellet, J., Szyf, M. and Stone, L. S. 2011. DNA methylation of SPARC and chronic low back pain. *Molecular Pain* 7: 65.

Tammimaki, A. and Mannisto, P. T. 2012. Catechol-O-methyltransferase gene polymorphism and chronic human pain: A systematic review and meta-analysis. *Pharmacogenetics and Genomics* 22: 673–91.

Tegeder, I., Costigan, M., Griffin, R. S., Abele, A., Belfer, I., Schmidt, H., Ehnert, C., Nejim, J., Marian, C., Scholz, J., et al. 2006. GTP cyclohydrolase and tetrahydrobiopterin regulate pain sensitivity and persistence. *Nature Medicine* 12: 1269–77.

Telese, F., Gamliel, A., Skowronska-Krawczyk, D., Garcia-Bassets, I. and Rosenfeld, M. G. 2013. 'Seq-ing' insights into the epigenetics of neuronal gene regulation. *Neuron* 77: 606–23.

Tennessen, J. A., Bigham, A. W., O'Connor, T. D., Fu, W., Kenny, E. E., Gravel, S., McGee, S., Do, R., Liu, X., Jun, G., et al. 2012. Evolution and functional impact of rare coding variation from deep sequencing of human exomes. *Science* 337: 64–9.

Thomas, E., Silman, A. J., Croft, P. R., Papageorgiou, A. C., Jayson, M. I. and Macfarlane, G. J. 1999. Predicting who develops chronic low back pain in primary care: A prospective study. *British Medical Journal* 318: 1662–7.

Tochiki, K. K., Cunningham, J., Hunt, S. P. and Geranton, S. M. 2012. The expression of spinal methyl-CpG-binding protein 2, DNA methyltransferases and histone deacetylases is modulated in persistent pain states. *Molecular Pain* 8: 14.

Tran, L., Chaloner, A., Sawalha, A. H., and Greenwood Van-Meerveld, B. 2012. Importance of epigenetic mechanisms in visceral pain induced by chronic water avoidance stress. *Psychoneuroendocrinology* 24 (5): 479–86.

Uchida, H., Matsushita, Y. and Ueda, H. 2013. Epigenetic regulation of BDNF expression in the primary sensory neurons after peripheral nerve injury: Implications in the development of neuropathic pain. *Neuroscience* 240: 147–54.

Valdes, A. M., Arden, N. K., Vaughn, F. L., Doherty, S. A., Leaverton, P. E., Zhang, W., Muir, K. R., Rampersaud, E., Dennison, E. M., Edwards, M. H., Jameson, K. A., Javaid, M. K., Spector, T. D., Cooper, C., Maciewicz, R. A. and Doherty, M. 2011. Role of the Nav1.7 R1150W amino acid change in susceptibility to symptomatic knee osteoarthritis and multiple regional pain. *Arthritis Care & Research* 63: 440–4.

Vargas-Alarcon, G., Alvarez-Leon, E., Fragoso, J. M., Vargas, A., Martinez, A., Vallejo, M., Martinez-Lavin, M. 2012. A SCN9A gene-encoded dorsal root ganglia sodium channel polymorphism associated with severe fibromyalgia. *BMC Musculoskeletal Disorders* 13: 23.

Walter, C. and Lotsch, J. 2009. Meta-analysis of the relevance of the OPRM1 118A>G genetic variant for pain treatment. *Pain* 146: 270–5.

Wang, Y., Liu, C., Guo, Q.-L., Yan J-Q, Zhu, X.-Y., Huang, C.-S. and Zou, W.-Y. 2011. Intrathecal 5-azacytidine inhibits global DNA methylation and methyl- CpG-binding protein 2 expression and alleviates neuropathic pain in rats following chronic constriction injury. *Brain Research* 1418: 64–9.

Wang, Z., Zang, C., Cui, K., Schones, D. E., Barski, A., Peng, W. and Zhao, K. 2009. Genome-wide mapping of HATs and HDACs reveals distinct functions in active and inactive genes. *Cell* 138: 1019–31.

Williams, F. M., Spector, T. D. and MacGregor, A. J. 2010. Pain reporting at different body sites is explained by a single underlying genetic factor. *Rheumatology* 49: 1753–5.

Williams, F. M., Bansal, A. T., van Meurs, J. B., Bell, J. T., Meulenbelt, I., Suri, P., Rivadeneira, F., Sambrook, P. N., Hofman, A., Bierma-Zeinstra, S., Menni, C., Kloppenburg, M., Slagboom, P. E., Hunter, D. J., Macgregor, A. J., Uitterlinden, A. G. and Spector, T. D. 2012. Novel genetic variants associated with lumbar disc degeneration in northern Europeans: A meta-analysis of 4600 subjects. *Annals of the Rheumatic Diseases* 72: 1141–48.

Wu, S. C. and Zhang, Y. 2010. Active DNA demethylation: Many roads lead to Rome. *Nat Nature Reviews Molecular Cell Biology* 11: 607–20.

Wu, H. and Zhang, Y. 2011. Mechanisms and functions of Tet protein-mediated 5-methylcytosine oxidation. *Genes & Development* 25: 2436–52.

Wu, H., Coskun, V., Tao, J., Xie, W., Ge, W., Yoshikawa, K., Li, E., Zhang and Y. and Sun, Y. E. 2010. Dnmt3a-dependent nonpromoter DNA methylation facilitates transcription of neurogenic genes. *Science* 329: 444–8.

Xie, W., Barr, C. L., Kim, A., Yue, F., Lee, A. Y., Eubanks, J., Dempster, E. L. and Ren, B. 2012. Base-resolution analyses of sequence and parent-of-origin dependent DNA methylation in the mouse genome. *Cell* 148: 816–31.

Yang, Y., Wang, Y., Li, S., Xu, Z., Li, H., Ma, L., Fan, J., Bu, D., Liu, B., Fan, Z., Wu, G., Jin, J., Ding, B., Zhu, X. and Shen, Y. 2004. Mutations in SCN9A, encoding a sodium channel alpha subunit, in patients with primary erythermalgia. *Journal of Medical Genetics* 41: 171–4.

Zee, B. M., Levin, R. S., Dimaggio, P. A. and Garcia, B. A. 2010. Global turnover of histone post-translational modifications and variants in human cells *Epigenetics & Chromatin* 3: 22.

Zhang, Z., Cai, Y. Q., Zou, F., Bie B. and Pan, Z. Z. 2011. Epigenetic suppression of GAD65 expression mediates persistent pain. *Nature Medicine* 17: 1148–55.

Zhu, X. Y., Huang, C. S., Li, Q., Chang, R. M., Song, Z. B., Zou, W. Y. and Guo, Q. L. 2012. p300 exerts an epigenetic role in chronic neuropathic pain through its acetyltransferase activity in rats following chronic constriction injury (CCI). *Molecular Pain* 8: 84.

Zhu, X. Y., Huang, C. S., Li, Q., Guo, Q. L., Wang, Y., He, X. and Liao, J. 2013. Temporal distribution of p300/CBP immunoreactivity in the adult rat spinal dorsal horn following chronic constriction injury (CCI). *Cellular and Molecular Neurobiology* 33: 197–204.

CHAPTER 6

The cannabinoid system and its role in nociception

Massimiliano Beltramo

A brief history of Cannabis

The cannabinoids owe their name to their presence, in significant amounts, in the plants of the genus *Cannabis*. Cannabinoids are produced from most parts of these plants, but they are highly concentrated in small droplets secreted by glands that are particularly abundant on the bracts of the female flower. Different preparations (known as kief, hashish, hash oil, Marijuana, etc.) containing a variable quantity of pharmacologically active cannabinoids can be obtained by processing the female flower.

Three putative varieties of cannabis have been described – *Cannabis sativa*, *Cannabis indica* and *Cannabis ruderalis* – all of them indigenous to Central and South Asia. Hence, not surprisingly, the first information to reach us about the use of cannabis was written down in ancient Chinese and Indian manuscripts. These manuscripts date back to the twenty-eighth century BC and relate to the use of hemp to produce fibre and textile. Actually hemp textile from around 4000–5000 BC have been collected from Northern China, confirming the authenticity of this report.

Archeological excavations in the Xinjiang-Uighur region unveiled a wooden bowl and a leather basket filled with *Cannabis sativa* achenes (seeds) in the 2,700-year-old grave of a shaman. The absence of plant parts from the male cannabis, which is less psychoactive, would suggest a possible pharmacological use of the plant (Russo et al., 2008). This would be the oldest record for such a use. Before the Greeks colonised the Euxine (Black Sea) area and got in contact with the Scythians, the ancient western world was unaware of *Cannabis sativa*'s existence and multiple uses. Cannabis seeds and hemp clothing are present in Scythian archeological sites of about 2,400–2,500 years ago. The presence of censers in which cannabis seeds would be burnt for their intoxicating properties closely matches the description of the use of cannabis made by Herodotus in his *Histories*, which was written in the fifth century BC (Herodotus' dates are unknown, but he is supposed to have died in the 420s BC). Surprisingly,

An Introduction to Pain and its Relation to Nervous System Disorders, First Edition. Edited by Anna A. Battaglia.
© 2016 John Wiley & Sons, Ltd. Published 2016 by John Wiley & Sons, Ltd.

Herodotus' description of the potentially psychoactive properties of cannabis remained for a long time the only one of its kind in Western Europe.

On the other hand, the use of cannabis for its medical properties is reported in various ancient sources coming from different eastern countries (India, Egypt, China, Persia, etc.) and goes back as far as 2000 BC. Different forms of cannabis preparation were used to treat a variety of conditions. Interestingly, in the Chinese herbal Pen-ts'ao cannabis, often mixed with a wine preparation, is mainly prescribed for its pain-relieving properties, even though it was also recommended for constipation, gout, malaria, rheumatism, and menstrual problems.

The therapeutic use of cannabis was mostly neglected in western medicine. It was only in 1839 that an article by O'Shaughnessy, which described the history of the use of cannabis products in the east, aroused interest in cannabis-based medical treatments (O'Shaugnessy, 1839). Experiments conducted by O'Shaughnessy led him to believe that cannabis could relieve pain and act as a muscle relaxant and anticonvulsant. O'Shaughnessy's work called attention to the subject, and between the end of the nineteenth century and the beginning of the twentieth century several over-the-counter medicines containing cannabis became available to treat a wide variety of symptoms and diseases (sleeplessness, neuralgia, dysmenorrhea, etc.). This blooming era of cannabis medical use was followed by a rapid decline and ended in the thirties, when laws were enforced banning the use of cannabis in medical preparations. It is only relatively recently, thanks to new compelling scientific results, that interest in the medical application of natural and synthetic cannabinoids was resurrected.

Discovery of the endocannabinoid system

Compared to the pharmacological effect of cannabinoids, which has been known for a long time, the discovery of an endogenous cannabinoid system constituted by endogenous ligands and their cognate receptors is extremely recent. The first step towards this discovery was the purification of the main active compound in marijuana: tetrahydrocannabinol (THC) (Mechoulam and Gaoni, 1965). After its identification, THC was used to perform a more precise assessment of the pharmacological effect of cannabinoids; in 1974 it was employed as a scaffold for the production of the first synthetic cannabinoids, CP55940, created by Pfizer Inc. The synthesis of CP55940 was instrumental in the discovery of the presence of selective cannabinoid binding sites in the brain. A series of *in vitro* experiments performed with the help of tritium-labelled CP55940 provided compelling evidence for the existence of a high-affinity, stereoselective and pharmacologically distinct cannabinoid receptor (Devane et al., 1988). This seminal discovery challenged the idea – common at the time – that, since cannabinoids are lipophylic molecules, they could insert themselves into the cell membrane and induce their effects by modifying the chemico-physical properties of the plasma membrane. Shortly afterwards, the first cannabinoid receptor (CB1) was cloned, definitively confirming that cannabinoids activate a classic G protein-coupled, seven-trans-membrane domain receptor (Matsuda et al., 1990). Three years later a second cannabinoid receptor (CB2) was identified and cloned from spleen macrophages. This promoted the hypothesis of two distinct cannabinoid receptors: CB1, mainly located in the central nervous system (CNS), and CB2, located in the immune system (Munro, Thomas and Abu-Shaar, 1993). More recent data portray a much less clear-cut situation, indicating a more widespread distribution of both receptors.

In 2004, a patent from Astra-Zeneca suggested that the orphan receptor GPR55 would be a third cannabinoid receptor. Later studies either indicated lysophosphatidylinositol as GPR55 ligand (Oka et al., 2007) or confirmed GPR55 as a cannabinoid receptor (Lauckner et al., 2008); this situation led to a confused pharmacological profile for this receptor. According to the classification of the International Union of Basic and Clinical Pharmacology (IUPHAR), GPR55 is still assigned to orphan receptors of class A, and the proposed ligand is lysophosphatidylinositol. Given the ambiguity of GPR55's status, this receptor will not be further considered in this chapter.

The discovery of cannabinoid receptors prompted more research for possible endogenous ligands. This research met with success in 1992, when a ligand named anandamide (arachidonoylethanolamine) was identified. The word 'anandamide' represents the fusion of the Sanskrit word *ananda* (bliss) with the English '-amide' – one of the chemical components of the endogenous ligand (Devane et al., 1992). A few years later another fatty acid derivative, 2-arachidonylglycerol (2-AG), was identified as a second endogenous cannabinoid ligand (Mechoulam et al., 1995) (see Figure 6.1). Additional lipid derivatives similar to anandamide have since been identified: virodhamine, N-arachydonoyldopamine, noladin ether, oleoylethanolamide and palmytoylethanolamide. However, they probably have a minor function in endocannabinoid signalling and some of them (oleoylethanolamide and palmytoylethanolamide) realise their main action through non-cannabinoid receptors. Additional components of the endocannabinoid system – the enzymes involved in the synthesis of anandamide (Di Marzo et al., 1994; Cadas et al., 1996; Hillard et al., 1997) and 2-AG (Stella, Schweitzer and Piomelli, 1997),

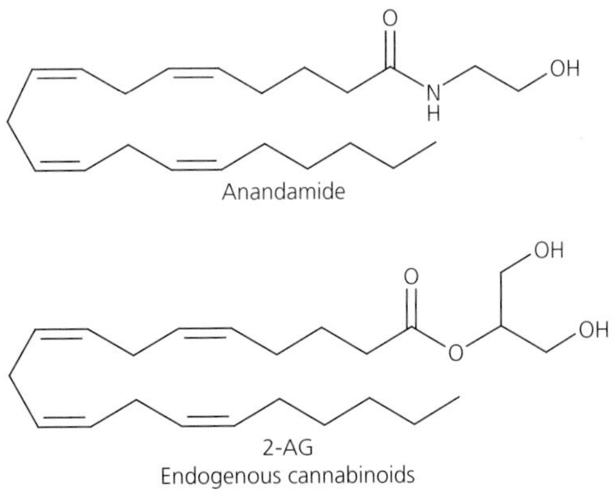

Anandamide

2-AG
Endogenous cannabinoids

Figure 6.1 Chemical structure of the two most studied endocannabinoids. Both molecules are derived from a lipid precursor and share an arachidonoyl moiety linked to either an ethanolamine (arachidonoylethanolamine, anandamide or AEA) or a glycerol (2-arachidonylglycerol, 2-AG).

as well as the systems involved in their inactivation (Di Marzo et al., 1994; Beltramo et al., 1997) – were discovered some years later. Thus, in less than ten years, all the relevant components of the endogenous cannabinoid system were identified and several pharmacological tools were made available for their study.

Cannabinoid receptor intracellular signalling

CB1 and CB2 are classic seven-transmembrane domain receptors that belong to class A of the G protein-coupled receptor (GPCR) superfamily. The main intracellular pathway triggered by both receptors is the activation of $G_{i/o}$ trimeric protein, followed by a decrease in cAMP intracellular level. Stimulation of the cannabinoid receptor elicits an overall inhibitory effect on neuronal activity. CB1 was reported to couple also to G_s protein (Glass and Felder, 1997), but this observation remains controversial (Mukhopadhyay et al., 2000). In addition to the main signalling pathway, other intracellular second-messenger cascades activated by CB1 have been described (Figure 6.2). Several calcium channel types (N-type, P/Q type and L-type) (Mackie et al., 1995; Mackie and Hille, 1992; Straiker et al., 1999) are inhibited and G protein-coupled inwardly rectifying potassium channels (GIRKs) are activated upon CB1 stimulation (Mackie et al., 1995), leading to a hyperpolarising effect (Figure 6.2). On the other hand, intracellular calcium mobilisation associated to CB1 stimulation by 2-AG or WIN55212–2 was also reported (Sugiura et al., 1996, 1997; Lauckner, Hille and Mackie, 2005). Pending on cell type, this effect was suggested to be mediated either by a Gi/o βγ subunit, through the activation of phospholipase C (PLC), or by a switch to a Gq/11 protein coupling of the receptor. The activation of this different pathway seems to be compound-specific, as it was observed with WIN55212–2 – but not with other agonists, which were unable to stimulate intracellular calcium mobilisation (Lauckner et al., 2005).

Another important pathway triggered by CB1 is that of MAP kinase leading to the phosphorylation of ERK1/2 (Howlett, 2005). It has been proposed that CB1-induced activation of MAP kinase is mediated by several mechanisms such as G_i βγ subunit, small G protein Raf-1, phosphatidylinositol 3-kinase, inhibition of adenylyl cyclase and protein kinase A. The stimulation of the MAP kinase cascade is involved, through modulation of gene expression, in regulating several important cellular functions such as differentiation, movement and proliferation.

There is evidence that β-arrestins could be involved in regulating CB1 signalling by acting on desensitisation; however, their role in the direct modulation of intracellular pathways is less clear.

Like CB1, CB2 has also been linked to the stimulation of multiple second-messenger pathways (Figure 6.2). An activation of ERK1/2 phosphorylation has been demonstrated in both recombinant and native expression systems (Shoemaker et al., 2005; Bouaboula et al., 1996; Correa et al., 2005; Carrier et al., 2004). Interestingly, this effect has been reported in microglia/macrophage cell lines and in neural progenitor cells (Palazuelos et al., 2006; Carrier et al., 2004), which suggests the presence of CB2 receptors in the central nervous system. CB2 also has an influence on NFkB, an important factor regulating inflammatory responses. Both the activation and the inhibition of this pathway have been reported after CB2 stimulation and, in some cases, a biphasic response has been

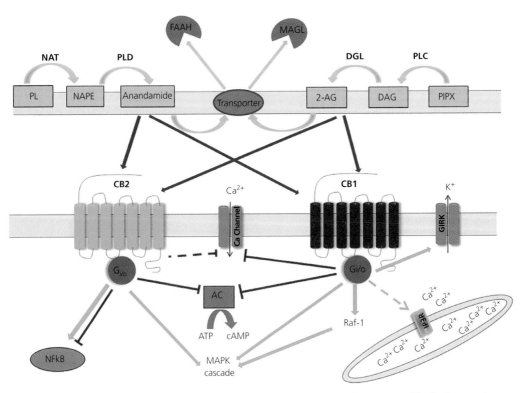

Figure 6.2 Schematic representation of the endocannabinoids system. This is a simplified scheme where only the main pathways are represented. The endocannabinoids anandamide and 2-AG are generated in a two-step process. Anandamide is produced through the action of a pospholipase D (PLD) that cleaves the N-arachydonoyl-phosphatidylethanolamine (NAPE) synthesized by an N-acyl transferase (NAT) from membrane phospholipids (PL). 2-AG is produce through the transformation of membrane phosphoinositides (PIPX) in 1,2-diacylglycerole (DAG) by the action of phospholipase C (PLC) and by the ensuing catalytic action of DAG-lipase (DLG), which produces arachidonic acid and 2-AG. Endocannabinoids are released into the extracellular space, where they could be taken up by the endocannabinoid transporter and inactivated by fatty acid amino hydrolase (FAAH, Anandamide) or by monoacylglyceride lipase (MAGL, 2-AG). Alternatively, endocannbinoids interact with the cannabinoid receptors (CB1 and CB2) and trigger various intracellular events. Activation of both receptors induces an inhibition of adenylate cyclase (AC) and a reduction of the intracellular cAMP level. Another common pathway activated by CB1 and CB2 is that of the MAPK cascade, where CB1 acts through different mechanisms to induce MAPK phosphorylation. CB1 activation reduces the extracellular calcium flux at various calcium channels, whereas for CB2 this is not firmly established. Conversely, the potassium efflux mediated by G protein-coupled inwardly rectifying potassium channel (GIRK) is enhanced through CB1 stimulation that also enhances intracellular calcium mobilization, even though this last process could be a compound specific effect and not a common characteristic of CB1 agonists. Stimulation or inhibition of NFkB could be triggered by CB2 activation and could play a role in modulating inflammatory responses.

observed it consists of an increase followed by a decrease (Derocq et al., 2000). Other pathways involved in CB2 intracellular signalling are AKT/PKB, PI3K/AKT, Ca^{2+}, JNK, PKA, NF-AT, CREB/ATF, JAK/STAT1, Caspase3 and 8, and ceramide; but our knowledge about their relevance in CB2 signalling is for the most part limited.

Endocannabinoids synthesis and degradation

As mentioned above, anandamide and 2-AG are the two main endocannabinoids. Anandamide originates in phosphatidyl lipids and is formed in a two-step process (Figure 6.2). The first step is the generation of N-arachydonoyl-phosphatidylethanolamine (NAPE) through a calcium- and cAMP-dependent process, catalysed by the enzyme N-acyltransferase (NAT) (Cadas et al., 1996). This step is followed by NAPE cleavage through a specific phospholipase D, which generates anandamide and phosphatidic acid. Anandamide is thought to be produced upon demand and immediately released (Piomelli et al., 1998). Like anandamide, 2-AG, too, is produced through the sequential activation of two enzymes (Figure 6.2). Membrane phosphoinositides are hydrolysed by phospholipase C (PLC) so as to produce 1,2-diacylglycerol (DAG), which, under the action of DAG-lipase, yields 2-AG (Stella, et al., 1997). Alternative pathways have also been proposed; these involve for example the activity of phospholipase A1 and lyso-PLC (Piomelli, 2003). 2-AG could have different roles, being either an end product or an intermediate in different lipid metabolic pathways.

Anandamide and 2-AG are highly lipophylic molecules. It has been suggested that they could interact with cannabinoid receptors by diffusion in the plasma membrane without leaving the cell where they are produced. Even though this remains a distinct possibility, experimental evidence indicates that endocannabinoids could cross the intercellular space and act on relatively distant receptors. The extracellular mobility of endocannabinoids would be facilitated by carrier proteins such as albumin (in the blood) or lipocaline (in the brain), which have been shown to bind anandamide (Bojesen and Hansen, 2003; Ruiz et al., 2013). Regardless of these hypotheses, the question of endocannabinoid transport in the extracellular fluids remains open.

The endocannabinoids' inactivation relies on a mechanism that couples internalisation and degradation (Figure 6.2). Anandamide is removed from the extracellular space by a process that has the characteristics of carrier-mediated facilitated diffusion in that it is saturable, selective (capable of recognising anandamide and 2-AG from other related molecules), cellular energy-independent, and inhibited in a competitive and stereoselective manner (Piomelli et al., 1999; Beltramo et al., 1997; Hillard, Edgemond and Campbell, 1995; Hillard and Jarrahian, 2003). Since the pharmacological characterisation of this process, the many attempts that have been made to clone endocannabinoid transporters have remained unsuccessful. In the absence of a molecular identity of these transporters, it has been proposed that the enzyme that inactivates anandamide, fatty acid amino hydrolase (FAAH), could be the driving force for anandamide internalisation (Deutsch et al., 2001). Hence the cellular mechanism of anandamide uptake remains controversial. Notwithstanding the internalisation mechanism, once inside the cell anandamide is cleaved by FAAH into ethanolamide and arachidonic acid, terminating its endocannabinoid activity. FAAH is distributed broadly into the CNS where it is often observed postsynaptically to CB1-expressing fibres. This suggests a role for it in mediating retrograde endocannabinoid signalling.

Studies on 2-AG are more limited but suggest that 2-AG and anandamide employ a common carrier-mediated internalisation mechanism (Beltramo and Piomelli, 2000). Following internalisation, 2-AG is converted into fatty acid and glycerol by a monoacylglyceride lipase that completes its deactivation (Goparaju et al., 1999; Dinh et al., 2002).

Pharmacological tools for investigating the endocannabinoid system

A brief account will be presented here of the extremely vast array of pharmacological tools that have been designed for modulating the activity of the cannabinoid system. Several companies and academic groups are engaged in the creation of selective pharmacological tools and a wide range of new chemical structures capable of binding and activating or inhibiting cannabinoid receptors or of interfering with the internalisation and degradation of endocannabinoids were discovered in the past twenty to twenty-five years. Only some of the most used molecules are mentioned here; for more extensive and detailed information the reader could consult several specific reviews (see Beltramo, 2009; Yang, Wang and Xie, 2012; Han et al., 2013; Riether, 2012).

Agonists

Natural cannabinoid, Δ^9-THC, and most of the first-generation synthetic cannabinoids (CP55940, WIN55212-2, HU210, etc.) have a limited selectivity and show a very high affinity for both cannabinoid receptors (Figure 6.3). Nevertheless, they are still largely used in cannabinoid studies and these molecules have a well-characterised pharmacological profile.

Different scaffolds were used as a starting point in the generation of selective ligands: THC, anandamide, or new synthetic structures (Figure 6.4). Selective CB2 agonists were obtained through the modification of both classic and non-classic cannabinoids. (Non-classic cannabinoids are bicyclic or tricyclic analogues of THC that lack the tricyclic ring constraint or possess a heterocyclic replacement of the pyranil ring.) JWH056, JWH133 (Huffman et al., 1999), L759633 and L759656 (Gareau et al., 1996) are all derived from classic cannabinoids and have a Ki for CB2 comprised between 3 and 32 nM (as compared to a CB1 Ki > 650 nM). Similar affinity and selectivity were obtained with non-classic cannabinoids such as L759632 and HU308 (Hanus et al., 1999). Several other selective agonists were the result of high throughput screening hit optimisation (L768242, Alias GW405833), GW842166X, AM1241, A796260, A836339, GRC10622, GRC10693, ADP371, etc.) and led to improved selectivity for CB2.

Conversely, modification of anandamide made possible the generation of molecules with a better affinity (more than 300 fold) for CB1 (Figure 6.4) such as arachidonyl-2'-chloroethylamide (ACEA) and arachidonylcyclopropylamide (ACPA) (Hillard et al., 1999). The effort to identify CB1-selective agonists has probably been limited by the consideration that most of cannabinoids psychotropic effects are attributed to CB1 activation.

Another interesting molecule that has been obtained by anandamide modification is (R)methanandamide, which has a modest selectivity for CB1 (about 40 time versus CB2) but has the advantage of being more resistant to hydrolytic degradation (Abadji et al., 1994). (R)methanandamide has been often used in *in vivo* experiments because of this feature.

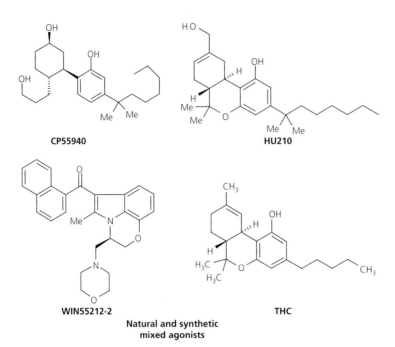

CP55940

HU210

WIN55212-2

THC

Natural and synthetic mixed agonists

Figure 6.3 Chemical structures of some mixed cannabinoid agonists. Tetrahydrocannabinol (THC) is a natural cannabinoid and the major psychoactive compound present in marijuana. HU210 is a synthetic derivative of THC with extremely high affinity for both CB1 and CB2 receptors. CP55940 is a non-tricyclic analogue of THC. WIN55212–2 is the prototype of cannabinoid ligands based on an aminoalkylindole scaffold.

Antagonists and inverse agonists

Soon after the discovery of CB1 and CB2, Sanofi identified two antagonists – SR141716A (Rinaldi-Carmona et al., 1994) and SR144528 (Rinaldi-Carmona et al., 1998) – selective for CB1 and CB2 respectively (Figure 6.5). These two molecules have been the reference antagonists in the field and have been broadly used to assign cannabinoid agonist effect to CB1 or CB2. Since then additional antagonists have been identified as being selective for either CB1: AM281 (Gifford et al., 1997), AM6545 (Tam et al., 2010), JD5037 (Tam et al., 2012); or CB2: AM630 (Pertwee et al., 1995), JTE907 (Iwamura et al., 2001), SCH336 (Lunn et al., 2006), and SCH 414319 (Lunn et al., 2008), but their use has been more limited. Noteworthy other CB1 antagonists (rimonabant, taranabant, otenabant, surinabant, AVE1625 and BMS646256) have been developed and used in clinical trials for obesity treatment, but not as a tool in preclinical studies on nociception.

It should be mentioned that most antagonists are actually inverse agonists. Inverse agonism is linked to a receptor's constitutive activity. Constitutive activity is the capacity of receptors to trigger intracellular signalling in the absence of an agonist. An inverse agonist is a molecule that inhibits constitutive activity, forcing the receptor into a neutral state and thus blocking its spontaneous signalling. By definition, a classic (neutral) antagonist has no effect per se on receptor activity but blocks the effect of agonists. CB1 or CB2 neutral antagonists have been described, but a general consensus is lacking about their pharmacological profile.

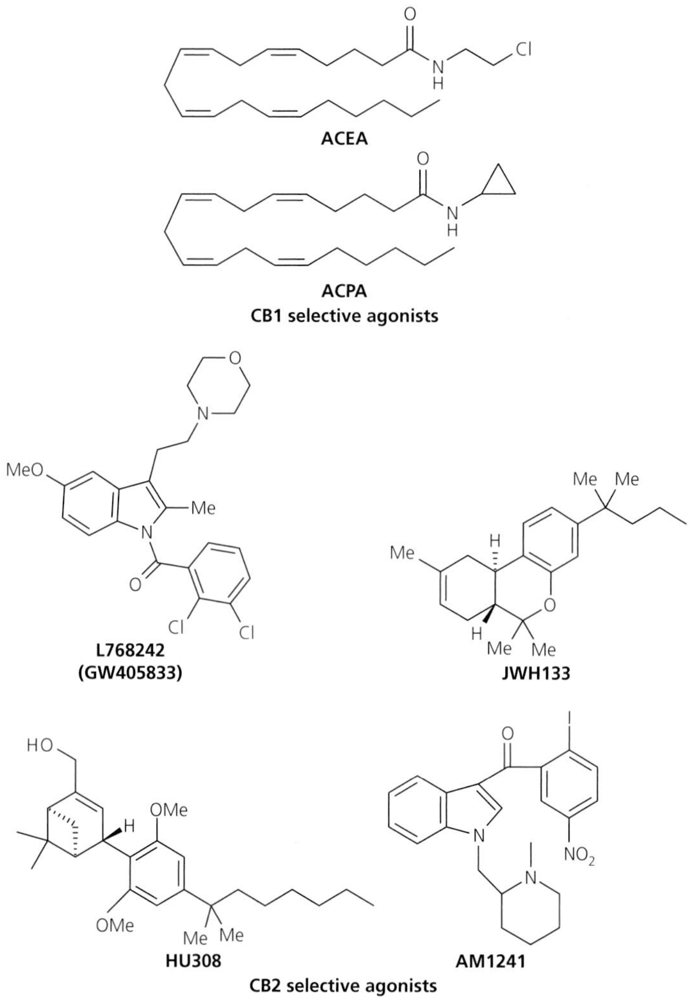

Figure 6.4 Chemical structures of some cannabinoid selective agonists. CB1-selective agonists were obtained through the modification of the ethanolamine portion of anandamide either by introducing a chlorine (arachidonyl-2′-chloroethylamide, ACEA) or a cyclopropyl (arachidonylcyclopropylamide, ACPA). The CB2-selective agonists presented in the figure were synthesized through the modification of THC (JWH133 and HU308) or aminoalkylindole scaffold (L768242 and AM1241).

Modulators of transport and degradation

Inhibitors of endocannabinoids' cellular internalisation can be divided into two main classes: aromatic acylamide derivatives such as AM404 (Figure 6.5), from which were later developed VDM11, OMDM1, OMDM2, UCM707, and AM1172; and carbamoyl-tetrazole such as LY2183240, which is, however, also a potent FAAH inhibitor. Inhibitors of endocannabinoid internalisation are active on both anandamide and 2-AG because of their common internalisation system. Conversely, specific inhibition of the endocannabinoid

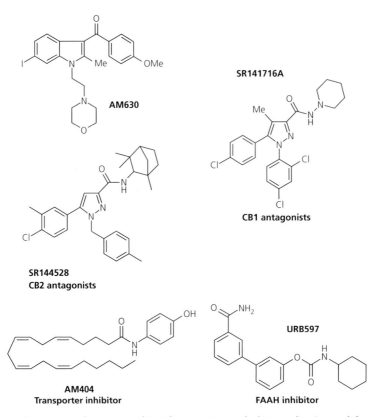

Figure 6.5 Chemical structures of some cannabinoid antagonists and of internalization and degradation inhibitors. SR141716A and SR144528 have a similar pyrazole-based structure, but they are selective for CB1 or CB2 respectively. Conversely, the CB2-,selective antagonist AM630 was obtained through selected modifications of an aminoalkylindole structure. The transporter inhibitor AM404 was created through the substitution of the ethanolamide of anandamide by an aromatic ring, whereas the FAAH inhibitor URB597 has a structure unrelated to anandamide.

degrading enzymes FAAH and MAG lipase should result in some degree of selectivity. Attention has focused mainly on FAAH, but a few MAG lipase inhibitors have also been developed. The first FAAH inhibitors were based on anandamide mimics capable of inhibiting the catalytic process. These are reversible inhibitors such as OL-135. Further modification led to the synthesis of URB597 (and later URB937), an almost irreversible inhibitor, considered nowadays the standard FAAH inhibitor (Figure 6.5). The most recent generation of FAAH inhibitors includes PF750, PF04457845 and JNJ1661010. For more details, the reader could consult recent reviews on the subject (Di Marzo, 2008; Roques, Fournie-Zaluski and Wurm, 2012).

The first MAG lipase inhibitor described was URB602, a molecule derived from FAAH inhibitor URB597, which is not, however, particularly selective. A second inhibitor (JZL-184), which showed good selectivity, was later developed from a completely different chemical structure (Long et al., 2009).

Cannabinoids and nociception

Endocannabinoid system localisation and its relation to the pain circuitry

The distribution of the different components of the cannabinoid system is an important element in our understanding of its involvement in nociception modulation. Information on the different components is, however, heterogeneous in quality and extent, due to the technical difficulties encountered in the study of lipid neurotransmitters.

Data on CB1 localisation, obtained using different techniques – binding, immunocytochemistry, in situ hybridisation and real-time polymerase chain reaction (PCR) – show that CB1 expression is quite widespread in the brain. The first data, obtained in binding experiments by using CP55950, indicated a high level of cannabinoid receptor in the basal ganglia, cerebral cortex, cerebellum and hippocampus, whereas medium to low levels were observed in the central gray, hypothalamus, spinal cord, reticular formation and pons (Herkenham et al., 1990). These results generally agree well with those of immunocytochemistry and messenger RNA distribution studies, which use *in situ* hybridisation. Together these data show that CB1 is present in several neuroanatomical structures involved in pain processing: the somatosensory cortex, the amygdala, rostral ventromedial medulla (RVM), the periaqueductal gray (PAG), the spinal trigeminal tract and nucleus, the dorsal horn and lamina X of the spinal cord, and dorsal root ganglia (DRG) (Tsou et al., 1998a; Hohmann and Herkenham, 1999; Salio et al., 2002; Agarwal et al., 2007). According to location and morphology, many, though not all, CB1-like immunoreactive neurons appear to be GABAergic.

The CB2 receptor was originally described as mainly localised in cells of the immune system (Munro et al., 1993). More recently its presence in the nervous system has been reported by different groups (Wotherspoon et al., 2005; Maresz et al., 2005; Zhang et al., 2003; Beltramo et al., 2006; Van Sickle et al., 2005; Onaivi et al., 2008). Nonetheless, the extent of CB2 distribution in the CNS is highly controversial, indications ranging from very widespread presence to complete absence. In this respect it should be mentioned that most antibodies directed against CB2 suffer from specificity problems (Baek et al., 2013; Brownjohn and Ashton, 2012). At present the most consistent data indicate a limited presence in a few localised regions such as the dorsal horn of the spinal cord and DRG (Sagar et al., 2005; Beltramo et al., 2006; Wotherspoon et al., 2005; Elmes et al., 2004).

Factors released by activated glia participate in the mediation of pathological pain; and the presence of the CB2 receptor in activated microglia in certain pathological conditions has been reported (Zhang et al., 2003; Beltramo et al., 2006; Stella, 2010) – as well as the expression of CB1 in astrocytes and microglia cells in culture (Zhang et al., 2000; Rodriguez, Mackie and Pickel, 2001; Waksman et al., 1999; Walter et al., 2003).

Due to the endocannabinoids' lipidic nature, their cellular localisation could not be studied through classic techniques such as in situ hybridisation or immunohistochemistry. Their localisation has been inferred from the expression of degrading enzymes and cannabinoid receptors, through microdyalisis studies and through the use of pharmacological tools capable of enhancing their concentration at the site of production by reducing internalisation or

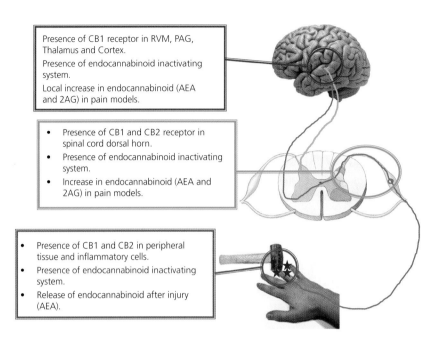

Presence of CB1 receptor in RVM, PAG, Thalamus and Cortex.
Presence of endocannabinoid inactivating system.
Local increase in endocannabinoid (AEA and 2AG) in pain models.

- Presence of CB1 and CB2 receptor in spinal cord dorsal horn.
- Presence of endocannabinoid inactivating system.
- Increase in endocannabinoid (AEA and 2AG) in pain models.

- Presence of CB1 and CB2 in peripheral tissue and inflammatory cells.
- Presence of endocannabinoid inactivating system.
- Release of endocannabinoid after injury (AEA).

Figure 6.6 Nociception and the endocannabinoid system. The figure highlights the main sites were components of the endocannabinoids system are present along the nociceptive pathway. From the scheme it is evident that the cannabinoid system is located in key areas for the modulation of the nociceptive input, from its origin at the periphery to its elaboration as conscious perception in the higher brain centres.

degradation. FAAH expression was observed in some brain regions related to nociception perception and processing, such as the superficial laminae of the spinal cord dorsal horn, the Lissauer tract and the ventral posterior lateral nucleus of the thalamus (Tsou et al., 1998b; Egertova, Cravatt and Elphick, 2003). FAAH has also been reported in peripheral tissues, where it could play a role in endocannabinoids inactivation in inflammatory processes. In situ hybridisation and immunocytochemistry for MAG lipase revealed the presence of FAAH in various brain regions, including the thalamus (Dinh et al., 2002). By coupling microdialysis with liquid chromatography or mass spectrometry, it was possible to measure an increase in anandamide in the PAG after formalin injection in the paw; this result indicated a direct connection between nociception and endocannabinoid production (Walker et al., 1999). Taken together, these data corroborate the idea that both anandamide and 2-AG are probably synthesised and released in sites that are critical to the control of nociception. Figure 6.6 represents schematically the presence of the different components of the cannabinoid system in relation to the main pathways and anatomical sites involved in pain perception and processing.

Evidence of cannabinoid antinociceptive effects

A large body of evidence concerning the antinociceptive effects of cannabinoids is available. These data were obtained in several pain models by using a wide array of pharmacological tools or genetic approaches. A comprehensive review of the related literature is out of the scope of the present chapter; to illustrate the antinociceptive effect of cannabinoids, only some examples will be presented.

A few general considerations are a necessary premise to this section. Initial studies on the behavioural and physiological effects of cannabinoids were performed using either the natural ligand Δ^9-THC or synthetic compounds that were acting at both CB1 and CB2 receptors. Peripheral or central administration of these compounds allowed the delineation of a series of functional effects: sedation, catalepsy, hypothermia, and antinociception. Since its description, this tetrad of effects has been considered a hallmark of cannabinoids action (Martin et al., 1991). The evaluation of nociception in most preclinical pain models requires the assessment of some kind of motor behaviour (paw or tail withdrawal, paw licking, etc.). Because of the cataleptic and sedating properties of cannabinoids, the evaluation of their analgesic effect is therefore all but a straightforward task. Side effects should be evaluated carefully when performing tests on cannabinoid antinociception; and, even when stringent experimental conditions are applied, the interpretation of results can be difficult and lead to controversy.

A second aspect to take into account is the pharmacological profile of ligands. Some agonists (e.g., WIN55212–2) could have dramatically different potency and efficacy in different species (Govaerts, Hermans and Lambert, 2004; Mukherjee et al., 2004), and differences in the pharmacological profile of the same molecule were observed in tests performed on tissues natively expressing CB2 by comparison to recombinant cells (Marini et al., 2013). Even though CB1- and CB2-selective agonists have been described, the degree of selectivity of these molecules remains equivocal. This is also true for antagonists that have often been used to confirm selective action on a single receptor subtype. In addition, as previously mentioned, the so-called 'antagonists' are actually inverse agonists and consequently could by themselves induce a physiological effect. Hence the use of genetically modified animal models, which bear deletion of one cannabinoid receptor, in combination with pharmacological treatments has proven highly valuable for understanding the relevance of CB1 versus CB2 in inducing antinociception. On the other hand the same approach used to investigate the FAAH and MAG lipase action suffers from the drawback that these enzymes are involved in the metabolism of lipids other than endocannabinoids and confounding effects could arise from this situation.

On another note, anandamide has been shown to be a ligand also for transient receptor potential vanilloid type 1 (TRPV1) (Van Der Stelt et al., 2005; Smart et al., 2000), a heat-, proton- and ligand-operated channel that bind capsaicin and is involved in nociception. Furthermore, cannabinoids – both endogenous and exogenous ones – could potentiate the action of the glycine receptor (Xiong et al., 2012), and it has been recently reported that some cannabinoid antagonists (AM251 and SR141716A) would be able to bind directly to the opioid receptor and inhibit morphine action (Seely et al., 2012). Finally, a series of effects attributed to cannabinoids could not be clearly assigned to an action on cannabinoid receptors and were attributed to additional receptors such as abnormal cannabidiol (Jarai et al., 1999). Thus a note of caution is always necessary when evaluating any antinociceptive effect attributed to cannabinoid action.

Soon after its identification, THC was shown to be analgesic in acute pain models (tail flick and hot plate), with an efficacy comparable to morphine (Bisher and Mecholam, 1968; Sofia, Vassar and Knobloch, 1975). These results attracted interest in the analgesic properties of cannabinoids, and Pfizer Inc. started a program that led to the synthesis of the cannabinoid analogue nantradol, which has antiemetic and analgesic properties (Milne, Koe and Johnson, 1979; Jacob, Ramabadran and Campos-Medeiros, 1981). Nabilone, another first-generation cannabinoid developed by Eli Lilly, has been reported to have analgesic effect in patients with multiple sclerosis, but it is presently commercialised under the name of Cesamet for the relief of nausea and vomiting caused by chemotherapy. Other synthetic cannabinoids, CP55940, WIN55212-2 and HU210, with mixed CB1–CB2 activity, have been widely used to study the cannabinoid system. These agonists produce analgesic response in the tail flick when applied peripherally (intravenously) or centrally (intrathecally in the lumbar region of the spinal cord) – an effect reversed by the selective CB1 antagonist SR141716A (Lichtman and Martin, 1991, 1997; Fox et al., 2001). Reversal of mechanical hyperalgesia in a neuropathic pain model of chronic pain was also reported for these three agonists, even though with a signifi-cant difference in potency, HU210 being the most potent (Fox et al., 2001). Assessment of various administration routes and antagonism with SR141716A suggest that these agonists trigger central and peripheral antinociception mediated by CB1. Similarly, in a persistent inflammatory pain model, complete Freund adjuvant (CFA), the anti-allodynic action of WIN55212–2 was blocked by SR141716A, corroborating a CB1-mediated effect (Martin et al., 1999b). Mixed cannabinoid agonists were also tested in several additional pain models that showed a wide spectrum of antinociceptive actions.

The first report showing that a CB2-selective agonist (HU308) could reduce nociception was published in 1999 (Hanuš et al., 1999). Since then a plethora of CB2-selective agonists based on different chemical structures have been designed and tested in various pain models. AM1241 and L768242 (GW405833) are probably the best characterised in a preclinical setting (Figure 6.4). These molecules showed an analgesic effect on a variety of pain models ranging from acute (tail flick and/or plantar test) to inflammatory (carrageenan, CFA, and formalin) and neuropathic (spinal nerve ligation and/or partial sciatic nerve ligation) (Yao et al., 2008; Bingham et al., 2007; Ibrahim et al., 2003; Nackley, Makriyannis and Hohmann, 2003; Gutierrez et al., 2007; Malan et al., 2002; Malan et al., 2001; Beltramo et al., 2006; Whiteside et al., 2005; Valenzano et al., 2005). The use of other CB2 agonists (e.g. JWH133, GW842166X, A796260, GRC10622, PRS211375, GSK554418A, LY2828360) substantially confirmed and extended these findings to additional pain models such as chronic constriction injury, knee joint osteoarthritis, acute visceral pain, intradermal capsaicin, chemotherapy induced pain, and so on (Beltramo, 2009; Han et al., 2013). The doses required to achieve antinociception are highly variable from compound to compound, for the same compound in different mod-els, using different routes of administration, and also for the same compound in the same model, pending on the behavioural read-out or experimental conditions. For example, antinociceptive concentration of AM1241 ranged from 2 to 0.033 mg/kg (Gutierrez et al., 2007; Quartilho et al., 2003).

Local knock-down of the receptor using antisense oligonucleotides injected in the spinal cord and directed against CB1 produced a reduction in the threshold to innocuous and noxious stimuli and an attenuation of agonist effect (Richardson, Aanonsen and Hargreaves, 1998; Dogrul et al., 2002). These data suggest that the spinal cord and the DRGs are impor-tant sites of cannabinoid antinociceptive action. Spontaneous responses to acute nociceptive stimuli (thermal, mechanical and chemical) were similar in CB1-/- and wild type mice. Considering that also in CB2-/- mice basal nociceptive response was not altered, the role of endocannabinoid in setting nociceptive thresholds remains ambiguous. Conversely, the antinociceptive effects of THC were absent or strongly reduced in CB1-/- (Ledent et al., 1999). The specific deletion of CB1 in peripheral nociceptors, combined with the administration of cannabinoid agonists, suggests a central role of CB1 in mediating cannabinoid antinociception in inflammatory and neuropathic pain (Agarwal et al., 2007). Studies on the antinociceptive effect of a peripherally restricted CB1 agonist (AZ11713908) in inflammatory pain using CB1-/- or CB2-/- mice indicate an action mainly on peripheral CB1, further confirming this hypothesis (Yu et al., 2010). On the other hand, the effect of putative CB2 agonists (GW405833, AM1241 and JWH133) was significantly lessened in neuropathic, inflammatory and acute pain models when tested in CB2-/- (Whiteside et al., 2005; Ibrahim et al., 2006; Yamamoto, Mikami and Iwamura, 2008). Concurrently, overexpression of CB2 significantly reduced mechanical allodynia in a joint pain model (La Porta et al., 2013). In conclusion, data from knock-out mice confirm that analgesia could be mediated by both CB1 and CB2 but do not provide a definitive clue regarding the relative importance of the two receptors.

Another approach to test cannabinoid antinociception is through the inhibition of endocannabinoids internalisation/degradation. This idea was fostered by the observation that electrical stimulation of the dorsal and lateral part of the PAG increased the anandamide level in PAG and produced CB1-mediated analgesia (Walker et al., 1999). The creation of FAAH-/- mice corroborated this idea, since in these mice an increase in anandamide level was associated to reduced pain sensitivity (Cravatt et al., 2001). Conversely, when delivered by intravenous injection in normal mice, anandamide has no effect on acute pain due to its rapid breakdown. However, co-administration of anandamide with the FAAH inhibitor resulted in a significant reduction of nociception in the tail flick test (Haller, Stevens and Welch, 2008). Similarly, the degradation-resistant anandamide analogue methanandamide is capable of reducing the mechanical allodynia and hyperalgesia caused by CFA injection (Potenzieri et al., 2008); and the administration of URB597, OL-135 (FAAH inhibitors), or JZL184 (MAG lipase inhibitor) reduced both hyperalgesia and allodynia in the neuropathic pain model (Kinsey et al., 2009; Russo et al., 2007; Desroches et al., 2008; Chang et al., 2006). Experiments performed with the FAAH inhibitor URB937, which does not cross the blood-brain barrier, suggest that endocannabinoid antinociception is chiefly due to a peripheral action (Clapper et al., 2010). Hence there is good agreement between the data obtained with FAAH-/- mice, with the degradation-resistant anandamide analogue, and with FAAH inhibitors. Nevertheless, it should be kept in mind that part of these effects could be mediated by action to receptors other than CB1/CB2. For instance, anandamide binds to TRPV1. Furthermore, N-palmitoyl ethanolamine and N-oleoyl ethanolamine, two lipid messengers, do not bind to a cannabinoid receptor but are hydrolyzed by FAAH, and bind to the peroxisome proliferator-activated receptor (PPAR), which is involved in nociception modulation. Thus multiple mechanisms could contribute to the analgesic effect observed after the blockade of endocannabinoid inactivation. Also, the administration of AM404, which is designed to block endocannabinoid internalisation, induces antinociception in various pain models: formalin test, CFA, and chronic constriction injury (Beltramo et al., 1997; La Rana et al., 2006; Palazzo et al., 2006; Mitchell et al., 2007). From this rapid survey it appears that blocking endocannabinoid inactivation is effective in preclinical models of both neuropathic and inflammatory pain, whereas its efficacy in acute pain needs to be further clarified.

Finally, we should consider the fact that the central tenet of cannabinoid antinocieptive effect, namely that stimulation of the cannabinoid system is producing antinociception, has been recently put in question. Evidence from a few studies points to a possible pronociceptive effect of cannabinoid agonists. In the spinal cord, CB1 is localised mainly on inhibitory GABAergic or glycinergic interneurons, which suggests a presynaptic relief of inhibitory input to C fibre nociceptors and an ensuing increase in nociception (Zhang et al., 2010; Pernia-Andrade et al., 2009). Furthermore, in certain models of inflammatory or activity-dependent hyperalgesia, CB1 antagonists were reported to be antinociceptive (Croci and Zarini, 2007; Pernia-Andrade et al., 2009) and CB1 knock-out mice to be hypoalgesic (Zimmer et al., 1999). To reconcile these results with the large body of evidence pointing to an antinociceptive action of cannabinoid receptor stimulation, two main hypotheses have been proposed. It has been suggested that a physiological state-dependent change in CB1 coupling to downstream effectors – such as G proteins and ion channels – in excitatory and inhibitory neurons that bear the cannabinoid receptor would promote the inhibition of nociceptor terminals under pathological conditions and their activation under physiological conditions. On the other hand, the presence of the cannabinoid receptor not only in neurons, but also on spinal cord astrocytes and microglia, which are known to contribute to pain modulation, could represent a point for a divergent control of cannabinoids on nociception (Zeilhofer, 2010). Interestingly, a recent report indicates that both CB agonist and CB antagonist suppress microglia activation (Ribeiro et al., 2013). Further studies are warranted to clarify the physiopathological significance of these unexpected and intriguing findings.

Mechanisms underpinning cannabinoid antinociception

The distribution of cannabinoid receptors clearly indicates that cannabinoids could act to mitigate pain not only inside the nervous system but also in the periphery (Figure 6.6). It is known that, upon injury, cells of the immune system could release pronociceptive molecules that sensitise the primary afferent neurons. The expression of CB2 in these cells is well documented (Munro et al., 1993; Galiegue et al., 1995; Klein et al., 1998). Hence cannabinoids could influence pain severity by reducing the release of pronociceptive mediators such as NGF, prostanoid, cytokines, ATP, histamine, and so on from cells of the immune system. This mechanism is probably especially relevant in conditions where inflammation has a foremost role in the genesis of pain but could be less significant in antinociception in neuropathic pain. Interestingly, both CB2 agonists and CB2 antagonists have been claimed to have immunosuppressive and anti-inflammatory properties, and further studies are needed to understand the complex actions of cannabinoids on the immune system.

Another peripheral mechanism that could account for CB2-mediated analgesia is an indirect one, via the stimulation of the release of antinociceptive factors: cannabinoids could trigger the secretion of β-endorphin from keratinocytes, producing an opioid-mediated cannabinoid analgesia (Ibrahim et al., 2005). This mechanism could, however, be compound-specific (AM1241), because it was not observed when using other CB2 agonists (A-796260 and L768242) (Yao et al., 2008; Whiteside et al., 2005).

CB1 is present in the DRG neurons, but its precise localisation and prevalence in the different neuronal populations is somehow controversial (Hohmann and Herkenham, 1999; Salio et al., 2002; Agarwal et al., 2007; Veress et al., 2013). DRG neurons are pseudo-unipolar and project their axons to both the periphery and the CNS. Hence peripheral and central terminals of DRG neurons express CB1. In the spinal cord dorsal horn, CB1 is localised in lamina I terminals that belong to both peptidergic and non-peptidergic nociceptors (Hegyi et al., 2009; Salio et al., 2002). The relative importance of the central versus the peripheral action of CB1 on DRG neurons was investigated using peripheral and central cannabinoid injections in inflammatory and neuropathic pain in mouse models that bear a general or a nociceptor-specific deletion of CB1. Specific deletion of CB1 from these neurons dramatically reduces the analgesic effect induced by the peripheral administration of WIN55212 in both inflammatory and neuropathic pain models (Agarwal et al., 2007). This suggests that CB1 expressed by peripheral and central terminals of nociceptors account for most of cannabinoids-induced analgesia. In addition to an immediate effect, it has been demonstrated recently that cannabinoids acting through CB1 play an important role in activity-dependent plasticity, inducing long-term depression at spinal cord dorsal horn synapses (Kato et al., 2012).

On the other hand, the presence of CB2 in DRGs and in the spinal cord dorsal horn has been proposed (Beltramo et al., 2006; Sagar et al., 2005; Elmes et al., 2004; Zhang et al., 2003; Svizenska et al., 2013). Capsaicine-stimulated CGRP release from primary afferent fibres was reduced through the application of the CB2-selective AM1241 or L768242 agonist to spinal cord slices. This effect was reverted by the CB2 antagonist SR144528 and conserved in CB1-/- mice, pointing to a localisation of CB2 on primary afferents, laminae I and II (Beltramo et al., 2006). On the other hand, the application of the CB2 agonist JWH133 directly to the spinal cord produced an effect on wide dynamic range (WDR) neurons, laminae V and VI, only on neuropathic rats (Sagar et al., 2005). A hypothesis able to reconcile these data would be that CB2 expression is regulated differently, superficial laminae constitutively expressing CB2 and deeper laminae expressing CB2 only under neuropathic pain conditions.

Cannabinoids action on spinal cord glia could also plays a role in cannabinoid-mediated analgesia by decreasing the release of pronociceptive mediators such cytokines (TNFα, IL6 and IL1-β) and nitric oxide (NO) that contribute to pain exacerbation. Indeed, at the spinal cord level, both the neuronal and the non-neuronal mechanism could participate in the antinociceptive effect of cannabinoids by blocking the transmission of noxious input to higher brain regions.

Supraspinal sites involved in the cannabinoid-induced analgesia were explored by using intracerebroventricular and localised cannabinoid injections. Cannabinoids induced antinociception when injected in the dorsolateral PAG, dorsal raphe nucleus RVM, amygdala, lateral posterior and submedius regions of the thalamus, superior colliculus and noradrenergic A5 region (Martin, Loo and Basbaum 1999a; Martin et al., 1996; Martin et al., 1995; Martin et al., 1998; Lichtman et al., 1996). Three of these sites were most intensely investigated: the PAG, the RVM and the amygdala.

Supraspinal modulation of pain engages neurons in the PAG that project to the RVM and from here to the dorsal horn of the spinal cord. Electrical stimulation of the PAG induces CB1-dependent analgesia, as demonstrated by the reversal obtained with the CB1 antagonist

SR141716A (Walker et al., 1999). Interestingly, it has been shown that cannabinoids inhibit both glutamate and GABA presynaptic release in the rat's PAG (Vaughan et al., 2000), which suggests a complex effect in modulating neuronal signalling.

The localised injection of cannabinoid in the RVM significantly increases the time latency in the tail flick test, indicating an antinociceptive effect (Martin et al., 1998). Neurons projecting from the RVM to the spinal cord belong to functionally distinct populations called 'off' and 'on' neurons. 'Off' neurons inhibit spinal nociceptive transmission; conversely, 'on' neurons have a facilitating effect on pain transmission. Microinjection of muscimol, a GABAA receptor agonist, in the RVM blocks the analgesic, but not the locomotor effect of intravenous cannabinoids. Extracellular single-unit recordings suggest that cannabinoids would reduce the 'off' cell inactivity and the 'on' cell burst, leading to an activation of the descending inhibitory pain pathway (Meng et al., 1998).

The amygdala is a brain structure with a well-known function in the coordination of fear and defense reactions, and its basolateral nucleus is involved in the formation and storage of aversive memories. Bilateral lesions of the amygdala in primate reduce the antinociceptive effect of WIN55212–2, and microinjection in the basolateral nucleus produces antinociception, indicating an involvement of this structure in the cannabinoid antinociceptive effect (Manning et al., 2001). CB1 is present in the basolateral complex, where it is localised in GABAergic interneurons. The implication of cannabinoids in the long-term depression of GABAergic inhibitory currents suggests that they could regulate aversive memory extinction (Marsicano et al., 2002).

Despite the role of the thalamus in processing nociceptive information, little is known about the cannabinoid effect on this structure. An increase of CB1 expression has been observed in the contralateral thalamus after unilateral axotomy of the tibial branch of the sciatic nerve (Siegling et al., 2001) and after spinal cord injury (Knerlich-Lukoschus et al., 2011). Interestingly, evidence was provided that CB2 plays a role in modulating thalamic nociceptive response in neuropathic, but not in sham-operated rats (Jhaveri et al., 2008).

It is clear that cannabinoids are involved in several key components of the pain pathways and could modulate nociception at different levels through various mechanisms: inhibition of the release of pronociceptive mediators from non-neuronal cells outside the central nervous system, modulation of nociceptive input into the spinal cord, modulation of descending inhibitory and excitatory pain pathways, and modulation of nociceptive input processing. The existence of these multiples mechanisms and sites of actions offers a wide array of opportunities for reducing nociception, but also clearly outlines the complexity of the system and the difficulties in obtaining a balanced action.

Which cannabinoid for which pain

The data gathered over the past thirty years on cannabinoid and antinociception constitute an impressive, but at times contradictory body of information. To extrapolate general indications about which intervention on the cannabinoid system would be the most appropriate for treating specific pain types is an interesting exercise that remains, however, highly speculative.

As a general consideration, cannabinoids are probably a more interesting target for chronic than for acute pain. Furthermore, on the basis of localisation and known physiological activity, one would predict that CB1 agonists would be more valuable in treating pain of neuropathic origin and CB2 agonists in mitigating inflammatory pain. The following sections illustrate the different strategies used to produce antinociception through cannabinoids, along with their possible advantages and drawbacks.

CNS-penetrating mixed agonists

The activation of the cannabinoid system in the brain has proven highly effective in inducing antinociception in a variety of preclinical and clinical pain models. Use of this approach in clinical setting results in antinociception that is, however, invariably associated to significant side effects, sometime severe enough to oblige patient to discontinue the treatment. In spite of this drawback, in some cases this strategy could still be an option. For example, in oncologic patients under chemotherapeutic treatment and suffering from pain, central acting cannabinoids could be a double-edge sword that reduces pain on the one hand and prevents nausea and vomiting induced by chemotherapeutic agents on the other. Another application could be in the treatment of multiple sclerosis patients, where, in addition to an antinociceptive effect, cannabinoids are also improving muscle stiffness and spasticity, sleep quality and bladder disturbances. (For a detailed overview of pain treatment in multiple sclerosis, see Chapter 14 in this volume). In these cases the balance between benefits and side effects could be acceptable. However, considering the limited market and the constraints, this approach is not particularly attractive to large pharmaceutical companies. Its application will continue to rely on cannabis cigarettes smoking or, eventually, on the extension of the application of cannabinoid drugs approved in the treatment of other pathologies. An example could be Sativex, a mixture of THC and cannabidiol produced by GW Pharmaceuticals, which has been recently approved in several countries for the symptomatic relief of spasticity in multiple sclerosis and is currently in phase 3 clinical trials, as an adjuvant analgesic in the treatment of advanced cancer patients.

Peripherally restricted CB agonists

As we have seen, the activation of CB1 in the CNS induces relevant side effects. To overcome this problem, peripherally restricted cannabinoids were conceived as a new approach. This strategy was based on the observation that CB1 expressed in the DRGs, which are outside the blood–brain barrier, would account for most of the cannabinoid-induced antinociception (Agarwal et al., 2007). Hence a peripherally restricted agonist should retain antinociceptive features without side effects. In addition, considering the potential benefit of stimulating simultaneously the CB2, the hurdle of designing a receptor-subtype selective compound could be avoided.

Astra-Zeneca has just completed phases I and II for their candidate molecule AZD1940, a peripherally restricted mixed CB1–CB2 agonist. In this respect AZD1940 efficacy was tested in capsaicin-induced pain (Kalliomaki et al., 2013), third-molar extraction pain and chronic low back pain. Unfortunately no significant analgesic effect was observed in primary end points, and at present there is no clear explanation for this failure. In addition, after multiple ascending-dose treatment, a significant increase in body weight and hypotension were observed. If one considers that Rimonabant, a CB1 antagonist, has been briefly on the market as a treatment for obesity and that cardiovascular effects of the CB1 agonist are well known (Pacher and Kunos, 2013), this is not entirely surprisingly. Nevertheless, other companies have ongoing programs based on this approach on preclinical (Cara Therapeutics) or clinical (JB Therapeutics) trials, and the results of these studies will help to understand whether this approach is still viable.

CB2-selective agonists

From the end of the 1990s on, CB2 attracted an increasing amount of interest in the cannabinoid field, and the antinociceptive effects triggered by its activation were reported in several models of pain in rodents. Due to a more restricted distribution in the central nervous system, targeting CB2 with a brain-penetrating selective agonist should have the distinct advantage of a more favourable psychotropic side effect profile than the one presented by CB1 agonists. In contrast to CB1 stimulation, CB2 stimulation seems to be cardioprotective (Steffens and Pacher, 2012) and without discernible effect on the metabolism. These considerations prompted several pharmaceutical companies to focus their attention on the development of CB2-selective agonists. Pharmos tested a moderately selective CB2 agonist on third-molar extraction and on capsaicin-induced pain in healthy volunteers. Efficacy was modest or absent, and this compound was discontinued. GlaxoSmithKline completed a clinical trial for dental and osteoarthritis pain with its selective CB2 agonist GW842166X, but data were not disclosed and the program seems to have been terminated. Glenmark Pharmaceuticals has reported the successful completion of phase I clinical trials for GRC1069; however, this molecule is no more in the company's pipeline. Several other companies (Kyowa Hakko Kirin, Abbott, Lilly, and Shionogi) have advanced CB2-selective agonists to clinical trials, but also in this case they are no more in the pipeline. Conversely, the compound APD371 from Arena is in phase I clinical trials. The limited information on clinical trial outcome makes any conclusion about the success of this approach premature.

Blockers of the endocannabinoid inactivation system

As described above, endocannabinoids are produced and released upon demand. Assuming that, in the presence of noxious stimuli, their synthesis would be enhanced in brain regions involved in processing noxious input, it was forecasted that blockade of their inactivation would enhance their concentration in these regions and would result in antinociception without side effects. In support of this view, several preclinical studies reported the analgesic effect of blocking endocannabinoid internalisation and breakdown. This prompted the initiation of drug-discovery projects aimed at the identification of FAAH inhibitors for the treatment of pain. The results of the first phase II clinical trials using an irreversible FAAH inhibitor, PF-04457845, to reduce osteoarthritic pain have been recently published (Huggins et al., 2012). On the positive side, there was no evidence of cannabinoid-type adverse events. However, despite a significant elevation of the plasma level in endogenous anandamide (tenfold) and in other related N-acylethanolamines, PF-04457845 failed to achieve any significant analgesic effect and the trial was stopped before completion. Two main possibilities could explain this failure. First, the treatment of a kind of pain with a stronger affective component, such as neuropathic pain, could be expected to be more positively affected on account of the anxiolytic- and antidepressant-like properties observed in preclinical models that use FAAH inhibition. Secondly, the enhancement of anandamide plasma levels could have increased its stimulatory activity on alternative targets such as TRPV1. This could have counteracted the beneficial effect of its action on cannabinoid receptors. On the basis of this hypothesis it has been postulated that a bifunctional molecule with FAAH inhibitory properties and antagonism at the TRPV1 receptor would probably be more apt to elicit antinociception.

Conclusions and perspectives

The past few years were full of promise for the delivery of a cannabinoid-based antinociceptive drug. Unfortunately, the many attempts made by different approaches (peripherally restricted cannabinoids, CB2-selective agonist, inhibitors of endocannabinoid internalisation and degradation) all fell short of the goal in clinical trials. Does this mean that the hope of a new, cannabinoid-based analgesic with an improved pharmacological and psychoactive profile is vanishing? Multiple reasons could be provided for the unexpected failure of all of these attempts, but two problems remain central: lack of efficacy and significant side effects. The lack of efficacy is worrisome, and is probably telling us that we are missing some essential piece of the puzzle or that the approaches used so far are not adequate. The undeniable fact that the cannabinoid system plays a role in nociception naturally leads to the conclusion that it should be possible to harness the system and to induce antinociception in humans. That pronociceptive activity of cannabinoid agonists, by direct action either on cannabinoid receptors or on other receptors (e.g. TRPV1), has also been reported and points to an unforseen complexity of the system that was probably underestimated.

Another aspect that has received little attention so far is the importance of the constitutive activity of native receptors. It has been shown that some CB2 ligands could behave as protean agonists, which means that they could be agonists, antagonists or inverse agonists, pending on the constitutive activity of the receptor (Mancini et al., 2009; Yao et al., 2006). A protean agonist is capable of either increasing or reducing the activity at the receptor, depending on that receptor's level of constitutive activity. Two of the most used CB2 agonists in preclinical studies (AM1241 and L768242) are protean agonists; but information on this feature is missing for most other ligands, and so are data on the constitutive activity of the receptor in the target tissue. It is therefore extremely difficult to estimate the importance of this feature on the overall outcome of a particular treatment. On the other hand, this trait could be eventually exploited in order to stabilise a situation and avoid an overstimulation or and understimulation of the system – for example, in order to reduce hyperalgesia without hampering the perception of normal pain sensations that have an adaptive function. Appealing as it is, this approach would have to overcome the difficulty of obtaining a protean agonist with the right balance between stimulation and inhibition, which could be different in the different types of pain.

The use of peripherally restricted cannabinoids was a promising strategy for circumventing the known psychotropic effect of cannabinoids. Unfortunately this strategy stumbled on cardiovascular and metabolic peripheral side effects. The existence of potentially harmful cannabinoid effects on the cardiovascular system and on metabolism was known, but probably underestimated when this strategy was designed. Actually most of this effect seems to be mediated by peripheral CB1, hence a peripherally restricted CB2 agonist could have a better safety profile. Another possible way out of this difficulty could be suggested by the case of β-adrenergic receptor (β-AR). Several β-AR blockers are available to treat heart failure; however, they show significant side effects. A new molecule, carvedilol, with a much improved safety profile has been discovered. Carvedilol is a functionally selective molecule (a biased agonist) capable of acting as an antagonist of the β-AR G protein-signalling pathway, which is cardiotoxic, and of simultaneously engaging through the same receptor cardioprotective β-arrestin signalling (Wisler et al., 2007). The existence of biased agonists has been reported also for the cannabinoid receptors. Intriguingly, *in vivo* data show that agonists such as Δ^9-THC and WIN55212-2 are more potent in inducing hypolocomotion than catalepsy or hypothermia, whereas WIN55212-2 analogues are more potent in inducing catalepsy and hypothermia – which suggests that different cannabinoid agonists may trigger different combinations of physiological responses (Bosier et al., 2010). These differences can tentatively be attributed to functional selectivity and exploited to produce more efficacious and safer molecules, capable of activating intracellular signalling that may lead to antinociception while avoiding side effects.

Two new endogenous cannabinoid ligands were described recently: a negative allosteric modulator (RVD-hemopressin) and an allosteric enhancer (lipoxin A4) (Bauer et al., 2012; Pamplona et al., 2012). Interestingly, a recent report connects biased agonism and allosteric modulator. ORG27569, an allosteric CB1 modulator that enhances the agonist binding affinity to CB1, actually inhibits the signalling efficacy of G protein-dependent agonists and induces a G protein-independent signalling (Ahn et al., 2013). These results point to a further layer of complexity in the cannabinoid signalling mechanism, which could be exploited in the design of new strategies for separating desired antinociceptive effects from unwanted metabolic and cardiovascular effects.

To conclude, while recent setbacks in clinical trials impose a serious reconsideration of approaches and goals, several new alternative strategies still appear possible, and there is therefore renewed hope that effective cannabinoid-based analgesics could be developed.

Acknowledgements

The author is grateful to Dr Hugues Dardente (Institut National de la Recherche Agronomique, UMR085 Physiologie de la Reproduction et de Comportements, France) for his critical reading of this manuscript.

References

Abadji, V., Lin, S., Taha, G., Griffin, G., Stevenson, L. A., Pertwee, R. G. and Makriyannis, A. 1994. (R)-meth-Anandamide: A chiral novel Anandamide possessing higher potency and metabolic stability. *Journal of Medicinal Chemistry* 37: 1889–93.

Agarwal, N., Pacher, P., Tegeder, I., Amaya, F., Constantin, C. E., Brenner, G. J., Rubino, T., Michalski, C. W., Marsicano, G., Monory, K., Mackie, K., Marian, C., Batkai, S., Parolaro, D., Fischer, M. J., Reeh, P., Kunos, G., Kress, M., Lutz, B., Woolf, C. J. and Kuner, R. 2007. Cannabinoids mediate analgesia largely via peripheral type 1 cannabinoid receptors in nociceptors. *Nature Neuroscience* 10: 870–9.

Ahn, K. H., Mahmoud, M. M., Shim, J. Y. and Kendall, D. A. 2013. Distinct roles of beta-arrestin 1 and beta-arrestin 2 in ORG27569-induced biased signaling and internalization of the cannabinoid receptor 1 (CB1). *The Journal of Biological Chemistry* 288: 9790–800.

Baek, J. H., Darlington, C. L., Smith, P. F. and Ashton, J. C. 2013. Antibody testing for brain immunohistochemistry: Brain immunolabeling for the cannabinoid CB2 receptor. *Journal of Neuroscience Methods* 216: 87–95.

Bauer, M., Chicca, A., Tamborrini, M., Eisen, D., Lerner, R., Lutz, B., Poetz, O., Pluschke, G. and Gertsch, J. 2012. Identification and quantification of a new family of peptide endocannabinoids (Pepcans) showing negative allosteric modulation at CB1 receptors. *The Journal of Biological Chemistry* 287: 36944–67.

Beltramo, M. 2009. Cannabinoid type 2 receptor as a target for chronic pain. *Mini Reviews in Medicinal Chemistry* 9: 11–25.

Beltramo, M. and Piomelli, D. 2000. Carrier-mediated transport and enzymatic hydrolysis of the endogenous cannabinoid 2-arachidonylglycerol. *Neuroreport* 11: 1231–5.

Beltramo, M., Stella, N., Calignano, A., Lin, S. Y., Makriyannis, A. and Piomelli, D. 1997. Functional role of high-affinity Anandamide transport, as revealed by selective inhibition. *Science* 277: 1094–7.

Beltramo, M., Bernardini, N., Bertorelli, R., Campanella, M., Nicolussi, E., Fredduzzi, S. and Reggiani, A. 2006. CB2 receptor-mediated antihyperalgesia: Possible direct involvement of neural mechanisms. *The European Journal of Neuroscience* 23: 1530–8.

Bingham, B., Jones, P. G., Uveges, A. J., Kotnis, S., Lu, P., Smith, V. A., Sun, S. C., Resnick, L., Chlenov, M., He, Y., Strassle, B. W., Cummons, T. A., Piesla, M. J., Harrison, J. E., Whiteside, G. T. and Kennedy, J. D. 2007. Species-specific in vitro pharmacological effects of the cannabinoid receptor 2 (CB2) selective ligand AM1241 and its resolved enantiomers. *British Journal of Pharmacology* 151: 1061–70.

Bisher, H. I. and Mecholam, R. 1968. Pharmacological effects of two active constituents of marihuana. *Archives Internationales de Pharmacodynamie et de Thérapie* 172: 24–31.

Bojesen, I. N. and Hansen, H. S. 2003. Binding of Anandamide to bovine serum albumin. *Journal of Lipid Research* 44: 1790–4.

Bosier, B., Muccioli, G. G., Hermans, E. and Lambert, D. M. 2010. Functionally selective cannabinoid receptor signalling: Therapeutic implications and opportunities. *Biochemical Pharmacology* 80: 1–12.

Bouaboula, M., Poinot-Chazel, C., Marchand, J., Canat, X., Bourrie, B., Rinaldi-Carmona, M., Calandra, B., Le Fur, G. and Casellas, P. 1996. Signaling pathway associated with stimulation of CB2 peripheral cannabinoid receptor: Involvement of both mitogen-activated protein kinase and induction of Krox-24 expression. *European Journal of Biochemistry* 237: 704–11.

Brownjohn, P. W. and Ashton, J. C. 2012. Spinal cannabinoid CB2 receptors as a target for neuropathic pain: An investigation using chronic constriction injury. *Neuroscience* 203: 180–93.

Cadas, H., Gaillet, S., Beltramo, M., Venance, L. and Piomelli, D. 1996. Biosynthesis of an endogenous cannabinoid precursor in neurons and its control by calcium and cAMP. *The Journal of Neuroscience* 16: 3934–42.

Carrier, E. J., Kearn, C. S., Barkmeier, A. J., Breese, N. M., Yang, W., Nithipatikom, K., Pfister, S. L., Campbell, W. B. and Hillard, C. J. 2004. Cultured rat microglial cells synthesize the endocannabinoid 2-arachidonylglycerol, which increases proliferation via a CB2 receptor-dependent mechanism. *Molecular Pharmacology* 65: 999–1007.

Chang, L., Luo, L., Palmer, J. A., Sutton, S., Wilson, S. J., Barbier, A. J., Breitenbucher, J. G., Chaplan, S. R. and Webb, M. 2006. Inhibition of fatty acid amide hydrolase produces analgesia by multiple mechanisms. *British Journal of Pharmacology* 148: 102–13.

Clapper, J. R., Moreno-Sanz, G., Russo, R., Guijarro, A., Vacondio, F., Duranti, A., Tontini, A., Sanchini, S., Sciolino, N. R., Spradley, J. M., Hohmann, A. G., Calignano, A., Mor, M., Tarzia, G. and Piomelli, D. 2010. Anandamide suppresses pain initiation through a peripheral endocannabinoid mechanism. *Nature Neuroscience* 13: 1265–70.

Correa, F., Mestre, L., Docagne, F. and Guaza, C. 2005. Activation of cannabinoid CB2 receptor negatively regulates IL-12p40 production in murine macrophages: Role of IL-10 and ERK1/2 kinase signaling. *British Journal of Pharmacology* 145: 441–8.

Cravatt, B. F., Demarest, K., Patricelli, M. P., Bracey, M. H., Giang, D. K., Martin, B. R. and Lichtman, A. H. 2001. Supersensitivity to Anandamide and enhanced endogenous cannabinoid signaling in mice lacking fatty acid amide hydrolase. *Proceedings of the National Academy of Sciences of the United States of America* 98: 9371–6.

Croci, T. and Zarini, E. 2007. Effect of the cannabinoid CB1 receptor antagonist rimonabant on nociceptive responses and adjuvant-induced arthritis in obese and lean rats. *British Journal of Pharmacology* 150: 559–66.

Derocq, J. M., Jbilo, O., Bouaboula, M., Segui, M., Clere, C. and Casellas, P. 2000. Genomic and functional changes induced by the activation of the peripheral cannabinoid receptor CB2 in the promyelocytic cells HL-60. Possible involvement of the CB2 receptor in cell differentiation. *The Journal of Biological Chemistry* 275: 15621–8.

Desroches, J., Guindon, J., Lambert, C. and Beaulieu, P. 2008. Modulation of the anti-nociceptive effects of 2-arachidonoyl glycerol by peripherally administered FAAH and MGL inhibitors in a neuropathic pain model. *British Journal of Pharmacology* 155: 913–24.

Deutsch, D. G., Glaser, S. T., Howell, J. M., Kunz, J. S., Puffenbarger, R. A., Hillard, C. J. and Abumrad, N. 2001. The cellular uptake of Anandamide is coupled to its breakdown by fatty-acid amide hydrolase. *The Journal of Biological Chemistry* 276: 6967–73.

Devane, W. A., Dysarz, F. A., 3rd, Johnson, M. R., Melvin, L. S. and Howlett, A. C. 1988. Determination and characterization of a cannabinoid receptor in rat brain. *Molecular Pharmacology* 34: 605–13.

Devane, W. A., Hanus, L., Breuer, A., Pertwee, R. G., Stevenson, L. A., Griffin, G., Gibson, D., Mandelbaum, A., Etinger, A. and Mechoulam, R. 1992. Isolation and structure of a brain constituent that binds to the cannabinoid receptor. *Science* 258: 1946–9.

Di Marzo, V. 2008. Targeting the endocannabinoid system: To enhance or reduce? *Nature reviews. Drug Discovery* 7: 438–55.

Di Marzo, V., Fontana, A., Cadas, H., Schinelli, S., Cimino, G., Schwartz, J. C. and Piomelli, D. 1994. Formation and inactivation of endogenous cannabinoid Anandamide in central neurons. *Nature* 372: 686–91.

Dinh, T. P., Carpenter, D., Leslie, F. M., Freund, T. F., Katona, I., Sensi, S. L., Kathuria, S. and Piomelli, D. 2002. Brain monoglyceride lipase participating in endocannabinoid inactivation. *Proceedings of the National Academy of Sciences of the United States of America* 99: 10819–24.

Dogrul, A., Gardell, L. R., Ma, S., Ossipov, M. H., Porreca, F. and Lai, J. 2002. 'Knock-down' of spinal CB1 receptors produces abnormal pain and elevates spinal dynorphin content in mice. *Pain* 100: 203–9.

Egertova, M., Cravatt, B. F. and Elphick, M. R. 2003. Comparative analysis of fatty acid amide hydrolase and cb(1) cannabinoid receptor expression in the mouse brain: Evidence of a widespread role for fatty acid amide hydrolase in regulation of endocannabinoid signaling. *Neuroscience* 119: 481–96.

Elmes, S. J., Jhaveri, M. D., Smart, D., Kendall, D. A. and Chapman, V. 2004. Cannabinoid CB2 receptor activation inhibits mechanically evoked responses of wide dynamic range dorsal horn neurons in naive rats and in rat models of inflammatory and neuropathic pain. *The European Journal of Neuroscience* 20: 2311–20.

Fox, A., Kesingland, A., Gentry, C., Mcnair, K., Patel, S., Urban, L. and James, I. 2001. The role of central and peripheral Cannabinoid1 receptors in the antihyperalgesic activity of cannabinoids in a model of neuropathic pain. *Pain* 92: 91–100.

Galiegue, S., Mary, S., Marchand, J., Dussossoy, D., Carriere, D., Carayon, P., Bouaboula, M., Shire, D., Le Fur, G. and Casellas, P. 1995. Expression of central and peripheral cannabinoid receptors in human immune tissues and leukocyte subpopulations. *European Journal of Biochemistry* 232: 54–61.

Gareau, Y., Dufresne, C., Gallant, M., Rochette, C., Sawyer, N., Slipetz, D. M., Tremblay, N., Weech, P. K., Metters, K. M. and Labelle, M. 1996. Structure activity relationships of tetrahydrocannabinol analogues on human cannabinoid receptors. *Bioorganic & Medicinal Chemistry Letters* 6: 189–194.

Gifford, A. N., Tang, Y., Gatley, S. J., Volkow, N. D., Lan, R. and Makriyannis, A. 1997. Effect of the cannabinoid receptor SPECT agent, AM 281, on hippocAMPal acetylcholine release from rat brain slices. *Neuroscience Letters* 238: 84–6.

Glass, M. and Felder, C. C. 1997. Concurrent stimulation of cannabinoid CB1 and dopamine D2 receptors augments cAMP accumulation in striatal neurons: Evidence for a Gs linkage to the CB1 receptor. *The Journal of Neuroscience* 17: 5327–33.

Goparaju, S. K., Ueda, N., Taniguchi, K. and Yamamoto, S. 1999. Enzymes of porcine brain hydrolyzing 2-arachidonoylglycerol, an endogenous ligand of cannabinoid receptors. *Biochemical Pharmacology* 57: 417–23.

Govaerts, S. J., Hermans, E. and Lambert, D. M. 2004. Comparison of cannabinoid ligands affinities and efficacies in murine tissues and in transfected cells expressing human recombinant cannabinoid receptors. *European Journal of Pharmaceutical Sciences* 23: 233–43.

Gutierrez, T., Farthing, J. N., Zvonok, A. M., Makriyannis, A. and Hohmann, A. G. 2007. Activation of peripheral cannabinoid CB1 and CB2 receptors suppresses the maintenance of inflammatory nociception: A comparative analysis. *British Journal of Pharmacology* 150: 153–63.

Haller, V. L., Stevens, D. L. and Welch, S. P. 2008. Modulation of opioids via protection of Anandamide degradation by fatty acid amide hydrolase. *European Journal of Pharmacology* 600: 50–8.

Han, S., Thatte, J., Buzard, D. J. and Jones, R. M. 2013. Therapeutic utility of cannabinoid receptor type 2 (CB2) selective agonists. *Journal of Medicinal Chemistry* 56: 8224–56.

Hanuš, L., Breuer, A., Tchilibon, S., Shiloah, S., Goldenberg, D., Horowitz, M., Pertwee, R. G., Ross, R. A., Mechoulam, R. and Fride, E. 1999. HU-308: A specific agonist for CB(2), a peripheral cannabinoid receptor. *Proceedings of the National Academy of Sciences of the United States of America* 96: 14228–33.

Hegyi, Z., Kis, G., Hollo, K., Ledent, C. and Antal, M. 2009. Neuronal and glial localization of the cannabinoid-1 receptor in the superficial spinal dorsal horn of the rodent spinal cord. *The European Journal of Neuroscience* 30: 251–62.

Herkenham, M., Lynn, A. B., Little, M. D., Johnson, M. R., Melvin, L. S., De Costa, B. R. and Rice, K. C. 1990. Cannabinoid receptor localization in brain. *Proceedings of the National Academy of Sciences of the United States of America* 87: 1932–6.

Hillard, C. J. and Jarrahian, A. 2003. Cellular accumulation of anandamide: Consensus and controversy. *British Journal of Pharmacology* 140: 802–8.

Hillard, C. J., Edgemond, W. S. and Campbell, W. B. 1995. Characterization of ligand binding to the cannabinoid receptor of rat brain membranes using a novel method: Application to Anandamide. *Journal of Neurochemistry* 64: 677–83.

Hillard, C. J., Edgemond, W. S., Jarrahian, A. and Campbell, W. B. 1997. Accumulation of N-arachidonoylethanolamine (anandamide) into cerebellar granule cells occurs via facilitated diffusion. *Journal of Neurochemistry* 69: 631–8.

Hillard, C. J., Manna, S., Greenberg, M. J., Dicamelli, R., Ross, R. A., Stevenson, L. A., Murphy, V., Pertwee, R. G. and Campbell, W. B. 1999. Synthesis and characterization of potent and selective agonists of the neuronal cannabinoid receptor (CB1). *The Journal of Pharmacology and Experimental Therapeutics* 289: 1427–33.

Hohmann, A. G. and Herkenham, M. 1999. Localization of central cannabinoid CB1 receptor messenger RNA in neuronal subpopulations of rat dorsal root ganglia: A double-label in situ hybridization study. *Neuroscience* 90: 923–31.

Howlett, A. C. 2005. Cannabinoid receptor signaling. *Handbook of Experimental Pharmacology* 168: 53–79.

Huffman, J. W., Liddle, J., Yu, S., Aung, M. M., Abood, M. E., Wiley, J. L. and Martin, B. R. 1999. 3-(1′,1′-Dimethylbutyl)-1-deoxy-delta8-THC and related compounds: Synthesis of selective ligands for the CB2 receptor. *Bioorganic and Medicinal Chemistry* 7: 2905–14.

Huggins, J. P., Smart, T. S., Langman, S., Taylor, L. and Young, T. 2012. An efficient randomised, placebo-controlled clinical trial with the irreversible fatty acid amide hydrolase-1 inhibitor PF-04457845, which modulates endocannabinoids but fails to induce effective analgesia in patients with pain due to osteoarthritis of the knee. *Pain* 153: 1837–46.

Ibrahim, M. M., Rude, M. L., Stagg, N. J., Mata, H. P., Lai, J., Vanderah, T. W., Porreca, F., Buckley, N. E., Makriyannis, A. and Malan, T. P., Jr. 2006. CB2 cannabinoid receptor mediation of antinociception. *Pain* 122: 36–42.

Ibrahim, M. M., Deng, H., Zvonok, A., Cockayne, D. A., Kwan, J., Mata, H. P., Vanderah, T. W., Lai, J., Porreca, F., Makriyannis, A. and Malan, T. P., Jr. 2003. Activation of CB2 cannabinoid receptors by AM1241 inhibits experimental neuropathic pain: Pain inhibition by receptors not present in the CNS. *Proceedings of the National Academy of Sciences of the United States of America* 100: 10529–33.

Ibrahim, M. M., Porreca, F., Lai, J., Albrecht, P. J., Rice, F. L., Khodorova, A., Davar, G., Makriyannis, A., Vanderah, T. W., Mata, H. P. and Malan, T. P., Jr. 2005. CB2 cannabinoid receptor activation produces antinociception by stimulating peripheral release of endogenous opioids. *Proceedings of the National Academy of Sciences of the United States of America* 102: 3093–8.

Iwamura, H., Suzuki, H., Ueda, Y., Kaya, T. and Inaba, T. 2001. In vitro and in vivo pharmacological characterization of JTE-907, a novel selective ligand for cannabinoid CB2 receptor. *The Journal of Pharmacology and Experimental Therapeutics* 296: 420–5.

Jacob, J. J., Ramabadran, K. and Campos-Medeiros, M. 1981. A pharmacological analysis of levonantradol antinociception in mice. *Journal of Clinical Pharmacology* 21: 327S–333S.

Jarai, Z., Wagner, J. A., Varga, K., Lake, K. D., Compton, D. R., Martin, B. R., Zimmer, A. M., Bonner, T. I., Buckley, N. E., Mezey, E., Razdan, R. K., Zimmer, A. and Kunos, G. 1999. Cannabinoid-induced mesenteric vasodilation through an endothelial site distinct from CB1 or CB2 receptors. *Proceedings of the National Academy of Sciences of the United States of America* 96: 14136–41.

Jhaveri, M. D., Elmes, S. J., Richardson, D., Barrett, D. A., Kendall, D. A., Mason, R. and Chapman, V. 2008. Evidence for a novel functional role of cannabinoid CB(2) receptors in the thalamus of neuropathic rats. *The European Journal of Neuroscience* 27: 1722–30.

Kalliomaki, J., Annas, P., Huizar, K., Clarke, C., Zettergren, A., Karlsten, R. and Segerdahl, M. 2013. Evaluation of the analgesic efficacy and psychoactive effects of AZD1940, a novel peripherally acting Cannabinoid agonist, in human capsaicin-induced pain and hyperalgesia. *Clinical and Experimental Pharmacology and Physiology* 40: 212–8.

Kato, A., Punnakkal, P., Pernia-Andrade, A. J., Von Schoultz, C., Sharopov, S., Nyilas, R., Katona, I. and Zeilhofer, H. U. 2012. Endocannabinoid-dependent plasticity at spinal nociceptor synapses. *The Journal of Physiology* 590: 4717–33.

Kinsey, S. G., Long, J. Z., O'neal, S. T., Abdullah, R. A., Poklis, J. L., Boger, D. L., Cravatt, B. F. and Lichtman, A. H. 2009. Blockade of endocannabinoid-degrading enzymes attenuates neuropathic pain. *The Journal of Pharmacology and Experimental Therapeutics* 330: 902–10.

Klein, T. W., Newton, C. and Friedman, H. 1998. Cannabinoid receptors and immunity. *Immunology Today* 19: 373–81.

Knerlich-Lukoschus, F., Noack, M., Von Der Ropp-Brenner, B., Lucius, R., Mehdorn, H. M. and Held-Feindt, J. 2011. Spinal cord injuries induce changes in CB1 cannabinoid receptor and C-C chemokine expression in brain areas underlying circuitry of chronic pain conditions. *Journal of Neurotrauma* 28: 619–34.

La Porta, C., Bura, S. A., Aracil-Fernandez, A., Manzanares, J. and Maldonado, R. 2013. Role of CB1 and CB2 cannabinoid receptors in the development of joint pain induced by monosodium iodoacetate. *Pain* 154: 160–74.

La Rana, G., Russo, R., Campolongo, P., Bortolato, M., Mangieri, R. A., Cuomo, V., Iacono, A., Raso, G. M., Meli, R., Piomelli, D. and Calignano, A. 2006. Modulation of neuropathic and inflammatory pain by the endocannabinoid transport inhibitor AM404 [N-(4-hydroxyphenyl)-eicosa-5,8,11,14-tetraenamide]. *The Journal of Pharmacology and Experimental Therapeutics* 317: 1365–71.

Lauckner, J. E., Hille, B. and Mackie, K. 2005. The Cannabinoid agonist WIN55,212–2 increases intracellular calcium via CB1 receptor coupling to Gq/11 G proteins. *Proceedings of the National Academy of Sciences of the United States of America* 102: 19144–9.

Lauckner, J. E., Jensen, J. B., Chen, H. Y., Lu, H. C., Hille, B. and Mackie, K. 2008. GPR55 is a cannabinoid receptor that increases intracellular calcium and inhibits M current. *Proceedings of the National Academy of Sciences of the United States of America* 105: 2699–704.

Ledent, C., Valverde, O., Cossu, G., Petitet, F., Aubert, J. F., Beslot, F., Bohme, G. A., Imperato, A., Pedrazzini, T., Roques, B. P., Vassart, G., Fratta, W. and Parmentier, M. 1999. Unresponsiveness to cannabinoids and reduced addictive effects of opiates in CB1 receptor knockout mice. *Science* 283: 401–4.

Lichtman, A. H. and Martin, B. R. 1991. Spinal and supraspinal components of cannabinoid-induced antinociception. *The Journal of Pharmacology and Experimental Therapeutics* 258: 517–23.

Lichtman, A. H. and Martin, B. R. 1997. The selective Cannabinoid antagonist SR 141716A blocks cannabinoid-induced antinociception in rats. *Pharmacology Biochemistry and Behavior* 57: 7–12.

Lichtman, A. H., Cook, S. A. and Martin, B. R. 1996. Investigation of brain sites mediating cannabinoid-induced antinociception in rats: Evidence supporting periaqueductal gray involvement. *The Journal of Pharmacology and Experimental Therapeutics* 276: 585–93.

Long, J. Z., Li, W., Booker, L., Burston, J. J., Kinsey, S. G., Schlosburg, J. E., Pavon, F. J., Serrano, A. M., Selley, D. E., Parsons, L. H., Lichtman, A. H. and Cravatt, B. F. 2009. Selective blockade of 2-arachidonoylglycerol hydrolysis produces cannabinoid behavioral effects. *Nature Chemical Biology* 5: 37–44.

Lunn, C. A., Reich, E. P., Fine, J. S., Lavey, B., Kozlowski, J. A., Hipkin, R. W., Lundell, D. J. and Bober, L. 2008. Biology and therapeutic potential of cannabinoid CB2 receptor inverse agonists. *British Journal of Pharmacology* 153: 226–39.

Lunn, C. A., Fine, J. S., Rojas-Triana, A., Jackson, J. V., Fan, X., Kung, T. T., Gonsiorek, W., Schwarz, M. A., Lavey, B., Kozlowski, J. A., Narula, S. K., Lundell, D. J., Hipkin, R. W. and Bober, L. A. 2006. A novel cannabinoid peripheral cannabinoid receptor-selective inverse agonist blocks leukocyte recruitment in vivo. *The Journal of Pharmacology and Experimental Therapeutics* 316: 780–8.

Mackie, K. and Hille, B. 1992. Cannabinoids inhibit N-type calcium channels in neuroblastoma-glioma cells. *Proceedings of the National Academy of Sciences of the United States of America* 89: 3825–9.

Mackie, K., Lai, Y., Westenbroek, R. and Mitchell, R. 1995. Cannabinoids activate an inwardly rectifying potassium conductance and inhibit Q-type calcium currents in AtT20 cells transfected with rat brain cannabinoid receptor. *The Journal of Neuroscience* 15: 6552–61.

Malan, T. P., Jr., Ibrahim, M. M., Vanderah, T. W., Makriyannis, A. and Porreca, F. 2002. Inhibition of pain responses by activation of CB(2) cannabinoid receptors. *Chemistry and Physics of Lipids* 121: 191–200.

Malan, T. P., Jr., Ibrahim, M. M., Deng, H., Liu, Q., Mata, H. P., Vanderah, T., Porreca, F. and Makriyannis, A. 2001. CB2 cannabinoid receptor-mediated peripheral antinociception. *Pain* 93: 239–45.

Mancini, I., Brusa, R., Quadrato, G., Foglia, C., Scandroglio, P., Silverman, L. S., Tulshian, D., Reggiani, A. and Beltramo, M. 2009. Constitutive activity of cannabinoid-2 (CB2) receptors plays an essential role in the protean agonism of (+)AM1241 and L768242. *British Journal of Pharmacology* 158: 382–91.

Manning, B. H., Merin, N. M., Meng, I. D. and Amaral, D. G. 2001. Reduction in opioid- and cannabinoid-induced antinociception in rhesus monkeys after bilateral lesions of the amygdaloid complex. *The Journal of Neuroscience* 21: 8238–46.

Maresz, K., Carrier, E. J., Ponomarev, E. D., Hillard, C. J. and Dittel, B. N. 2005. Modulation of the cannabinoid CB2 receptor in microglial cells in response to inflammatory stimuli. *Journal of Neurochemistry* 95: 437–45.

Marini, P., Cascio, M. G., King, A., Pertwee, R. G. and Ross, R. A. 2013. Characterization of cannabinoid receptor ligands in tissues natively expressing cannabinoid CB2 receptors. *British Journal of Pharmacology* 169: 887–99.

Marsicano, G., Wotjak, C. T., Azad, S. C., Bisogno, T., Rammes, G., Cascio, M. G., Hermann, H., Tang, J., Hofmann, C., Zieglgansberger, W., Di Marzo, V. and Lutz, B. 2002. The endogenous cannabinoid system controls extinction of aversive memories. *Nature* 418: 530–4.

Martin, B. R., Compton, D. R., Thomas, B. F., Prescott, W. R., Little, P. J., Razdan, R. K., Johnson, M. R., Melvin, L. S., Mechoulam, R. and Ward, S. J. 1991. Behavioral, biochemical, and molecular modeling evaluations of cannabinoid analogs. *Pharmacology Biochemistry and Behavior* 40: 471–8.

Martin, W. J., Hohmann, A. G. and Walker, J. M. 1996. Suppression of noxious stimulus-evoked activity in the ventral posterolateral nucleus of the thalamus by a Cannabinoid agonist: Correlation between electrophysiological and antinociceptive effects. *The Journal of Neuroscience* 16: 6601–11.

Martin, W. J., Loo, C. M. and Basbaum, A. I. 1999a. Spinal cannabinoids are anti-allodynic in rats with persistent inflammation. *Pain* 82: 199–205.

Martin, W. J., Tsou, K. and Walker, J. M. 1998. Cannabinoid receptor-mediated inhibition of the rat tail-flick reflex after microinjection into the rostral ventromedial medulla. *Neuroscience Letters* 242: 33–6.

Martin, W. J., Patrick, S. L., Coffin, P. O., Tsou, K. and Walker, J. M. 1995. An examination of the central sites of action of cannabinoid-induced antinociception in the rat. *Life Sciences* 56: 2103–9.

Martin, W. J., Coffin, P. O., Attias, E., Balinsky, M., Tsou, K. and Walker, J. M. 1999b. Anatomical basis for cannabinoid-induced antinociception as revealed by intracerebral microinjections. *Brain Research* 822: 237–42.

Matsuda, L. A., Lolait, S. J., Brownstein, M. J., Young, A. C. and Bonner, T. I. 1990. Structure of a cannabinoid receptor and functional expression of the cloned cDNA. *Nature* 346: 561–4.

Mechoulam, R. and Gaoni, Y. 1965. A total synthesis of Dl-Delta-1-ssh. *Journal of the American Chemical Society* 87: 3273–5.

Mechoulam, R., Ben-Shabat, S., Hanus, L., Ligumsky, M., Kaminski, N. E., Schatz, A. R., Gopher, A., Almog, S., Martin, B. R., Compton, D. R. and Et Al. 1995. Identification of an endogenous 2-monoglyceride, present in canine gut, that binds to cannabinoid receptors. *Biochemical Pharmacology* 50: 83–90.

Meng, I. D., Manning, B. H., Martin, W. J. and Fields, H. L. 1998. An analgesia circuit activated by cannabinoids. *Nature* 395: 381–3.

Milne, G. M., Koe, B. K. and Johnson, M. R. 1979. Stereospecific and potent analgetic activity for nantradol: A structurally novel, cannabinoid-related analgetic. *NIDA Research Monograph* 27: 84–92.

Mitchell, V. A., Greenwood, R., Jayamanne, A. and Vaughan, C. W. 2007. Actions of the endocannabinoid transport inhibitor AM404 in neuropathic and inflammatory pain models. *Clinical and Experimental Pharmacology and Physiology* 34: 1186–90.

Moldrich, G. and Wenger, T. 2000. Localization of the CB1 cannabinoid receptor in the rat brain. An immuno-histochemical study. *Peptides* 21: 1735–42.

Mukherjee, S., Adams, M., Whiteaker, K., Daza, A., Kage, K., Cassar, S., Meyer, M. and Yao, B. B. 2004. Species comparison and pharmacological characterization of rat and human CB2 cannabinoid receptors. *European Journal of Pharmacology* 505: 1–9.

Mukhopadhyay, S., Mcintosh, H. H., Houston, D. B. and Howlett, A. C. 2000. The CB(1) cannabinoid receptor juxtamembrane C-terminal peptide confers activation to specific G proteins in brain. *Molecular Pharmacology* 57: 162–70.

Munro, S., Thomas, K. L. and Abu-Shaar, M. 1993. Molecular characterization of a peripheral receptor for cannabinoids. *Nature* 365: 61–5.

Nackley, A. G., Makriyannis, A. and Hohmann, A. G. 2003. Selective activation of cannabinoid CB(2) receptors suppresses spinal fos protein expression and pain behavior in a rat model of inflammation. *Neuroscience* 119: 747–57.

O'Shaugnessy, W. B. 1839. On the Preparations of the Indian Hemp, or Gunjah (Cabbanis indica); Their effect on the animal system in health, and their utility in the treatement of tetanus and other convulsive diseases *Transactions of the Medical and Physical Society of Bengal*, 421–461.

Oka, S., Nakajima, K., Yamashita, A., Kishimoto, S. and Sugiura, T. 2007. Identification of GPR55 as a lysophosphatidylinositol receptor. *Biochemical and Biophysical Research Communications* 362: 928–34.

Onaivi, E. S., Ishiguro, H., Gong, J. P., Patel, S., Meozzi, P. A., Myers, L., Perchuk, A., Mora, Z., Tagliaferro, P. A., Gardner, E., Brusco, A., Akinshola, B. E., Liu, Q. R., Chirwa, S. S., Hope, B., Lujilde, J., Inada, T., Iwasaki, S., Macharia, D., Teasenfitz, L., Arinami, T. and Uhl, G. R. 2008. Functional expression of brain neuronal CB2 cannabinoid receptors are involved in the effects of drugs of abuse and in depression. *Annals of the New York Academy of Sciences* 1139: 434–49.

Pacher, P. and Kunos, G. 2013. Modulating the endocannabinoid system in human health and disease: Successes and failures. *The FEBS Journal* 280: 1918–43.

Palazuelos, J., Aguado, T., Egia, A., Mechoulam, R., Guzman, M. and Galve-Roperh, I. 2006. Non-psychoactive CB2 Cannabinoid agonists stimulate neural progenitor proliferation. *FASEB Journal* 20: 2405–7.

Palazzo, E., De Novellis, V., Petrosino, S., Marabese, I., Vita, D., Giordano, C., Di Marzo, V., Mangoni, G. S., Rossi, F. and Maione, S. 2006. Neuropathic pain and the endocannabinoid system in the dorsal raphe: Pharmacological treatment and interactions with the serotonergic system. *The European Journal of Neuroscience* 24: 2011–20.

Pamplona, F. A., Ferreira, J., Menezes De Lima, O., Jr., Duarte, F. S., Bento, A. F., Forner, S., Villarinho, J. G., Bellocchio, L., Wotjak, C. T., Lerner, R., Monory, K., Lutz, B., Canetti, C., Matias, I., Calixto, J. B., Marsicano, G., Guimaraes, M. Z. and Takahashi, R. N. 2012. Anti-inflammatory lipoxin A4 is an endogenous allosteric enhancer of CB1 cannabinoid receptor. *Proceedings of the National Academy of Sciences of the United States of America* 109: 21134–9.

Pernia-Andrade, A. J., Kato, A., Witschi, R., Nyilas, R., Katona, I., Freund, T. F., Watanabe, M., Filitz, J., Koppert, W., Schuttler, J., Ji, G., Neugebauer, V., Marsicano, G., Lutz, B., Vanegas, H. and Zeilhofer, H. U. 2009. Spinal endocannabinoids and CB1 receptors mediate C-fiber-induced heterosynaptic pain sensitization. *Science* 325: 760–4.

Pertwee, R., Griffin, G., Fernando, S., Li, X., Hill, A. and Makriyannis, A. 1995. AM630, a competitive cannabinoid receptor antagonist. *Life Sciences* 56: 1949–55.

Piomelli, D. 2003. The molecular logic of endocannabinoid signalling. *Nature Reviews Neuroscience* 4: 873–84.

Piomelli, D., Beltramo, M., Giuffrida, A. and Stella, N. 1998. Endogenous cannabinoid signaling. *Neurobiology of Disease* 5: 462–73.

Piomelli, D., Beltramo, M., Glasnapp, S., Lin, S. Y., Goutopoulos, A., Xie, X. Q. and Makriyannis, A. 1999. Structural determinants for recognition and translocation by the Anandamide Transporter. *Proceedings of the National Academy of Sciences of the United States of America* 96: 5802–7.

Potenzieri, C., Brink, T. S., Pacharinsak, C. and Simone, D. A. 2008. Cannabinoid modulation of cutaneous Adelta nociceptors during inflammation. *Journal of Neurophysiology* 100: 2794–806.

Quartilho, A., Mata, H. P., Ibrahim, M. M., Vanderah, T. W., Porreca, F., Makriyannis, A. and Malan, T. P., Jr. 2003. Inhibition of inflammatory hyperalgesia by activation of peripheral CB2 cannabinoid receptors. *Anesthesiology* 99: 955–60.

Ribeiro, R., Wen, J., Li, S. and Zhang, Y. 2013. Involvement of ERK1/2, cPLA2 and NF-kappaB in microglia suppression by cannabinoid receptor agonists and antagonists. *Prostaglandins and Other Lipid Mediators* 100–1: 1–14.

Richardson, J. D., Aanonsen, L. and Hargreaves, K. M. 1998. Hypoactivity of the spinal cannabinoid system results in NMDA-dependent hyperalgesia. *The Journal of Neuroscience* 18: 451–7.

Riether, D. 2012. Selective cannabinoid receptor 2 modulators: A patent review 2009–present. *Expert Opinion on Therapeutic Patents* 22: 495–510.

Rinaldi-Carmona, M., Barth, F., Millan, J., Derocq, J. M., Casellas, P., Congy, C., Oustric, D., Sarran, M., Bouaboula, M., Calandra, B., Portier, M., Shire, D., Breliere, J. C. and Le Fur, G. L. 1998. SR 144528, the first potent and selective antagonist of the CB2 cannabinoid receptor. *The Journal of Pharmacology and Experimental Therapeutics* 284: 644–50.

Rinaldi-Carmona, M., Barth, F., Heaulme, M., Shire, D., Calandra, B., Congy, C., Martinez, S., Maruani, J., Neliat, G., Caput, D. et al. 1994. SR141716A, a potent and selective antagonist of the brain cannabinoid receptor. *FEBS Letters* 350: 240–4.

Rodriguez, J. J., Mackie, K. and Pickel, V. M. 2001. Ultrastructural localization of the CB1 cannabinoid receptor in mu-opioid receptor patches of the rat Caudate putamen nucleus. *The Journal of Neuroscience* 21: 823–33.

Roques, B. P., Fournie-Zaluski, M. C. and Wurm, M. 2012. Inhibiting the breakdown of endogenous opioids and cannabinoids to alleviate pain. *Nature reviews. Drug Discovery* 11: 292–310.

Ruiz, M., Sanchez, D., Correnti, C., Strong, R. K. and Ganfornina, M. D. 2013. Lipid-binding properties of human ApoD and Lazarillo-related lipocalins: Functional implications for cell differentiation. *The FEBS Journal* 280: 3928–43.

Russo, E. B., Jiang, H. E., Li, X., Sutton, A., Carboni, A., Del Bianco, F., Mandolino, G., Potter, D. J., Zhao, Y. X., Bera, S., Zhang, Y. B., Lu, E. G., Ferguson, D. K., Hueber, F., Zhao, L. C., Liu, C. J., Wang, Y. F. and Li, C. S. 2008. Phytochemical and genetic analyses of ancient Cannabis from Central Asia. *Journal of Experimental Botany* 59: 4171–82.

Russo, R., Loverme, J., La Rana, G., Compton, T. R., Parrott, J., Duranti, A., Tontini, A., Mor, M., Tarzia, G., Calignano, A. and Piomelli, D. 2007. The fatty acid amide hydrolase inhibitor URB597 (cyclohexylcarbamic acid 3'-carbamoylbiphenyl-3-yl ester) reduces neuropathic pain after oral administration in mice. *The Journal of Pharmacology and Experimental Therapeutics* 322: 236–42.

Sagar, D. R., Kelly, S., Millns, P. J., O'shaughnessey, C. T., Kendall, D. A. and Chapman, V. 2005. Inhibitory effects of CB1 and CB2 receptor agonists on responses of DRG neurons and dorsal horn neurons in neuropathic rats. *The European Journal of Neuroscience* 22: 371–9.

Salio, C., Fischer, J., Franzoni, M. F. and Conrath, M. 2002. Pre- and postsynaptic localizations of the CB1 cannabinoid receptor in the dorsal horn of the rat spinal cord. *Neuroscience* 110: 755–64.

Seely, K. A., Brents, L. K., Franks, L. N., Rajasekaran, M., Zimmerman, S. M., Fantegrossi, W. E. and Prather, P. L. 2012. AM-251 and rimonabant act as direct antagonists at mu-opioid receptors: Implications for opioid/cannabinoid interaction studies. *Neuropharmacology* 63: 905–15.

Shoemaker, J. L., Ruckle, M. B., Mayeux, P. R. and Prather, P. L. 2005. Agonist-directed trafficking of response by endocannabinoids acting at CB2 receptors. *The Journal of Pharmacology and Experimental Therapeutics* 315: 828–38.

Siegling, A., Hofmann, H. A., Denzer, D., Mauler, F. and De Vry, J. 2001. Cannabinoid CB(1) receptor upregulation in a rat model of chronic neuropathic pain. *European Journal of Pharmacology* 415, R5–7.

Smart, D., Gunthorpe, M. J., Jerman, J. C., Nasir, S., Gray, J., Muir, A. I., Chambers, J. K., Randall, A. D. and Davis, J. B. 2000. The endogenous lipid Anandamide is a full agonist at the human vanilloid receptor (hVR1). *British Journal of Pharmacology* 129: 227–30.

Sofia, R. D., Vassar, H. B. and Knobloch, L. C. 1975. Comparative analgesic activity of various naturally occurring cannabinoids in mice and rats. *Psychopharmacologia* 40: 285–95.

Steffens, S. and Pacher, P. 2012. Targeting cannabinoid receptor CB(2) in cardiovascular disorders: Promises and controversies. *British Journal of Pharmacology* 167: 313–23.

Stella, N. 2010. Cannabinoid and cannabinoid-like receptors in microglia, astrocytes, and astrocytomas. *Glia* 58: 1017–30.

Stella, N., Schweitzer, P. and Piomelli, D. 1997. A second endogenous cannabinoid that modulates long-term potentiation. *Nature* 388: 773–8.

Straiker, A., Stella, N., Piomelli, D., Mackie, K., Karten, H. J. and Maguire, G. 1999. Cannabinoid CB1 receptors and ligands in vertebrate retina: Localization and function of an endogenous signaling system. *Proceedings of the National Academy of Sciences of the United States of America* 96: 14565–70.

Sugiura, T., Kodaka, T., Kondo, S., Tonegawa, T., Nakane, S., Kishimoto, S., Yamashita, A. and Waku, K. 1996. 2-Arachidonoylglycerol, a putative endogenous cannabinoid receptor ligand, induces rapid, transient elevation of intracellular free Ca2+ in neuroblastoma x glioma hybrid NG108–15 cells. *Biochemical and Biophysical Research Communications* 229: 58–64.

Sugiura, T., Kodaka, T., Kondo, S., Nakane, S., Kondo, H., Waku, K., Ishima, Y., Watanabe, K. and Yamamoto, I. 1997. Is the cannabinoid CB1 receptor a 2-arachidonoylglycerol receptor? Structural requirements for triggering a Ca2+ transient in NG108–15 cells. *Journal of Biochemistry* 122: 890–5.

Svizenska, I. H., Brazda, V., Klusakova, I. and Dubovy, P. 2013. Bilateral changes of cannabinoid receptor type 2 protein and mRNA in the dorsal root ganglia of a rat neuropathic pain model. *The Journal of Histochemistry and Cytochemistry* 61: 529–47.

Tam, J., Vemuri, V. K., Liu, J., Batkai, S., Mukhopadhyay, B., Godlewski, G., Osei-Hyiaman, D., Ohnuma, S., Ambudkar, S. V., Pickel, J., Makriyannis, A. and Kunos, G. 2010. Peripheral CB1 cannabinoid receptor blockade improves cardiometabolic risk in mouse models of obesity. *The Journal of Clinical Investigation* 120: 2953–66.

Tam, J., Cinar, R., Liu, J., Godlewski, G., Wesley, D., Jourdan, T., Szanda, G., Mukhopadhyay, B., Chedester, L., Liow, J. S., Innis, R. B., Cheng, K., Rice, K. C., Deschamps, J. R., Chorvat, R. J., Mcelroy, J. F. and Kunos, G. 2012. Peripheral cannabinoid-1 receptor inverse agonism reduces obesity by reversing leptin resistance. *Cell Metabolism* 16: 167–79.

Tsou, K., Brown, S., Sanudo-Pena, M. C., Mackie, K. and Walker, J. M. 1998a. Immunohistochemical distribution of cannabinoid CB1 receptors in the rat central nervous system. *Neuroscience* 83: 393–411.

Tsou, K., Nogueron, M. I., Muthian, S., Sanudo-Pena, M. C., Hillard, C. J., Deutsch, D. G. and Walker, J. M. 1998b. Fatty acid amide hydrolase is located preferentially in large neurons in the rat central nervous system as revealed by immunohistochemistry. *Neuroscience Letters* 254: 137–40.

Valenzano, K. J., Tafesse, L., Lee, G., Harrison, J. E., Boulet, J. M., Gottshall, S. L., Mark, L., Pearson, M. S., Miller, W., Shan, S., Rabadi, L., Rotshteyn, Y., Chaffer, S. M., Turchin, P. I., Elsemore, D. A., Toth, M., Koetzner, L. and Whiteside, G. T. 2005. Pharmacological and pharmacokinetic characterization of the cannabinoid receptor 2 agonist, GW405833, utilizing rodent models of acute and chronic pain, anxiety, ataxia and catalepsy. *Neuropharmacology* 48: 658–72.

Van Der Stelt, M., Trevisani, M., Vellani, V., De Petrocellis, L., Schiano Moriello, A., Campi, B., Mcnaughton, P., Geppetti, P. and Di Marzo, V. 2005. Anandamide acts as an intracellular messenger amplifying Ca2+ influx via TRPV1 channels. *The EMBO Journal* 24: 3026–37.

Van Sickle, M. D., Duncan, M., Kingsley, P. J., Mouihate, A., Urbani, P., Mackie, K., Stella, N., Makriyannis, A., Piomelli, D., Davison, J. S., Marnett, L. J., Di Marzo, V., Pittman, Q. J., Patel, K. D. and Sharkey, K. A. 2005. Identification and functional characterization of brainstem cannabinoid CB2 receptors. *Science* 310: 329–32.

Vaughan, C. W., Connor, M., Bagley, E. E. and Christie, M. J. 2000. Actions of cannabinoids on membrane properties and synaptic transmission in rat periaqueductal gray neurons in vitro. *Molecular Pharmacology* 57: 288–95.

Veress, G., Meszar, Z., Muszil, D., Avelino, A., Matesz, K., Mackie, K. and Nagy, I. 2013. Characterisation of cannabinoid 1 receptor expression in the perikarya, and peripheral and spinal processes of primary sensory neurons. *Brain Structure and Function* 218: 733–50.

Waksman, Y., Olson, J. M., Carlisle, S. J. and Cabral, G. A. 1999. The central cannabinoid receptor (CB1) mediates inhibition of nitric oxide production by rat microglial cells. *The Journal of Pharmacology and Experimental Therapeutics* 288: 1357–66.

Walker, J. M., Huang, S. M., Strangman, N. M., Tsou, K. and Sanudo-Pena, M. C. 1999. Pain modulation by release of the endogenous cannabinoid Anandamide. *Proceedings of the National Academy of Sciences of the United States of America* 96: 12198–203.

Walter, L., Franklin, A., Witting, A., Wade, C., Xie, Y., Kunos, G., Mackie, K. and Stella, N. 2003. Nonpsychotropic cannabinoid receptors regulate microglial cell migration. *The Journal of Neuroscience* 23: 1398–405.

Whiteside, G. T., Gottshall, S. L., Boulet, J. M., Chaffer, S. M., Harrison, J. E., Pearson, M. S., Turchin, P. I., Mark, L., Garrison, A. E. and Valenzano, K. J. 2005. A role for cannabinoid receptors, but not endogenous opioids, in the antinociceptive activity of the CB2-selective agonist, GW405833. *European Journal of Pharmacology* 528: 65–72.

Wisler, J. W., Dewire, S. M., Whalen, E. J., Violin, J. D., Drake, M. T., Ahn, S., Shenoy, S. K. and Lefkowitz, R. J. 2007. A unique mechanism of beta-blocker action: Carvedilol stimulates beta-arrestin signaling. *Proceedings of the National Academy of Sciences of the United States of America* 104: 16657–62.

Wotherspoon, G., Fox, A., Mcintyre, P., Colley, S., Bevan, S. and Winter, J. 2005. Peripheral nerve injury induces cannabinoid receptor 2 protein expression in rat sensory neurons. *Neuroscience* 135: 235–45.

Xiong, W., Wu, X., Li, F., Cheng, K., Rice, K. C., Lovinger, D. M. and Zhang, L. 2012. A common molecular basis for exogenous and endogenous cannabinoid potentiation of glycine receptors. *The Journal of Neuroscience* 32: 5200–8.

Yamamoto, W., Mikami, T. and Iwamura, H. 2008. Involvement of central cannabinoid CB2 receptor in reducing mechanical allodynia in a mouse model of neuropathic pain. *European Journal of Pharmacology* 583: 56–61.

Yang, P., Wang, L. and Xie, X. Q. 2012. Latest advances in novel cannabinoid CB(2) ligands for drug abuse and their therapeutic potential. *Future Medicinal Chemistry* 4: 187–204.

Yao, B. B., Hsieh, G. C., Frost, J. M., Fan, Y., Garrison, T. R., Daza, A. V., Grayson, G. K., Zhu, C. Z., Pai, M., Chandran, P., Salyers, A. K., Wensink, E. J., Honore, P., Sullivan, J. P., Dart, M. J. and Meyer, M. D. 2008. In vitro and in vivo characterization of A-796260: A selective cannabinoid CB2 receptor agonist exhibiting analgesic activity in rodent pain models. *British Journal of Pharmacology* 153: 390–401.

Yao, B. B., Mukherjee, S., Fan, Y., Garrison, T. R., Daza, A. V., Grayson, G. K., Hooker, B. A., Dart, M. J., Sullivan, J. P. and Meyer, M. D. 2006. In vitro pharmacological characterization of AM1241: A protean agonist at the cannabinoid CB2 receptor? *British Journal of Pharmacology* 149: 145–54.

Yu, X. H., Cao, C. Q., Martino, G., Puma, C., Morinville, A., St-Onge, S., Lessard, E., Perkins, M. N. and Laird, J. M. 2010. A peripherally restricted cannabinoid receptor agonist produces robust anti-nociceptive effects in rodent models of inflammatory and neuropathic pain. *Pain* 151: 337–44.

Zeilhofer, H. U. 2010. Spinal cannabinoids: A double-edged sword? (Commentary on Zhang et al.). *The European Journal of Neuroscience* 31: 223–4.

Zhang, G., Chen, W., Lao, L. and Marvizon, J. C. 2010. Cannabinoid CB1 receptor facilitation of substance P release in the rat spinal cord, measured as neurokinin 1 receptor internalization. *The European Journal of Neuroscience* 31: 225–37.

Zhang, J., Hoffert, C., Vu, H. K., Groblewski, T., Ahmad, S. and O'Donnell, D. 2003. Induction of CB2 receptor expression in the rat spinal cord of neuropathic but not inflammatory chronic pain models. *The European Journal of Neuroscience* 17: 2750–4.

Zimmer, A., Zimmer, A. M., Hohmann, A. G., Herkenham, M. and Bonner, T. I. 1999. Increased mortality, hypoactivity, and hypoalgesia in cannabinoid CB1 receptor knockout mice. *Proceedings of the National Academy of Sciences of the United States of America* 96: 5780–5.

CHAPTER 7

EphB receptors and persistent pain

Isabella Gavazzi

As already highlighted in Chapter 2 by A. Latremolière, hypersensitivity to noxious and non-noxious stimuli in persistent pain is, at least in part, due to a change in the strength of the synapses between primary sensory afferents and neurons in the *dorsal horn* of the spinal cord, including projection neurons – a change usually due to previous strong activation of sensory afferents. This activity-dependent increase in the efficacy of sensory afferent synapses is known as *central sensitisation*. An understanding of the specific molecular mechanisms responsible for the onset and maintenance of central sensitisation would thus provide potential targets for therapeutic strategies in all forms of chronic pain, in which central sensitisation plays an important role. Crucially, these forms allegedly involve *neuropathic pain*, a common complaint for which effective analgesics are particularly scarce. In this chapter we shall present in some detail evidence linking a specific family of molecules, the *EphBRs* (*EphB receptors*, thus called because they were first identified on erythropoietin-producing hepatocellular carcinoma cells), to central sensitisation. EphBRs are by no means the only, or arguably even the most important, molecules involved in this process, but they present some features that make them a quite promising therapeutic target. They are also discussed here as an example of the type of experiments that must be performed in order to demonstrate the involvement of any molecule in the mechanisms underlying central sensitisation.

A brief introduction to Eph receptors and their ligands, the ephrins

Eph receptor tyrosine kinases were first identified in the late 1980s and were originally 'orphan' receptors, in other words their ligands were unknown. Over the following decade, the family expanded to include 14 members, which makes it the largest subgroup in the receptor tyrosine kinases family. Eight Eph receptor ligands were identified; they are collectively named *ephrins*. The Eph family is divided into two subclasses, A and B, and currently comprises, in mammals, nine EphA receptors (EphA1–A8, EphA10) and five EphBRs (EphB1–B4

An Introduction to Pain and its Relation to Nervous System Disorders, First Edition. Edited by Anna A. Battaglia.

and EphB6). EphA and EphBRs differ in terms of sequence homology and ligand-binding specificity. EphA receptors preferentially bind ephrinAs (A1–A5); EphBRs usually bind one of three ephrinBs (B1–3). Within subclasses, binding is highly promiscuous (some limited interclass binding also exists). Both ephrinAs and ephrinBs are membrane-tethered molecules; but, while ephrinAs are anchored to the cell membrane via a glycosylphosphatidylinositol (GPI) anchor, ephrinBs are transmembrane molecules. The fact that both Eph receptors and ephrins are linked to the membrane has some interesting consequences, particularly in the case of EphBRs and ephrinBs, where both receptors and ligands are transmembrane molecules. The most striking feature of these molecules, which differentiates them from other receptor tyrosine kinases and their ligands, is that receptor-ligand interactions may lead to signalling both in the receptor and in the ligand-expressing cell (bidirectional signalling: forward signalling in the receptor-expressing cell, reverse signalling in the ligand-expressing cell; see Pasquale, 2008). This bidirectional signalling has a variety of implications, both in terms of the functions that these molecules can have and for their amenability to experimental studies. To add a further level of complexity, it has fairly recently been discovered that Eph receptor-ephrin interaction may take place not only in trans – that is, when receptor and ligand are expressed on neighbouring cells – but also in cis – that is, between a receptor and a ligand expressed on the same cell (see, e.g., Antion et al., 2010). The precise biological significance of the expression, on a single cell or on neighbouring cells, of different combinations of receptors or ligands is still unclear, but could represent one way of regulating signalling. The binding of ephrins to Eph receptors causes receptor homo- or heterodimerisation (or oligomerisation), which leads to receptor autophosphorylation, thus enabling binding with proteins that contain the Src homology 2 domain. EphrinBs on the other hand can interact, among others, with Src non-receptor tyrosine kinases and with group I metabotropic glutamate receptors. Many EphBRs and all ephrinBs contain PDZ-binding domains at their cytoplasmic C terminus, which allows them to interact with postsynaptic PDZ domain-containing proteins; this would be crucial for their synaptogenic role (see below). It is thought that the size of clusters of receptor–ligand complexes, and therefore their signalling potential, may reflect the level of expression of the receptor, which would thus modulate the level – and probably the type – of response. To date, no specific agonists or antagonists exist for EphBRs and ephrinBs. Most functional studies rely therefore on the use of EphB-Fc or ephrinB-Fc chimeric molecules. EphB-Fc chimeric molecules consist of the extracellular portion of two EphBRs, fused with the Fc portion of an IgG molecule. This dimeric form of the EphBR can bind to endogenous EphrinBs, preventing them from binding to the endogenous EphBRs and thus effectively acting as a receptor antagonist. Conversely, the dimeric ephrinB2-Fc chimera can bind to and dimerise endogenous EphBRs, thus acting as an agonist and activating intracellular signalling pathways. However, EphB-Fc chimera could also potentially activate ephrinB2 reverse signalling, and ephrinB2-Fc could act as an antagonist for endogenous ephrins, so that experimental findings obtained with these molecules need to be interpreted with caution.

What is the biological role of this large class of molecules? They have been shown to have a huge variety of functions in most tissues and throughout development and adulthood. Because both Eph receptors and ephrins are membrane-bound, they are ideal to mediate interactions that require cell–cell contact, such as patterning, segmentation, cell migration and axonal guidance.

Therefore it is not surprising that in the nervous system they were initially described as fundamental players in controlling neuronal development and neural circuits formation (for review, see Pasquale, 2008; Feldheim and O'Leary, 2010; Klein, 2012). Initially they were identified as primarily chemorepellent molecules; later on it became clear that they can also regulate adhesive events. How EphBR–ephrinB binding can switch from leading to cell repulsion to stabilising cell–cell contact is still not clear, but probably it is crucial whether the receptor becomes internalised (leading to repulsion) or remains localised to the membrane, where the strong affinity of EphBRs for their ligands would lead to cell adhesion. Eph receptors and ephrins appear to be important for regionalisation at earlier stages and for axonal guidance at later stages (this is reviewed in Sefton and Nieto, 1997). At the turn of the present millennium, a new function started to emerge, particularly for Eph receptors of the B subclass, as mediators of excitatory synaptogenesis (Dalva et al., 2000). A variety of studies, carried out in primary neuronal cultures or in EphB2 receptor knock-out mice, demonstrated an important role for EphB2 receptors, in particular in ensuring the localisation of the *ionotropic glutamate NMDA receptors* at glutamatergic synapses, and also for the trafficking of these receptors' NR2B subunits to the synapse. Furthermore, EphB2 receptors can regulate *AMPA receptors'* clustering and trafficking to the synapse in cultured cortical neurons, as well as dendritic spine formation, maintenance and plasticity (for a review of the role of EphBRs and ephrins in synaptogenesis, and more generally for their role on synaptic plasticity, see Klein, 2009; Lai and Ip, 2009; Hruska and Dalva, 2012; Sheffler-Collins and Dalva, 2012; Sloniowski and Ethell, 2012). It was awareness of this role of EphBRs at glutamatergic synapses – and in particular of their interaction with NMDA receptors, which play a crucial role in the onset of central sensitisation (see Chapter 2 in this volume) – that spurred a group of researchers at King's College London to investigate the involvement of these molecules in animal models of pain (Battaglia et al., 2003) and led them to the surprising discovery that in adult animals EphBRs can rapidly modulate synaptic function at spinal cord level, thus altering the organism's sensitivity to noxious stimulation.

EphB receptors and hypersensitivity in inflammatory and neuropathic pain

At the beginning of the last decade, Battaglia and colleagues were studying the distribution of Eph receptors and ephrin ligands in primary sensory neurons and in the spinal cord of adult rats. In the developing nervous system, ephrins often act as negative guidance cues, and these researchers were exploring the possibility that, in adulthood, ephrins may be implicated in failure of regeneration within and into the central nervous system (CNS) after injury. Localisation studies of these molecules are difficult on account of the paucity of specific antibodies, which work well in immunohistochemistry. However, these researchers were struck by the strong immunoreactivity for ephrinB2 they discovered – specifically on small primary sensory neurons, which are likely to be nociceptors. One of the receptors for ephrinB2, EphB1, was found on neurons in the dorsal horn of the spinal cord (Figure 7.1). The localisation of these molecules, respectively on the pre- and postsynaptic side of the first, glutamatergic, synapse of the so-called 'pain' pathways, together with the findings mentioned above in relation to the important function of these receptors at

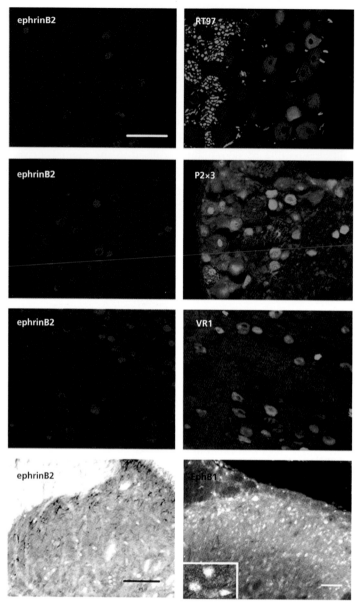

Figure 7.1 Expression of EphB1 receptors and ephrinB1 in sensory pathways: EphrinB2 ligands are present on nociceptors in DRGs, and EphB1 is expressed in neurons in the adult rat spinal cord. Battaglia et al. (2003) examined the distribution of EphB1 receptors and ephrinB ligands in the DRGs and spinal cord of adult rats. The top six panels are micrographs of sections of lumbar DRGs, double-stained with antibodies anti-ephrinB2 (left), and with markers for different subclasses of DRG neurons (right): RT97 (a marker for large and intermediate neurons, largely mechanoceptive and proprioceptive), $P2X_3$ (an ATP receptor expressed on a subset of nociceptive neurons) and VR1 (the vanilliod receptor, now named TRPV1; a marker of a subset of nociceptive neurons). EphrinB2 staining was not observed on RT97 positive, non-nociceptive neurons. On the other hand, all TRPV1-expressing neurons and nearly 80 per cent of $P2X_3$ immunoreactive neurons were positive for ephrinB2 immunofluorescence, indicating that a large proportion of nociceptive neurons express ephrinB2. The bottom two panels are sections of spinal cord (dorsal horn). EphrinB2 labelling (left panel) in the dorsal horn was associated with nerve terminals (presumably sensory afferents) in laminae I and II. In the dorsal horn, EphB1 immunoreactivity (right panel) was predominantly associated with laminae I–III neurons. In the inset: EphB1 staining on a motoneuron. Bars: 100 m; all DRG micrographs were taken at the same magnification. Source: Battaglia 2003.

developing glutamatergic synapses in the brain, led Battaglia and colleagues (2003) to formulate the hypothesis that EphBR–ephrinB interactions may be important for the plastic changes known to occur at these synapses in chronic pain states. Their initial hypothesis, though, was that these molecules would be important for the late structural changes at these synapses that accompany prolonged central sensitisation. They were therefore quite surprised to discover that the activation of EphB1 receptors, achieved by injecting intrathecally an ephrinB2-Fc chimera into the lumbar region of the spinal cord of adult rats, led rapidly (within 15–30 minutes) to behavioural hypersensitivity to thermal stimuli in the hind paws. No mechanical hyperalgesia appears to develop in the rat, while in the mouse some researchers observed a development of mechanical hyperalgesia as well (Yu et al., 2012). Subsequently a fairly large number of behavioural studies, some of which successfully replicated in independent laboratories, have shown how EphBRs present in the spinal cord contribute to the increased sensitivity to thermal and mechanical stimuli that accompany inflammatory and neuropathic pain. In the absence of specific agonists or antagonists for any of the Eph receptors, these studies have had to employ primarily one of two alternative methods: either they have made use of EphBR and ephrinB ligand-Fc chimeric molecules; or they have made use of knock-out mice (lacking a functional EphB1 receptor or ephrinB2 ligand).

Using EphB1-Fc and ephrinB2-Fc molecules as agonists and antagonists for EphB receptors

Both chimeric EphB1-Fc and ephrinB2-Fc molecules can be delivered to the dorsal spinal cord via intrathecal injection and thus act on endogenous EphBRs/ephrinB ligands.

Using this technique, it was established that EphB–ephrinB interactions probably do not to play a role in acute nociception, since injecting EphB1–Fc chimeras did not affect the response to heat or mechanical stimulation in intact animals or the response in the first phase of the *formalin model* (a model of tissue injury), which is considered akin to acute nociception (Battaglia et al., 2003; one exception is represented by the work of Yu et al., 2012, who found reduced response in mice in phase I as well). However, experiments conducted in several independent laboratories showed that, in both rats and mice, the injection of ephrinB2-Fc, acting as an EphBR agonist, did sensitise the animals to thermal stimulation – and, as mentioned above, in mice at times also to mechanical stimuli: see Battaglia et al., 2003; Song, Zheng, et al., 2008; Yu et al., 2012; in some experiments ephrinB2-Fc was replaced by ephrinB1-Fc, which led, however, to undistinguishable results.

Conversely, blocking the EphBR activation through injection of EphB1-Fc or EphB2-Fc prevented the development of thermal and mechanical hyperalgesia, or spontaneous pain, in a number of models of persistent pain (formalin injection in the hind paw, a model of tissue injury; carrageenan injection, a model of *inflammatory pain*; *chronic constriction injury* (CCI), a model of neuropathic pain; see Battaglia et al., 2003; Song, Zheng. et al., 2008; Dong et al., 2011). This suggested that EphBR activation is indeed required for the development of hypersensitivity and spontaneous pain in these models. The results obtained with EphB1-Fc in pain models exclude the possibility that the hyperalgesic effect of ephrinB2-Fc was purely pharmacological and not linked to a physiological mechanism.

It was mentioned above how EphBRs are quite promiscuous in their binding to the three ephrinB ligands; hence these studies with chimeric molecules cannot block or activate a specific EphBR. The immunolocalisation studies of Battaglia and colleagues (2003) suggested that EphB1

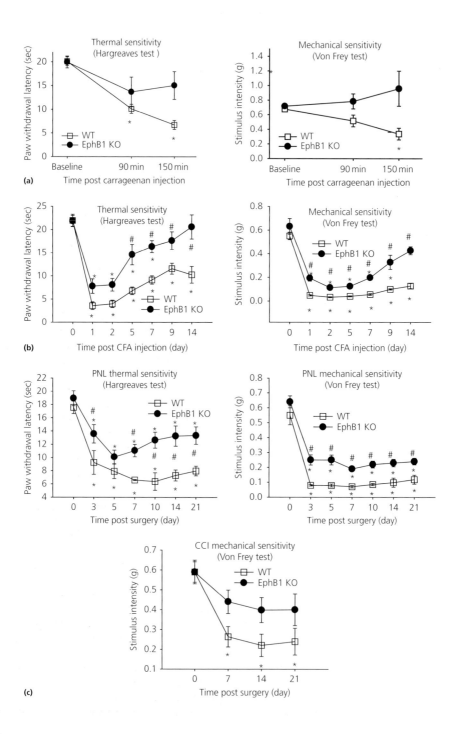

receptors and ephrin B2 ligands, present on dorsal horn neurons and in peripheral nociceptors respectively, could be the primary molecules involved in the modulation of synaptic efficacy in the dorsal horn. Other immunolocalisation as well as western blot studies (which, however, do not allow specific localisation at the cellular level) seemed to confirm the involvement of EphB1 in neurons of the dorsal horn of the spinal cord, while some contradictory findings were reported for the ephrinB involved (B1 vs B2; see below for details). In the absence of specific agonists and antagonists, and given the lack of adequately specific antibodies for immunolocalisation for all receptors and ligands, a univocal identification of the molecules involved has, for the moment, primarily relied on studies of genetically modified animals.

Inflammatory and neuropathic pain in EphB1 knock-out mice and nociceptor-specific ephrinB2 knock-out mice

There is some unavoidable degree of ambiguity in interpreting data from transgenic knock-out animals, since the knocked-out molecule is absent throughout the animal's life, hence the difficulty to tease out its adult function versus its possible roles in development. This is of course potentially a major problem with molecules such as EphBRs and ephrinBs, which have a prominent function in nervous system development. Despite this, or possibly because the fundamental role of these molecules in development ensures the presence of compensatory mechanisms in knock-out mice, *EphB1 knock-out* animals appear largely normal, and in particular their sensitivity to thermal and mechanical stimuli is undistinguishable from that of their wild-type littermates (Cibert-Goton et al., 2013). However, when examined in a wide variety of models of persistent pain (formalin, carrageenan, Selzer and Bennett models of neuropathic pain, cancer pain: see Han et al., 2008; Liu et al., 2011; Cibert-Goton et al., 2013) (Figure 7.2), EphB1 knock-out mice demonstrate blunted responses (significantly smaller or no development

Figure 7.2 EphB1 knock-out (KO) mice display reduced thermal and mechanical hyperalgesia in inflammatory pain models and accelerated recovery from mechanical and thermal hyperalgesia following nerve injury. The behavioural responses of EphB1 KO and wild-type (WT) mice were examined in two tests of inflammatory pain (**a** and **b**) and two tests of neuropathic pain (**c**). For the testing of inflammatory pain, either carrageenan (**a**) or Complete Freud's Adjuvant (CFA) (**b**) were injected in a hind paw, and responses to thermal (Hargreaves) and mechanical (von Frey) stimuli were recorded up to a maximum of 150 minutes (**a**) or 14 days (**b**). Following carrageenan injection, all WT mice developed thermal and mechanical hyperalgesia, whereas EphB1 KO mice did not develop either. Following CFA injection, which elicits a greater and more prolonged inflammatory response than carrageenan, both WT and EphB1 KO mice developed thermal and mechanical hyperalgesia. However, EphB1 KO mice displayed a faster recovery in terms of thermal hyperalgesia and a significantly smaller development of mechanical allodynia. Behavioural responses following peripheral nerve lesion were examined in EphB1 KO mice and in WT mice using two models of nerve injury: partial nerve lesion (PNL) and CCI. Development of both thermal and mechanical hyperalgesia was studied after PNL; mechanical hyperalgesia only was examined after CCI. WT mice developed significant thermal hyperalgesia in both models. In WT mice with PNL, hyperalgesia developed about three days after injury and persisted throughout the duration of the experiment. This reduction in paw withdrawal latency was significantly smaller in EphB1 KO mice at all time points, except in the five days after injury. WT mice also developed mechanical allodynia, with a marked decrease in the stimulus intensity required to trigger a lifting response of the injured paw by comparison to baseline. EphB1 KO mice displayed a less pronounced mechanical allodynia when compared to WT at all time points (p < 0.05). In the CCI model, mice responses to mechanical stimuli were tested at 7, 14 and 21 days after injury. WT mice developed a significant mechanical allodynia by 7 days (p < 0.05), and this persisted throughout the experiment, while EphB1 KO mice failed to develop a significant mechanical allodynia. * p < 0.05 vs respective baseline, # p < 0.05 comparing WT and EphB1 KO mice at equivalent time points. Source: Cibert-Goton 2013.

of mechanical and thermal hyperalgesia and spontaneous pain, or faster recovery), which support the hypothesis that EphB1 receptors are involved in the development of hypersensitivity and spontaneous pain in these models.

A similar, even if somewhat less striking, outcome was obtained with mice in which ephrinB2 was removed, specifically, from a subset of sensory neurons that was thought to represent the majority of nociceptive afferents (ephrinB2 knock-out mice do not survive, due to lethal defects in their cardiovascular development) (Zhao et al., 2010). The ephrinB2 gene is removed from nociceptors with the help of the cre-lox system, cre being under the promoter for Nav1.8 – a sodium channel thought to be present exclusively in nociceptors. The residual hypersensitivity in nociceptor-specific ephrinB2 knock-out mice could be explained by any of the following, alone or in combination: (1) ephrinB2 is removed in nociceptors during development, therefore compensatory mechanisms could have partially overcome its absence; (2) it has recently been discovered that the cre-lox system mentioned above, where the cre is under the Nav1.8 promoter, does not inactivate the target genes in all nociceptive neurons (and also inactivates genes in non-nociceptive neurons (Shields et al., 2012) – therefore there could still be significant levels of ephrinB2, particularly after injury or inflammation: the presence of ephrinB2 protein in injured animals was not examined in detail in conditional ephrinB2 knock-outs; (3) other ephrinBs, in particular ephrinB1, may contribute to the development of hypersensitivity.

In favour of a predominant involvement of ephrinB2 is a study conducted by using RNA interference techniques, which knocked down ephrinB2 with small interfering RNAs and found that *mechanical allodynia* was reduced in rats that had undergone an L5 spinal nerve crush (a model of neuropathic pain) (Kobayashi et al., 2007). The authors of this study could not detect ephrinB1 or B3 in dorsal root ganglion (DRG) neurons. In this case, since ephrinB2 was knocked out in adult animals, compensatory mechanisms are unlikely to have been activated.

In conclusion, studies carried out both in adult rats with 'agonist-like' and 'antagonist-like' molecules and in transgenic animals, using a variety of pain models, have demonstrated that EphBRs, and in particular EphB1, and ephrinBs (possibly both ephrinB1 and ephrinB2) are required for the onset and for the maintenance of chronic hypersensitivity (or for both). But what is the underlying mechanism?

EphB receptors may modulate NMDA receptors function

Battaglia and colleagues (2003) were initially induced to examine the role of EphBRs in pain processing by studies that showed that EphBRs associate with NMDA receptors during development; and later studies on knock-out mice had shown abnormalities in synaptic plasticity in the hippocampus of EphB2 knock-out animals, a phenomenon known to involve NMDA receptors (Grunwald et al., 2001; Henderson et al., 2001). It was therefore plausible that a link between EphBR activation and NMDA receptor function could be the element that explained the results obtained in the behavioural studies described above, indicating an involvement of EphB1–ephrinB interactions in the onset of chronic pain syndromes. This idea would also be supported by an electrophysiological study from the Song laboratory in Dallas (Song, Zheng, et al., 2008). It is known that nerve injury can cause NMDA-dependent sensitisation of wide dynamic range (WDR) neurons in the spinal cord. Song and co-workers demonstrated that WDR neurons from rats with a nerve injury (CCI) have increased responses to innocuous brush and pressure and to painful pinch and that this increased responses can be suppressed by repeated treatment with EphB1-Fc.

The direct demonstration of the role of EphBRs in regulating NMDA receptor function *in vivo* came only with studies on adult rats; in these animals the hypersensitivity induced by activating EphBRs with ephrinB2-Fc could be prevented by injecting the non-competitive NMDA receptor antagonist MK801, prior to injecting ephrinB2-Fc (Battaglia et al., 2003). This suggests that EphBR activation may indeed modulate NMDA receptor function. But how could this happen? EphBRs have been shown to bind to and phosphorylate non-receptor tyrosine kinases of the Src family. Tyrosine kinases of the Src family are known to bind to and phosphorylate the NR2B subunit of the NMDA receptor, and this has been linked to the onset of long-term potentiation in the hippocampus (Rostas et al., 1996). In relation to processing of nociceptive inputs in the spinal cord, several lines of evidence support the hypothesis that ephrinB2-Fc injection increases the amount of phospho-Src bound to the EphB1 receptor. Intrathecal ephrinB2-Fc injection and a number of models of persistent pain (carrageenan or formalin injection, CCI and partial ligation) also cause an increase in NR2B phosphorylation (see e.g. Slack et al., 2008) (Figure 7.3). This increase in p-NR2B was prevented in some models by using the Src inhibitor PP2, in others by knocking out EphB1 or nociceptors-expressed ephrinB2 (Slack et al., 2008; Wu et al., 2011; Cibert-Goton et al., 2013) (Figure 7.3a and 7.3b); and this supports the hypothesis that Eph-ephrin binding leads to phosphorylation of Src, followed by phosphorylation of NR2B.

The most economical explanation of these findings is that a variety of insults to peripheral nociceptors (inflammation or nerve injury, for example) can lead to the activation of EphB1 receptors on secondary sensory neurons in the spinal cord (Figure 7.4). Ligand binding induces a clustering of EphBRs, which is followed by the phosphorylation of tyrosine residues on the intracellular portion of the receptors. This recruits signalling molecules that contain SH2 domains, such as molecules belonging to the Src family of non-receptor tyrosine kinases. One member of such a family, Fyn, has been shown to be associated with both EphB and NMDA receptors (see, e.g., Murai and Pasquale, 2004), and it can phosphorylate NR2B (Nakazawa et al., 2001). This phosphorylation would lead to increased activity of the NMDA receptor, as shown in cultures of cortical neurons, with greater Ca^{++} influx (Takasu et al., 2002). Downstream of this increased activity one would find, among other things, increased levels of the product of the immediate early gene c-fos. Indeed EphB1-Fc treatment, or EphB1 knock-out, can prevent the increase in levels of c-fos that is normally observed in persistent pain models (Battaglia et al., 2003; Cibert-Goton et al., 2013). A similar intracellular signalling pathway had been proposed for EphBRs expressed on hippocampal and cortical neurons; this was carefully examined in culture studies (Takasu et al., 2002), where ephrinB2-Fc induced the activation of the EphBR, which in turn modulated a NMDA receptor-dependent influx of calcium, presumably via phosphorylation of NR2B by a Src family kinase. Ca^{++} increases led to phosphorylation of the Ca^{++}/cAMP responsive element-binding protein (CREB). NMDA-dependent gene expression was also enhanced (in particular c-fos, BDNF and cpg15).

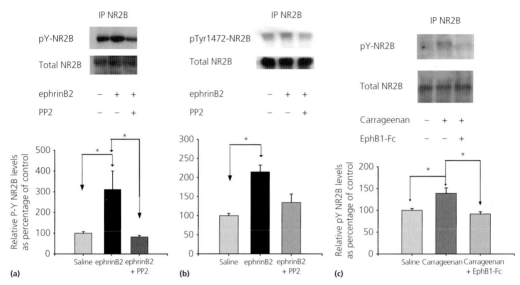

Figure 7.3 EphrinB2-Fc treatment and inflammation induce an increase in the phosphorylation of the NR2B subunit of the NMDA receptor, mediated by the non-receptor tyrosine kinase Src. In order to investigate the intracellular signalling pathways activated by EphB1 receptors in dorsal horn neurons, Slack et al. (2008) examined the phosphorylation state of the NR2B subunit of the NMDA receptor following intrathecal (i.e., through injection into the cerebrospinal fluid, in this case at lumbar level) treatment with EphrinB2-Fc in adult rats. The involvement of the non-receptor tyrosine kinase Src in NR2B phosphorylation was examined by administering the Src family kinases inhibitor PP2. The top panels are representative immunoblots that examine the phosphorylation state of NR2B (using anti-phosphotyrosine antibodies followed by anti-NR2B antibodies in immunoprecipitates obtained from lysates of the dorsal portion of the lumbar spinal cord using NR2B antibodies). In panels **a** and **b**, the rats, from which the spinal cords were dissected, had received two intrathecal injections separated by a five-minute interval. The first injection was either vehicle or the Src family kinases inhibitor PP2, and the second injection was either ephrinB2-Fc or saline solution. For panel **c**, rats received an intraplantar injection of carrageenan, preceded by intrathecal injection with either the soluble EphB1 receptor chimera EphB1-Fc or Human-Fc in saline. After 40 minutes the lumbar portion of their spinal cords was dissected and NR2B was immunoprecipitated from the dorsal portion of the cords. By comparison to control levels, levels of tyrosine phosphorylation of NR2B were significantly increased following ephrinB2-Fc treatment in the spinal cord or carrageenan injection in the paw. The increase induced by ephrinB2-Fc was prevented through the prior administration of PP2, which indicates that the non-receptor tyrosine kinase Src is responsible for this phosphorylation. In panel B it is shown that Tyr1472 residue was a target of phosphorylation on NR2B following ephrinB2-Fc treatment. Intrathecal administration of Eph-B1-Fc prior to the carrageenan injection significantly reduced the phosphorylation levels of NR2B. This suggests that the endogenous ephrin in the dorsal horn was prevented from binding to EphBs in the spinal cord by the soluble receptor, hence preventing the phosphorylation of NR2B induced by carrageenan. In all bar charts the relative phosphoproteins level (means +/− SEM) are expressed as a percentage of the saline controls and are normalised to the respective NR2B band. In all experiments, the levels of total NR2B remained unchanged. Test: ANOVA on ranks.* $p < 0.05$. Source: Adapted from Slack et al., 2008.

Some studies have tried to uncover further details about the intracellular signalling pathways activated through EphBR activation on dorsal horn neurons, presumably downstream of the increased Ca^{++} entry (but the specific link between EphB1 activation and the pathways described below still needs to be clarified). EphrinB1-Fc intrathecal injection leads to a time- and dose-dependent increase in expression of PI3K-p110γ (which is notoriously involved in central sensitisation), a phosphorylation of AKT, and increased c-fos expression

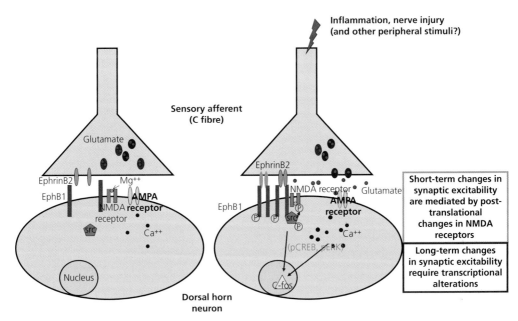

Figure 7.4 Current hypothesis on the mechanisms linking EphB1 receptor activation with central sensitisation of primary afferent synapses onto dorsal horn neurons. In the non-sensitised state, ephrinB2 (and/or B1?) ligands and EphB1 receptors would be present pre- and postsynaptically, but would not form clusters. Inflammation, nerve injury and possibly other sensitising signals are likely to cause a clustering of ephrinB2 on the presynaptic cell (possibly linked to an increase in expression of ephrinB2). Dimerisation (and possibly the formation of oligomers) of EphB1 receptors leads to autophosphorylation and recruitment of a Src family kinase to the membrane. Src would then phosphorylate the NR2B subunit of the NMDA receptor, with consequent increase of Ca++ entry. This would mediate short-term alterations in synaptic efficacy. The signalling pathways downstream of Src phosphorylation and Ca++ entry may vary, depending on the specific model examined (inflammatory, neuropathic or cancer pain), and can involve CamKII, PI3K, and MAPKs (not illustrated here). Ultimately, ERK and/or CREB are phosphorylated, translocate to the nucleus and lead to the transcription of genes, including c-fos. This would lead to long-term, persistent changes in synaptic efficacy.

(Yu et al., 2012). The inhibition of spinal PI3K inhibited the activation of spinal AKT, the increase of c-fos, and the mechanical and thermal hyperalgesia induced by ephrinB1-Fc injection in mice. Interestingly, both the inhibition that preceded ephrinB1-Fc injection and the inhibition that followed the injection were effective, indicating that PI3K is involved both in the onset and in the maintenance of increased sensitivity. The injection of EphB1-Fc reduced the pain behaviours associated with formalin injection and with CCI (in the latter case it had this effect both before and after injury); it also decreased expression of PI3K-p110 γ, phosphorylation of AKT and expression of c-fos. Furthermore, pre-treatment with a PI3K inhibitor prevented the activation of extracellular signal-regulated kinase (ERK) in the spinal cord induced by ephrinB1. Similarly, ephrinB1-Fc intrathecal injection in mice also caused the activation of the mitogen-activated protein kinases (MAPKs) pathway (Ruan et al., 2010); some activated MAPKs (p-p38, p-JNK) were observed in both neurons and astrocytes, p-ERK only in neurons. Blocking MAPK before and after ephrinB1-Fc treatment respectively prevented and reversed both the hyperalgesia (both thermal and mechanical: these experiments were conducted in mice) and the increase in c-fos expression induced by ephrinB1-Fc injection. Formalin injection in the paw and CCI caused increases in p-MAPK, and the injection of EphB1-Fc prevented spontaneous pain (formalin) and the thermal and mechanical hyperalgesia induced by formalin or CCI, as well as the increase in p-MAPK. The NMDA receptor antagonist MK-801, given prior to ephrinB1-Fc injection, prevented both the onset of hyperalgesia and the MAPK activation, supporting the hypothesis that this pathway is indeed downstream of increases in Ca^{++} influx.

EphB receptors and spinal LTP

There is an ongoing discussion, sometimes in the shape of an open argument, on the role of long-term potentiation (LTP) in central sensitisation. The discussion appears at times to be largely semantic in nature and it is, anyway, beyond the scope of this chapter. The activity-dependent potentiation of the synapses of primary sensory afferents onto dorsal horn neurons does share some similarities with the LTP described in other regions, primarily the hippocampus. Some forms of LTP are NMDA-dependent and involve up-regulation of p-CAMKII, p-ERK and phosphorylation of CREB, leading to increased expression of c-Fos (Miyamoto, 2006). It has been found that high-frequency stimulation of primary afferents (the same stimulus normally used to stimulate LTP in the hippocampus) leads to up-regulation of p-CAMKII, p-ERK and p-CREB in the spinal cord as well as to up-regulation of c-fos in the mouse and in the rat, and this can be blocked by pre-treatment with the NMDA receptor antagonist MK801; by injection of EphB1-Fc; or by knocking out the EphB1 gene (that is, in EphB1 KO mice) (Liu, Han, et al., 2009). Furthermore, in naïve rats, LTP induced by four trains of high-frequency stimulation was blocked through pre- (but not post-) treatment with EphB1-Fc. EphrinB1-Fc treatment did not induce LTP, but, if it was followed by two trains of high-frequency stimulation (normally ineffective), LTP was induced. The precise physiological significance of these experiments is still unclear, partly because the significance of LTP induced through high-frequency stimulation of C fibres is debated; but these findings do anyway provide some further pieces in the jigsaw of the potential mechanisms that underlie the EphB1 receptor modulation of the synaptic function in the spinal cord, via supporting LTP in WDR.

However, there is one important point that none of the experiments mentioned above addresses: How do primary nociceptors signal to the postsynaptic EphB1 receptor, to induce its activation? EphrinBs are not secreted by the presynaptic neuron; as far as we know, they are instead inserted in its membrane.

How does inflammation or nerve injury induce the activation of spinal EphB receptors?

Oddly, none of the numerous studies and reviews now available on the role of Eph receptors and ephrins in modulating synaptic efficacy in the adult brain has addressed the question of why presynaptic ephrinBs activate postsynaptic EphBs (or, in some cases, vice versa) only when an intense barrage of stimuli reaches the presynaptic terminal, since both molecules are expressed at all times on either side of the synaptic cleft. The only study trying to address this issue remains the one by Battaglia et al. (2003). These researchers cultured primary sensory neurons *in vitro* and stimulated them with ATP. They had previously shown that about 80 per cent of ephrinB2-expressing nociceptors expressed the ATP receptor $P2X_3$ (Figure 7.1). After five minutes of exposure to ATP, they fixed the cells and compared the staining for ephrinB2 in non-stimulated versus stimulated cells. They noticed that the intensity of ephrinB2 staining in the cell membrane increased in the treated neurons and the staining itself appeared more punctate. It would therefore seem plausible that the activation of peripheral nociceptors causes both an insertion of more ephrinB2 in the membrane (rapid trafficking of membrane molecules from a pool of vesicles has been clearly demonstrated for other proteins, including trkB receptors and AMPA receptors (Meyer-Franke et al., 1998; Galan, Laird and Cervero, 2004) and a clustering of transmembrane ephrins. As mentioned above, the ligand needs to be at least dimerised in order to induce a response (and some responses even require oligomerisation). The suggestion from this study would be that activation of nociceptors leads to increased levels of ephrinB2 and to its clustering on presynaptic terminals. Due to the short duration of the experiment, *de novo* synthesis of ephrinB2 can be excluded.

Other studies have looked at longer time courses, to see whether there were changes in EphB or ephrinB levels in models of persistent pain, which could explain the increased activation of the receptors; the hypothesis would be that an increased level of the receptor or the ligand molecule (or of both) would make clustering more likely. The results so far are not conclusive, but they suggest that levels of EphB1 or ephrinB2 (or ephrinB1) do change in some pain models, even if no study has specifically looked at them in the presynaptic or postsynaptic membrane (Table 7.1). Most studies were conducted on models of neuropathic pain (nerve injury). So far as the expression of the EphB1 receptor in the spinal cord is concerned, there appears to be agreement on its presence, but there is disagreement on changes in its expression after injury. There are reports that this expression is unchanged (see Kobayashi et al., 2007 – a study that also excludes the presence of EphB2 and B4), as well as reports that its levels are increased (see Song, Cao, et al., 2008). For EphBRs, there is thus agreement on the identity of the receptor, disagreement on whether its expression levels change, whereas for the ligand its levels are generally reported as increased after injury, both in the DRG and in the spinal cord, but there is disagreement as to the identity of the ligand (ephrinB1 or ephrinB2; see Table 7.1). The reported spinal cord up-regulation of the ligand could potentially be in sensory afferent terminals or in cells (glia or neurons). In Song, Cao, et al.'s (2008) experiments, the up-regulation of ephrinB1 in the spinal cord was prevented by dorsal rhizotomy after CCI; therefore it was hypothesised that most of the ephrinB1 in the spinal cord is actually on nerve terminals (predominantly IB4 positive). It has to be noticed that protein levels were always determined with biochemical techniques (western blot on tissue lysates in Kobayashi et al., 2007; immunoprecipitates in Song, Cao, et al., 2008) that have a degree of ambiguity, since the up-regulation may be on glial cells (or on other cells, for example in blood vessels), and not on neurons.

The contradictory findings in relation to changes in the levels of the receptor could be explained by differences in the models and in the time points analysed. It is somewhat harder to invoke differences of protocol to explain the disagreement concerning the identity of the ligand. When the specific ligand present is identified by methods that rely on the use of antibodies, one could argue that it is difficult to be absolutely certain about the identity of the

Table 7.1 Table summarising the changes in expression levels of EphBRs and ephrinBs in DRGs and the spinal cord, as described by various authors in experimental animal models of chronic pain.

Author	DRGs	Spinal cord	Injury model	Technique
Kobayashi et al. 2007	EphrinB2 up at 7 d (B1 and B3 not detected)	EphB1 no change ephrinB2 up at 7 d	Crush of L5 spinal nerve Rat	WB
Song et al. 2008a Song et al. 2008b	EphB1 up ephrinB1 up (small to medium size neurons)	EphB1 up ephrinB1 up (predominantly in IB4+ terminals)	Chronic constriction injury Rat	IP followed by WB
Uchida et al. 2009	EphrinB1 up		Lysophosphatidic acid Mouse	Microarrays real-time quantitative PCR
Liu et al. 2011		EphB1 up ephrinB2 up	Bone cancer pain (Tibial bone cavity tumour cell implantation) Rat	IP followed by WB
Dong et al. 2011	EphrinB1 down EphB1 no change	EphB1 up ephrinB1 up	Bone cancer pain (intra-tibial inoculation of cancer cells) Rat	WB
Orikawa et al. 2010	Increase in ephrinB1 No change in EphB1		Cancer pain (transplant of melanoma cells in plantar hind paw) Mouse	Quantitative real time RT-PCR
Peng et al. 2010		EphrinB2 up	Colon-urethra cross-organ sensitisation Rat	WB
Liu et al. 2009b		EphB1 up ephrinB1 and B2 no change	Morphine withdrawal Mouse	IP followed by WB

WB, Western blotting; IP, immunoprecipitation.
To facilitate comparison, the experimental models employed, the species and the techniques utilised to assess levels of expression of EphBRs and ephrinBs (protein or mRNAs) are indicated in the last two columns.

molecule visualised, because, as mentioned before, the specificity of antibodies to EphBRs and ephrins is usually questionable. This is probably due to the fact that most of these receptors and ligands have high sequence homology, so it is difficult to create good antibodies, which will bind only the variable portion of the molecule. However, this disagreement is also present in studies where the problem should be more easily controlled. Uchida, Matsumoto and Ueda (2009) studied alterations in gene expression induced by lysophosphatic acid with the help of microarrays followed by confirmatory real-time quantitative polymerase chain reaction (PCR) and identified an up-regulation of ephrinB1 in mice: an antisense oligonucleotide for ephrinB1 reversed the thermal and mechanical hyperalgesia induced by lysophosphatic acid (oddly, in Uchida et al.'s experiments ephrinB1-Fc caused both thermal and mechanical hyperalgesia, but with a time lag of hours, in contrast to the 30 minutes consistently reported elsewhere).

However, we have unpublished data from in situ hybridisation experiments indicating that, in the rat, approximately 50 per cent of DRG neurons express ephrinB2, while only 12 per cent express ephrinB1, and this percentage is actually halved after sciatic nerve lesion. Also, Kobayashi et al. (2007) could ameliorate pain hypersensitivity after nerve crush by injecting ephrinB2 siRNAs, which further supports the hypothesis of a role for this specific ligand.

Regardless of the contradictions still present in these studies, overall they seem to confirm that injury leads to an increase in the expression of either the ligand or the receptor or both, and this may lead to an increase in EphB1 activation.

EphB receptors and glial activation

In recent years it has become apparent that neurons are not the only cells involved in the onset and maintenance of chronic pain states: *glial cells, microglia* (see Chapter 4 in this volume) and in particular *astrocytes* also play a role, which is progressively being clarified, even if many questions on the specific mechanisms remain unanswered (see, e.g., Gao and Ji, 2010). Both in EphB1 knock-out mice and in nociceptor-specific ephrinB2 knock-out mice, microglia activation was reduced in a model of neuropathic pain by comparison to wild type littermates (Zhao et al., 2010; Cibert-Goton et al., 2013). This difference between knock-outs and wild type may be secondary to differences in the activation of EphB1 in postsynaptic neurons with consequent lack of activation of NMDA receptors. However, it is also possible that EphBRs or ligands on glial cells are involved. Ruan et al. (2010), showed how intrathecal injection of ephrinB1-Fc induced MAPK activation in astrocytes and behavioural pain; immunolocalisation studies indicated that both EphB1 receptors and ephrinB2 are associated with neurons and astrocytes, whereas microglia displays little EphB1 receptors staining and no ephrinB2 staining (Liu et al., 2011). Further investigations on this topic are warranted.

A peripheral site of action?

So far this chapter has focused predominantly on the role of postsynaptic EphBRs in modulating spinal processing of pain, versus any presynaptic mechanism. In some brain areas, presynaptic increase in neurotransmitter release also contributes to synaptic LTP. *In vivo* studies on the optic tectum of Xenopus suggest that the activation of presynaptic ephrinBs (obtained through infusion of EphB2-Fc) may lead to increased neurotransmitter release, which in turn leads, after 30 minutes, to a NMDA-dependent postsynaptic enhancement of transmission (Lim, Matsuda and Poo, 2008). Whether similar mechanisms may be involved in pain is still unknown.

Nonetheless, some evidence exists for a role of EphB1 receptors expressed on DRG neurons in pain hypersensitivity, at least in neuropathic pain, even if it is still unclear whether EphB1 receptors expressed at the central terminals or elsewhere in the neuron are important. Nerve injury may lead to an increased expression of EphB1 receptors on primary afferent neurons, causing increased excitability. Furthermore, no nerve injury-induced hyperexcitability of medium-size DRG neurons is present in EphB1 knock-out mice (Han et al., 2008). In particular, in mice with CCI, there is a depolarisation of resting membrane potential, a decrease in the threshold of the action potential and an increase in repetitive discharge. None of these was present in EphB1 knock-out mice dissociated DRG neurons (Han et al., 2008). Similarly, the increased excitability of small-sized DRG neurons from rats exposed to CCI was blocked by bath application of EphB1-Fc; their excitability was further increased by EphrinB1-Fc, which had no effect on naïve neurons (Song, Zheng, et al., 2008). Due to the nature of these experiments, which were performed on DRGs *in vitro*, it is still unclear whether the EphB1 receptors responsible for this increased excitability are expressed on the peripheral or on the central terminals of neurons.

Some studies conducted on mice show how EphBR–ephrinB interaction in peripheral tissues, rather than within the spinal cord, can play a role in the development of hypersensitivity to somatosensory stimulation. They will be mentioned here for completeness, but these experiments are more relevant to the issue of peripheral sensitisation, therefore they lie beyond the scope of this chapter. Studies by Cao et al. (2008) demonstrated that thermal and mechanical hyperalgesia could be induced by peripheral (paw) injection of ephrinB1–Fc. Spontaneous pain (as measured by licking and biting of the injected paw) was also induced. P-MAPK p38, p-ERK1/2 and p-JNK expression were significantly increased in the skin of the treated paw. Pre-treatment with MAKP inhibitors prevented the development of hyperalgesia and spontaneous pain; post-treatment could reverse established hyperalgesia ad spontaneous pain. EphrinB1 also induced c-fos increase in the spinal cord, which was prevented by injecting MAPK inhibitors before or after treatment. EphB1-Fc injection could decrease spontaneous pain and MAPK phosphorylation if given 30 minutes before a formalin injection. By analogy to what has been found in the spinal cord, phosphorylation of MAPK would be mediated by peripheral NMDA receptors, since it could be blocked by pre-treatment with MK801. This study could not demonstrate that ephrinB1 did indeed act on the nerve terminals and was not acting instead on non-neuronal cells, increasing the levels of inflammatory mediators.

In this same mouse model, the activation of PI3K and AKT was also shown in peripheral sensory neurons, following ephrinB1-Fc intraplantar injection. Activated AKT was found to be present in both peptidergic and non-peptidergic terminals, as well as in the cell body of small and medium-size DRG neurons. Injection of PI3K inhibitors before or after the ephrinB1 injection prevented or reversed behavioural and thermal hyperalgesia, as well as preventing the phosphorylation of AKT and spinal increases in the expression of c-fos (Guan et al., 2010).

EphB receptors in visceral and cancer pain

Most of the studies on the role of EphBRs in pain hypersensitivity have focused on models of neuropathic or somatic inflammatory pain. However, some more recent studies have also utilised models of visceral or cancer pain and have highlighted how this receptor–ligand system is crucially involved in the onset and maintenance of hypersensitivity in these models.

EphB1 receptors in cancer pain
Cancer pain remains a considerable clinical challenge, and its underlying mechanisms are still incompletely understood. However, the current consensus is that most forms of cancer pain are likely to have both a neuropathic and an inflammatory component and to have multiple

triggers, which activate multiple pathways. On the basis of what has been presented so far, an involvement of EphBRs and their ligands in the onset and maintenance of this most intractable type of pain would therefore not be unexpected.

A first group of studies aimed at investigating the involvement of the EphB–ephrin system in cancer pain used models of *bone cancer*. Bone-cancer pain, as well as the morphine tolerance associated with treating it, remains a significant clinical problem. As was the case with studies in other pain models, the experiments conducted on models of bone-cancer pain led to reports of up-regulation of EphB1 receptors in the spinal cord (Table 7.1; Dong et al., 2011; Liu et al., 2011), but to contradictory results concerning changes in the levels of EphB1 in DRGs and the identity and regulation of expression of ephrinBs. However, intrathecal injection of the blockers EphB1-Fc (Dong et al., 2011) or Ephb2-Fc (Liu et al., 2011) after cancer pain induction consistently and significantly reduced the mechanical allodynia induced by the bone cancer. The intracellular signalling mechanisms triggered by EphB1 receptors activation in cancer pain appear to be similar to the ones identified in other models, since injection of the blocker EphB2-Fc prevented and reversed c-fos induction and the activation of astrocytes and microglia; it also prevented the phosphorylation of NR1 and NR2B and the activation of Src, p-ERK, p-CaMKII, and p-CREB increases – but not of p-NR2A, which is not affected by cancer or by EphB2-Fc (Liu et al., 2011). EphB1 activation has been shown to lead to the activation of MAPKs cell signalling (see above), and this could result in a pro-inflammatory response and an increase in the production of inflammatory cytokines. EphB1-Fc injection suppressed the increase in mRNA for IL-1beta, IL-6 and TNF-alpha (Dong et al., 2011). In the study by Liu et al. (2011), ephrin B2 appeared to co-localise with calcitonin gene-related peptide (CGRP)-containing terminals but not with IB4-containing terminals, in disagreement with what was reported by Song et al. (Song, Cao, et al., 2008). The reason behind these contradictory findings and their significance are for the moment unknown.

Another group (Orikawa et al., 2010) provided some correlative evidence for a role of the EphB1 receptors in a different model of severe cancer pain: the orthotopic transplantation of melanoma cells in the plantar region of the hind paw in mice. They examined (1) variations in ephrinB1 expression by quantitative real-time PCR and (2) EphB1 receptors in DRGs, and they found an increase in ephrinB1 and no change for EphB1 receptors. NR2B phosphorylation is induced in the spinal cord of experimental mice and IL1beta production is increased in the cancer-injected paw. All of these increases are prevented by a CCK1 receptor antagonist and by an orally active cholecystokinin-2/gastrin receptor antagonist, Z-360, in this severe, opioid-resistant cancer model in mice. Z-360 was used because, in a study in combination with gemcitabine and analgesics, it had displayed a trend towards reduced pain in patients with advanced pancreatic cancer in a phase Ib/IIa clinical trial. Z-360 also prolongs survival in pancreatic cancer sufferers. Neither Z-360 nor the CCK1 receptor antagonist affected tumour size or EphB1 gene expression. The proposed cascade of events is an increase in IL1beta in cancer-inoculated regions; this causes an ephrinB1 up-regulation, which would induce NR2B phosphorylation via EphBRs.

A purely correlative study in human patients with pancreatic cancer showed that the overexpression of EphB2 and ephrinB2 mRNAs was markedly associated with abdominal and/or back pain, as well as being a poor prognostic factor – like in breast cancer, but unlike in colonorectal cancer (Lu et al., 2012). Whether the expression of EphB2 and ephrinB2 in pancreatic cancer is in any way mechanistically related to the pain is still unknown.

EphB1 receptors in visceral pain

Where *visceral pain* is at stake, the number of studies available is even more limited than for cancer pain, but they seem yet again to suggest that the same mechanisms that have been proposed in the other models of pain discussed so far are at play here as well. Peng et al. (2010) discovered a role for EphBR–ephrin interactions in viscero-visceral referred pain, which leads to cross-sensitisation between urethra and colon. Cross-organ sensitisation is due to central sensitisation. The physiological and pathological significance of this phenomenon, which leads to sensitisation of the urethra after the activation of nociceptors in the colon, is still unknown. Intracolonic instillation of the irritant mustard oil plus test stimulation of the colon in order to induce acute irritation caused an increase in the level of ephrin B and of phosphorylation of EphB1/B2 in the lumbosacral spinal cord of rats, Src phosphorylation, and NMDA receptor phosphorylation (Tyr1336 and Tyr1472 of NR2B). Reflex sensitisation of the urethra, as well as ephrinB2 and EphB1/B2 phosphorylation and Src phosphorylation, were reduced by treatment with EphB1 or EphB2-Fc. Sensitisation was also reduced by PP2 (a Src blocker) and by the NMDA receptor antagonist APV. However, neither PP2 nor APV prevented an ephrinB2 increase or an increase in EphB1/B2 phosphorylation – which indicates that these are upstream of Src activation and NMDA receptor function regulation. Wu et al. (2011) found that the intrathecal treatment of the lumbosacral cord with ephrinB2-Fc potentiates the pelvic–urethra reflex, as well as Src and NR2B phosphorylation. All of these findings, as mentioned above, would suggest a similar signalling cascade to that described for inflammatory and neuropathic pain models and for cancer pain.

EphB receptors and morphine tolerance

There is a further feature of the function of EphBRs in the spinal cord that needs to be mentioned here, since it may have quite a significant impact on the therapeutic potential of these receptors for chronic pain. Opiates, including morphine, are still used extensively in the management of pain, and in some cases they are the only effective analgesic drugs available. There are, however, issues linked to their use, an important one of them being the development of physical dependence and tolerance. The cellular and molecular mechanisms responsible for the development of *opioid tolerance and dependence* are complex and have not been fully clarified. It seems, however, that the NMDA receptor/nitric oxide pathway plays a fundamental role in the development and maintenance of opioid tolerance and withdrawal (see Inturrisi, 2002; Pasternak, 2007). In order to study the development and maintenance of opioid dependence, withdrawal symptoms are induced by treating mice or rats with escalating doses of morphine and then giving them an injection of naloxone, to precipitate withdrawal symptoms. Surprisingly, EphB1 receptor protein (but not ephrinB1 or B2) appears to be significantly up-regulated in the spinal cord of mice after escalating morphine treatment (Liu, Li, et al., 2009). Chronic exposure to morphine, followed by withdrawal, causes a significant increase in the phosphorylation of the NR2B subunit of the NMDA receptor as well as in the activated forms of ERK and CREB, and increases in c-fos expression. Injection with EphB2-Fc reduces the behavioural symptoms of withdrawal, as well as the increase in phosphorylation in the aforementioned molecules and in c-fos expression. It has also been shown that morphine withdrawal symptoms in morphine-treated mice are significantly reduced in EphB1 knock-out mice or by using EphB-Fc chimeric molecules in mice and rats (Han et al., 2008; Liu, Li, et al., 2009). Even simple EphB1 knock-down (EphB +/−) has analgesic effects in neuropathic pain and reduces morphine withdrawal symptoms. Also, low doses of EphB2-Fc could prevent the development of morphine tolerance in bone-cancer rats (Liu et al., 2011). Interestingly, chronic morphine treatment and withdrawal also cause alterations in CGRP levels in afferent terminals in the spinal cord. EphB2-Fc treatment prevents these changes (Liu, Li, et al., 2009). This would suggest that EphB–ephrinB signalling has an impact on presynaptic as well as on postsynaptic neurons in this context. Whether this is mediated by retrograde signalling via ephrinBs, or EphB1 receptors present on sensory afferents are responsible, remains to be determined.

A few open questions

Since the function of EphBRs–ephrinBs interaction in persistent pain was first described, much progress has been made in clarifying the mechanisms involved, and also in establishing the extent to which these interactions are important in different forms and at different stages of chronic pain syndromes. However, many questions remain unanswered, some of which have already been mentioned.

A first set of questions is linked to the observations, described above, on the expression and alterations of levels of EphBRs and ligands in the DRGs and spinal cord. These questions could be summarised as follows: (1) What is the ligand, expressed on primary afferents central terminals, for EphB1 receptors on spinal cord neurons? (2) What is the nature and the significance of changes in expression levels of EphB1 and ephrinBs in different pain models?

The presence of EphB1 on dorsal horn neurons has been reported by several independent laboratories, and therefore appears well established. But which molecule expressed on nociceptors binds to and activates EphB1 in dorsal horn neurons? Both the evidence supporting a role for ephrinB1 and the evidence supporting a role for ephrinB2 are quite convincing, and some of it does not rely on potentially unreliable antibodies, but on arguably more reliable in situ hybridisation and PCR. Further studies are needed to establish the reasons for these contradictory findings. This is particularly important for the potential development of analgesic drugs: it would establish which specific molecules should be targeted. With the widespread distribution and wide variety of functions of these molecules, specificity is essential. EphrinB3 appears not to be present in adult DRGs, which limits the potential candidates to two. Curiously, Song and colleagues reported ephrinB1 expression on IB4 positive (therefore non peptidergic) terminals (Song, Cao, et al., 2008), while Liu et al. (2011) reported that ephrinB2 was expressed in CGRP-containing sensory afferents. The possibility that different subsets of nociceptors express a different ligand ought to be considered. As for the contradictory findings related to regulations of the levels of expression of EphBRs and ligands, their significance needs to be explored further. The reported differences may be due to the methods employed for quantifications, but this is unlikely to be the only explanation. Liu et al. (2011) reported that both ephrinB-Fc injection and bone-cancer pain-induced hyperalgesia in mice could be prevented through EphB2-Fc treatment; however, ephrinB2-Fc did not affect overall receptor expression but increased its level of phosphorylation, and also down-regulated both ephrinB2 and p-ephrinB2 – while bone cancer increases EphB1 and p-EphB1, ephrinB2 and its phosphorylated form. Differences in regulations of expression in pain states of different aetiology were therefore reported also from the same laboratory, and this excluded methodological differences. How these differences in the regulation of expression levels of receptors and ligands affect downstream signalling needs to be explored.

In relation to the downstream signalling activated by ephrinB–EphBR binding, a number of further questions are important: (1) Is reverse signalling via ephrinB2 required, in addition to the better studied forward signalling? (2) What other events are downstream of EphB1 activation, besides NR2B phosphorylation? (3) What is the functional significance of the reported up-regulation of EphB1 receptors on injured DRG neurons?

There is little evidence either in favour or against a role of signalling triggered by ephrinB2 or B1 in presynaptic cells. Changes in levels of expression are no proof of a role for reverse signalling. In one study reported above in a model of cancer pain (Liu et al., 2011), EphB2-Fc treatment prevented the increase of EphB1, but not that of ephrinB2, induced by bone-cancer pain, and this reduced hyperalgesia. This was interpreted as suggesting that, at least in this type of pain, forward signalling rather than reverse signalling is important; but this evidence is not conclusive. In the brain (hippocampus, mossy fibres–CA3 synapses), LTP and LTD rely on ephrinB reverse signalling and postsynaptic EphB2 receptors serve as ligands, while at CA3-CA1 synapses ephrinB reverse signalling (possibly leading to phosphorylation of NR2A) would take place in postsynaptic neurons, and synaptic plasticity is dependent on NMDA receptors (Grunwald et al., 2004; Armstrong et al., 2006). On the basis of studies on genetically modified mice, it has been proposed (Klein, 2009; Lai and Ip, 2009) that the kinase activity of EphBRs may not be necessary for LTP and LTD in the hippocampus, and EphBRs would act as ligands for ephrinBs, or by direct cis interactions with molecules such as NMDA receptors. However, compensatory mechanisms during development to replace the lost function cannot be excluded in any of these studies; moreover, in the mutants used to demonstrate that EphB forward signalling is not required, the mutated EphB2 receptors retain part of their cytoplasmic tail, which could be phosphorylated and could lead to the activation of intracellular signalling pathways. We found increases of p-Src associated with EphBRs, which would directly implicate them, rather than ephrinBs, in activating intracellular signalling (Battaglia et al., 2003).

Regarding the consequences of EphBR activation, further to its modulation of NMDA receptor function by phosphorylation of the NR2 subunit, EphB1 receptors could act on glutamate receptor trafficking at the synapse. Changes in the number and type of AMPA and NMDA receptors underlie synaptic plasticity as much as the regulation of function. Interactions of glutamate receptors with scaffolding proteins and with synaptic adhesion molecules such as EphBRs are known to be important for receptor trafficking, but it is still not clear whether this happens in pain (Sheffler-Collins and Dalva, 2012). EphBs may regulate synaptic AMPA receptor subunit composition (Sheffler-Collins and Dalva, 2012). Deficits in trafficking have been linked with Alzheimer's disease, addiction and schizophrenia. Eph–ephrins interactions regulate the function of *metabotropic glutamate (mGlu) receptors* in other systems (see, e.g., Calò et al., 2005), and mGlu receptors are also important in pain, but no information is available so far on whether some of the behavioural consequences of EphB1 receptor (or also ephrinB) activation are mediated via mGlu receptors.

Song et al. reported an increased expression of EphB1 receptors in the DRGs after CCI and have also demonstrated that the increased excitability of DRG neurons in this model can be blocked by EphB-Fc chimeric molecules (Song, Cao, et al., 2008; Song, Zheng, et al., 2008). Conversely, EphrinB1-Fc can increase the excitability of these neurons. This would suggest that forward signalling can be important in primary sensory afferents, as well as in postsynaptic dorsal horn neurons. However, it cannot be for the moment determined whether EphB1 receptors exert their role at the central or at the peripheral terminal of DRG neurons, since the experiments on DRG neurons excitability were not conducted *in vivo*.

Another important issue is linked to the fact that Eph and ephrins are also expressed on glial cells; particularly astrocytic expression may be important to regulate synaptic function, but this area is completely unexplored. We know that ephrinB1-Fc treatment in mice causes an increase of p-p38 and p-JNK in astrocytes, but we do not know whether this is a direct effect (Ruan et al., 2010).

Another area that may be worth investigating, particularly in relation to long-term structural plasticity in chronic pain, is based on the knowledge that the dynamic formation and retractions of spines are thought to be one of the factors contributing to *synaptic plasticity* and possibly memory to formation in the hippocampus. The significance of spine formation to central sensitisation is not known, but it is known that the EphB–ephrin interaction modulates spine formation (see Henkemeyer et al., 2003).

Some intriguing results by Liu et al. (2011) suggest that EphB2-Fc, applied intrathecally, may not simply act as an EphB1 receptor antagonist, but may also contribute to regulating the level of expressions of EphB1 itself, via the metalloproteases MMP-2 and MMP-9. Considering that it is thought that levels of expression of the receptor are important in regulating its function, a better understanding of the interactions metalloproteases-EphBRs would be essential.

Potential for translation

EphBRs and their ephrin ligands have a wide distribution both inside and outside the CNS, and in the adult their involvement has been demonstrated in a variety of cancers, as well as in Alzheimer's disease and in other neurological diseases such as epilepsy (Hruska and Dalva, 2012). As a consequence, these molecules are considered potential targets for therapy in a variety of neurological and non-neurological diseases. This should hopefully lead to a concerted effort in the development of drugs that can interfere with their function. This may help the development of analgesic drugs targeting EphB1 receptor function. Should the blocking of EphB1 receptor activation lead to a significant side effect, precisely for their widespread functions in adulthood, and in particular for their important roles in cancer, drugs that target more specifically their function to chronic pain could be developed on the basis of our understanding of the signalling pathways downstream of EphB1 receptor activation. In particular, the interaction of EphB1 receptors with NMDA receptors could be targeted, as this appears to be a common mechanism through with EphB1 receptors lead to hyperalgesia in the models examined so far.

Overview

Studies carried out in recent years have highlighted how the release of neurotransmitters is not the only way in which neurons can communicate; transmembrane proteins at pre- and postsynaptic sites can play a role not only in synaptogenesis, but also in modulating synaptic plasticity in the adult. Both Eph receptors and their ephrin ligands have been shown to play a role in this respect (see Hruska and Dalva, 2012). Several features of this class of molecules, and aspects of their signalling mechanisms, make them difficult to study: these features include their high sequence homology, their binding promiscuity, their ability to interact both in cis and in trans, and their ability to activate both reverse and forward signalling. Despite these difficulties, a fairly large body of evidence now exists, clearly implicating Eph–ephrins interactions in the onset and maintenance of hypersensitivity and spontaneous pain in a variety of pain models, including inflammatory, neuropathic, cancer and visceral pain. Interaction with postsynaptic NMDA receptors, leading to their phosphorylation and subsequently to an increase of Ca^{++} influx in the postsynaptic neuron, is likely to be a common mechanism through which EphB1 receptor activation leads to hypersensitivity, even if other mechanisms are likely to be involved, at least in some of these models. Interestingly, EphB1 receptors are also needed for the development of morphine tolerance, one of the major problems linked to the use of opioids as analgesic drugs. Despite questions still unanswered on the cellular and molecular mechanisms that mediate the involvement of EphBRs in persistent pain, it is clear that these molecules represent a promising molecular target for the development of new therapeutic strategies for treating chronic pain.

References

Antion, M. D., Christie, L. A., Bond, A. M., Dalva, M. B. and Contractor, A. 2010. Ephrin-B3 regulates glutamate receptor signaling at hippocampal synapses. *Molecular and Cellular Neuroscience* 45 (4): 378–88.

Armstrong, J. N., Saganich, M. J., Xu, N. J., Henkemeyer, M., Heinemann, S. F. and Contractor, A. 2006. B-ephrin reverse signaling is required for NMDA-independent long-term potentiation of mossy fibers in the hippocampus. *Journal of Neuroscience* 26 (13): 3474–81.

Battaglia, A. A., Sehayek, K., Grist, J., McMahon, S. B. and Gavazzi, I. 2003. EphB receptors and ephrin-B ligands regulate spinal sensory connectivity and modulate pain processing. *Nature Neuroscience* 6 (4): 339–40.

Calò, L., Bruno, V., Spinsanti, P., Molinari, G., Korkhov, V., Esposito, Z., Patane, M., Melchiorri, D., Freissmuth, M. and Nicoletti, F. 2005. Interactions between ephrin-B and metabotropic glutamate 1 receptors in brain tissue and cultured neurons. *Journal of Neuroscience* 25 (9): 2245–54.

Cao, J. L., Ruan, J. P., Ling, D. Y., Guan, X. H., Bao, Q., Yuan, Y., Zhang, L. C., Song, X. J. and Zeng, Y. M. 2008. Activation of peripheral ephrinBs/EphBs signaling induces hyperalgesia through a MAPKs-mediated mechanism in mice. *Pain* 139 (3): 617–31.

Cibert-Goton, V., Yuan, G., Battaglia, A., Fredriksson, S., Henkemeyer, M., Sears, T. and Gavazzi, I. 2013. Involvement of EphB1 receptors signalling in models of inflammatory and neuropathic pain. *PLoSOne* 8 (1): e53673.

Dalva, M. B., Takasu, M. A., Lin, M. Z., Shamah, S. M., Hu, L., Gale, N. W. and Greenberg, M. E. 2000. EphB receptors interact with NMDA receptors and regulate excitatory synapse formation. *Cell* 103 (6): 945–56.

Dong, Y., Mao-Ying, Q. L., Chen, J. W., Yang, C. J., Wang, Y. Q. and Tan, Z. M. 2011. Involvement of EphB1 receptor/ephrinB1 ligand in bone cancer pain. *Neuroscience Letters* 496 (3): 163–7.

Feldheim, D. A. and O'Leary, D. D. 2010. Visual map development: Bidirectional signaling, bifunctional guidance molecules, and competition. *Cold Spring Harbor Perspective in Biology* 2 (11): a001768.

Galan, A., Laird, J. M. and Cervero, F. 2004. In vivo recruitment by painful stimuli of AMPA receptor subunits to the plasma membrane of spinal cord neurons. *Pain* 112 (3): 315–23.

Gao, Y. J. and Ji, R. R. 2010. Chemokines, neuronal-glial interactions, and central processing of neuropathic pain. *Pharmacology and Therapeutics* 126 (1): 56–68.

Grunwald, I. C., Korte, M., Wolfer, D., Wilkinson, G. A., Unsicker, K., Lipp, H. P., Bonhoeffer, T. and Klein, R. 2001. Kinase-independent requirement of EphB2 receptors in hippocampal synaptic plasticity. *Neuron* 32 (6): 1027–40.

Grunwald, I. C., Korte, M., Adelmann, G., Plueck, A., Kullander, K., Adams, R. H., Frotscher, M., Bonhoeffer, T. and Klein, R. 2004. Hippocampal plasticity requires postsynaptic ephrinBs. *Nature Neuroscience* 7 (1): 33–40.

Guan, X. H., Lu, X. F., Zhang, H. X., Wu, J. R., Yuan, Y., Bao, Q., Ling, D. Y. and Cao, J. L. 2010. Phosphatidylinositol 3-kinase mediates pain behaviors induced by activation of peripheral ephrinBs/EphBs signaling in mice. *Pharmacology Biochemistry and Behavior* 95 (3): 315–24.

Han, Y., Song, X. S., Liu, W. T., Henkemeyer, M. and Song, X. J. 2008. Targeted mutation of EphB1 receptor prevents development of neuropathic hyperalgesia and physical dependence on morphine in mice. *Molecular Pain* 4, 60.

Henderson, J. T., Georgiou, J., Jia, Z., Robertson, J., Elowe, S., Roder, J. C. and Pawson, T. 2001. The receptor tyrosine kinase EphB2 regulates NMDA-dependent synaptic function. *Neuron* 32 (6): 1041–56.

Henkemeyer, M., Itkis, O. S., Ngo, M., Hickmott, P. W. and Ethell, I. M. 2003. Multiple EphB receptor tyrosine kinases shape dendritic spines in the hippocampus. *Journal of Cell Biology* 163 (6): 1313–26.

Hruska, M. and Dalva, M. B. 2012. Ephrin regulation of synapse formation, function and plasticity. *Molecular and Cellular Neuroscience* 50 (1): 35–44.

Inturrisi, C. E. 2002. Clinical pharmacology of opioids for pain. *Clinical Journal of Pain* 18 (4): S3–S13.

Klein, R. 2009. Bidirectional modulation of synaptic functions by Eph/ephrin signaling. *Nature Neuroscience* 12 (1): 15–20.

Klein, R. 2012. Eph/ephrin signalling during development. *Development* 139 (22): 4105–9.

Kobayashi, H., Kitamura, T., Sekiguchi, M., Homma, M. K., Kabuyama, Y., Konno, S., Kikuchi, S. and Homma, Y. 2007. Involvement of EphB1 receptor/EphrinB2 ligand in neuropathic pain. *Spine* 32 (15): 1592–8.

Lai, K. O. and Ip, N. Y. 2009. Synapse development and plasticity: Roles of ephrin/Eph receptor signaling. *Current Opinion in Neurobiology* 19 (3): 275–83.

Lim, B. K., Matsuda, N. and Poo, M. M. 2008. Ephrin-B reverse signaling promotes structural and functional synaptic maturation in vivo. *Nature Neuroscience* 11 (2): 160–69.

Liu, W. T., Li, H. C., Song, X. S., Huang, Z. J. and Song, X. J. 2009a. EphB receptor signaling in mouse spinal cord contributes to physical dependence on morphine. *FASEB Journal* 23 (1): 90–8.

Liu, S., Liu, W. T., Liu, Y. P., Dong, H. L., Henkemeyer, M., Xiong, L. Z. and Song, X. J. 2011. Blocking EphB1 receptor forward signaling in spinal cord relieves bone cancer pain and rescues analgesic effect of morphine treatment in rodents. *Cancer Research* 71 (13): 4392–402.

Liu, W. T., Han, Y., Li, H. C., Adams, B., Zheng, J. H., Wu, Y. P., Henkemeyer, M. and Song, X. J. 2009b. An in vivo mouse model of long-term potentiation at synapses between primary afferent C-fibers and spinal dorsal horn neurons: Essential role of EphB1 receptor. *Molecular Pain* 5 (29). doi: 10.1186/1744-8069-5-29.

Lu, Z., Zhang, Y., Li, Z., Yu, S., Zhao, G., Li, M., Wang, Z., Wang, Q. and Yang, Y. 2012. Overexpression of the B-type Eph and ephrin genes correlates with progression and pain in human pancreatic cancer. *Oncology Letters* 3 (6): 1207–12.

Meyer-Franke, A., Wilkinson, G. A., Kruttgen, A., Hu, M., Munro, E., Hanson, M. G., Jr., Reichardt, L. F. and Barres, B. A. 1998. Depolarization and cAMP elevation rapidly recruit TrkB to the plasma membrane of CNS neurons. *Neuron* 21 (4): 681–93.

Miyamoto, E. 2006. Molecular mechanism of neuronal plasticity: Induction and maintenance of long-term potentiation in the hippocampus. *Journal of Pharmacological Sciences* 100 (5): 433–42.

Murai, K. K. and Pasquale, E. B. 2004. Eph receptors, ephrins, and synaptic function. *Neuroscientist* 10 (4): 304–14.

Nakazawa, T., Komai, S., Tezuka, T., Hisatsune, C., Umemori, H., Semba, K., Mishina, M., Manabe, T. and Yamamoto, T. 2001. Characterization of Fyn-mediated tyrosine phosphorylation sites on GluR epsilon 2 (NR2B) subunit of the N-methyl-D-aspartate receptor. *Journal of Biological Chemistry* 276 (1): 693–9.

Orikawa, Y., Kato, H., Seto, K., Kobayashi, N., Yoshinaga, K., Hamano, H., Hori, Y., Meyer, T. and Takei, M. 2010. Z-360, a novel therapeutic agent for pancreatic cancer, prevents up-regulation of ephrin B1 gene expression and phosphorylation of NR2B via suppression of interleukin-1 beta production in a cancer-induced pain model in mice. *Molecular Pain* 6 (72). doi: 10.1186/1744-8069-6-72.

Pasquale, E. B. 2008. Eph-ephrin bidirectional signaling in physiology and disease. *Cell* 133 (1): 38–52.

Pasternak, G. W. 2007. When it comes to opiates, just say NO. *Journal of Clinical Investigation* 117 (11): 3185–7.

Peng, H. Y., Chen, G. D., Lai, C. H., Tung, K. C., Chang, J. L. and Lin, T. B. 2010. Endogenous ephrinB2 mediates colon-urethra cross-organ sensitization via Src kinase-dependent tyrosine phosphorylation of NR2B. *American Journal of Physiology: Renal Physiology* 298 (1): F109–F117.

Rostas, J. A., Brent, V. A., Voss, K., Errington, M. L., Bliss, T. V. and Gurd, J. W. 1996. Enhanced tyrosine phosphorylation of the 2B subunit of the N-methyl-D-aspartate receptor in long-term potentiation. *Proceedings of the National Academy of Sciences of the United States of America* 93 (19): 10452–6.

Ruan, J. P., Zhang, H. X., Lu, X. F., Liu, Y. P. and Cao, J. L. 2010. EphrinBs/EphBs signaling is involved in modulation of spinal nociceptive processing through a mitogen-activated protein kinases-dependent mechanism. *Anesthesiology* 112 (5): 1234–49.

Sefton, M. and Nieto, M. A. 1997. Multiple roles of Eph-like kinases and their ligands during development. *Cell and Tissue Research* 290 (2): 243–50.

Sheffler-Collins, S. I. and Dalva, M. B. 2012. EphBs: An integral link between synaptic function and synaptopathies. *Trends in Neuroscience* 35 (5): 293–304.

Shields, S. D., Ahn, H. S., Yang, Y., Han, C., Seal, R. P., Wood, J. N., Waxman, S. G. and Dib-Hajj, S. D. 2012. Nav1.8 expression is not restricted to nociceptors in mouse peripheral nervous system. *Pain* 153 (10): 2017–30.

Slack, S., Battaglia, A., Cibert-Goton, V. and Gavazzi, I. 2008. EphrinB2 induces tyrosine phosphorylation of NR2B via Src-family kinases during inflammatory hyperalgesia. *Neuroscience* 156 (1): 175–83.

Sloniowski, S. and Ethell, I. M. 2012. Looking forward to EphB signaling in synapses. *Seminars in Cell & Developmental Biology* 23 (1): 75–82.

Song, X. J., Cao, J. L., Li, H. C., Zheng, J. H., Song, X. S. and Xiong, L. Z. 2008a. Upregulation and redistribution of ephrinB and EphB receptor in dorsal root ganglion and spinal dorsal horn neurons after peripheral nerve injury and dorsal rhizotomy. *European Journal of Pain* 12 (8): 1031–9.

Song, X. J., Zheng, J. H., Cao, J. L., Liu, W. T., Song, X. S. and Huang, Z. J. 2008b. EphrinB-EphB receptor signaling contributes to neuropathic pain by regulating neural excitability and spinal synaptic plasticity in rats. *Pain* 139 (1): 168–180.

Takasu, M. A., Dalva, M. B., Zigmond, R. E. and Greenberg, M. E. 2002. Modulation of NMDA receptor-dependent calcium influx and gene expression through EphB receptors. *Science* 295 (5554): 491–5.

Uchida, H., Matsumoto, M. and Ueda, H. 2009. Profiling of BoNT/C3-reversible gene expression induced by lysophosphatidic acid: EphrinB1 gene up-regulation underlying neuropathic hyperalgesia and allodynia. *Neurochemistry International* 54 (3–4): 215–21.

Wu, H. C., Chang, C. H., Peng, H. Y., Chen, G. D., Lai, C. Y., Hsieh, M. C. and Lin, T. B. 2011. EphrinB2 induces pelvic-urethra reflex potentiation via Src kinase-dependent tyrosine phosphorylation of NR2B. *American Journal of Physiology: Renal Physiology* 300 (2): F403–F411.

Yu, L. N., Zhou, X. L., Yu, J., Huang, H., Jiang, L. S., Zhang, F. J., Cao, J. L. and Yan, M. 2012. PI3K contributed to modulation of spinal nociceptive information related to ephrinBs/EphBs. *PLoSOne* 7 (8): e40930.

Zhao, J., Yuan, G., Cendan, C. M., Nassar, M. A., Lagerstrom, M. C., Kullander, K., Gavazzi, I. and Wood, J. N. 2010. Nociceptor-expressed ephrin-B2 regulates inflammatory and neuropathic pain. *Molecular Pain* 6 (77). doi: 10.1186/1744-8069-6-77.

SECTION II
Pain in the brain

CHAPTER 8

Brain imaging in experimental pain

Massieh Moayedi and Tim V. Salomons

Traditional theories of pain have depicted pain as a 'bottom-up' phenomenon: a direct reflection of peripheral injury or pathology. Descartes, often considered the father of Specificity theory (but see Moayedi and Davis, 2013), posited that the pain system runs on a direct line from the skin to a pain centre in the brain. Activation of this line was viewed as necessary and sufficient for the experience of pain, with the implication of a one-to-one correspondence between peripheral stimulation and the pain experience. Such theories allow only a minimal role for the brain, viewing it largely as a relay centre for messages from the periphery, as evidenced by Descartes' metaphor of the brain as a bell to be rung by tugging on a string in the periphery (Descartes, 1664).

The Specificity account of pain was advanced by the discovery of nociceptors: these are nerve endings in the periphery that respond exclusively to stimulation that is strong enough to signal potential damage. Sherrington (1903) used the term 'nocicipient' to describe cutaneous end organs specific to noxious stimuli – a framework that was validated by Perl's discovery of primary afferents responsive only to mechanical noxious stimulation (Burgess and Perl, 1967), as well as of polymodal nociceptors and high-threshold mechanoreceptors (Bessou and Perl, 1969). These discoveries supported one aspect of the bottom-up account, namely that pain is associated with the activation of specialised receptors in the periphery. It is critical to note, however, that pain and nociception are not synonymous. Nociception, according to the 2011 taxonomy of the International Association for the Study of Pain (IASP), is 'the neural process of encoding noxious stimuli' (Merskey and Bogduk, 1994: 209) – a process that is initiated by activation of nociceptors through noxious stimulation. In contrast to nociception, however, pain requires a conscious perception.

The clinical literature provides multiple examples of what Melzack and Wall (2004) referred to as the 'variable link' between nociception and the conscious perception of pain. Beecher (1946) famously documented soldiers who sustained grave injuries on the battlefield but experienced no pain until they were safe from direct threat, highlighting the fact that nociception is not sufficient for the experience of pain. Furthermore, many chronic pain disorders occur without apparent injury or measurable pathology, or persist long after an initial injury has healed; this suggests that the relationship between nociception and pain might be less exclusive than

An Introduction to Pain and its Relation to Nervous System Disorders, First Edition. Edited by Anna A. Battaglia.
© 2016 John Wiley & Sons, Ltd. Published 2016 by John Wiley & Sons, Ltd.

depicted by early theorists. Observation of the centrally mediated facilitation of pain in many of these chronic pain disorders (Edwards, 2005; Arendt-Nielsen and Yarnitsky, 2009) suggests that one explanation for the complex relationship between nociception and pain is that the brain plays a far more prominent role than that initially hypothesised by Descartes.

Even in the presence of nociceptive input, the experience of pain can be transformed by the cognitive and emotional context in which nociception occurs. Pain that is believed to signal a grave health risk is experienced differently from a similar level of pain that is believed to be innocuous. Cognitive and affective responses to pain are critical to clinical presentation, as belief-based schemas such as catastrophising are repeatedly shown to affect outcomes in chronic pain (Campbell and Edwards, 2009). These top-down modulatory influences provide further demonstration of the 'variable link' between nociception and pain perception.

As our understanding of the peripheral events that give rise to nociception develops, it is clear that the depiction of specific channels or 'labelled lines' for pain was prescient. Nevertheless, it is equally clear that nociception alone cannot account for the entirety of pain experience. The brain plays an active role in modulating the conscious perception of pain in response to the sensory, emotional and cognitive context in which nociception occurs. Understanding how the brain instantiates this active role, however, remains a challenge. While work in non-human species continues to provide insight into key processes, it has one critical limitation: pain is inherently a subjective experience and impossible to fully understand without subjective report. Thus, while work in non-human species allows for more invasive and directly interventional studies, understanding how the brain shapes the human experience of pain requires techniques that can combine the direct *in vivo* examination of neural processes with self-report. Modern brain-imaging techniques provide this opportunity and therefore represent a critical tool for understanding the brain's contribution to human pain experience. Now, more than twenty years after the first neuroimaging studies were used to investigate pain (Jones et al., 1991; Talbot et al., 1991) the field of pain neuroimaging has expanded exponentially – but how far has it advanced our understanding of how the brain contributes to the pain experience? This chapter reviews the contribution of neuroimaging to our understanding of the brain's role in this experience and assesses some of the methodological and inferential challenges that have limited progress.

Brain-imaging methods

Functional brain imaging

There are various imaging modalities that can be used to investigate brain mechanisms of pain. As indicated in Figure 8.1, methodologies for examining brain function differ in their temporal and spatial specificity. Additionally, imaging techniques vary in terms of their suitability for examining evoked or basal states or in terms of their ability to capture specific aspects of pain transmission or response. Electroencephalography (EEG) is a non-invasive brain-imaging technique for measuring the brain's intrinsic electrical activity through the skull with a single or an array of electrodes. It offers high temporal resolution, allowing for the measurement of spontaneous neural activity related to spontaneous pain or for event-related neural activity (stimulus-evoked potentials like pulsed laser heat stimuli) on a millisecond time scale. However, the spatial resolution of these techniques is comparatively low, because each electrode records

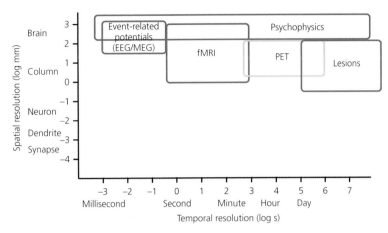

Figure 8.1 Schematic of the spatial and temporal resolution of different techniques to elucidate brain functions. The y-axis represents the spatial resolution, and the x-axis represents the temporal resolution. The colour of the boxes represent the invasiveness of the technique: yellow is invasive, and green is non-invasive.

the sum of activities from large swathes of brain (large, that is, by comparison to the size of individual neurons). Positron emission tomography (PET) imaging is based on the spatial distribution of radioactivity emitted by radioactive isotopes (injected into the subject as part of a molecule) that emit positrons as they decay. The system investigated relies on the radionuclide used. While PET offers relatively poor temporal resolution, it (along with a similar but more recent method, arterial spin labelling) provides quantitative images, making it possible to study basal states. A further advantage of PET is the ability to use radiolabelled molecules in order to target the activity of specific neurotransmitter systems during a given paradigm. Therefore PET can be used to investigate the role of neurotransmitters like dopamine, serotonin and opiates in nociceptive processing and in pain modulation.

While other functional techniques offer advantages in terms of temporal resolution, receptor specificity or the ability to provide quantitative (rather than relative) measures, functional magnetic resonance imaging has been the most widely used measure, primarily because it is non-invasive and is perceived to have a good balance of temporal and spatial resolution relative to other imaging modalities (see Figure 8.1). This chapter will therefore focus primarily on fMRI.

Functional Magenetic Resonance Imaging

Functional magnetic resonance imaging measures a hemodynamic response, usually the blood oxygenation level-dependent (BOLD) signal. The hemodynamic response is used as a proxy for neuronal activity, on the assumption that oxygenated blood flows preferentially to regions where neuronal populations are more active. This variation can be detected by the MRI scanner due to differences in the magnetic properties of oxygenated and deoxygenated blood.

There are several ways to exploit blood oxygenation level-dependent (BOLD) fMRI so as to glean information on brain function. The first is to investigate brain function in response to a stimulus or a task or related to a perception. Another method is to investigate brain regions that are temporally synchronised in the absence of a task – in other words, to investigate resting-state functional connectivity.

Stimulus-evoked fMRI

Most fMRI studies evaluate brain responses related to an experimental stimulus by comparing them to a baseline or control stimulus. This method can be used to assess brain regions related to noxious stimuli and individual differences in painful perception. It allows the investigation of brain mechanisms underlying acute pain, and also stimulus-evoked pains (such as pain induced through topical capsaicin application) in experimental models of sensitisation.

The BOLD hemodynamic response function (HRF) has a delay with respect to the stimulus onset: it is maximal at about 4–6 seconds after the stimulus onset and lasts between 10 and 14 seconds. Furthermore, the time to scan the full brain (and thus to return to re-sampling the activation change in a given region) tends to be in the 2–3-second range. Given these temporal constraints, fMRI studies are largely focused on global changes in signal that occur over longer periods, rather than changes that might occur over short periods (e.g., through the activation of different nociceptors). Studies that have investigated brain responses to nociceptive stimulation in this way have identified a set of brain regions – the so-called 'pain matrix' – that are consistently activated. The findings related to these regions will be described in fuller detail below.

Percept-related fMRI

Sensation and perception are dynamic processes: the intensity of a sensation can fluctuate in the absence of changes to the stimulus. These changes are due to bottom-up and top-down modulation related to various physiological and psychological factors. For example, some acute pain qualities can vary over time (Davis and Pope, 2002), either attenuating or increasing during a tonic stimulation (Hashmi and Davis, 2008). Given the temporal constraints mentioned above, fluctuations in pain percepts over time may not be captured in traditional stimulus-fMRI.

Therefore, to identify the neural basis of these temporal fluctuations in perception, several groups developed percept-related fMRI. This method relies on the selection of a clear, definable percept that the subject is experiencing, the precise timing when s/he is experiencing it, and the temporal profile of the intensity of percept. These measurements are acquired through some method of continuous online perceptual ratings during fMRI acquisition. The BOLD response recorded during the fMRI session is correlated to the online ratings provided by the subject.

Percept-related fMRI has been used to identify and dissect brain responses linked to various pain-related percepts, such as noxious tonic chemically induced stinging and burning pain (Porro et al., 1998), noxious heat (Apkarian et al., 1999; Gelnar et al., 1999; Chen et al., 2002), cold-evoked prickle sensations (Davis et al., 2002), and noxious cold-evoked paradoxical heat sensation (Davis et al., 2004). Importantly, some of the studies compared these pain-related brain activity to innocuous stimuli and control tasks (e.g., Apkarian et al., 1999; Gelnar et al., 1999; Chen et al., 2002).

Functional connectivity

In addition to traditional, evoked-response approaches that identify brain regions commonly co-activated during a task, it is possible to investigate 'functional connectivity' – that is, the degree to which these brain regions are meaningfully linked during a task or at rest. There are several types of functional connectivity methods during task that can be used for the purpose of assessing hierarchical processing in the brain (in other words the order of processing) – for example effective connectivity, mediation analysis, regional coherence – or to determine how connectivity is moderated by a particular behavioural or psychophysical condition (e.g. reaction time or pain intensity), using a psychophysiological interaction (PPI) analysis. These methods are covered in detail elsewhere (see Friston et al., 1997; Friston, 2002; Friston, 2011; Smith, 2012). This chapter will focus on resting state functional connectivity. Resting state fMRI (rs-fMRI) data are collected in the absence of any task – the subject is instructed not to think of anything in particular. Rs-fMRI data consist of activity with ultra-low frequency (<0.1 Hz) oscillations between brain regions (Friston, 1994). The 'connectivity' measured with rs-fMRI includes brain regions that are connected through direct connections, indirection connections, or a common input.

Rs-fMRI has been used to examine intrinsic brain networks that putatively reflect brain architecture. The first intrinsic brain network identified to be functionally connected was the so-called 'default mode network' (DMN) (Gusnard and Raichle, 2001; Mazoyer et al., 2001; Raichle et al., 2001). In addition to the DMN, various networks have been identified on the basis of the spatial pattern of regions that are functionally connected (Smith et al., 2009), and their co-activation during a task (Fox et al., 2005; De Luca et al., 2006; Weissman-Fogel et al., 2010); such networks include the salience, executive control, sensorimotor, cognitive and emotional networks. New theories are now incorporating the role of these networks in pain perception and modulation (Farmer, Baliki and Apkarian, 2012; Kucyi and Davis, 2014).

Brain imaging of experimental acute pain

Imaging acute pain

The first two brain-imaging studies to investigate brain regions involved in nociception and pain perception and modulation in healthy subjects were performed using PET (Jones et al., 1991; Talbot et al., 1991). One study reported that heat pain, when compared to an innocuous warm stimulus, evoked an increase in regional cerebral blood flow (rCBF) in the contralateral thalamus, in the lentiform nucleus and in the cingulate cortex (ACC) – or rather in the mid-cingulate cortex (MCC), with the current nomenclature (Jones et al., 1991). The other study reported increased rCBF in S1, S2 and the MCC (Talbot et al., 1991). The first fMRI study to investigate the neural correlates of pain perception found that painful electrical stimulation increased the BOLD signal in S1 and MCC (Davis et al., 1995). Many studies have since investigated pain-related brain activation, and some have specifically attempted to identify the neural correlates of the different dimensions of pain. Several reviews and meta-analyses have summarised these findings (Treede et al., 1999; Casey, 2000; Davis, 2000; Peyron, Laurent and Garcia-Larrea, 2000; Price, 2000; Rainville, 2002; Apkarian et al., 2005; Farrell, Laird and Egan, 2005; Bingel and Tracey, 2008; Duerden and Albanese, 2011). In general, brain-imaging studies of pain have reported the activation of S1, S2, MCC, insula, prefrontal and motor regions. As mentioned above, these regions have collectively been referred to as the 'pain matrix' – a putative network that has been supposed to integrate nociceptive signals into the experience of pain. The putative roles of these regions in pain-related processing, as determined by neuroimaging studies, are described below.

S1

It is well known, from classic brain stimulation studies in humans and from animal electro-physiology studies, that S1 is somatotopically organised (Penfield and Boldrey, 1937). This S1 tactile homunculus has been largely confirmed in human brain-imaging studies (for a review, see Apkarian et al., 2005). It generally shows an inverted organisation: the face is represented ventrolaterally, the upper limb is represented more dorsolaterally, and the lower limb is represented within the medial wall. Furthermore, it is noteworthy that nociresponsive cells have been identified in the primate S1 (Kenshalo and Isensee, 1983; Kenshalo et al., 1988; Chudler et al., 1990; Kenshalo et al., 2000).

Although nearly all neuroimaging studies have confirmed that S1 responds to innocuous A-beta fibre stimulation, these methods have not provided a clear role for S1 responses to A-delta (nociceptive) stimulation (Bushnell et al., 1999). Therefore, on the basis of neuroimaging findings, it remains unclear whether S1 is implicated in pain perception.

Furthermore, imaging studies have obtained vastly different findings in S1 activation in pain paradigms. Specifically, some studies have identified significant contralateral S1 activation to the stimulation of the arm (e.g., Talbot et al., 1991), whereas others did not find S1 activity in response to noxious stimulation (e.g., Jones et al., 1991). In a later study, Apkarian and colleagues (1992) found that S1 actually showed deactivation in response to noxious thermal stimuli. However, in a meta-analysis, Apkarian and colleagues (2005) reported that 69 per cent of PET, 76 per cent of fMRI, 10 per cent of EEG and 70 per cent of magnetoen-cephalography (MEG) studies reported an S1 activation in acute pain paradigms. More recently, MEG (Timmermann et al., 2001; Ploner et al., 2006), EEG (Valentini et al., 2012; Hu et al., 2014) and diffuse optical tomography (Becerra et al., 2009) studies have reported S1 responses evoked by painful stimuli. Also, a recent meta-analysis identified S1 as a region that is activated across 140 neuroimaging studies of acute experimental heat and cold pain (Duerden and Albanese, 2011). Several human neuroimaging studies have found graded responses in S1 (and other cortical areas) (Derbyshire et al., 1997; Porro et al., 1998; Coghill et al., 1999; Moulton et al., 2005). Further evidence for a role for S1 in pain perception comes from a study that used hypnotic suggestion to dissociate pain intensity from pain unpleasant-ness (Rainville et al., 1999) and found that S1 and possibly S2 responses correlated with pain intensity, whereas responses in the ACC correlated with pain unpleasantness. In sum, while a large number of neuroimaging studies have supported a role for S1 in nociception and pain processing, there is considerable inconsistency.

These inconsistencies may be related to the sensory stimuli used in eliciting pain. Specifically, there is a clear role for the activation of S1 to innocuous tactile stimulation, and thus it is possible that noxious stimuli become contaminated by tactile stimuli; for example, when a heat probe is used to stimulate the arm, the probe places some pressure on the arm, and fibres that encode innocuous warm temperatures or touch may also be firing. Many of the studies use stimulators that activate both A-delta and A-beta fibres; and they do not control for these inputs. The use of a laser to elicit noxious stimuli gets around this potential confound, as it specifically activates nociceptors – and not tactile sensory afferent. More recent and more sensitive methods have clearly identified, in S1, activity specific to laser-evoked potential (LEP), and have clearly demonstrated that S1 processes nociceptive information at early stages and at the latest stages of nociceptive processing (Lee, Mouraux and Iannetti, 2009; Valentini et al., 2012; Hu et al., 2014) and that LEP-specific gamma-band activity in S1 is a direct and obligatory correlate of pain intensity (Zhang et al., 2012). Therefore the combination of LEPs, study designs that could parse nociceptive activity from general, salience-related brain activity, and novel analysis methods has demonstrated that S1 does indeed process nociceptive information – encoding pain magnitude and location.

Further evidence for the role of S1 in localising nociceptive stimuli on the body comes from the identification of pain somatotopic maps in this brain region. The first functional investigations in humans revealed a gross somatotopy of nociceptive responses in S1 (Lamour, Willer and Guilbaud, 1983; Andersson et al., 1997; Kenshalo et al., 2000; Bingel et al., 2004). Recent advances in functional neuroimaging techniques allow for the detection of surprisingly finer somatotopic maps of nociceptive signals. For example, phase-encoded MRI analysis was used to reveal that nociceptive somatotopic maps in S1 are fine-grained (Mancini et al., 2012). Furthermore, body regions where spatial acuity is very high, such as fingers (Weinstein, 1968; Mancini et al., 2014), are magnified in S1 independently of their intra-epidermal innervation density (Mancini et al., 2013). In addition, somatotopic maps for nociceptive inputs in S1 are co-aligned to maps of tactile, innocuous mechanical input to the same body part (Mancini et al., 2012).

Therefore neuroimaging studies have replicated many of the previous findings from electrophysiological and brain stimulation about the role of S1 in nociception and pain perception. Specifically, neuroimaging studies have confirmed the somatotopic organisation of S1 and have shown that S1 responses are graded to noxious stimulus intensity. Together, these findings suggest that S1 is implicated in the sensory–discriminative dimension of pain, as it encodes the location and intensity of a noxious stimulus.

S2/parietal operculum

Although many imaging studies report that S2 is activated during painful stimuli, the findings have not clarified the role of S2 in pain processing. According to Apkarian et al. (2005), 68 per cent of PET, 81 per cent of fMRI, 60 per cent of EEG and 95 per cent of MEG studies reported S2 activation in investigations of the neural correlates of acute pain in healthy controls. Also, Duerden and Albanese's (2011) meta-analysis identified S2 as one of the regions that are most commonly activated in neuroimaging studies of experimental pain. However, it is not clear how S2 activations reflect various aspects of the pain experience. It is also unclear whether S2 receives nociceptive information from the thalamus via a relay through S1 (i.e., serial processing) (Allison et al., 1989; Hari et al., 1993), or directly, in parallel to information being sent to S1 (Pons, Garraghty and Mishkin, 1992; Ploner et al., 2006; Liang, Mouraux and Iannetti, 2011). Furthermore, evidence from LEP studies has demonstrated that the early electrophysiological nociceptive response (N2) is caused by activity in S2/the posterior insular cortex (e.g., Frot et al., 2007), which suggests that there is parallel processing in S1 and S2. This is based on the timing of the observed electrophysiological responses – serial activity would lead to waveforms that occur serially, whereas parallel activity would result in a waveform with activity from both S1 and S2. This latter possibility can be resolved through source localisation methods. Using these methods, recent findings have opposed a parallel processing model with regard to the sensory–discriminative dimension of pain (Valentini et al., 2012).

One study that investigates painful LEP with subdural electrodes has demonstrated somatotopic organisation along the anterior–posterior axis, the foot being anterior to the face, within S2 (Vogel et al., 2003). This organisation is different from that of tactile input, and it has been suggested that there may be a pain somatotopic map independent of the tactile somatotopic map (Apkarian et al., 2005).

The insula

The insula is a large brain area consisting of three subregions: the anterior, the mid and the posterior insula. The insula has been consistently activated in most experimental studies of pain. Apkarian et al. (2005) reported that 88 per cent of PET, 100 per cent of fMRI, 30 per cent of EEG and 20 per cent of MEG studies of acute pain in healthy volunteers reported insular activation. A more recent meta-analysis of 140 experimental pain neuroimaging studies also identified the insula as one of the regions most likely to be activated (Duerden and Albanese, 2011).

Given the ubiquity of insula activation in imaging studies, many groups have postulated a central role for the insula in pain perception. For example, the insula responds in a graded fashion to increasing intensity of noxious stimuli (Coghill et al., 1999). In line with this, Garcia-Larrea has posited that the posterior insula is akin to the tertiary somatosensory cortex (S3), which receives direct input from the spinothalamic system and produces pain through network interaction with other brain regions (Garcia-Larrea, 2012). Similarly, Craig has suggested that the posterior insula is the site of sensory–discriminative nociceptive processing (Craig, 2002). Support for this hypothesis comes from the presence of somatotopic maps in the insula (Brooks et al., 2005; Hua et al., 2005; Henderson et al., 2007); in fact there are several somatotopic maps for different noxious stimuli and within different subregions (Baumgartner et al., 2010). In these somatotopic maps, the face is anterior to the foot.

Brooks and Tracey (2007) have suggested that the insula is a multidimensional integration site for pain. This hypothesis is based on the convergence of the afferent pathways underlying the sensory–discriminative and the cognitive–affective dimensions of pain (the lateral and the medial spinothalamic tracts, respectively) at the insula (Treede et al., 1999); on the fact that electrophysiological stimulation of the insula elicits painful percepts (Ostrowsky et al., 2002); and on the fact that lesions to this region are related to abnormal pain percepts (Greenspan et al., 1999; Starr et al., 2009). In contrast, Baliki and colleagues (2009) have advanced the hypothesis that, in addition to integrating the dimensions of pain perception, the insula is a central, multimodal magnitude estimator and a nociceptive-specific magnitude estimator. However, Craig (2003b, 2003a, 2009, 2010) and others (Devinsky, Morrell and Vogt, 1995; Critchley, 2004; Critchley et al., 2004; Pollatos et al., 2005; Pollatos et al., 2007) have proposed that the insula is involved in integrating multimodal information important for sensorimotor, emotional, allostatic/homeostatic and cognitive functions and should be called a limbic sensory cortex. In fact Craig reclassifies pain as a homeostatic emotion rather than regarding it as a submodality of the somatosensory system.

The insula has also been implicated in a network responsible for encoding, evaluating and responding to salient sensory stimuli (Downar et al., 2001; Downar et al., 2002; Downar et al., 2003; Mouraux and Iannetti, 2009; Iannetti and Mouraux, 2010; Legrain et al., 2011; Mouraux et al., 2011). In a meta-analysis of the function of the insula, Kurth and colleagues (2010) investigated 79 studies that showed insular activation in imaging studies of acute pain stimulation paradigms in healthy controls, and found that all of the subregions of the insula were activated. It is, however, noteworthy that this same study identified different subregions of the insula that showed responses to paradigms that tested the neural correlates of processes related to pain, such as somatosensation, motor output, attention, interoception, and emotion, and that these activations generally overlapped with that of pain, underscoring the more general function of the insula in processing salient stimuli.

In sum, because the insula receives multimodal input, it is an ideal site for integrating information from the various dimensions of pain. However, it is activated in many other paradigms and across different modalities (Downar et al., 2003; Yarkoni et al., 2011). Therefore the insula may best be considered as a region that encodes behaviourally salient stimuli, including pain.

The cingulate cortex

The cingulate cortex is a large, heterogeneous brain region that can be subdivided into several subregions. There are many classification systems that are used in order to differentiate between these subregions, and this adds to the complexity of comparing studies that use different species or different nomenclatures. In this chapter, the nomenclature developed by Vogt and colleagues (2005) will be used whenever possible. It is noteworthy that the region previously referred to as the ACC has now been divided into two regions: the MCC and the ACC. The ACC can be further subdivided into the subgenual ACC (sgACC) and the perigenual ACC (pgACC). The MCC can be further subdivided into the anterior MCC (aMCC) and the posterior MCC (pMCC). Various regions of the cingulate cortex are involved in nociceptive processing and pain modulation (for a review, see Vogt and Sikes, 2000; Bushnell, Ceko and Low, 2013), although this chapter will largely focus on the roles of the MCC, as this is the cingulate region most consistently activated in neuroimaging studies of pain (Shackman et al., 2011). Brain-imaging studies have consistently shown MCC activity during noxious stimulation. For example, the MCC showed graded responses to the increasing intensity of noxious stimuli (Coghill et al., 1999). Furthermore, according to Apkarian et al. (2005), 94 per cent of PET, 81 per cent of fMRI, 100 per cent of EEG and 25 per cent of MEG studies reported cingulate activation in imaging studies of acute pain in healthy volunteers.

On the basis of converging evidence across domains, the aMCC has often been considered the seat of the affective dimension of pain (or unpleasantness, as opposed to pain discriminability) (Vogt, Sikes and Vogt, 1993). For example, tracing studies have shown that the spinothalamic tract projects directly to the MCC (Dum, Levinthal and Strick, 2009); electrophysiological studies have demonstrated the presence of neurons with increased activity during pain anticipation and escape from a noxious stimulus (Hutchison et al., 1999; Koyama et al., 2001; Iwata et al., 2005); and cingulotomy and cingulectomy – that is, the surgical ablation and disruption of the cingulate cortex in cases of intractable pain – have been somewhat effective in reducing the affective dimension of pain in patients (for a review, see Fuchs et al., 2014, and also Davis et al., 1994).

Early neuroimaging evidence for the role of the ACC/MCC in the processing of the affective dimension of pain came from an influential study where hypnotic suggestion was used in order to manipulate the perception of pain intensity and unpleasantness. In this study, Rainville and colleagues (1997) found that the modulation of pain unpleasantness (but not of its intensity) affected the CBF in the ACC.

More recent evidence has demonstrated that the MCC may be something more complex than a node in the limbic circuit. It has been proposed that, rather than simply encode affect, the cingulate may serve as a hub between the affective processing and the motor planning involved in processing and planning affectively motivated motor outputs (Vogt and Sikes, 2009; Shackman et al., 2011). Support for the affective premotor cingulate concept comes from studies that have shown that the stimulation of a region analogous to aMCC is related to the urge to escape a threatening stimulus (Iwata et al., 2005) and has direct projections to primary motor regions, as well as corticobulbar/corticospinal projections (Dum and Strick, 1992).

In addition to the role of the cingulate cortex as an affective–motor interface, the aMCC's role in cognitive processing has been well documented; influential early models even suggest that this region is specialised for cognition, as affective processing occurs in more ventral and anterior regions (Devinsky et al., 1995; Bush, Luu and Posner, 2000). It is noteworthy that the evidence cited above in support of a role for the aMCC in pain-related affect was a key piece of evidence against this segregated model of cingulate function (Shackman et al., 2011). Indeed, there is ample evidence that, in addition to its role in affective and motor responses, the MCC is involved in the cognitive–evaluative processing of pain (for a comprehensive review, see Seminowicz and Davis, 2007; Bushnell et al., 2013).

Like the insula in nociceptive processing and pain integration, the MCC is well situated to integrate the cognitive–evaluative and affective dimensions of pain. Its connectivity with motor regions suggests that it is a hub for mounting an appropriate motor response to pain (Morrison, Perini and Dunham, 2013). Therefore a working model for the role of the cingulate cortex is that of a hub for the integration of cognitive and affective information towards the initiation of appropriate defensive actions.

Other brain regions

A set of other brain regions are also often – but less consistently – activated in acute pain-imaging paradigms, for example the prefrontal cortex (PFC) and the cortical and subcortical motor regions. Given the size and heterogeneity of the PFC, its role in pain modulation is beyond the scope of this chapter. Briefly, there is no evidence suggesting that the PFC receives nociceptive information directly from the thalamus. In contrast to tract tracing studies, PET and fMRI studies often identified PFC regions activated through noxious stimulation (Apkarian et al., 2005). Interestingly, no EEG or MEG studies reported PFC activation. The PFC does, however, have reciprocal connections with a number of nociresponsive brain regions such as the cingulate cortex, the insula, and the somatosensory cortex (Nieuwenhuys, Voogd and van Huijzen, 2008; for more in-depth reviews, see Bushnell et al., 2013; Colloca et al., 2013).

Motor regions are activated in acute pain paradigms in healthy controls, but less reliably than the aforementioned brain regions (Apkarian et al., 2005). In the context of pain, it is believed that motor regions serve two purposes: to orient the body toward the source of pain and to initiate nocifensive behaviour (e.g., to avoid the stimulus). Similarly, another study reported that hypertonic saline injection into the masseter muscle, an experimental acute pain model, was related to deactivation in the face region of M1, as measured by fMRI (Nash et al., 2010). It is, however, noteworthy that some studies have not observed activation in M1 during noxious stimulation (Romaniello et al., 2000; Halkjaer et al., 2006).

The basal ganglia are a set of subcortical nuclei that have been associated with motor function. However, it is noteworthy that the basal ganglia do receive input from many cortical and subcortical regions and form several functional loops (Nieuwenhuys et al., 2008). It has been proposed that this extensive wiring to the brain suggests that the basal ganglia may be implicated in more than just motor functions (Haber and Knutson, 2009). For instance, the basal ganglia are often activated in neuroimaging studies of experimental pain (Borsook et al., 2010). It has been suggested that the cortico-thalamo-basal ganglia–cortical loops may provide a unique anatomical substrate for the integration of the various dimensions of pain (Borsook et al., 2010). However, more work is required to further investigate this possibility.

Other motor regions, including the primary motor cortex (PMC) and the supplementary motor area (SMA), are also commonly activated in brain-imaging studies of experimental pain in healthy volunteers (Apkarian et al., 2005; Duerden and Albanese, 2011). The PMC has been implicated in the cognitive modulation of motor output, and the SMA has been implicated in motor planning and in the temporal organisation of movements (Picard and Strick, 2001). Therefore their activation during experimental pain is probably related to the cognitive and motivational dimensions of pain with regard to nocifensive behaviours.

In sum, many studies have found that acute pain stimuli evoke activities in widespread brain regions such as the S1, the S2, the ACC, the MCC, and the insula (Apkarian et al., 2005). Some fMRI studies have also reported that noxious stimuli elicit activity in additional brain areas such as the PFC, the motor cortex and the SMA, and also subcortically – in the basal ganglia, thalamus and brainstem (Apkarian et al., 2005; Duerden and Albanese, 2011). In line with the concept that pain is multidimensional, these findings demonstrate that pain is associated with activity not only in regions that are traditional parts of pain pathways, but also in areas implicated in innocuous somatosensory, cognitive and motor functions.

Salience, pain and specificity

Neuroimaging provides a unique window for viewing how the human brain processes pain in real time. After more than twenty years of compiling functional neuroimaging data, it is worth assessing the utility of these data in terms of furthering our understanding of how the brain processes pain. As reviewed above, fMRI studies consistently identify a pattern of activation (referred to as the "pain matrix") commonly involving the ACC, the insula, the thalamus and the somatosensory cortices.

A key assumption underlying the term 'pain matrix' is that this pattern of activation is specific to pain. Given how consistent this spatial pattern is while subjects are experiencing pain, it is unsurprising that it has been considered as a potential objective marker or 'neurosignature' for pain. Furthermore, the activation of 'pain matrix' regions has been used as evidence of perceptual similarity between pain and social isolation (Eisenberger, Lieberman and Williams, 2003; Macdonald and Leary, 2005), The notion that pain matrix activation in these studies indicates that these aversive psychological states are experienced as physically painful has pervaded both the scientific literature and popular press. A 2005 review article claimed that psychological distress is 'experienced as painful because reactions to rejection are mediated by aspects of the physical pain system' (Macdonald and Leary, 2005: 202), and a recent article in the *New York Times* stated that 'being socially rejected doesn't just feel bad, it hurts' (Paul, 2011). In the wake of these studies, a host of others have inferred that other evoked negative affective states (including distress from completing mathematical problems: Lyons and Beilock, 2012) are experienced as painful, on the basis of observed activation in the pain matrix.

The argument that underlies these studies relies on the assumption that the pattern of brain activity elicited by a nociceptive stimulus actually reflects the neural mechanism that gives rise to pain. Judging the presence of physical pain to be based on a pattern of neural activity relies on *reverse inference*. This argument takes the following form:

- If Tom is feeling pain, the pain matrix is activated.
- The pain matrix is activated.
- Therefore Tom is feeling pain.

Given the consistency of pain matrix activation during pain, the ubiquity of this type of argument is unsurprising. The logical fallacy becomes more apparent, however, if we apply the same inferential structure to a different example:

- If Tom is in Moscow, he is in Russia.
- Tom is in Russia.
- Therefore Tom is in Moscow.

The latter example exposes the logical flaw of reverse inference. There are many places in Russia that Tom could be other than Moscow, so the fact he is in Russia is not sufficient for inferring that he is in Moscow. Poldrack (2006) has argued that, while reverse inferences are logically flawed, they should be treated probabilistically, as the likelihood of the inference being true depends on the degree of exclusivity between the antecedent and the consequent. In the above example, the reverse inference is unlikely to be true, since Moscow is a relatively small part of Russia. The inference would be far more likely in the case of Singapore, where the relationship between city and state is exclusive (one can't be in the country without being in the state). To apply a similar logic in the case of neuroimaging, we would have higher confidence in a reverse inference if a given pattern of activation were exclusively associated with a given mental state. Thus, to evaluate the reverse inference that pain matrix activation signals the presence of pain, we must evaluate whether pain matrix activation is exclusively associated with the experience of pain.

In an early series of fMRI experiments, Downar and colleagues (2000, 2001, 2002) called the specificity of these responses into question by demonstrating that a network of regions largely overlapping the pain matrix were associated with the relevance and novelty of environmental stimuli rather than with their painfulness. Similarly. Mouraux and colleagues (Mouraux and Iannetti, 2009; Mouraux et al., 2011) have shown that salient (but non-painful) stimuli in other sensory modalities (visual, auditory, non-painfully tactile) can elicit patterns of activation remarkably similar to those generated in response to nociceptive stimuli. These studies demonstrate that pain matrix activation is not sufficient for the experience of pain; and we have recently demonstrated that it is not necessary for it either. Using cold-pressor and nociceptive heat, we demonstrated intact sensory and emotional pain responses in an individual with extensive damage to the pain matrix (including near-complete ablation of the cingulate and insular cortices) (Feinstein et al., 2015).

On the basis of these studies, it has been concluded that the 'pain matrix' is better characterised as a 'salience matrix' associated with detecting, processing and reacting to salient sensory events (Iannetti et al., 2013; Uddin, 2015). Clearly this interpretation (and the empirical findings that give rise to it) indicates that observed activation in this matrix of regions is not sufficiently specific to pain to justify reverse inferences. This non-specificity calls into question the conclusion of experiential similarity between pain and other negative affective states (e.g., social isolation), which is based on nothing more than overlapping neuroimaging findings.

Perhaps more importantly from the perspective of basic pain research, however, are the implications for neuroimaging's ability to elucidate the brain's role in pain. If, on the one hand, the neural response to pain really is no more than an acknowledgement by the brain that incoming nociceptive input is salient and requires appropriate defensive routines, then Descartes' original conceptualisation of the brain as a bell to be rung by a string in the periphery might be an accurate characterisation of the brain's limited role in the perceptual experience of pain. If, on the other hand, the brain plays a far more active role in shaping the unique perceptual experience of pain (as seems almost certain, given the wide variety of pain experiences and the relative lack of variety of the nociceptive input), then the non-specificity of the 'pain matrix' suggests that neuroimaging research on pain has failed in identifying neural responses that make pain perceptually distinct from other salient sensory and emotional experiences.

One possible explanation for the lack of specificity is that technical limitations make fMRI poorly suited to detect responses specific to pain. As previously discussed, the temporal resolution is of the order of seconds, far below the timescale of the nociceptive input. Furthermore, fMRI relies on oxygenated blood flow as a proxy for neuronal activation, and this introduces temporal delays and further measurement imprecision. Finally, while the spatial resolution (in cubic millimetres) is superior to the spatial resolution obtained through techniques like PET or EEG, it is too coarse to differentiate between smaller populations of more specialised neurons. This might be particularly important in regions like the insula and the ACC, whose integrative role means that several types of neurons might exist in close proximity to each other. Nocireceptive neurons have been detected in both regions by using techniques with higher spatial resolution (Hutchison et al., 1999; Frot et al., 2008; Frot, Faillenot and Mauguiere, 2014), which suggests that fMRI might simply not be able to distinguish between activation in these regions driven by nociceptive responses and activation in regions driven by other types of input.

Another possibility is that the analytical techniques being used are not optimised for the detection of differences. Analyses typically looking to localise function to a particular region (or collection of regions) commonly determine whether a sufficient number of voxels within that region reach a threshold level of activation under a particular condition. Regions meeting these criteria after data are smoothed and spatially averaged are considered to be activated by that condition. Such an approach largely ignores (and, in the case of smoothing and spatial averaging, obscures) the spatial pattern that might give rise to overall activation within a given cluster. Thus two activation maps might identify a common cluster, but the fine-grained spatial pattern that gives rise to this cluster might be different. This possibility is the basis for a newer technique called multivoxel pattern analysis (MVPA), which seeks patterns of activation that distinguish between conditions.

MVPA is the basis for a new 'neurosignature', proposed by Wager et al. (2013). In a comprehensive set of experiments, Wager and colleagues first establish a neurosignature that differentiates between four levels of perceived pain. They then demonstrate that this neurosignature can distinguish between painful and non-painful stimuli in an independent data set. In a third experiment they use the neurosignature to distinguish between social and physical pain. Finally, they demonstrate the sensitivity of this neurosignature to an analgesic agent (reminfentanil).

This study represents an important step towards elucidating neural responses specifically associated with pain. The paper represents a substantial technical step forward, and its importance in directing pain neuroimaging towards improved specificity should not be overlooked. Nevertheless, a number of limitations of this initial step need to be made clear, particularly where they highlight further challenges for the field of pain neuroimaging:

1 Moving towards clinical utility: despite publication in an important clinical journal, the clinical utility of this neurosignature remains to be seen. Some immediate uses – such as providing an alternative dependent measure for use in drug discovery studies – are possible, but many of the clinical benefits of such a neurosignature would seem to derive from their application to clinical pain states. FMRI designs generally require evoked stimuli with specific timings (such that the evoked state can be compared to another condition, or to baseline). Thus the study describes responses to acute thermal pain and, as stated by the authors, 'has not been validated for clinical pain and cannot currently be used in clinical tests' (Wager et al., 2013: 1396). The challenges of imaging chronic pain are endemic to the field rather than a specific flaw of this work. Combining this analytical approach with percept-related MRI or incorporating techniques like resting state fMRI, ASL or PET, which are less reliant on discrete evoked stimuli, might be helpful in this regard.

2 Ability to distinguish current pain status: the paper by Wager and colleagues establishes the robustness of the neurosignature in distinguishing between coarse categories. In their first study they collapse data across four levels of perceived pain and establish the model on its ability to distinguish these combined maps (rather than categorising individual trials for the presence or absence of pain). In subsequent studies, the model is tested for its ability to distinguish between high (painful) and low (non-painful) temperatures, or between discrete categories (social vs physical pain). While in their

second study the researchers examine the model's ability to distinguish painful temperatures on a trial-by-trial basis, future work should focus on improving the power to detect pain on single trials (i.e., to detect whether an individual is in pain at any given point in time).

3 Distinguishing between equally salient states: while the pattern of activation traditionally referred to as the 'pain matrix' is more accurately characterised as a 'salience matrix', this term has little utility from the perspective of cognitive neuroscience. It tells us very little about what individual regions are contributing without explaining why broad swaths of cortex evolved only to solve the most rudimentary adaptive discrimination (does this stimulus require attention?) – a function that remains intact in lower species without cortical development. Nevertheless, no analysis of pain-specific responses can be complete without acknowledging the confound between pain and salience. What is more painful is always more salient. The converse, however (that what is more salient is always more painful), is clearly not the case. Thus the same caution needs to be applied to spatial-based algorithms as to more traditional methods. Wager and colleagues' model differentiates social from physical pain, but it is not clear whether these discriminations are made possible by differences in the degree to which the respective stimuli draw attention to, or require, immediate responses. Matching for salience is difficult, as even designs that have explicitly attempted to do so rely on relatively crude measures. Nevertheless, providing such a measure, even if incomplete, should be fundamental to specificity studies.

4 Are differences detected at the level of the brain critical to perceptual distinctions? A further step in determining the brain changes that give rise to perceptual changes is establishing their necessity or their sufficiency for those perceptual differences. It is entirely possible that the spatial patterns that best differentiate between perceptual states are entirely epiphenomenal and play no important role in instantiating those states (just as the observation of puddles might indicate a rainy day, though puddles result from rain rather than vice versa).

This final issue is related to an obstacle that lingers along the path of obtaining greater specificity: the types of design commonly employed in fMRI studies. Traditionally, functional imaging studies have employed single-session designs in which a particular physical or cognitive state is evoked and the resultant activations are attributed to that state. These designs, while useful, are strictly correlational and do not allow for inferences about causal mechanisms or about the neural operations that are necessary or sufficient for a particular state. Advancing our understanding of the neural operations that underlie specific states requires the development of better designs for inferring causality. These include longitudinal designs where one scan might be before a particular intervention and another afterwards. Such designs are challenging, given the high signal-to-noise ratio in fMRI (which results in reduced power to detect differences between time points), but recent studies have suggested good test–retest reliability in fMRI designs that use thermal pain (Letzen et al., 2014; Quiton et al., forthcoming), supporting the feasibility of longitudinal designs in the study of pain with fMRI. Similarly, techniques like transcranial magnetic stimulation (TMS), which can disrupt or enhance neural processes, can be used with fMRI to provide stronger causal inferences.

Concluding remarks

There have been many advances in our understanding of nociception and pain modulation since the advent of neuroimaging. Neuroimaging research, however, is costly and has temporal and spatial limitations that restrict its ability to examine some aspects of pain, including the identification of activation patterns specific to pain. Despite these challenges, however, neuroimaging remains our only method of collecting non-invasive, *in vivo* information about brain function in humans. This compels technical innovation to move towards stronger inferences and improved specificity. Longitudinal, interventional designs and multimodal approaches can help move the field towards stronger causal inferences, while new analytical techniques have begun to improve our ability to identify neural responses specific to pain. Nevertheless, it will remain best practice to supplement neuroimaging data with findings from lesion and animal studies and other neuroscientific techniques that allow for more direct inferences.

References

Allison, T., McCarthy, G., Wood, C. C., Darcey, T. M., Spencer, D. D. and Williamson, P. D. 1989. Human cortical potentials evoked by stimulation of the median nerve. I: Cytoarchitectonic areas generating short-latency activity. *Journal of Neurophysiology* 62: 694–710.

Andersson, J. L. R., Lilja, A., Hartvig, P., Lüngstrîm, B., Gordh, T., Handwerker, H. and Torebjîrk, E. 1997. Somatotopic organization along the central sulcus, for pain localization in humans, as revealed by positron emission tomography. *Experimental Brain Research* 117: 192–9.

Apkarian, A. V., Bushnell, M. C., Treede, R. D. and Zubieta, J. 2005. Human brain mechanisms of pain perception and regulation in health and disease. *European Journal of Pain* 9: 463–84.

Apkarian, A. V., Darbar, A., Krauss, B. R., Gelnar, P. A. and Szeverenyi, N. M. 1999. Differentiating cortical areas related to pain perception from stimulus identification: Temporal analysis of fMRI activity. *Journal of Neurophysiology* 81: 2956–63.

Apkarian, A. V., Stea, R. A., Manglos, S. H., Szeverenyi, N. M., King, R. B. and Thomas, F. D. 1992. Persistent pain inhibits contralateral somatosensory cortical activity in humans. *Neuroscience Letters* 140: 141–7.

Arendt-Nielsen, L. and Yarnitsky, D. 2009. Experimental and clinical applications of quantitative sensory testing applied to skin, muscles and viscera. *Journal of Pain* 10: 556–72.

Baliki, M. N., Geha, P. Y. and Apkarian, A. V. 2009. Parsing pain perception between nociceptive representation and magnitude estimation. *Journal of Neurophysiology* 101: 875–87.

Baumgartner, U., Iannetti, G. D., Zambreanu, L., Stoeter, P., Treede, R. D. and Tracey, I. 2010. Multiple somatotopic representations of heat and mechanical pain in the operculo-insular cortex: A high-resolution fMRI study. *Journal of Neurophysiology* 104: 2863–72.

Becerra, L., Harris, W., Grant, M., George, E., Boas, D., Borsook, D. 2009. Diffuse optical tomography activation in the somatosensory cortex: Specific activation by painful vs. non-painful thermal stimuli. *PLoS One* 4: e8016.

Beecher, H. K. 1946. Pain in men wounded in battle. *Annals of Surgery* 123: 96–105.

Bessou, P. and Perl, E. R. 1969. Response of cutaneous sensory units with unmyelinated fibers to noxious stimuli. *Journal of Neurophysiology* 32: 1025–43.

Bingel, U. and Tracey, I. 2008. Imaging CNS modulation of pain in humans. *Physiology* 23: 371–80.

Bingel, U., Lorenz, J., Glauche, V., Knab, R., Glascher, J., Weiller, C. and Buchel, C. 2004. Somatotopic organization of human somatosensory cortices for pain: A single trial fMRI study. *Neuroimage* 23: 224–32.

Borsook, D., Upadhyay, J., Chudler, E. H. and Becerra, L. 2010. A key role of the basal ganglia in pain and analgesia: Insights gained through human functional imaging. *Molecular Pain* 6 (27). doi:10.1186/1744-8069-6-27.

Boyd, D., Butler, M., Carr, D. B., Cohen, M. J., Devor, M., Dworkin, R., Greenspan, J. D., Jensen, T. S., King, S., Koltzenburg, M., Loeser, J. D., Merskey, H., Okifuji, A., Paice, J., Serra, J., Treede, R. D. and Woda, A. 2011. IASP pain terminology: An update on the IASP taskforce on taxonomy. Part III: Pain terms, a current list with definitions and notes on usage. In *IASP Taxonomy*, 3rd edn, ed. by J. D. Loeser. Seattle, WA: IASP Press, pp. 209–14.

Brooks, J. C. and Tracey, I. 2007. The insula: A multidimensional integration site for pain. *Pain* 128: 1–2.

Brooks, J. C., Zambreanu, L., Godinez, A., Craig, A. D. and Tracey, I. 2005. Somatotopic organisation of the human insula to painful heat studied with high resolution functional imaging. *Neuroimage* 27: 201–9.

Burgess, P. R., Perl, E. R. 1967. Myelinated afferent fibres responding specifically to noxious stimulation of the skin. *Journal of Physiology* 190: 541–62.

Bush, G., Luu, P. and Posner, M. I. 2000. Cognitive and emotional influences in anterior cingulate cortex. *Trends in Cognitive Sciences* 4: 215–22.

Bushnell, M. C., Ceko, M. and Low, L. A. 2013. Cognitive and emotional control of pain and its disruption in chronic pain. *Nature Reviews Neuroscience* 14: 502–11.

Bushnell, M. C., Duncan, G. H., Hofbauer, R. K., Ha, B., Chen, J. I. and Carrier, B. 1999. Pain perception: Is there a role for primary somatosensory cortex *Proceedings of the National Academy of Sciences of the United States of America* 96: 7705–9.

Campbell, C. M. and Edwards, R. R. 2009. Mind–body interactions in pain: The neurophysiology of anxious and catastrophic pain-related thoughts. *Translational Research: The Journal of Laboratory and Clinical Medicine* 153: 97–101.

Casey, K. L. 2000. Concepts of pain mechanisms: The contribution of functional imaging of the human brain. *Progress in Brain Research* 129: 277–87.

Chen, J. I., Ha, B., Bushnell, M. C., Pike, B. and Duncan, G. H. 2002. Differentiating noxious- and innocuous-related activation of human somatosensory cortices using temporal analysis of fMRI. *Journal of Neurophysiology* 88: 464–74.

Chudler, E. H., Anton, F., Dubner, R. and Kenshalo, D. R., Jr. 1990. Responses of nociceptive S1 neurons in monkeys and pain sensation in humans elicited by noxious thermal stimulation: Effect of interstimulus interval. *Journal of Neurophysiology* 63: 559–69.

Coghill, R. C., Sang, C. N., Maisog, J. M. and Iadarola, M. J. 1999. Pain intensity processing within the human brain: A bilateral, distributed mechanism. *Journal of Neurophysiology* 82: 1934–43.

Colloca, L., Klinger, R., Flor, H. and Bingel, U. 2013. Placebo analgesia: Psychological and neurobiological mechanisms. *Pain* 154: 511–14.

Craig, A. D. 2002. How do you feel? Interoception: The sense of the physiological condition of the body. *Nature Reviews Neuroscience* 3: 655–66.

Craig, A. D. 2010. Once an island, now the focus of attention. *Brain Structure and Function* 124: 395–6.

Craig, A. D. B. 2003a. A new view of pain as a homeostatic emotion. *Trends in Neurosciences* 26: 303–7.

Craig, A. D. B. 2003b. Pain mechanisms: Labeled lines versus convergence in central processing. *Annual Reviews in Neuroscience* 26: 1–30.

Craig, A. D. B. 2009. How do you feel – now? The anterior insula and human awareness. *Nature Reviews Neuroscience* 10: 59–70.

Critchley, H. D. 2004. The human cortex responds to an interoceptive challenge. *Proceedings of the National Academy of Sciences of the United States of America* 101: 6333–4.

Critchley, H. D., Wiens, S., Rotshtein, P., Ohman A. and Dolan, R. J. 2004. Neural systems supporting interoceptive awareness. *Nature Neuroscience* 7: 189–95.

Davis, K. D. 2000. Studies of pain using fMRI. In *Pain Imaging*, ed. by M. C. Bushnell and K. L. Casey. Seattle, WA: IASP Press, pp. 195–210.

Davis, K. D. and Pope, G. E. 2002. Noxious cold evokes multiple sensations with distinct time courses. *Pain* 98: 179–85.

Davis, K. D., Hutchison, W. D., Lozano, A. M. and Dostrovsky, J. O. 1994. Altered pain and temperature perception following cingulotomy and capsulotomy in a patient with schizoaffective disorder. *Pain* 59: 189–99.

Davis, K. D., Pope, G. E., Crawley, A. P. and Mikulis, D. J. 2002. Neural correlates of prickle sensation: A percept-related fMRI study. *Nature Neuroscience* 5: 1121–2.

Davis, K. D., Pope, G. E., Crawley, A. P. and Mikulis, D. J. 2004. Perceptual illusion of 'paradoxical heat' engages the insular cortex. *Journal of Neurophysiology* 92: 1248–51.

Davis, K. D., Wood, M. L., Crawley, A. P. and Mikulis, D. J. 1995. fMRI of human somatosensory and cingulate cortex during painful electrical nerve stimulation. *Neuroreport* 7 (1): 321–5.

De Luca, M., Beckmann, C. F., De Stefano, N., Matthews, P. M. and Smith, S. M. 2006. fMRI resting state networks define distinct modes of long-distance interactions in the human brain. *Neuroimage* 29: 1359–67.

Derbyshire, S. W. G., Jones, A. K. P., Gyulai, F., Clark, S., Townsend, D. and Firestone, L. L. 1997. Pain processing during three levels of noxious stimulation produces differential patterns of central activity. *Pain* 73: 431–45.

Descartes, R. 1664. L'homme *de René Descartes et un traitté* De la formation du foetus *du mesme autheur, avec les remarques de Louis de La Forge, docteur en medecine … sur le* Traité de l'homme *de René Descartes & sur les figures par luy inventées,* ed. by C. Clerselier, with a Preface by F. Schuyl. Paris: Charles Angot.

Devinsky, O., Morrell, M. J. and Vogt, B. A. 1995. Contributions of anterior cingulate cortex to behavior. *Brain* 118: 279–306.

Downar, J., Crawley, A. P., Mikulis, D. J. and Davis, K. D. 2000. A multimodal cortical network for the detection of changes in the sensory environment. *Nature Neuroscience* 3: 277–83.

Downar, J., Mikulis, D. J. and Davis, K. D. 2003. Neural correlates of the prolonged salience of painful stimulation. *Neuroimage* 20: 1540–51.

Downar, J., Crawley, A. P., Mikulis, D. J. and Davis, K. D. 2001. A cortical network for the detection of novel events across multiple sensory modalities. *Neuroimage* 13: S310.

Downar, J., Crawley, A. P., Mikulis, D. J. and Davis, K. D. 2002. A cortical network sensitive to stimulus salience in a neutral behavioral context across multiple sensory modalities. *Journal of Neurophysiology* 87: 615–20.

Duerden, E. G. and Albanese, M. C. 2011. Localization of pain-related brain activation: A meta-analysis of neuroimaging data. *Human Brain Mapping* 34: 109–49.

Dum, R. and Strick, P. 1992. Medial wall motor areas and skeletomotor control. *Current Opinion in Neurobiology* 2: 836–39.

Dum, R., Levinthal, D. and Strick, P. 2009. The spinothalamic system targets motor and sensory areas in the cerebral cortex of monkeys. *Journal of Neuroscience* 29: 14223–35.

Edwards, R. R. 2005. Individual differences in endogenous pain modulation as a risk factor for chronic pain. *Neurology* 65: 437–43.

Eisenberger, N. I., Lieberman, M. D. and Williams, K. D. 2003. Does rejection hurt? An FMRI study of social exclusion. *Science* 302: 290–2.

Farmer, M. A., Baliki, M. N. and Apkarian, A. V. 2012. A dynamic network perspective of chronic pain. *Neuroscience Letters* 520: 197–203.

Farrell, M. J., Laird, A. R. and Egan, G. F. 2005. Brain activity associated with painfully hot stimuli applied to the upper limb: A meta-analysis. *Human Brain Mapping* 25: 129–39.

Feinstein, J. S., Khalsa, S. S., Salomons, T. V., Prkachin, K. M., Frey-Law, L. A., Lee, J. E., Tranel and D., Rudrauf, D. 2015. Preserved emotional awareness of pain in a patient with extensive bilateral damage to the insula, anterior cingulate, and amygdala. *Brain Structure & Function*. doi: 10.1007/s00429-014-0986-3.

Fox, M. D., Snyder, A. Z., Vincent, J. L., Corbetta, M., Van, E. D. C. and Raichle, M. E. 2005. The human brain is intrinsically organized into dynamic, anticorrelated functional networks. *Proceedings of the National Academy of Sciences of the United States of America* 102: 9673–8.

Friston, K. J. 1994. Functional and effective connectivity in neuroimaging: A synthesis. *Human Brain Mapping* 2: 56–78.

Friston, K. 2002. Beyond phrenology: What can neuroimaging tell us about distributed circuitry? *Annual Review of Neuroscience* 25: 221–50.

Friston, K. J. 2011. Functional and effective connectivity: A review. *Brain Connectivity* 1: 13–36.

Friston, K. J., Buechel, C., Fink, G. R., Morris, J., Rolls, E. and Dolan, R. J. 1997. Psychophysiological and modulatory interactions in neuroimaging. *Neuroimage* 6: 218–29.

Frot, M., Faillenot, I. and Mauguiere, F. 2014. Processing of nociceptive input from posterior to anterior insula in humans. *Human Brain Mapping* 35 (11): 5486–99.

Frot, M., Magnin, M., Mauguiere, F. and Garcia-Larrea, L. 2007. Human SII and posterior insula differently encode thermal laser stimuli. *Cerebral Cortex* 17: 610–20.

Frot, M., Mauguiere, F., Magnin, M. and Garcia-Larrea, L. 2008. Parallel processing of nociceptive, A-delta inputs in SII and midcingulate cortex in humans. *Journal of Neuroscience* 28: 944–52.

Fuchs, P. N., Peng, Y. B., Boyette-Davis, J. A. and Uhelski, M. L. 2014. The anterior cingulate cortex and pain processing. *Frontiers in Integrative Neuroscience* 8 (35). doi: 10.3389/fnint.2014.00035.

Garcia-Larrea, L. 2012. The posterior insular–opercular region and the search of a primary cortex for pain. *Neurophysiologie Clinique* 42: 299–313.

Gelnar, P. A., Krauss, B. R., Sheehe, P. R., Szeverenyi, N. M. and Apkarian, A. V. 1999. A comparative fMRI study of cortical representations for thermal painful, vibrotactile, and motor performance tasks. *Neuroimage* 10: 460–82.

Greenspan, J. D., Lee, R. R. and Lenz, F. A. 1999. Pain sensitivity alterations as a function of lesion location in the parasylvian cortex. *Pain* 81: 273–82.

Gusnard, D. A. and Raichle, M. E. 2001. Searching for a baseline: Functional imaging and the resting human brain. *Nature Reviews Neuroscience* 2: 685–94.

Haber, S. N. and Knutson, B. 2009. The reward circuit: Linking primate anatomy and human imaging. *Neuropsychopharmacology* 35: 4–26.

Halkjaer, L., Melsen, B., McMillan, A. S. and Svensson, P. 2006. Influence of sensory deprivation and perturbation of trigeminal afferent fibers on corticomotor control of human tongue musculature. *Experimental Brain Research* 170: 199–205.

Hari, R., Karhu, J., Hamalainen, M., Knuutila, J., Salonen, O., Sams, M. and Vilkman, V. 1993. Functional organization of the human first and second somatosensory cortices: A neuromagnetic study. *European Journal of Neuroscience* 5: 724–34.

Hashmi, J. A. and Davis, K. D. 2008. Effect of static and dynamic heat pain stimulus profiles on the temporal dynamics and interdependence of pain qualities, intensity, and affect. *Journal of Neurophysiology* 100: 1706–15.

Henderson, L. A., Gandevia, S. C. and Macefield, V. G. 2007. Somatotopic organization of the processing of muscle and cutaneous pain in the left and right insula cortex: A single-trial fMRI study. *Pain* 128: 20–30.

Hu, L., Valentini, E., Zhang, Z. G., Liang, M. and Iannetti, G. D. 2014. The primary somatosensory cortex contributes to the latest part of the cortical response elicited by nociceptive somatosensory stimuli in humans. *Neuroimage* 84: 383–93.

Hua, l. H., Strigo, I. A., Baxter, L. C., Johnson, S. C. and Craig, A. D. 2005. Anteroposterior somatotopy of innocuous cooling activation focus in human dorsal posterior insular cortex. *American Journal of Physiology Regulatory, Integrative and Comparative Physiology* 289: R319–R325.

Hutchison, W. D., Davis, K. D., Lozano, A. M., Tasker, R. R. and Dostrovsky, J. O. 1999. Pain-related neurons in the human cingulate cortex. *Nature Neuroscience* 2: 403–5.

Iannetti, G. D. and Mouraux, A. 2010. From the neuromatrix to the pain matrix (and back). *Experimental Brain Research* 205: 1–12.

Iannetti, G. D., Salomons, T. V., Moayedi, M., Mouraux, A. and Davis, K. D. 2013. Beyond metaphor: Contrasting mechanisms of social and physical pain. *Trends in Cognitive Sciences* 17: 371–8.

Iwata, K., Kamo, H., Ogawa, A., Tsuboi, Y., Noma, N., Mitsuhashi, Y., Taira, M., Koshikawa, N. and Kitagawa, J. 2005. Anterior cingulate cortical neuronal activity during perception of noxious thermal stimuli in monkeys. *Journal of Neurophysiology* 94: 1980–91.

Jones, A. K. P., Brown, W. D., Friston, K. J., Qi, L. Y. and Frackowiak, R. S. J. 1991. Cortical and subcortical localization of response to pain in man using positron emission tomography. *Proceedings of the Royal Society of London, Series B: Biological Sciences* 244: 39–44.

Kenshalo, D. R., Iwata, K., Sholas, M. and Thomas, D. A. 2000. Response properties and organization of nociceptive neurons in area 1 of monkey primary somatosensory cortex. *Journal of Neurophysiology* 84: 719–29.

Kenshalo, D. R., Jr. and Isensee, O. 1983. Responses of primate, SI cortical neurons to noxious stimuli. *Journal of Neurophysiology* 50: 1479–96.

Kenshalo, D. R., Jr., Chudler, E. H., Anton, F. and Dubner, R. 1988. SI nociceptive neurons participate in the encoding process by which monkeys perceive the intensity of noxious thermal stimulation. *Brain Research* 454: 378–82.

Koyama, T., Kato, K., Tanaka, Y. Z. and Mikami, A. 2001. Anterior cingulate activity during pain-avoidance and reward tasks in monkeys. *Neuroscience Research* 39: 421–30.

Kucyi, A. and Davis, K. D. 2014. The dynamic pain connectome. Trends Neurosci.

Kurth, F., Zilles, K., Fox, P. T., Laird, A. R. and Eickhoff, S. 2010. A link between the systems: Functional differentiation and integration within the human insula revealed by meta-analysis. *Brain Structure and Function* 214: 519–34.

Lamour, Y., Willer, J. C. and Guilbaud, G. 1983. Rat somatosensory (SmI) cortex. I: Characteristics of neuronal responses to noxious stimulation and comparison with responses to non-noxious stimulation. *Experimental Brain Research* 49: 35–45.

Lee, M. C., Mouraux, A. and Iannetti, G. D. 2009. Characterizing the cortical activity through which pain emerges from nociception. *Journal of Neuroscience* 29: 7909–16.

Legrain, V., Iannetti, G. D., Plaghki, L. and Mouraux, A. 2011. The pain matrix reloaded: A salience detection system for the body. *Progress in Neurobiology* 93: 111–24.

Letzen, J. E., Sevel, L. S., Gay, C. W., O'Shea, A. M., Craggs, J. G., Price, D. D. and Robinson, M. E. 2014. Test–retest reliability of pain-related brain activity in healthy controls undergoing experimental thermal pain. *Journal of Pain* 15: 1008–14.

Liang, M., Mouraux, A. and Iannetti, G. D. 2011. Parallel processing of nociceptive and non-nociceptive somatosensory information in the human primary and secondary somatosensory cortices: Evidence from dynamic causal modeling of functional magnetic resonance imaging data. *Journal of Neuroscience* 31: 8976–85.

Lyons, I. M. and Beilock, S. L. 2012. When math hurts: Math anxiety predicts pain network activation in anticipation of doing math. *PLoS One* 7: e48076.

Macdonald, G. and Leary, M. R. 2005. Why does social exclusion hurt? The relationship between social and physical pain. *Psychological Bulletin* 131: 202–23.

Mancini, F., Haggard, P., Iannetti, G. D., Longo, M. R. and Sereno, M. I. 2012. Fine-grained nociceptive maps in primary somatosensory cortex. *Journal of Neuroscience* 32: 17155–62.

Mancini, F., Sambo, C. F., Ramirez, J. D., Bennett, D. L., Haggard, P. and Iannetti, G. D. 2013. A fovea for pain at the fingertips. *Current Biology* 23: 496–500.

Mancini, F., Bauleo, A., Cole, J., Lui, F., Porro, C. A., Haggard, P. and Iannetti, G. D. 2014. Whole-body mapping of spatial acuity for pain and touch. *Annals of Neurology* 75: 917–24.

Mazoyer, B., Zago, L., Mellet, E., Bricogne, S., Etard, O., Houde, O., Crivello, F., Joliot, M., Petit, L. and Tzourio-Mazoyer, N. 2001. Cortical networks for working memory and executive functions sustain the conscious resting state in man. *Brain Research Bulletin* 54: 287–98.

Melzack, R. and Wall, P. D. 2004. *The Challenge of Pain*. London: Penguin Books.

Merskey, H. and Bogduk, N., eds. 1994. Pain terms: A current list with definitions and notes on usage. In *Classification of Chronic Pain*, 2nd edn. Seattle, WA: IASP Task Force on Taxonomy, IASP Press, pp. 209–13.

Moayedi, M. and Davis, K. D. 2013. Theories of pain: From specificity to gate control. *Journal of Neurophysiology* 109: 5–12.

Morrison, I., Perini, I. and Dunham, J. 2013. Facets and mechanisms of adaptive pain behavior: Predictive regulation and action. *Frontiers in Human Neuroscience* 7 (755). doi: 10.3389/fnhum.2013.00755.

Moulton, E. A., Keaser, M. L., Gullapalli, R. P. and Greenspan, J. D. 2005. Regional intensive and temporal patterns of functional MRI activation distinguishing noxious and innocuous contact heat. *Journal of Neurophysiology* 93: 2183–93.

Mouraux, A. and Iannetti, G. D. 2009. Nociceptive laser-evoked brain potentials do not reflect nociceptive-specific neural activity. *Journal of Neurophysiology* 101: 3258–69.

Mouraux, A., Diukova, A., Lee, M. C., Wise, R. G. and Iannetti, G. D. 2011. A multisensory investigation of the functional significance of the 'pain matrix'. *Neuroimage* 54: 2237–49.

Nash, P. G., Macefield, V. G., Klineberg, I. J., Gustin, S. M., Murray, G. M. and Henderson, L. A. 2010. Changes in human primary motor cortex activity during acute cutaneous and muscle orofacial pain. *Journal of Oral & Facial Pain* 24: 379–90.

Nieuwenhuys, R., Voogd, J. and van Huijzen, C. 2008. *The Human Central Nervous System*, 4th edn. Berlin: Springer.

Ostrowsky, K., Magnin, M., Ryvlin, P., Isnard, J., Guenot, M. and Mauguiere, F. 2002. Representation of pain and somatic sensation in the human insula: A study of responses to direct electrical cortical stimulation. *Cerebral Cortex* 12: 376–85.

Paul, P. 2011. Rejection may hurt more than feelings. *New York Times*, 14 May. At http://www.nytimes.com/2011/05/15/fashion/is-rejection-painful-actually-it-is-studied.html?_r=1 (accessed 27 September 2015).

Penfield, W., Boldrey, E. 1937. Somatic motor and sensory representation in the cerebral cortex of man as studied by electrical stimulation. *Brain* 60: 389–443.

Peyron, R., Laurent, B. and Garcia-Larrea, L. 2000. Functional imaging of brain responses to pain: A review and meta-analysis. *Neurophysiologie Clinique/Clinical Neurophysiology* 30: 263–88.

Picard, N. and Strick, P. L. 2001. Imaging the premotor areas. *Current Opinion in Neurobiology* 11: 663–72.

Ploner, M., Gross, J., Timmermann, L. and Schnitzler, A. 2006. Pain processing is faster than tactile processing in the human brain. *Journal of Neuroscience* 26: 10879–82.

Poldrack, R. A. 2006. Can cognitive processes be inferred from neuroimaging data? *Trends in Cognitive Sciences* 10: 59–63.

Pollatos, O., Gramann, K. and Schandry, R. 2007. Neural systems connecting interoceptive awareness and feelings. *Human Brain Mapping* 28: 9–18.

Pollatos, O., Kirsch, W. and Schandry, R. 2005. Brain structures involved in interoceptive awareness and cardioafferent signal processing: A dipole source localization study. *Human Brain Mapping* 26: 54–64.

Pons, T. P., Garraghty, P. E. and Mishkin, M. 1992. Serial and parallel processing of tactual information in somatosensory cortex of rhesus monkeys. *Journal of Neurophysiology* 68: 518–27.

Porro, C. A., Cettolo, V., Francescato, M. P. and Baraldi, P. 1998. Temporal and intensity coding of pain in human cortex. *Journal of Neurophysiology* 80: 3312–20.

Price, D. D. 2000. Psychological and neural mechanisms of the affective dimension of pain. *Science* 288: 1769–72.

Quiton, R. L., Keaser, M. L., Zhuo, J., Gullapalli, R. P. and Greenspan, J. D. forthcoming. Intersession reliability of fMRI activation for heat pain and motor tasks. *Neuroimage Clinical*.

Raichle, M. E., MacLeod, A. M., Snyder, A. Z., Powers, W. J., Gusnard, D. A. and Shulman, G. L. 2001. A default mode of brain function. *Proceedings of the National Academy of Sciences of the United States of America* 98: 676–82.

Rainville, P. 2002. Brain mechanisms of pain affect and pain modulation. *Current Opinion in Neurobiology* 12: 195–204.

Rainville, P., Carrier, B., Hofbauer, R. K., Bushnell, M. C. and Duncan, G. H. 1999. Dissociation of sensory and affective dimensions of pain using hypnotic modulation. *Pain* 82: 159–71.

Rainville, P., Duncan, G. H., Price, D. D., Carrier, B. and Bushnell, M. C. 1997. Pain affect encoded in human anterior cingulate but not somatosensory cortex. *Science* 277: 968–71.

Romaniello, A., Cruccu, G., McMillan, A. S., Arendt-Nielsen, L. and Svensson, P. 2000. Effect of experimental pain from trigeminal muscle and skin on motor cortex excitability in humans. *Brain Research* 882: 120–7.

Seminowicz, D. A. and Davis, K. D. 2007. A re-examination of pain–cognition interactions: Implications for neuroimaging. *Pain* 130: 8–13.

Shackman, A. J., Salomons, T. V., Slagter, H. A., Fox, A. S., Winter, J. J. and Davidson, R. J. 2011. The integration of negative affect, pain and cognitive control in the cingulate cortex. *Nature Reviews Neuroscience* 12: 154–67.

Sherrington, C. S. 1903. Science and medicine in the modern university. *Science* 18: 675–84.

Smith, S. M. 2012. The future of FMRI connectivity. *Neuroimage* 62: 1257–66.

Smith, S. M., Fox, P. T., Miller, K. L., Glahn, D. C., Fox, P. M., Mackay, C. E., Filippini, N., Watkins, K. E., Toro, R., Laird, A. R. and Beckmann, C. F. 2009. Correspondence of the brain's functional architecture during activation and rest. *Proceedings of the National Academy of Sciences of the United States of America* 106: 13040–5.

Starr, C., Sawaki, L., Wittenberg, G., Burdette, J., Oshiro, Y., Quevedo, A. and Coghill, R. 2009. Roles of the insular cortex in the modulation of pain: Insights from brain lesions. *Journal of Neuroscience* 29: 2684–94.

Talbot, J. D., Marrett, S., Evans, A. C., Meyer, E., Bushnell, M. C. and Duncan, G. H. 1991. Multiple representation of pain in human cerebral cortex. *Science* 251: 1355–8.

Timmermann, L., Ploner, M., Haucke, K., Schmitz, F., Baltissen, R. and Schnitzler, A. 2001. Differential coding of pain intensity in the human primary and secondary somatosensory cortex. *Journal of Neurophysiology* 86: 1499–503.

Treede, R. D., Kenshalo, D. R., Gracely, R. H. and Jones, A. K. P. 1999. The cortical representation of pain. *Pain* 79: 105–11.

Uddin, L. Q. 2015. Salience processing and insular cortical function and dysfunction. *Nature Reviews Neuroscience* 16: 55–61.

Valentini, E., Hu, L., Chakrabarti, B., Hu, Y., Aglioti, S. M. and Iannetti, G. D. 2012. The primary somatosensory cortex largely contributes to the early part of the cortical response elicited by nociceptive stimuli. *Neuroimage* 59: 1571–81.

Vogel, H., Port, J. D., Lenz, F. A., Solaiyappan, M., Krauss, G. and Treede, R. D. 2003. Dipole source analysis of laser-evoked subdural potentials recorded from parasylvian cortex in humans. *Journal of Neurophysiology* 89: 3051–60.

Vogt, B. A. and Sikes, R. W. 2000. The medial pain system, cingulate cortex, and parallel processing of nociceptive information. *Progress in Brain Research* 122: 223–35.

Vogt, B. A. and Sikes, R. W. 2009. Cingulate nociceptive circuitry and roles in pain processing: The cingulate premotor pain model. In *Cingulate Neurobiology and Disease*, ed. by B. A. Vogt. New York: Oxford University Press, pp. 312–38.

Vogt, B. A., Sikes, R. W. and Vogt, L. T. 1993. Anterior cingulate cortex and the medial pain system. In *Neurobiology of Cingulate Cortex and Limbic Thalamus: A Comprehensive Handbook*, ed. by B. A. Vogt and M. Gabriel. Boston: Birkhauser, pp. 313–44.

Vogt, B. A., Vogt, L., Farber, N. B. and Bush, G. 2005. Architecture and neurocytology of monkey cingulate gyrus. *Journal of Comparative Neurology* 485: 218–39.

Wager, T. D., Atlas, L. Y., Lindquist, M. A., Roy, M., Woo, C. W. and Kross, E. 2013. An fMRI-based neurologic signature of physical pain. *New England Journal of Medicine* 368: 1388–97.

Weinstein, S. 1968. Intensive and extensive aspects of tactile sensitivity as a function of body part, sex, and laterality. In *The Skin Senses*, ed. by D. R. Kenshalo. Springfield, IL: Thomas, pp. 195–222.

Weissman-Fogel, I., Moayedi, M., Taylor, K. S., Pope, G. and Davis, K. D. 2010. Cognitive and default-mode resting state networks: Do male and female brains 'rest' differently? *Human Brain Mapping* 31: 1713–26.

Zhang, Z. G., Hu, L., Hung, Y. S., Mouraux, A. and Iannetti, G. D. 2012. Gamma-band oscillations in the primary somatosensory cortex: A direct and obligatory correlate of subjective pain intensity. *Journal of Neuroscience* 32: 7429–38.

Yarkoni, T., Poldrack, R. A., Nichols, T. E., Van Essen, D. C. and Wager, T. D. 2011. Large-scale automated synthesis of human functional neuroimaging data. *Nature Methods* 8: 665–70.

CHAPTER 9
Placebo effects in pain

Luana Colloca, Adam P. Horin and Damien Finniss

Introduction

Placebo effects are an important aspect of medical practice, with broad implications for routine clinical care and clinical trial design (Colloca and Benedetti, 2005; Colloca, Flaten, and Meissner, 2013; Finniss and Benedetti, 2005; Finniss et al., 2010). This topic area is both interesting and controversial; recent literature promotes a better understanding of placebo effects, with a view to ethically harnessing them to clinical practice (Colloca and Miller, 2011a). It is important to remember that medical professionals adhere to the standard that they are obliged to treat their patients by using the most effective known methods (Enck et al., 2013). Keeping this in mind, finding the most ethical and beneficial ways to implicate knowledge from placebo research can definitely be valuable to medical practice (Brody, 1980; Brody, Colloca and Miller, 2012). There has been a host of research over the past several decades that has expanded our knowledge of the mysterious phenomenon we know as 'the placebo effect'. Understanding its mechanisms from both a psychological and a neurobiological viewpoint has permitted a better understanding of how we can use placebo effects to our benefit.

From a psychological viewpoint, placebo effects have been studied by using several experimental designs such as verbal suggestions, conditioning, and social learning (Colloca and Miller, 2011b; Price, Finniss and Benedetti, 2008). All of these methods have been used in some way in order to induce the expectancy of a positive outcome. Expectation remains one of the key aspects of exhibiting a placebo effect (Benedetti, 2008). The psychological mechanisms discussed later in this chapter represent critical elements in the psychosocial environment around the patient that are known to generate and modulate placebo effects. Therefore selected psychological determinants can be manipulated and placebo effects measured. The neurobiological mechanisms of placebo effects in the brain and in the spinal cord are being extensively studied to help us understand the interaction between psychosocial context and underlying biological processes in general (Eippert, Bingel, et al., 2009; Eippert, Finsterbusch, et al., 2009). These neurobiological mechanisms have been studied through

An Introduction to Pain and its Relation to Nervous System Disorders, First Edition. Edited by Anna A. Battaglia.
© 2016 John Wiley & Sons, Ltd. Published 2016 by John Wiley & Sons, Ltd.

several techniques such as brain imaging and pharmacological manipulation, and using agonists and antagonists to elucidate the pathways involved (Benedetti, 2013). Several systems have been linked to the exhibition of a placebo effect – including the opioid, cannabinoid, dopamine, oxytocin and cholesystokinin (CCK) pathways (Colloca, Flatten, et al., 2013; Benedetti, 2013; Kessner et al., 2013). The mechanisms of placebo effects sparked the interest of the scientific community in further research, which aimed to evaluate the prospects of using these effects to enhance treatments and medical practice. The significance of placebo effects is now being recognised. The psychosocial environment (or context) around a patient has the ability to affect that patient's mind, brain and body. Therefore the doctor/clinician–patient relationship represents an important aspect of health care – one that may actually augment response to routine therapies.

This chapter will present recent research on the topic of placebo effects. There is a tremendous amount of data about psychological and neurobiological processes that support prospective mechanisms for placebo effects.

Definitions

Before discussing this research in any detail, it is important to show how we define a placebo and the placebo effect. A placebo is traditionally defined as an 'inert' form of treatment (such as a sugar pill) that elicits a response similar to that of an active treatment. The placebo effect is the response observed in the patient – a response attributable to the placebo (de la Fuente-Fernandez, Schulzer and Stoessl, 2002). One of the paradoxes of this traditional formulation is that, by definition, something that is inert has no active property that can elicit an effect. New conceptualisations have advanced our definitions. We now understand that a placebo is actually a simulation of a therapeutic ritual, and placebo effects are the effects of the psychosocial context on the patient's brain and body (Price et al., 2008). It is also important to note that there has been research to support the existence of a nocebo effect. This is a response too, just like the placebo effect; but is a negative response elicited by negative expectations (Barsky et al., 2002; Benedetti et al., 2007; Colloca and Miller, 2011c; Hahn, 1997; Kennedy, 1961). The nocebo effect has not been studied as extensively as the placebo effect because of the obvious bioethical implications of knowingly giving negative expectations to a patient or to a subject.

True placebo effects

In clinical practice and in uncontrolled trials, the reported 'success rate' may be due to one or more of the three phenomena described below: natural history, regression to the mean, and subject bias (for reviews, see Colloca and Miller, 2011a and Colloca, Benedetti and Porro, 2008). Many symptoms (e.g., pain) show spontaneous variations and fluctuations over time; their unchecked, natural development is known as 'natural history'. If a subject takes a placebo just before his/her pain starts decreasing, she may believe that the placebo is effective, although that decrease would have occurred anyway. This is an example of subject bias. Relapses and remissions can occur in the absence of any treatment or therapeutic strategies and are not genuine placebo effects. To demonstrate a placebo effect, it is necessary to identify the difference between the natural history group and the placebo group.

Another example of potential bias is related to regression to the mean. This phenomenon refers to the fact that a variable will tend to move closer to the centre of the distribution from initial to later measurements, even if no biologically or psychologically mediated placebo effects are present. If patients tend to receive their initial clinical assessment when their pain is near its greatest intensity, then their pain level is likely to be lower at the second pain assessment. This kind of improvement cannot be attributed to any intervention (active or placebo). Regression to the mean often appears together with placebo effects in clinical trials, and the only reliable way to see what proportion of an observed improvement might actually be attributable to the placebo is, again, to compare a group receiving a placebo to a natural history group (that is, a group receiving no treatment).

Other sources of confusion include error in symptom assessment (false positive errors) and co-interventions. A patient can report that a treatment makes her feel better, detecting by mistake symptomatic relief. This phenomenon is termed 'false positive errors' and is commonly observed in medical practice; it is related to both the physician who diagnoses a patient's symptoms and the patient who reports symptom severity. The success rate may also be due to unidentified co-interventions, which produce parallel effects on the observed benefit, and to the effect of being under study (the Hawthorne effect). In conclusion, the placebo effect is a real psychobiological effect, attributable to specific psychological and environmental determinants with equally specific physiological processes.

Psychological mechanisms

Placebo and nocebo effects have been widely studied from a psychological perspective (Price et al., 2008). We know that there are many ways of inducing placebo effects in a subject; and many of these mechanisms have been extensively studied. The psychology behind a placebo effect can be looked at in three groups of determinants or mechanisms: verbal suggestions, conditioning, and social learning (Colloca, Flaten, et al., 2013; Colloca, Lopiano, et al., 2013). All of these mechanisms have the ability to influence expectations in a subject or in a patient and therefore a degree of overlay can be present. This is important, because expectancy is known to be the most important psychological determinant of placebo effects.

Verbal suggestion

Subjects can be influenced by verbal suggestions, which induce expectations (Colloca and Miller, 2011b). Many studies have proven how verbal suggestions can significantly affect a patient's expectations of a treatment's randomisation and effectiveness. A simple example of verbal suggestion is demonstrated in an experiment performed in 1988 by Kirsch and Weixel. In this study, decaffeinated coffee was administered to two groups under two different verbal suggestions. One group was told that they would receive either real or decaffeinated coffee, while the other group was told that they would receive real coffee. The subjects in the latter group exhibited a higher placebo effect than the first group (Kirsch and Weixel, 1988). This shows us how verbal suggestion can be used to elicit an expected response even when there is no mechanistic trigger to cause that response. A placebo effect can be elicited just by telling a patient or a subject that something is supposed to happen.

In two earlier studies, scientists demonstrated this idea of using verbal suggestions to influence patient outcomes. Luparello and colleagues (1970) were able to produce verbally induced placebo responses in subjects with asthma. The subjects were given either isoproterenol or carbachol. Both drugs were administered in one of two ways: the subjects were told either that the drug was a bronchodilator or that it was a bronchoconstrictor. Those who were given the correct information exhibited greater responsiveness to the drug than those who were misinformed and made to believe that the drug had opposite effects to the ones it actually had. This study shows how just being told that a treatment is going to exhibit properties that it does not really possess can cause it to act less effectively.

In a similar study, McFadden and colleagues (1969) used verbal suggestion with asthmatic patients in order to increase and decrease airway resistance. The researchers were able to use verbal suggestion to influence response in about 50 percent of their patients. This finding shows us how verbal suggestion can be used to alter patients' expectations and how these expectations can in turn, influence the effectiveness of a treatment.

The studies mentioned so far are early examples of using verbal suggestion to create expectations in a patient. More modern studies have now been able to use this knowledge to help manage illnesses such as irritable bowel syndrome (IBS) (Kaptchuk et al., 2010; Kaptchuk et al., 2008), pain (Colloca et al., 2004), depression (Leuchter et al., 2002; Leuchter et al., 2004; Leuchter et al., 2008; Leuchter et al., 2009), and Parkinson's disease (Pollo et al., 2002).

For example, IBS patients can benefit from being administered placebos. Patients were randomised to one of three groups: no treatment (a natural history control); rectal placebo; and rectal lidocaine. There was a higher placebo response in patients who were told that they would be receiving an active treatment, as opposed to those who were told that there was a chance that they would be receiving a placebo (Vase et al., 2005). Similar results had already been found in an earlier study (Verne et al., 2003). In these studies, certain suggestions of benefit were associated with larger placebo effects, demonstrating the importance of information, perception, and patients' expectation.

Studies on pain have been the basis of many placebo experiments. For example, placebo effects have been explored for the purpose of reducing the patient's need for painkillers. In a study by Pollo and colleagues in 2001, patients were allocated to three different verbal suggestion groups. One group was not told what they would be receiving as a painkiller. Another group was told that they would be receiving either a powerful drug or a placebo. The final group was told that they would be receiving a powerful drug. All of the patients received an intravenous saline solution and were also able to take additional buprenorphine on request. In fact it was the total additional buprenorphine dose that was the primary outcome in the study. The subjects were treated for three days and were treated upon request. The researchers measured the amounts of drug requested in order to determine the beneficial placebo responsiveness. There were fewer requests in the group that was told that they were receiving the powerful drug than in the group that was told that they may or may not receive the drug

(Pollo et al., 2001). This study shows that verbal suggestion can have an impact on the patient's expectation of a treatment being effective and can modulate the magnitude of the placebo component in a treatment. There have been many studies that show how a patient's expectation of a successful outcome has been influenced by the clinician's interaction with that patient. In a study, patients received a placebo pain medication after surgery. There were two groups in the experiment. One group was told that they were being given a powerful painkiller, while the other was given hidden administration of the drug from a machine. The patients who were given hidden administration of the painkiller were more likely to request more pain relief than those in the first group (Amanzio et al., 2001).

Thus, verbal suggestion has been extensively studied and, while this method seems to have – and it does indeed have – many implications for medicine, it does not, however, elicit as strong a response as other ways of creating expectations in subjects, such as conditioning or social learning (Colloca, Klinger et al., 2013).

Conditioning

Studies have shown that classical conditioning can also be an effective way of modulating a placebo effect (Colloca and Benedetti, 2009). Classical conditioning, a form of learning, has been a well-studied concept in psychology; it originates in Ivan Petrovich Pavlov's work with dogs. In Pavlov's experiment the dogs' salivation is the conditioned response (CR) to the ringing of a bell, which is the conditioned stimulus (CS). The CS was previously paired with the presentation of food – an unconditioned stimulus (US) that caused the dogs to exhibit the CR, salivation (Pavlov and Anrep, 1927). Classical conditioning has been demonstrated to be a modulator of placebo effects (Colloca and Miller, 2011b). In these studies a real drug is given repeatedly (and paired with an effect). The drug is then substituted with a placebo, and a conditioned response occurs. This conditioned response is a placebo effect and is highly dependent on the conditioning protocol and on psychosocial context. Furthermore, studies have revealed placebo responses from sources not endowed with volition, such as in the immune system, the endocrine system (Colloca and Miller, 2011b) and the respiratory system (Benedetti et al., 1998; Benedetti et al., 1999). This demonstrates that placebo effects can occur even in the absence of conscious awareness and therefore can operate both independently from or in combination with expectancy mechanisms (Colloca and Miller, 2011d).

Initially conditioned placebo responses were observed in animals (Ader and Cohen, 1975; Herrnstein, 1962). In Herrnstein's (1962) study the administration of scopolamine to rats was paired with that of sweetened milk. After several pairings, rats were able to exhibit scopolamine-like responses just after the administration of sweetened milk. Ader and Cohen (1975) were able to induce an immunosuppressant response in rats after administration of the immunosuppressant drug cyclophosphamide. Interestingly, this drug was paired with a sweet liquid (saccharine), and, when the cyclophosphamide was removed, a conditioned immune suppression was achieved to the placebo saccharine solution. The two researchers found that the response was greater in rats who had received two doses rather than just one. This demonstrates that conditioning protocols can be used to induce a placebo effect and that by varying the administration we can affect the intensity of the placebo effect.

More recent studies have been able to extend this work on conditioning, demonstrating placebo responses in human subjects with experimental pain and in a clinical population. Benedetti and colleagues were able to produce placebo effect that involved different bodily systems that reply to unconscious processes such as immune and endocrine functions (Benedetti et al., 2003).

Social learning

Individuals exhibiting a placebo effect can acquire it by social vicarious learning, whereby they observe a certain response to a treatment and then exhibit a similar response (Colloca and Miller, 2011b). This idea of observational learning was first described in a study by Colloca and Benedetti (2009). In the study, a subject observed another subject undergoing a series of administered shocks. The researchers then assessed how the first subject was influenced by the response of the subject s/he had observed (Colloca and Benedetti, 2009). They found that subjects with a high empathy scores on their questionnaire were more likely to be influenced by the responses of another subject. However, empathy plays a role (or has an effect) when interpersonal interactions are involved (Colloca and Benedetti, 2009; Swider and Babel, 2013), since placebo and nocebo effects after the observation of a video reply are not linked to the level of individual of empathy (Vogtle, Barke and Kroner-Herwig, 2013). A recent study suggests that men exhibit a stronger placebo response induced by observational learning than women (Swider and Babel, 2013).

Observational or social learning introduces the notion of the importance of a patient's vicarious experience on shaping future responses to treatments.

Regarding the nocebo effect, studies have been performed that show how placebos can be used to induce negative effects on a patient. In a 2010 study, Mazzoni and colleagues were able to induce negative effects in healthy volunteers through social learning and modelling. The subjects were asked to inhale a sample of clean air, which they were told contained a toxin that would elicit specific symptoms. Half of the subjects watched an actor inhale the air and exhibit symptoms of the toxin. These subjects proved to be more likely to exhibit the symptoms, even though they were not actually inhaling a toxin (Mazzoni et al., 2010).

Social learning may also be involved in the patient–clinician relationship and in placebo effects. Further research is needed to clarify to what extent social interactions and vicarious learning influence clinical outcomes.

Expectation

It is clear that expectation is one of the key determinants in placebo effects. All the mechanisms mentioned above are based on the ability of creating expectations in the patients to the point where they exhibit placebo effects. Expectations can play a large role in determining the strength of a placebo or nocebo effect and can be influenced by all the psychological mechanisms of a placebo response discussed so far (Colloca and Miller, 2011b). A patient's first major expectation is about the therapeutic outcome of a treatment (Benedetti, Carlino and Pollo, 2011). This expectation can be affected by the patient's own medical history or by his/her observation of similar treatments in other individuals. This is a good reason why clinicians should be aware of a patient's expectations and prior medical history before starting a treatment. If a treatment did not work for a patient in the past, it may not work effectively in the present either, on account of that patient's previous history with it. If the patient has a preconceived expectation of benefit from a treatment, s/he is more likely to respond to it positively. If the patient is pessimistic about the treatment, s/he may not respond to it very well. Pessimism may be caused by his/her worrying about potential side effects of a treatment s/he is about to receive. This would produce a nocebo effect in a treatment.

The patient could also be influenced by his/her own anxiety (Benedetti et al., 2011). If s/he is anxious about medical treatment, this may not allow him/her to respond to it properly. Finding ways to reduce anxiety in patients would be pivotal in helping to increase the efficacy of treatments. This can be achieved through various methods that will be presented later in the chapter, including through the patient–clinician relationship.

Character traits can be a great determinant of placebo effects. Many studies on the subject have shown what types of character traits will exhibit the strongest placebo responses. Understanding what character traits will respond best to a treatment could be a way to further personalise medicine for patients. Traits that have been looked at are empathy, trust, anxiety, fear, and optimism. Emotions like these can dramatically change an individual's placebo response (Tracey, 2010).

Neurobiological mechanisms

All these psychological mechanisms are crucial to the formation of a placebo response and allow us to disentangle the primary components of generating a placebo effect in human beings – healthy volunteers and patients alike – and notably the specific neurobiological mechanisms that underlie the psychological aspects of this response. So far, research has suggested the involvement of several neurobiological pathways, for example the opioid, cannabinoid, dopamine, oxytocin, vasopressin and cholesystokinin (CCK) pathways (Benedetti, 2013; Colagiuri et al., 2015).

Molecular aspects

Placebo analgesia, from a neurobiological perspective, has classically been attributed to opioid and non-opioid mechanisms (Finniss et al., 2010). Pioneering studies suggested that a placebo response can be blocked by naxolone, an opioid antagonist – which therefore indicates the involvement of the opioid system (Colloca and Benedetti, 2005).

An early study from 1978 opened up the research line supporting that endogenous opioids can be crucial modulators of placebo analgesia. By using a model of dental postoperative pain, the authors gave either a placebo or naxolone to reduce patients' pain at three and four hours after the surgery. The patients who received the placebo first and were placebo responders had an increase in pain when they received naxolone as their second drug, This suggested that the placebo responses could be mediated by endorphin release (Levine, Gordon and Fields, 1978). Studies have also shown that placebo responses induced by verbal suggestion are partially blocked by naloxone and that other non-opioid systems can play a role (Colloca and Benedetti, 2005). In a 1999 study by Amanzio and colleagues, the team used a combination of some of the psychological mechanisms, including expectation and conditioning (Amanzio and Benedetti, 1999). The authors studied the mechanisms of a placebo response that involved the activation of endogenous opioids. They conditioned their patients with one of two drugs: opioid agonist morphine hydrochloride or the nonopioid ketorolac tromethamine. The expectation cues that they used, which produced placebo responses, as well as the expectation cues paired with morphine, were all blocked by naxolone (Amanzio and Benedetti, 1999). Since they were able to block a placebo effect with naxolone, the study showed that a placebo effect is likely to be associated with opioid system.

Another pathway mentioned earlier was the cannabinoid pathway. Benedetti and colleagues (2011) were recently able to use a specific CB1 cannabinoid receptor antagonist, rimonabant, to block non-opioid placebo responses, but not opioid placebo effects.

Dopaminergic pathways have been studied and found to have some potential involvement in placebo analgesia (Schweinhardt et al., 2009). In a recent genetic study by Hall and colleagues, placebo responses in IBS patients have been linked to the dopaminergic pathway and to the polymorphisms of catechol-O-methyltransferase (COMT), an enzyme involved in the catabolism of dopamine (Hall et al., 2012). Cholesystokinin (CCK) has been considered a major neurotransmitter involved in the nocebo response. An early study showed how this endogenous neuropeptide was able to reverse opiate analgesia (Faris et al., 1983). CCK has been used in research for both placebo and nocebo effects (Benedetti et al., 2006).

More recently, Colloca and colleagues demonstrated that arginine vasopressin induces an enhancement of placebo analgesic effects when given intranasally so that it can cross the brain-blood-barrier and that these effects depend upon a sexual dimorphism (e.g., women responded with larger effects in the presence of a woman experimenter) (Colloca et al., 2015).

Neurophysiological aspects

We know that the placebo response and the psychological mechanisms that produce it are associated with the central nervous system (Colloca, Klinger, et al., 2013). Responses such as heart rate and skin conductance can be altered in placebo responses induced through the administration of an inert substance. Understanding the neurophysiological reactions and how they can be altered in a placebo response can help us better understand the above-mentioned pathways. Since these neurophysiological pathways are controlled by the autonomic nervous system, a branch of the central nervous system, we know that they can be recorded without bias from a participant, because we have no voluntary control over them.

There have been studies to show how placebo analgesia can be localised in different parts of the body. In one study patients were injected with capsaicin so that a burning sensation was produced in them. They were injected in the right and left hand and foot. A placebo cream was applied to one of the body parts and they were told that it was just a local anaesthetic. The study found that the subjects had exhibited placebo responses only in the treated body parts (Benedetti, Arduino and Amanzio, 1999). This shows us how the placebo effect can be concentrated in just one area of the body and can have a specific somatotopy. The mechanism for this localisation is yet to be understood.

The neurophysiological aspects of the placebo response are not entirely known, but there are proposed mechanisms that suggest potential pathways in the body. A study by Benedetti and colleagues compared verbally induced placebo responses with conditioned placebo responses in the endocrine system of the body. They used sumatriptan, a serotonin agonist of the 5-HT 1B/1D receptors that stimulates the production of growth hormone and inhibits the secretion of cortisol, together with a placebo drug, to condition a subject to exhibit the effect of sumatriptan when that subject was eventually given just the placebo drug (Benedetti et al., 2003). They were able to induce the placebo response through conditioning, but not through verbal suggestion related to the effects of the drug. This shows us that the placebo response can physiologically be produced by an unconscious awareness of the pathways that should be activated in the body.

Brain imaging

The brain is a complex puzzle that needs to be put together through trial and error in research. There has been a lot of progress in mapping out areas of the brain that control different responses and emotions (for a detailed evaluation of both the progress and the limitations of brain imaging in experimental pain, see Massieh and Salomons's contribution in this volume). We have recently started to understand how the brain might be involved in a placebo effect and to identify the underpinning areas and pathways. Studies that used a range of brain-mapping technologies have been able to show that the descending pain modulatory system plays a crucial role in uncovering much of the mystery of the placebo mechanisms (Tracey, 2010).

Bingel and colleagues (2006) tested the connection between the rostral anterior cingulate cortex (rACC) and subcortical areas. It was already known that the rACC was involved in the formation of a placebo response, but they demonstrated that the rACC interacts with other areas of the brain that are rich in opioid receptors, such as the bilateral amygdalae and the periaqueductal gray (PAG).

In another study performed a few years later, Eippert and colleagues found more evidence to suggest the interactions between various areas of the brain (see Eippert, Bingel, et al., 2009). They based their study on the understanding that placebo analgesia is controlled by the endogenous opioid system by using fMRI in combination with the administration of naxolone. Interestingly, several areas of the brain were modulated by naxolone, such as the hypothalamus, the PAG, and the rostral ventromedial medulla (RVM). They also found, in support of this study, that the administration of naxolone completely stopped interactions between the rACC and the PAG during a placebo response.

In the same year Eippert and colleagues performed another study in order to investigate the involvement of the spinal cord in placebo analgesia (Eippert, Finsterbusch, et al., 2009). They used spinal fMRI to analyze placebo effects at the level of the spinal cord. The spinal cord's pain-related activity was decreased when a placebo response was exhibited. These findings suggest that a placebo effect may in some ways inhibit spinal cord processing.

More recently Wager and colleagues (2011) have provided evidence for how different brain activations observed in subjects can be used to predict the effectiveness of a placebo response. They combined different fMRI studies successfully, creating brain-imaging tools to predict brain placebo effects in healthy volunteers.

Petrovic and colleagues (2002), using positron emission tomography (PET), suggested a link between opioids and placebo analgesia. Their study showed that the rACC exhibited increased activity in both forms of analgesia; thus they revealed a link between the neurobiological mechanisms of the opioid and placebo systems. Similarly, Zubieta and colleagues (2005) also used PET techniques to corroborate the hypothesis that placebo effects were controlled by the endogenous opioid system.

Genetics

There has been a growing interest in the identification of genetic variants associated with placebo responsiveness (for a review, see Colagiuri et al., 2015). While the research carried out on the genetics that underlies the placebo effect has not been as extensive as research on other mechanisms, several links between genes and placebo responsiveness have been discovered.

Furmark and colleagues (2008) studied involving patients with social anxiety disorder. Subjects were asked to perform a public speaking task while their brain activity was recorded via PET. Furmark and colleagues genotyped all of the patients for specific genes related to serotonin. They found that different carriers for the serotonin-related gene influence the placebo response in social anxiety, at both the behavioral and the neural level. Polymorphisms modulating monoaminergic tone have been linked to the degree of placebo responsiveness in patients with major depressive disorder (Leuchter et al., 2009). Leuchter and colleagues hypothesised that, if placebo responses were controlled by the central reward pathways, which are modulated by monoamines and these are under strong genetic control, 'common functional polymorphisms modulating monoaminergic tone would be related to degree of improvement during placebo treatment of subjects with major depressive disorder' (Leuchter et al., 2009). They found that subjects with monoamine oxidase A G/T polymorphisms had a smaller placebo effect than subjects with a different genotype.

Patients with IBS have also been studied from a genetic perspective. After being randomized to 'waitlist' sham acupuncture delivered according to a business-like doctor–patient interaction and sham acupuncture performed in the context of an enhanced and supportive doctor–patient relationship, patients were studied for pain outcome, which was indicated by changes from baseline in the IBS Symptom Severity Scale after three weeks of treatment. Interestingly, the number of methionine alleles in the *COMT* Val158Met polymorphism (rs4633) was considered in relation to the clinical outcomes. Patients with Met/Met alleles exhibited strong placebo analgesic effects and benefited from the enhanced and supportive doctor–patient relationship. Patients with Val/Val alleles benefited minimally from placebo responses and from the doctor–patient relationship, potentially opening the way to personalised therapeutic approaches (Hall et al., 2012).

Recently Pecina et al. (2013) indicated that a single-nucleotide polymorphism in the fatty acid amide hydrolase (*FAAH*) gene, C385A (rs324420), which regulates the release of endogenous cannabinoids, represents a good predictor of opioid-mediated placebo analgesia. In fact the activation of the μ-opioid neurotransmission observed in areas associated with reward-motivated learning and memory processing significantly correlated with placebo analgesia and *FAAH* Pro129/Pro129 homozygosity. Those with Pro129/Pro129 homozygosity showed larger psychophysical placebo analgesia and regional μ-opioid activation during placebo administration than those with Pro129/Pro129 heterozygotes.

Clinical implications

Placebo effects are an integral part of routine care and there have been recent attempts to evaluate the literature on placebo, with a view to understanding its implications for clinical trial design and clinical practice. As placebo effects are the result of the whole therapeutic context, it is important to note that one does not need to administer a placebo in order to elicit a placebo effect. In fact placebo effects occur in routine clinical care when 'real' treatments are given. The overall outcome of a treatment is therefore due to both mechanisms of the treatment and placebo mechanisms. This has implications for both clinical trial design and clinical practice,

Clinical trials

Placebo effects have been a source of interest and a nuisance in clinical trials since the placebo-controlled trial was adopted. Large placebo responses reduce the demonstrable effect of the therapy in question. Furthermore, these trials do not actually represent a normal therapeutic context, for in real life patients do not have a 50 per cent chance of receiving a real or a sham treatment when seeking medical care. On this basis, new work is promoting progression in clinical trial design. For example, a recent paper has described a possible new format for clinical trials; this format is meant to help the patient benefit more from the treatment (Enck et al., 2013). The first important step is to minimise the placebo response during clinical trials. It is important to be able to analyse the drug being tested without any interference from a placebo response. This would improve the efficacy of treatments being tested. Once a treatment is approved and put into use, the placebo effect should be studied and maximized in patients. This is in the patients' best interest and will only increase the effects of the drugs used in their treatment. It has also been suggested that placebo effects are individual and that attempts should be made to harness individualized placebo effects. Unfortunately, both from a clinical trial and from a practice perspective, identifying 'the placebo responder' has been a difficult task; but much work is in progress on this front. Many studies have been dedicated to investigating how genetics, character traits, and prior experience and expectations from a treatment can alter an individual's placebo response. Using these data as stepping stones towards personalizing treatments would be – again – in best interest of patients as specific individuals (Enck et al., 2013).

Clinical practice: Communication

The consultation between a patient and a clinician is an important aspect of medical practice. If one imagines one's own consultation with a physician, one realises that features like the physician's facial expressions, his/her tone of voice, the way in which s/he delivers information, the way in which s/he listens, and many more influence how one will feel about the diagnosis and the prognosis being delivered, no matter how minor or serious it may be. Studies on placebo effects have used the patient–clinician consultation as a means of enhancing the effectiveness of a treatment; such studies demonstrate the importance of thinking about the best way for a clinician to relay information to a patient. Exhibiting a placebo effect depends in very large measure on the expectation of a certain outcome, which we know can be influenced by verbal suggestions such as those made by the clinician during consultation with a patient.

One example came from a study by Thomas in 1987 where patients were split into two groups: a positive consultation group and a negative consultation group. All of the patients were suffering from minor illnesses that would have healed in time. Within these two groups half of the patients received a placebo treatment and the other half received treatment. Thomas found that the patients who received positive consultations healed faster than those who received negative consultations. He also found that there was no significant difference between the placebo group and the no-treatment group, which demonstrated the impact of the clinician's consultation with the patients (Thomas, 1987).

While it is important to have a warm and positive consultation between a patient and clinician, this also means that a small amount of ethical deception may be important if the patient is to benefit. Not disclosing some adverse side effects to a patient can help to reduce the nocebo effect that occurs frequently with medication administration (Colloca and Finniss, 2012). In this instance, the removal of the expectancy of minor side effects to medication can reduce their frequency, while important side effects can still be disclosed to the patient. This practical application of placebo and nocebo research represents an interesting future direction in clinical practice.

Patient–clinician relationship

The patient–clinician relationship does not only play a key role in the effectiveness of a placebo response, but it can also affect the potential outcomes of a nocebo effect. Studies have shown how a clinician's description of the outcome can induce or reduce nocebo effects (Varelmann et al., 2010).

As reported above, Kaptchuk and colleagues (2008) performed a study in order to determine the degree to which a warm patient–clinician relationship can increase patients' responses to treatments for IBS. This study had divided the participants into three different treatment groups. One group was put on a waiting list; it was to act as a control and observe the natural progression of the disorder. Another group had minimal interaction with the clinician and received a placebo treatment. In the third group, participants had a warm consultation with the clinician while also receiving a placebo treatment of sham acupuncture (Kaptchuk et al., 2008). Researchers found a graded effect in response to placebo, whereby the additional interaction with the therapist (with standardised elements of the consultation, such as warmth and enquiry about symptoms) resulted in larger treatment effects for placebo acupuncture.

The previously mentioned study from Thomas also has a great deal to do with the clinician–patient relationship. His patients who received warmer optimistic consultations had higher success rates than those who received a negative consultation (Thomas, 1987).

Expectancy: A critical element of placebo in practice

As mentioned many times, expectation plays an important role in modulating a placebo response. The studies discussed here have shown how a patient's expectation of results can affect the effectiveness of a treatment in both negative and positive ways. There are many studies that suggest the placebo effect to be an important aspect of medical treatment (Colloca and Miller, 2011d). The patient's expectations play a key role in is/her exhibiting a placebo response; hence it is important to understand how to induce a placebo effect in a patient.

Colloca and Benedetti (2006) have discussed how a patient's prior experiences with different treatments can contribute to that patient's expectation how s/he will respond to a treatment. The patient's medical history can be an important factor in inducing a placebo effect.

A 2008 study by Sneed and colleagues demonstrated a patient's expectation could play a large role in how that patient responds to treatment. They looked at the difference in responses between participants in active-controlled trials and placebo-controlled trials. They found that patients who expected to receive at least some form of treatment responded to treatments better than those who thought they had a chance of receiving only a placebo. This demonstrates that the degree of expectancy in the patient can alter his/her response to a treatment (Sneed et al., 2008).

The clinician's expectation has been found to be just as important as the patient's. In a study by Jensen and colleagues (2013), the team used brain imaging to analyze the activity in the clinician's brain during a consultation. Jensen and colleagues' study showed brain activation in areas of the brain involved in the expectation of pain relief. Thus they determined that the clinician's ability to empathize and relate to the patient was an important factor in successful treatment.

Controversy

There are many obvious ethical implications regarding the use of placebos in medicine. Not only do we need to question whether or not the use of placebos can be sufficient or beneficial enough to replace active treatments that known to be effective; we also need to consider the use of deception in medicine. From an ethical standpoint, one would think that it is wrong to use deceit with your patients; but, if this is for their benefit, could it be deemed appropriate? Placebo research has been a major topic of discussion in bioethics committees and institutional review boards. In a recent review, Miller and Colloca (2009) focused on two questions in order to evaluate the ethical implications of using placebo effects in medicine. They asked (1) whether a placebo effect could be used to produce clinical benefits for the patient and (2) whether a placebo effect could be formed without using deception (Miller and Colloca, 2009). Potentially, a placebo effect could be triggered without the use of deception. If a clinician gives a patient a placebo as treatment, discloses that it is a placebo and tells the patient that it will have beneficial effects, that clinician would not be lying to the patient, because research has proved these placebos to activate physiological pathways in the body. This would still instil in the patient the expectation of a positive outcome from the treatment.

One of the aspects involved in producing a placebo effect is how this affects the trust gained by the clinician from his/her patient. Using deception can break this trust, and this in turn could diminish the placebo effect (Bok, 1974).

In conclusion, it is important to understand fully the implications of using placebos as treatments. While placebos have been able to yield real physiological changes in patients, this does not mean that a placebo could be used to replace an effective treatment. Understanding the mechanisms can nevertheless help to enhance forms of treatment already known to be effective.

On a more general note, placebo effects are genuine psychobiological effects that are a part of routine clinical care, and therefore understanding how they operate can only lead to the improvement of medical practice.

Acknowledgements

This research was supported by the University of Maryland Baltimore School of Nursing.

References

Ader, R. and Cohen, N. 1975. Behaviorally conditioned immunosuppression. *Psychosomatic Medicine* 37 (4): 333–40.

Amanzio, M. and Benedetti, F. 1999. Neuropharmacological dissection of placebo analgesia: Expectation-activated opioid systems versus conditioning-activated specific subsystems. *Journal of Neuroscience* 19 (1): 484–94.

Amanzio, M., Pollo, A., Maggi, G. and Benedetti, F. 2001. Response variability to analgesics: A role for non-specific activation of endogenous opioids. *Pain* 90 (3): 205–15.

Barsky, A. J., Saintfort, R., Rogers, M. P. and Borus, J. F. 2002. Nonspecific medication side effects and the nocebo phenomenon. *Journal of the American Medical Association* 287 (5): 622–7.

Benedetti, F. 2008. Mechanisms of placebo and placebo-related effects across diseases and treatments. *Annual Review of Pharmacology and Toxicology* 48: 33–60.

Benedetti, F. 2013. Placebo and the new physiology of the doctor-patient relationship. *Physiological Reviews* 93 (3): 1207–46.

Benedetti, F., Arduino, C. and Amanzio, M. 1999. Somatotopic activation of opioid systems by target-directed expectations of analgesia. *Journal of Neuroscience* 19 (9): 3639–48.

Benedetti, F., Carlino, E. and Pollo, A. 2011. How placebos change the patient's brain. *Neuropsychopharmacology* 36 (1): 339–54.

Benedetti, F., Amanzio, M., Rosato, R. and Blanchard, C. 2011. Nonopioid placebo analgesia is mediated by CB1 cannabinoid receptors. *Nature Medicine* 17 (10): 1228–30.

Benedetti, F., Amanzio, M., Vighetti, S. and Asteggiano, G. 2006. The biochemical and neuroendocrine bases of the hyperalgesic nocebo effect. *Journal of Neuroscience* 26 (46): 12014–22.

Benedetti, F., Lanotte, M., Lopiano, L. and Colloca, L. 2007. When words are painful: Unraveling the mechanisms of the nocebo effect. *Neuroscience* 147 (2): 260–71.

Benedetti, F., Amanzio, M., Baldi, S., Casadio, C. and Maggi, G. 1999. Inducing placebo respiratory depressant responses in humans via opioid receptors. *European Journal of Neuroscience* 11 (2): 625–31.

Benedetti, F., Pollo, A., Lopiano, L., Lanotte, M., Vighetti, S. and Rainero, I. 2003. Conscious expectation and unconscious conditioning in analgesic, motor, and hormonal placebo/nocebo responses. *Journal of Neuroscience* 23 (10): 4315–23.

Benedetti, F., Amanzio, M., Baldi, S., Casadio, C., Cavallo, A., Mancuso, M., Ruffini, E., Oliaro, A. and Maggi, G. 1998. The specific effects of prior opioid exposure on placebo analgesia and placebo respiratory depression. *Pain* 75 (2–3): 313–9.

Bingel, U., Lorenz, J., Schoell, E., Weiller, C. and Buchel, C. 2006. Mechanisms of placebo analgesia: rACC recruitment of a subcortical antinociceptive network. *Pain* 120 (1–2): 8–15.

Bok, S. 1974. The ethics of giving placebos. *Scientific American* 231 (5): 17–23.

Brody, H. 1980. *Placebo and the Philosophy of Medicine*. Chicago: University of Chicago Press.

Brody, H., Colloca, L. and Miller, F. G. 2012. The placebo phenomenon: Implications for the ethics of shared decision-making. *Journal of General Internal Medicine* 27 (6): 739–42.

Colagiuri, B., Schenk, L. A., Kessler, M. D., Dorsey, S. G. and Colloca, L. 2015. The placebo effect: From concepts to genes. *Neuroscience* 307:171–90.

Colloca, L. and Benedetti, F. 2005. Placebos and painkillers: Is mind as real as matter? Nat Rev Neurosci 6 (7): 545–52.

Colloca, L. and Benedetti, F. 2006. How prior experience shapes placebo analgesia. *Pain* 124 (1–2): 126–33.

Colloca, L. and Benedetti, F. 2009. Placebo analgesia induced by social observational learning. *Pain* 144 (1–2): 28–34.

Colloca, L. and Finniss, D. 2012. Nocebo effects, patient–clinician communication, and therapeutic outcomes. *Journal of the American Medical Association* 307 (6): 567–8.

Colloca, L. and Miller, F. G. 2011a. Harnessing the placebo effect: The need for translational research. *Philosophical Transactions of the Royal Society of London, Series B: Biological Sciences* 366 (1572): 1922–30.

Colloca, L. and Miller, F. G. 2011b. How placebo responses are formed: A learning perspective. *Philosophical Transactions of the Royal Society of London, Series B: Biological Sciences* 366 (1572): 1859–69.

Colloca, L. and Miller, F. G. 2011c. The nocebo effect and its relevance for clinical practice. *Psychosomatic Medicine* 73 (7): 598–603.

Colloca, L. and Miller, F. G. 2011d. Role of expectations in health. *Current Opinion in Psychiatry* 24 (2): 149–55.

Colloca, L., Benedetti, F. and Porro, C. A. 2008. Experimental designs and brain mapping approaches for studying the placebo analgesic effect. *European Journal of Applied Physiology* 102 (4): 371–80.

Colloca, L., Flaten, M. A. and Meissner, K. 2013. *Placebo and Pain: From Bench to Bedside*. Oxford: Elsevier.

Colloca, L., Lopiano, L., Lanotte, M. and Benedetti, F. 2004. Overt versus covert treatment for pain, anxiety, and Parkinson's disease. *Lancet Neurology* 3 (11): 679–84.

Colloca, L., Klinger, R., Flor, H. and Bingel, U. 2013. Placebo analgesia: Psychological and neurobiological mechanisms. *Pain* 154 (4): 511–4.

Colloca, L., Pine, D. S., Ernst, M., Miller, F. G. and Grillon, C. 2015. Vasopressin boosts placebo analgesic effects in women: A randomized trial. *Biological Psychiatry* pii: S0006-3223(15)00638-1. doi: 10.1016/j.biopsych.2015.07.019.

de la Fuente-Fernandez, R., Schulzer, M. and Stoessl, A. J. 2002. The placebo effect in neurological disorders. *Lancet Neurology* 1 (2): 85–91.

Eippert, F., Finsterbusch, J., Bingel, U. and Büchel, C. 2009. Direct evidence for spinal cord involvement in placebo analgesia. *Science* 326 (5951). doi: 10.1126/science.1180142.

Eippert, F., Bingel, U., Schoell, E. D., Yacubian, J., Klinger, R., Lorenz, J. and Büchel C. 2009. Activation of the opioidergic descending pain control system underlies placebo analgesia. *Neuron* 63 (4): 533–43.

Enck, P., Bingel, U., Schedlowski, M. and Rief, W. 2013. The placebo response in medicine: Minimize, maximize or personalize? *Nature Reviews Drug Discovery* 12 (3): 191–204.

Faris, P. L., Komisaruk, B. R., Watkins, L. R. and Mayer, D. J. 1983. Evidence for the neuropeptide cholecystokinin as an antagonist of opiate analgesia. *Science* 219 (4582): 310–2.

Finniss, D. G. and Benedetti, F. 2005. Mechanisms of the placebo response and their impact on clinical trials and clinical practice. *Pain* 114 (1–2): 3–6.

Finniss, D. G., Kaptchuk, T. J., Miller, F. and Benedetti, F. 2010. Biological, clinical, and ethical advances of placebo effects. *Lancet* 375 (9715): 686–95.

Furmark, T., Appel, L., Henningsson, S., Ahs, F., Faria, V., Linnman, C., Pissiota, A., Frans, O., Bani, M., Bettica, P., et al. 2008. A link between serotonin-related gene polymorphisms, amygdala activity, and placebo-induced relief from social anxiety. *Journal of Neuroscience* 28 (49): 13066–74.

Hahn, R. A. 1997. The nocebo phenomenon: Concept, evidence, and implications for public health. *Preventive Medicine* 26 (5 Pt 1): 607–11.

Hall, K. T., Lembo, A. J., Kirsch, I., Ziogas, D. C., Douaiher, J., Jensen, K. B., Conboy, L. A., Kelley, J. M., Kokkotou, E. and Kaptchuk, T. J. 2012. Catechol-O-methyltransferase val158met polymorphism predicts placebo effect in irritable bowel syndrome. *PLoS One* 7 (10): e48135.

Herrnstein, R. J. 1962. Placebo effect in the rat. *Science* 138 (3541): 677–8.

Jensen, K. B., Petrovic, P., Kerr, C. E., Kirsch, I., Raicek, J., Cheetham, A., Spaeth, R., Cook, A., Gollub, R. L., Kong, J., and Kaptchuk, T. J. 2014. Sharing pain and relief: Neural correlates of physicians during treatment of patients. *Molecular Psychiatry* 19(3): 392–398.

Kaptchuk, T. J., Kelley, J. M., Conboy, L. A., Davis, R. B., Kerr, C. E., Jacobson, E. E., Kirsch, I., Schyner, R. N., Nam, B. H., Nguyen, L. T., Park, M., et al. 2008. Components of placebo effect: Randomised controlled trial in patients with irritable bowel syndrome. *British Medical Journal* 336 (7651): 999–1003.

Kaptchuk, T. J., Friedlander, E., Kelley, J. M., Sanchez, M. N., Kokkotou, E., Singer, J. P., Kowalczykowski, M., Miller, F. G., Kirsch, I. and Lembo, A. J. 2010. Placebos without deception: A randomized controlled trial in irritable bowel syndrome. *PLoS One* 5 (12): e15591.

Kennedy, W. P. 1961. The nocebo reaction. *Medical World* 95: 203–5.

Kessner, S., Sprenger, C., Wrobel, N., Wiech, K. and Bingel, U. 2013. Effect of oxytocin on placebo analgesia: A randomized study. *Journal of the American Medical Association* 310 (16): 1733–5.

Kirsch, I. and Weixel, L. J. 1988. Double-blind versus deceptive administration of a placebo. *Behavioral Neuroscience* 102 (2): 319–23.

Leuchter, A. F., Cook, I. A., Witte, E. A., Morgan, M. and Abrams, M. 2002. Changes in brain function of depressed subjects during treatment with placebo. *American Journal of Psychiatry* 159 (1): 122–9.

Leuchter, A. F., Morgan, M., Cook, I. A., Dunkin, J., Abrams, M. and Witte, E. 2004. Pretreatment neurophysiological and clinical characteristics of placebo responders in treatment trials for major depression. *Psychopharmacology (Berlin)* 177 (1–2): 15–22.

Leuchter, A. F., Cook, I. A., DeBrota, D. J., Hunter, A. M., Potter, W. Z., McGrouther, C. C., Morgan, M. L., Abrams, M. and Siegman, B. 2008. Changes in brain function during administration of venlafaxine or placebo to normal subjects. *Clinical EEG and Neuroscience* 39 (4): 175–81.

Leuchter, A. F., McCracken, J. T., Hunter, A. M., Cook, I. A. and Alpert, J. E. 2009. Monoamine oxidase a and catechol-o-methyltransferase functional polymorphisms and the placebo response in major depressive disorder. *Journal of Clinical Psychopharmacology* 29 (4): 372–7.

Levine, J. D., Gordon, N. C. and Fields, H. L. 1978. The mechanism of placebo analgesia. *Lancet* 2 (8091): 654–7.

Luparello, T. J., Leist, N., Lourie, C. H. and Sweet, P. 1970. The interaction of psychologic stimuli and pharmacologic agents on airway reactivity in asthmatic subjects. *Psychosomatic Medicine* 32 (5): 509–13.

Mazzoni, G., Foan, L., Hyland, M. E. and Kirsch, I. 2010. The effects of observation and gender on psychogenic symptoms. *Health Psychology* 29 (2): 181–5.

McFadden, E. R., Jr., Luparello, T., Lyons, H. A. and Bleecker, E. 1969. The mechanism of action of suggestion in the induction of acute asthma attacks. *Psychosomatic Medicine* 31 (2): 134–43.

Miller, F. G. and Colloca, L. 2009. The legitimacy of placebo treatments in clinical practice: Evidence and ethics. *American Journal of Bioethics* 9 (12): 39–47.

Pavlov, I. P. 1927. *Conditioned Reflexes: An Investigation of the Physiological Activity of the Cerebral Cortex*, trans. G. V. Anrep. London: Oxford University Press.

Pecina, M., Martinez-Jauand, M., Hodgkinson, C., Stohler, C.S., Goldman, D., Zubieta, J.K. 2013. FAAH selectively influences placebo effects. *Molecular Psychiatry* 19(3):385–91

Petrovic, P., Kalso, E., Petersson, K. M. and Ingvar, M. 2002. Placebo and opioid analgesia: Imaging a shared neuronal network. *Science* 295 (5560): 1737–40.

Pollo, A., Amanzio, M., Arslanian, A., Casadio, C., Maggi, G., Benedetti, F. 2001. Response expectancies in placebo analgesia and their clinical relevance. *Pain* 93 (1): 77–84.

Pollo, A., Torre, E., Lopiano, L., Rizzone, M., Lanotte, M., Cavanna, A., Bergamasco, B. and Benedetti, F. 2002. Expectation modulates the response to subthalamic nucleus stimulation in Parkinsonian patients. *Neuroreport* 13 (11): 1383–6.

Price, D. D., Finniss, D. G. and Benedetti, F. 2008. A comprehensive review of the placebo effect: Recent advances and current thought. *Annual Review of Psychology* 59: 565–90.

Schweinhardt, P., Seminowicz, D. A., Jaeger, E., Duncan, G. H. and Bushnell, M. C. 2009. The anatomy of the mesolimbic reward system: A link between personality and the placebo analgesic response. *Journal of Neuroscience* 29 (15): 4882–7.

Sneed, J. R., Rutherford, B. R., Rindskopf, D., Lane, D. T., Sackeim, H. A. and Roose, S. P. 2008 Design makes a difference: A meta-analysis of antidepressant response rates in placebo-controlled versus comparator trials in late-life depression. *American Journal of Geriatric Psychiatry* 16 (1): 65–73.

Swider, K. and Babel, P. 2013. The effect of the sex of a model on nocebo hyperalgesia induced by social observational learning. *Pain* 154 (8): 1312–7.

Thomas, K. B. 1987. General practice consultations: Is there any point in being positive? *British Medical Journal (Clinical Research edition)* 294 (6581): 1200–2.

Tracey, I. 2010. Getting the pain you expect: Mechanisms of placebo, nocebo and reappraisal effects in humans. *Nature Medicine* 16 (11): 1277–83.

Varelmann, D. Pancaro, C., Cappiello, E. C. and Camann, W. R. 2010. Nocebo-induced hyperalgesia during local anesthetic injection. *Anesthesia & Analgesia* 110 (3): 868–70.

Vase, L., Robinson, M. E., Verne, G. N. and Price, D. D. 2005. Increased placebo analgesia over time in irritable bowel syndrome (IBS) patients is associated with desire and expectation but not endogenous opioid mechanisms. *Pain* 115 (3): 338–47.

Verne, G. N., Robinson, M. E., Vase, L., Price, D. D. 2003. Reversal of visceral and cutaneous hyperalgesia by local rectal anesthesia in irritable bowel syndrome (IBS) patients. *Pain* 105 (1–2): 223–30.

Vogtle, E., Barke, A. and Kroner-Herwig, B. 2013. Nocebo hyperalgesia induced by social observational learning. *Pain* 154 (8): 1427–33.

Wager, T. D., Atlas, L. Y., Leotti, L. A., Rilling, J. K. 2011. Predicting individual differences in placebo analgesia: Contributions of brain activity during anticipation and pain experience. *Journal of Neuroscience* 31 (2): 439–52.

Zubieta, J. K., Bueller, J. A., Jackson, L. R., Scott, D. J., Xu, Y., Koeppe, R. A., Nichols, T. E. and Stohler, C. S. 2005. Placebo effects mediated by endogenous opioid activity on mu-opioid receptors. *Journal of Neuroscience* 25 (34): 7754–62.

CHAPTER 10

Psychology and pain

Lance M. McCracken

Introduction

There are a number of different perspectives or points of view that one can take in trying to understand pain. Some of these are relatively narrow, some are broader. Some people spend their entire research careers looking at pain as a neurophysiological problem, or as a brain problem, or as a genetic problem. Certainly these are legitimate points of view and they may one day produce more effective treatments. At the same time it is often claimed that we should adopt a biopsychosocial point of view on pain (Engel, 1977; Gatchel et al., 2007), or even a biopsychosocioeconomic one (Waddell and Aylward, 2010). It can be argued that, once our interest in pain enters into the realm of human experience, pain is certainly no longer a problem of physiology, neurology, or genes alone. As a human experience, pain is also a fundamentally psychosocial problem. The point of this chapter is to look at this aspect of pain.

The chapter will include several separate sections, each designed to examine separate psychosocial aspects of pain, especially chronic pain. These parts of the chapter will help the reader to examine and understand (1) the loose relationship between pain and pathophysiology, (2) the 'nocebo' response, (3) our current multidimensional definition of the pain experience, (4) the wider impacts of pain, (5) models of pain that incorporate or focus on psychosocial aspects, (6) the evidence for psychological treatment approaches, and (7) processes or mechanisms of pain treatment.

Pain experiences and physical causation

It is natural to assume that, if there is pain, then there must be a discernible physical cause for that pain. Certainly pain is often a signal that needs to be heeded and investigated. At the same time it is remarkable how ordinary it is for pain to appear without a discernible root cause and without a sign of the underlying pathology. With pain, particularly chronic pain, the links between the experience and the presence of an underlying physical cause or disease is often very weak indeed.

An Introduction to Pain and its Relation to Nervous System Disorders, First Edition. Edited by Anna A. Battaglia.
© 2016 John Wiley & Sons, Ltd. Published 2016 by John Wiley & Sons, Ltd.

Consider as an example low back pain. Low back pain is extremely common, one of the most common chronic or recurrent types of pain disorder encountered in clinical practice. In the past it was routinely assumed that low back pain is a reflection of some underlying spinal pathology and that this pathology needed to be found and treated. Studies of the spine, however, reveal a more complicated picture. In one such study 98 people without symptoms of back pain underwent magnetic resonance imaging (MRI) examinations of their lumbosacral spine (Jensen et al., 1994). The images produced were subsequently read by two neuroradiologists who were blind to the clinical status of the study participants, and the results were intriguing. Sixty-four percent of the scans showed some type of disc pathology, including what they defined as a bulge, protrusion, or extrusion, and 38 per cent had an abnormality at more than one level of the participants' spine. The researchers also found other abnormalities, such as herniation of the disk into the vertebral body in 19 per cent and defects in the outer fibrous cover of the disc in 14 per cent of cases. At this point they concluded that this high rate of apparent 'pathology' in asymptomatic people meant that disc pathology found in people with low back pain in many cases ought to be regarded as coincidental and not the root of the problem (Jensen et al., 1994).

In the mid-1990s a task force was set up by the International Association for the Study of Pain (IASP) to examine the problem of back pain and disability at work (Fordyce et al., 1995). One of the main conclusions of this group of 25 experts in healthcare, research, and work settings was that more than 85 per cent of low back pain sufferers could not, on the basis of the state of medical knowledge at the time, be characterised as having a specific, known pathology. They termed the condition experienced by the large majority of low back pain sufferers 'nonspecific low back pain'. Quite radically, they advised that it be addressed not as a medical problem; in their view, treatment approaches ought to pay greater attention to disability management. There does not appear to be any evidence that this situation has changed considerably in the last twenty years. Current advice for back pain is to rule out so-called 'red flags', regarded as possible indicators of serious spinal pathology – as these are likely to be present in only a very small percentage of patients – and then to encourage resumptions of normal activities (e.g., Samanta, Kendall and Samanta, 2003).

'Nocebo': An interesting example of pain appearing without a painful stimulus

A placebo is generally defined as a simulated treatment that is assumed not to have a specific therapeutic mechanism, for instance a pharmacological mechanism. Colloca, Horin and Finniss provide a comprehensive overview of placebo effects in Chapter 9. There is also a term 'nocebo' that can be said to designate a pain experience that occurs in the absence of any direct painful stimulus (Schweiger and Parducci, 1981).

In a typical paper that demonstrates the nocebo response, subjects were invited to participate in a study of the effects of a mild electric current, applied via electrodes attached to the forehead, on reaction time. They were told that the electrical current is safe but often painful (Bayer, Baer and Early, 1991). Participants were subsequently asked to complete a series of reaction-time tasks and a device with a dial on it was adjusted before each trial; it was ostensibly related to an increase in electrical current. Pain ratings were collected after each trial, on a scale ranging from 0 to 15, and 10 on this scale was designated as the point where similar pain would typically require analgesic medication. The device with the dial and the description of the role of the electrical current were, however, deceptions – a sham. No electrical current was ever delivered, even though participants were led to believe that there would be. In those exposed to this deception, 33 out of 60, or 55 per cent, reported pain. Seven participants reported pain at or above the analgesic threshold, and one asked for medication for headache and for the stimulator to be turned down. Interestingly, pain ratings also significantly increased as the settings on the sham stimulator dial were increased, although – again – no actual painful stimulus was present during any of the trials.

Pain defined

Beginning in 1979 it was recognised by the IASP that there was a need for agreed-upon definitions for key pain-related terms and for a classificatory scheme to support accurate communication and further understanding of pain as a problem. A first version of these terms was published in 1979 and subsequently updated and revised in 1986 and 1994. One part that has remained consistent is the definition of pain offered there. According to this definition, pain is 'an unpleasant sensory and emotional experience associated with actual or potential tissue damage, or described in terms of such damage' (Merskey and Bogduk, 1994: 210). The authors go on to offer a few further clarifications of this definition. They highlight the following points: (1) inability to communicate does not mean a person does not feel pain; (2) pain is always unpleasant and therefore has an emotional component; (3) 'many people report pain in the absence of tissue damage or any likely pathophysiological cause; usually this happens for psychological reasons' (1994: 210), and (4) if any person 'regards their experience as pain, and if they report it in the same ways as pain caused by tissue damage, it should be accepted as pain' (1994: 210). The only point here that some experts might question is the dichotomy implicit in point (3). It can be misleading to imply that pain is either physical or psychological, as it is always both – in fact the formal definition here says this. Probably we ought to be humble about our ability to find the pathophysiology underlying many cases of chronic pain. Hence it is probably best to be open to psychological influences on pain in all cases of people who come for services and to assume that the vast majority of these cases will include both pathophysiology that may be difficult to detect and these other important psychological influences.

The impact of chronic pain

Chronic pain, pain lasting more than three or six months, is reported by about one in eight people in the United Kingdom (Breivik et al., 2006; Smith et al., 2001). About half of them report severe chronic pain and, perhaps not surprisingly, this is more common in older people (Smith et al., 2001). Most people who develop chronic pain, around 80 per cent, will still have chronic pain when surveyed four years later, and those with the greatest pain and the worst self-rated general health are the least likely to recover (Elliott et al., 2002). According to this same community survey, people with relatively lower social and physical functioning are the most likely to develop chronic pain (when they do not first suffer from it in the first place). Data such as these highlight the importance of overall health and functioning as predictors of later chronic pain outcomes.

There is no doubt that chronic pain can have a significant impact on the functioning of those who suffer with it. In a large survey of pain in Europe (N = 4,839), significant percentages of the respondents reported that their pain left them less able to sleep (56 per cent), exercise (50 per cent), lift objects (49 per cent), complete household chores (42 per cent), walk (40 per cent), engage in social activities (34 per cent), or have sex (24 per cent) (Breivik et al., 2006). Respondents reported a mean of 7.8 days of missed work in the past six months, 19 per cent said that they had lost jobs due to pain, and 21 per cent reported that they had been diagnosed with depression.

Chronic 'physical' conditions such as chronic pain and psychiatric conditions are highly comorbid. A World Health Organization population survey (N = 245,404) carried out in 60 countries showed that 9 to 23 per cent of participants with one or more chronic physical conditions also met the criteria for depression, and the highest rate occurred in those with two or more physical conditions (Moussavi et al., 2007). A large study in Germany (N = 4,181) showed very similar rates of anxiety disorders in people with physical conditions (8.9 per cent to 20.3 per cent; Sareen et al., 2006). In these surveys the physical health conditions included cancer, chronic pain, diabetes, heart disease, and respiratory disease. In the World Mental Health Survey – another large population-based survey (N = 85,088), carried out in 17 countries that included Europe but not the United Kingdom – the rate of depression and anxiety disorders turned out to be roughly twice higher in people with chronic neck and back pain than in people without these kinds of pains (Demyttenaere et al., 2007).

It is common for researchers and clinicians to wonder whether the high rates of co-occurrence of chronic pain and psychiatric disorders imply some shared underlying pathophysiology. (Iannitelli and Tirassa have contributed an inspiring chapter on the co-morbidity of pain with depression: see Chapter 13 in this volume.) Another way of looking at this is to appreciate that pain is a threatening and uncertain experience; it can impose restrictions on functioning and it can cause losses, for instance of activities, relationships, jobs, confidence, and security. In such circumstances fear, anxiety, sadness, or even anger, are rather natural responses (e.g., Cassell, 2004). From there the potential for these natural reactions to develop into problems in their own right, and to add to the overall impacts of pain, is significant, particularly while the pain continues.

The gate control theory of pain

There was a time when conditions like chronic pain, particularly in the absence of clear physical pathology, would have been attributed to some processes of emotional conflict, or might have been called 'psychogenic' or 'psychosomatic' (Alexander, 1950); but this is less often the case now, and most often it is entirely inappropriate to make such attributions. Today it is most apt to hold a view of pain that does not dichotomise its experience as either physical or psychological, one that appreciates that pain is an interaction of these influences. One of the important milestones in the history of pain studies that set the stage firmly for this interactive view was what is called the 'gate control theory' (Melzack and Wall, 1965).

Following research on pain in the modern age, four facts have stood out that need to be accommodated within a useful model of pain: (1) pain can occur without any clear cause; (2) a clear injury may occur and yet there may be no pain; (3) an injury can heal, and yet pain may persist; and (4) treatments that ought to block the experience of pain, such as anaesthetising or removing an affected body part, often do not put an end to it.

In essence the gate control theory postulated the existence of a gating mechanism in the spinal cord, a mechanism that operates through an interaction of inputs from actual or potential tissue injury, other sensory inputs, and inputs from the brain, including psychological influences. Through the action of this hypothetical gate, the pain experience was tied not only to the apparently affected tissues but also to other factors, which could either increase or decrease the experience. The introduction of this theory fuelled many studies, and in particular it promoted an appreciation of psychosocial influences on the pain construct. For many years this model was presented to those who sought treatment for chronic pain as a way to help them to understand the role played in their pain experience by such factors as attention, emotions, thoughts, and behaviour patterns. Even so, it gradually became clear that many of the specific details of this hypothesis could not be supported through evidence; and eventually a new theory was needed. For most researchers and clinicians, the gate control theory is now a piece of history, relevant to the field but no longer operating as a guide for research.

Psychosocial approaches to treatment for pain

One important step toward understanding the role of psychosocial factors in chronic pain was taken in what is called the 'operant approach' to chronic pain (Fordyce, 1976). This approach first developed in the 1960s. At that time it provided a highly practical alternative to more limited prevailing approaches that might have attributed pain to personality problems or personal traits. Within the operant approach it was recognised that the behaviour of people with pain can be influenced by pain, obviously, but also by events in their environment, such as other feelings, effects of medication, and responses from others. From this basic idea there emerged a focus on overt actions of the person with pain that represent 'pain behaviour' – for example complaints, postures, other non-verbal expressions, avoidance, and other behaviour patterns that constitute disability – and on 'feel-well behaviour'. As well as paying heed to the behaviour patterns themselves, researchers placed a focus on the various reinforcing or punishing consequences these patterns meet at home, at work and in healthcare situations. Here, regardless of the underlying pain problem, displays of pain and disability can be shaped, elaborated, and maintained through contingent reinforcement, usually while patterns of healthy functioning are either not reinforced or punished. This model provided a way of understanding chronic pain and disability and of developing methods for reversing it, primarily by altering the reinforcing consequences for related behaviour patterns. Although this is probably an incomplete way to look at pain, the processes emphasised within the operant approach remain relevant to the management of chronic pain and illness today.

The operant approach had a large impact on the design of pain management treatment programs when it was introduced, mostly beginning in the 1970s. After a short time, however, the operant approach was expanded into a more comprehensive approach. This involved a greater focus on cognitive processes, coping, and self-management. This expansion became what is now called the cognitive behavioural approach, or cognitive behavioural therapy (CBT) (Turk, Meichenbaum and Genest, 1983). The intention of the cognitive behavioural approach was to integrate the focus on environmental influences from the operant approach with the focus on thoughts, expectations, and beliefs from the then rapidly developing methods of cognitive therapy. One of the primary principles in CBT was the notion that the person's emotional reactions and behaviour are largely determined by how they view themselves, their world or their circumstances, and their future. The treatment methods that emerged from this combination of behavioural and cognitive factors include such things as pain education, relaxation, distraction, cognitive restructuring, pacing methods, goal setting, and various other forms of skills training.

The fear–avoidance model of chronic pain

In more recent times the general cognitive behavioural model has given rise to some more specific models of the suffering and disability associated with chronic pain. One of the currently more prominent examples is what is called the fear-avoidance model (Vlaeyen and Linton, 2000). Within this model, disabling chronic pain is a result of a number of separate but related psychological processes, including catastrophic thinking, experiences of fear, patterns of avoidance, pain-related hypervigilance, and potential physical deconditioning. Essentially this model proposes a self-perpetuating process of appraisals that pain is a threat, and then a series of emotional, cognitive, and behavioural responses that feed a cycle of distress, 'deactivation', and disability. Within this model, those who respond to pain as if it were a catastrophe enter this cycle of disability; those who respond otherwise enter a process of recovery. Data from many studies routinely show that fear and avoidance-related processes appear significantly more important than pain intensity itself as determiners of patient functioning, wellbeing, and healthcare use (Crombez et al., 1999; McCracken, Evon and Karapas, 2002; McCracken, Gross and Eccleston, 2002; McCracken and Samuel, 2007). In the more than twelve years since this model was introduced, it has led to a highly productive period of research (Vlaeyen and Linton, 2012), to specific 'exposure'-based treatments, and to a number of related treatment outcome studies that support the benefits of these types of treatments (Bailey et al., 2010). One issue that has certainly remained a key issue in chronic pain research from the fear-avoidance model is the important role of avoidance (McCracken and Samuel, 2007). At the same time it has been noted that this model has a few limitations and would benefit from some modifications and expansion, for instance by focusing on motivation and specific processes of recovery (Crombez et al., 2012).

The newest model of pain and suffering: Psychological flexibility

From just a couple studies in the 1990s another shift of focus in psychological approaches to chronic pain has started to develop. This shift moves somewhat away from concepts of 'coping' and from emphasis on controlling the content of feelings and thoughts related to pain, toward 'acceptance' and emphasis on changing the ways in which we experience feelings and thoughts (McCracken, 2005, 2011). The model that guides this current approach is called the psychological flexibility model (Hayes et al., 2011). Psychological flexibility is the capacity to persist in, or change, behaviour patterns in a way that is guided by a person's chosen goals and is sensitive to what the situation at hand allows without being unnecessarily restricted by psychological experiences, especially those that are based on cognitive processes (Hayes et al., 2011; McCracken and Morley, 2014). This is admittedly a definition that sounds very complex. Another, simpler and less technical way to describe this process is to say that it is the ability to remain open, aware, and engaged. The particular psychological processes that make up this overall process are called acceptance, cognitive defusion, flexible present-focused awareness, values, committed action, and self as context (Hayes et al., 2011; McCracken and Morley, 2014). It is not crucial to understand what each of these unusual terms mean, but perhaps just to get a general sense of the overall approach.

In a nutshell, it is very common (and basically normal) for humans to get stuck in life's difficulties and to suffer. This is true both for people with pain problems and for the rest of us. The psychological flexibility model proposes that people suffer because they encounter things they do not like and they, quite naturally, struggle to eliminate or avoid these things. For many of the things with which people struggle, this struggling either does not eliminate the experience or does not yield successful functioning – as when we struggle with our own thoughts, urges, and feelings. Another problem is our tendency to get caught in our own thoughts and worries and to lose touch with our current situation, what we are doing, and whom we are with. Our minds are so busy evaluating, interpreting, and judging events and other people around us that we lose touch, and we do not even realise at the time that our evaluations, interpretations, and judgments are not the same thing as the world itself. From this position of struggling with our experiences and of being stuck in our own thinking, we are less able to remain clear on our goals and on what is important for us to do, and we become less able to do what these goals require. We call these processes experiential

Table 10.1 Typical treatment methods used in traditional cognitive behavioural treatments for chronic pain.

Reconceptualization (pain can be self-managed)	Pain education
Coping skills training (e.g., attention management)	Communication skills training
Relaxation	Pacing/activity management
Cognitive restructuring	Problem solving skills
Graded physical activation	Physical exercise
Exposure treatment (facing situations avoided)	Other health education

avoidance, cognitive fusion, and failure in goal-directed or values-based and committed action. The therapeutic approach designed to address these problems, and to increase psychological flexibility, is a form of cognitive behavioural therapy (CBT) called acceptance and commitment therapy (ACT) (Hayes, Strosahl and Wilson, 2013).

Evidence for CBT, ACT, and multidimensional approaches to chronic pain

The types of treatment method typically included in CBT are presented in Table 10.1. There are now more than ten or twelve systematic reviews of psychological treatments for pain, most of them focused on packages of treatment that include these types of methods. One of the recent reviews produced through the system of Cochrane reviews went through 42 randomised controlled trials (RCTs) of cognitive or behavioural therapy (Williams, Eccleston and Morley, 2012). These authors extracted and analysed data gathered from these trials both immediately after treatment and at follow-up periods of six months or more. They examined comparisons both with treatment as usual and with waiting-list conditions and with other active treatment comparison conditions, and they looked at outcomes that included pain, disability, depression, and catastrophising. They found that (1) CBT produced small improvements in disability and catastrophising, but not in pain or mood, in comparison to active control conditions; (2) it also produced small to moderate improvements on all outcomes in comparison to treatment as usual or waiting-list conditions; but (3) some of the improvements were lost at follow-up. The authors described CBT as useful in approaching the management of chronic pain, but they also highlighted the needs for finding which components of the treatment have the greatest impact; for matching treatments to patients' needs; and for focusing more on treatment mechanisms.

The evidence base for ACT and chronic pain is still developing. The first RCT of ACT-related to pain was a small trial that successfully provided a brief ACT-based treatment for people at risk of missing work due to stress or pain (Dahl, Wilson and Nilsson, 2004). This study was conducted in Sweden. It was, however, not purely a study of pain. The first clearly pain-focused RCT was not published until 2008. This was a trial of treatment for chronic pain due to whiplash, also conducted in Sweden (Wicksell et al., 2008). A systematic review of 'acceptance-based' treatment for chronic pain was published recently but only less than one third of the studies included ACT; most of them were studies of mindfulness-based treatments (Veehof et al., 2011). There are now at least eight published RCTs of ACT for chronic pain (McCracken and Morley, 2014). Some of them are preliminary or pilot studies and many have small sample sizes; however, the rate of published trials seems to be increasing, and more definitive studies may soon be seen. One interesting feature of the current trials of ACT for chronic pain is their diversity. Some included individual psychological treatment, and some included group-based treatment. Some of the treatments were very short, of just a few hours, while some were longer and possibly more intensive, for instance they went on for four weeks at a rate of five days per week. Some were delivered face to face and some through the internet.

Sometimes when treatment for chronic pain is delivered on the basis of a psychosocial model and contains psychological methods, it is delivered by a psychologist who works alone. This may be done in general practice or primary-care settings, or it may be done in secondary-care pain centres. Sometimes such treatment is delivered by a multidisciplinary or interdisciplinary team in highly specialist centres (here 'interdisciplinary' means that the team is not merely made up of specialists in more than one discipline, but these disciplines also work together in a highly integrated fashion). The smallest multidisciplinary teams delivering psychologically based treatments might consist of a psychologist and a physiotherapist – and maybe a nurse, too. Larger interdisciplinary teams would certainly include physicians, nurses, psychologists and physiotherapists – and possibly others too: occupational therapists, vocational counsellors, pharmacists. Smaller teams with reduced specialist medical input are often able to provide services for the larger number of people with common chronic musculoskeletal pain problems in primary or secondary care. The larger and more integrated teams, which have higher specialist medical input, are able to address the needs of people with complex chronic pain problems, complex co-morbid medical conditions and high levels of distress and disability, and people who may have failed to respond to less intensive treatments. Systematic reviews clearly show that multidisciplinary or interdisciplinary treatments, especially the ones that deliver higher intensity treatment, are able to help people with chronic pain to improve their pain, emotional functioning, daily activities, and in some cases their work status (Guzmán et al., 2001; Hoffman et al., 2007; Scascighini et al., 2008).

This may come as a surprise to some, but when researchers examine both the effectiveness of various treatments for chronic pain and the costs required they find that intensive multidisciplinary or interdisciplinary pain treatments based on a general cognitive behavioural model are the most efficient and cost-effective option, in comparison to other routinely used treatments, such as operations and medications (Turk and Burwinkle, 2005; Gatchel and Okifuji, 2006).

The role of psychology in the outcome of treatment trials

A meeting was held in November 2002 with 27 people from academia, government agencies, and pharmaceutical industries. The purpose of the group formed at this meeting was to provide recommendations for the core outcome domains that should be considered by those planning clinical trials for chronic pain treatments. The results of that meeting were the first recommendations from the Initiative on Methods, Measurement, and Pain Assessment in Clinical Trials (IMMPACT) (Turk et al., 2003). The outcome domains specified in the IMMPACT's initial report were pain, physical functioning, emotional functioning, ratings of improvement and satisfaction, adverse events, and disposition (adherence or withdrawal). These recommendations are important, as they reflect the complexity of chronic pain as a problem and make room for a clear role of psychological outcomes.

Another key issue related to outcomes for clinical trials of chronic pain treatment is that of the domains that people with pain say are important. The IMMPACT group addressed this issue in a follow-up set of recommendations (Turk et al., 2008). For this purpose the group conducted a two-phase study. First it conducted focus groups from clinical settings and generated a pool of 19 items reflecting aspects of daily life affected by pain. These items were then used in an online survey of diverse people with chronic pain. Items with the highest mean ratings of importance were enjoyment of life in general, fatigue, emotional well-being, weakness, and staying asleep at night. Another general finding was that all of the 19 items were rated as significantly affected: all were above 6.5 out of 10, on a scale from 'not at all' to 'extremely important'. The authors concluded that this meant that 'most if not all aspects of their lives are significantly affected by chronic pain' (Turk et al., 2008: 281).

Treatment mechanisms in chronic pain

Treatment mechanisms, sometimes called treatment processes, are those changes made directly, through treatment, that produce shifts in outcome variables. A commonsense model of pain treatment might propose, for example, that pain reduction is the mechanism by which people improve in their daily functioning, emotional well-being and degree of reliance on health care (outcomes). Unfortunately or interestingly, things are not quite this simple. We know from studies of multimodal rehabilitation of chronic pain that people do experience significant improvements in pain-related interference with functioning, daily activities, and emotional functioning (McCracken, Gross, et al., 2002). We also know that, when we examine other changes that correlate with those improvements and could function as their mechanisms, we find that reductions in pain-related anxiety and avoidance are significantly correlated with outcome improvements – much more than are improvements in physical capacity (reflected in directly assessed flexibility, or in lifting and carrying capacity) or in pain. Even in terms of more subjective experiences of treatment, it has been shown that people seeking treatment for pain are more satisfied with their treatment six months later, when they feel that their assessment was complete, that treatment procedures were well explained, and that their functioning has improved (McCracken, Evon, et al., 2002). Each of these factors was more important than pain reduction itself; and, once all this was taken into account for a prediction model, pain reduction turned out to play no significant role in predicting treatment satisfaction.

Over the years, there has not always been a great deal of attention to precise mechanisms of psychological treatments. This is unfortunate, because this kind of attention is now considered our best means of securing progress in the field of therapy development (Eccleston, Williams and Morley, 2009; Williams et al., 2012; McCracken and Morley, 2014). For many years, the tacit assumption underlying psychological approaches to chronic pain was that successful pain management is based on the learning of a certain set of skills – physical exercise, stretching, pacing, and cognitive restructuring – and then on adherence to using these skills over time. When the use of these skills was examined recently in a large study in London (N = 2345), it was found that adherence to these skills accounted for only a small amount of variance in people's treatment outcome; therefore, according to these data, such adherence cannot constitute a substantial mechanism of treatment (Curran, Williams and Potts, 2009).

The model of treatment underlying ACT proposes a direct role for psychological flexibility in relation to treatment outcome. Increasing amounts of data and increasingly sophisticated designs are steadily providing support for the role of acceptance, values-based action, and general psychological flexibility as the means by which ACT creates it positive impact on treatment outcome measures (McCracken and Gutiérrez-Martínez, 2011). In fact the role of psychological flexibility appears to be more important than that of more traditional pain management strategies such as distraction, cognitive restructuring, relaxation and pacing (Vowles and McCracken, 2010), and also more important than that of other commonly considered processes such as pain, distress, self-efficacy or kinesiophobia (Wicksell, Olsson and Hayes, 2010).

Summary

Chronic pain is one of the most prevalent and most expensive problems in all of health care. Psychosocial influences and impacts are integral to the experience of chronic pain. It is right therefore that these factors are integrated into the best current treatments and that we continue to research them.

There is sometimes a tendency in health care to dichotomise physical and mental aspects of health. We label and categorise these as separate issues and then reinforce the distinction by training treatment providers in one or in the other, but not in both, and by separating the hospital departments and other treatment delivery systems for these separate problems. Implicitly or explicitly, we, as a wider society, sometimes prioritise physical health problems and stigmatise those with mental health needs. It ought to be clear by now that this should not be done; chronic pain as a clinical problem is a key example of where an integration of psychosocial and biological or physiological factors is needed.

Today cognitive behavioural approaches to chronic pain reflect a number of different models (Jensen, 2011); they can appear confusing to those outside the field, but they are certainly worth studying and understanding – both because they represent an important part of the care of a large part of our population and because they are not simple common sense.

Regardless of the specific models they follow, psychological treatments have a reliable evidence base and they continue to evolve. In comparison to other commonly available alternatives, interdisciplinary treatments based on cognitive behavioural models are widely regarded as the most clinically efficient and cost-effective treatments for chronic pain. And it is likely that they will improve.

References

Alexander, F. 1950. *Psychosomatic Medicine: Its Principles and Applications*. New York: W. W. Norton.

Bailey, K. M., Carleton, N., Vlaeyen, J. W. and Asmundson, G. J. 2010. Treatments addressing pain-related fear and anxiety in patients with chronic musculoskeletal pain: A preliminary review. *Cognitive and Behavior Therapy* 39: 46–63.

Bayer, T. L., Baer, P. E. and Early, C. 1991. Situational psychophysiological factors in psychologically induced pain. *Pain* 44: 45–50.

Breivik, H., Collett, B., Ventafridda, V., Cohen, R. and Gallacher, D. 2006. Survey of chronic pain in Europe: Prevalence, impact on daily life, and treatment. *European Journal of Pain* 10: 287–333.

Cassell, E. J. 2004. *The Nature of Suffering and the Goals of Medicine*, 2nd edn. Oxford: Oxford University Press.

Crombez, G., Vlaeyen, J. W. S., Heuts, P. H. T. G. and Lysens, R. 1999. Pain-related fear is more disabling than pain itself: Evidence on the role of pain-related fear in chronic back pain disability. *Pain* 80: 329–39.

Crombez, G., Eccleston, C., Van Damme, S., Vlaeyen, J. W. and Karoly, P. 2012. Fear-avoidance model of chronic pain: The next generation. *Clinical Journal of Pain* 28: 475–83.

Curran, C., Williams, A. C. de C. and Potts, H. W. W. 2009. Cognitive–behavioral therapy for persistent pain: Does adherence after treatment affect outcome? *European Journal of Pain* 13: 178–88.

Dahl, J., Wilson, K. G. and Nilsson, A. 2004. Acceptance and commitment therapy and the treatment of persons at risk for long-term disability resulting from stress and pain symptoms: A preliminary randomized trial. *Behavior Therapy* 35: 785–801.

Demyttenaere K., Bruffaerts R., Lee S., Posada-Villa, J., Kovess, V., Angermeyer, M. C., Levinson, D., de Girolamo, G., Nakane, H., Mneimneh, Z., et al. 2007. Mental disorders among persons with chronic back or neck pain: Results from the world mental health surveys. *Pain* 129: 332–42.

Eccleston, C., Williams, A. C. and Morley, S. 2009. Psychological therapies for the management of chronic pain (excluding headache) in adults (Review). *Cochrane Database of Systematic Reviews* 11. doi: 10.1002/14651858. CD007407.pub2.

Elliott, A. M., Smith, B. H., Hannaford, P. C., Smith, W. C. and Chambers, W. A. 2002. The course of chronic pain in the community: Results of a 4-year follow-up study. *Pain* 99: 299–307.

Engel, G. L. 1977. The need for a new medical model: A challenge for biomedicine. *Science* 196 (4286): 129–36.

Fordyce, W. E. 1976. *Behavioral Methods for Chronic Pain and Illness*. St Louis, MO: CV Mosby.

Fordyce, W. E. 1995. *Back Pain in the Workplace: Management of Disability in Nonspecific Conditions*. Seattle, WA: IASP Press.

Gatchel, R. J. and Okifuji, A. 2006.. Evidence-based scientific data documenting the treatment and cost-effectiveness of comprehensive pain programs for chronic non-malignant pain. *The Journal of Pain* 7: 779–93.

Gatchel, R. J., Peng, Y. B., Peters, M. L., Fuchs, P. N., Turk, D. C. 2007. The biopsychosocial approach to chronic pain: Scientific advances and future directions. *Psychological Bulletin* 133 (4): 581–624.

Guzmán, J., Esmail, R., Karjalainen, K., Malmivaara, A., Irvin, E. and Bombardier, C. 2001. Multidisciplinary rehabilitation for chronic low back pain: Systematic review. *British Medical Journal* 322: 1511–16.

Hayes, S. C., Strosahl, K. D. and Wilson, K. G. 2013. *Acceptance and Commitment Therapy: The Process and Practice of Mindful Change*. New York: Guildford Press.

Hayes, S. C., Villatte, M., Levin, M. and Hildebrandt, M. 2011. Open, aware and active: Contextual approaches as an emerging trend in the behavioural and cognitive therapies. *Annual Review of Clinical Psychology* 7: 141–68.

Hoffman, B. M., Papas, R. K., Chatkoff, D. K. and Kerns, R. D. 2007. Meta-analysis of psychological interventions for chronic low back pain. *Health Psychology* 26: 1–19.

Jensen, M. P. 2011. Psychosocial approaches to pain management: An organizational framework. *Pain* 152: 717–25.

Jensen, M. C., Brant-Zawadzki, M. N., Obuchowski, N., Modic, M. T., Malkasian, D. and Ross, J. S. 1994. Magnetic resonance imaging of the lumbar spine in people without back pain. *The New England Journal of Medicine* 331: 68–73.

McCracken, L. M. 2005. *Contextual Cognitive Behavioral Therapy for Chronic Pain*. Progress in Pain Eesearch and Management 33. Seattle, WA: IASP Press.

McCracken, L. M., ed. 2011. *Mindfulness and Acceptance in Behavioral Medicine: Current Theory and Practice*. Oakland, CA: New Harbinger Press.

McCracken, L. M. and Gutiérrez-Martínez, O. 2011. Processes of change in psychological flexibility in an interdisciplinary group-based treatment for chronic pain based on Acceptance and Commitment Therapy. *Behaviour Research & Therapy* 49: 267–74.

McCracken, L. M. and Morley, S. 2014. The psychological flexibility model: A basis for integration and progress in psychological approaches to chronic pain management. *The Journal of Pain* 15. doi: 10.1016/j. jpain.2013.10.014.

McCracken, L. M. and Samuel, V. M. 2007. The role of avoidance, pacing, and other activity patterns in chronic pain. *Pain* 130: 119–25.

McCracken, L. M., Evon, D. and Karapas, E. T. 2002a. Satisfaction with treatment for chronic pain in a specialty service: Preliminary prospective results. *European Journal of Pain* 6: 387–93.

McCracken, L. M., Gross, R. T. and Eccleston, C. 2002b. Multimethod assessment of treatment process in chronic low back pain: Comparison of reported pain-related anxiety with directly measured physical capacity. *Behaviour Research and Therapy* 40: 585–94.

Melzack, R. and Wall, P. D. 1965. Pain mechanisms: A new theory. *Science* 150: 971–9.

Mersky, H. and Bogduk, N. 1994. *Classification of Chronic Pain*, 2nd edn. Seattle, WA: IASP Task Force on Taxonomy, IASP Press.

Moussavi, S., Chatterji, S., Verdes, E., Tandon, A., Patel, V. and Ustun, B. 2007. Depression, chronic diseases, and decrements in health: Results from the World Health Survey. *Lancet* 370: 851–8.

Samanta, J., Kendall, J. and Samanta, A. 2003. Chronic low back pain. *British Medical Journal* 326 (535). doi: 10.1136/bmj.327.7406.107-a.

Sareen, J., Jacobi, F., Cox, B. J., Belik, S. L., Clara, I. and Stein, M. B. 2006. Disability and poor quality of life associated with comorbid anxiety disorders and physical conditions. *Archives Internal Medicine* 166 (19): 2109–16.

Scascighini, L., Toma, V., Dober-Speilmann, S. and Sprott, H. 2008. Multidisciplinary treatment for chronic pain: S systematic review. *Rheumatology* 47: 670–8.

Schweiger, A. and Parducci, A. 1981. Nocebo: The psychological induction of pain. *Pavlovian Journal of Biological Science* 16: 140–3.

Smith, B. H., Elliott, A. M., Chambers, W. A., Smith, W. C., Hannaford, P. C. and Penny, K. 2001. The impact of chronic pain in the community. *Family Practice* 18: 292–9.

Turk, D. C. and Burwinkle, T. M. 2005. Clinical outcomes, cost-effectiveness, and the role of psychology in treatments for chronic pain sufferers. *Professional Psychology: Research and Practice* 36: 602–10.

Turk, D. C., Meichenbaum, D. and Genest, M. 1983. *Pain and Behavioral Medicine: A Cognitive–Behavioral Perspective*. New York: Guilford Press.

Turk, D. C., Dworkin, R. H., Allen, R. R., Bellamy, N., Branderburg, N., Carr, D. B., Cleeland, C., Dionne, R., Farrar, J. T., Galer, B. S., et al. 2003. Core outcome domains for chronic pain clinical trials: IMMPACT recommendations. *Pain* 106: 337–45.

Turk, D. C., Dworkin, R. H., Revicki, D., Harding, G., Burke, L. B., Cella, D., Cleeland, C. S., Cowan, P., Farrar, J. T., Hertz, S., et al. 2008. Identifying important outcome domains for chronic pain clinical trials: An IMMPACT survey of people with pain. *Pain* 137: 276–85.

Veehof, M. M., Oskam, M. J., Schreurs, K. M. G. and Bohlmeijer, E. T. 2011. Acceptance-based interventions for the treatment of chronic pain: A systematic review and meta-analysis. *Pain* 152: 533–42.

Vlaeyen, J. W. S. and Linton, S. J. 2000. Fear-avoidance and its consequences in chronic musculoskeletal pain: A state of the art. *Pain* 85: 317–32.

Vlaeyen, J. W. S. and Linton, S. J. 2012. Fear-avoidance of chronic musculoskeletal pain: 12 years on. *Pain* 153: 1144–7.

Vowles, K. E. and McCracken, L. M. 2010. Comparing the role of psychological flexibility and traditional pain management coping strategies in chronic pain treatment outcomes. *Behaviour Research & Therapy* 48: 141–6.

Waddell, G. and Aylward, M. 2010. *Models of Sickness and Illness Applied to Common Health Problems*. London: Royal Society of Medicine Press.

Wicksell, R. K., Olsson, G. L. and Hayes, S. C. 2010. Psychological flexibility as a mediator of improvement in Acceptance and Commitment Therapy for patients with chronic pain following whiplash. *European Journal of Pain* 14: 1059.e1–1059.e11.

Wicksell, R. K., Ahlqvist, J., Bring, A., Melin, L., Olsson, G. L. 2008. Can exposure and acceptance strategies improve functioning and life satisfaction in people with chronic pain and whiplash-associated disorders (WAD)? A randomized controlled trial. *Cognitive Behavioural Therapy* 37: 169–82.

Williams, A. C. de C., Eccleston, C. and Morley, S. 2012. Psychological therapies for the management of chronic pain (excluding headache) in adults (Review). *Cochrane Database of Systematic Reviews* 11. doi: 10.1002/14651858. CD007407.pub3.

SECTION III
Pain in the lifecycle and in nervous system disorders

CHAPTER 11

Pain in neonates and infants

Fiona Moultrie, Segzi Goksan, Ravi Poorun and Rebeccah Slater

Infant pain assessment

It is widely accepted that infants display physiological, behavioural, and biochemical responses to noxious stimulation and numerous pain assessment tools have been developed to use these indicators in order to measure pain in infants (Maxwell, Malavolta and Fraga, 2013). Given that infants admitted to intensive care experience a high number of painful procedures per day – estimates suggest that some infants may experience as many as 50 painful procedures per day (Carbajal et al., 2008) – accurate pain assessment in this population is extremely important. Pain assessment is a fundamental component of effective pain management; it facilitates the identification of pain, determines the degree of intervention required, demonstrates the effectiveness of the chosen intervention, and indicates the need for further intervention. Self-reporting of pain is regarded as the gold standard of assessment, but of course this is not applicable to patients under three years of age. The recent acceptance, within the medical community, that a lack of verbal communication in neonates and infants does not negate the potential experience of pain has brought with it the challenge of accurate assessment in this unique patient population.

Biochemical responses to noxious stimuli – responses that include rises in cortisol, adrenaline, noradrenaline, and endorphins (Schmeling and Coran, 1991) – formed the cornerstone of key research conducted in the 1980s. These measures provided clear evidence that infants manifest a biological response to noxious events; however, such approaches cannot practically be incorporated into bedside clinical pain assessment. Instead, changes in infant behaviour and physiology form the basis of most clinical pain assessment tools. Despite significant advances in pain research over the past three decades, which have led to the development of over forty neonatal pain assessment tools of varying degrees of feasibility and clinical utility (Lee and Stevens, 2013), there remains no gold standard for the assessment of pain in infants (Ranger, Johnston and Anand, 2007). Most of the tools that have been designed are multidimensional or composite, incorporating the scoring of both physiological and behavioural responses, in an effort to provide the most sensitive, specific and honest representation of pain.

An Introduction to Pain and its Relation to Nervous System Disorders, First Edition. Edited by Anna A. Battaglia.
© 2016 John Wiley & Sons, Ltd. Published 2016 by John Wiley & Sons, Ltd.

The composite tools are thought to offer a thorough representation of the infant's pain experience and to have the most established psychometric properties. Crying, changes in facial expression and characteristic body movements such as limb withdrawal or agitation are examples of the complex behavioural responses that occur during the experience of pain; among these, facial expression is regarded as the most sensitive and specific measure in neonates (Craig et al., 1993; Grunau, Johnston and Craig, 1990; Slater et al., 2008). These indicators of pain are subjectively assessed and change with age, particularly from extremely premature to term infants, which limits their usefulness.

Physiological responses that include changes in heart rate, respiratory rate, oxygen saturation and blood pressure have the advantage of being recorded objectively; but they are limited by their lack of nociceptive specificity (Raeside, 2011). Indeed both physiological and behavioural responses can be non-specifically elicited by a variety of other conditions, such as hunger and distress, and can be influenced by other factors such as gestational age, postnatal age, sleep state, neurological impairment, time since last procedure, and overall past experience of pain. It is therefore not surprising that behavioural and physiological measures are relatively poorly correlated (Craig et al., 1993).

Neonates as premature as 25 weeks gestation have been shown to mount behavioural, physiological and cortical responses to noxious stimulation (Bartocci et al., 2006; Slater et al., 2006; Stevens et al., 1996; Stevens et al., 2014). Evidence however suggests that up to 35 per cent of infants do not appear to respond behaviourally to tissue-damaging procedures such as venepuncture and heel lancing (Johnston et al., 1999; Slater et al., 2009). The interpretation of this phenomenon remains controversial; a lack of response could represent a reduction in or an absence of pain, or simply the infant's inability to mount an energetically costly response. One study has demonstrated that cortical pain processing, as evidenced by recording somatosensory cortical activity using near-infrared spectroscopy, can occur in the absence of linked behavioural changes such as facial expression; this fact presents the clinician with further challenges for pain assessment and management (Slater et al., 2008). Indeed, if infants are processing the nociceptive stimuli in the central nervous system but not mounting a behavioural response, this leads to the possibility that infant pain is being underestimated. Furthermore, studies of evoked brain activity have demonstrated that very premature infants may not discriminate between noxious and innocuous stimulation, which further complicates the assessment of pain (Fabrizi et al., 2011).

Infant pain management

Neonates in intensive care undergo an average of 11 acute painful procedures per day and it has been estimated that only 36 per cent of neonates receive analgesia for these frequent procedures (Roofthooft et al., 2014). There has been a growing emphasis on incorporating pain management into routine clinical care, which has prompted the emergence of a myriad of guidelines attempting to synthesise and rationalise the evidence from randomised controlled trials, observational studies and systematic reviews. There remains great controversy, however, regarding the efficacy, safety, and long-term effects of most analgesic agents, and a dramatic variation continues to exist in practice. A multimodal approach has been promoted that includes minimising the number of noxious invasive procedures, the adoption of environmental stress-reducing interventions, and of course the regular use of pharmacological as well as non-pharmacological methods of pain relief (Anand and the International Evidence-Based Group for Neonatal, 2001). Efforts have been made to limit the number of invasive and potentially painful procedures through the implementation of care clustering, the reduction of unnecessary procedures and attempts, and the introduction of transcutaneous measurements when appropriate (Anand and the International Evidence-Based Group for Neonatal, 2001). Nevertheless, in clinical practice pain management interventions remain underutilised; this is probably due to a failure to assess pain, an underestimation of pain and a fear of side effects.

Pharmacological interventions
Non-opioids

A number of pharmacological interventions can be prescribed to provide pain relief to infants in acute, chronic and postoperative pain. Paracetamol has recently been licensed for use in patients who weigh under 10 kilograms, and it is the most commonly prescribed non-opioid for the relief of mild to moderate pain in neonates. It is also valuable as an opioid-sparing agent for moderate postsurgical pain (Ceelie et al., 2013). Hepatotoxicity is the most serious risk and occurs typically through errors in prescribing or administration of the very small doses required. There have been many pharmacokinetic studies of paracetamol in neonates; however, further prospective pharmacodynamic studies evaluating its safety in the extreme premature population are required before the use of this drug in routine clinical practice can be recommended (Pacifici and Allegaert, 2015). Non-steroidal anti-inflammatories such as ibuprofen are not prescribed for their analgesic properties in neonatal care and are associated with significant side effects, including impaired platelet aggregation, renal dysfunction and gastric bleeding, which limits their potential use in neonates (Parry, 2008).

Sucrose

There is a growing body of evidence supporting the administration of oral sucrose before noxious events, as a form of preemptive analgesia (Stevens et al., 2013). With over 30 randomised controlled trials, sucrose is the most studied pharmacological intervention for neonatal pain. Non-nutritive sucking with sucrose prior to minor procedures reduces pain-related behaviour and physiological responses in neonates to a greater degree than breastfeeding (Carbajal et al., 2003), expressed breast milk (Skogsdal, Eriksson and Schollin, 1997) and non-nutritive sucking (Carbajal, 1999). The dosage, the safety of repeated use, and the effectiveness of sucrose in extremely premature infants are yet to be determined (Stevens et al., 2013), and there is evidence that administration of sucrose does not reduce the resultant hyperalgesia caused by noxious events (Taddio, Shah and Katz, 2009). Furthermore, while sucrose does reduce pain scores that are based on behavioural and physiological measures, it does not alter the evoked nociceptive-specific brain activity (Slater, Cornelissen, et al., 2010). The administration of sucrose therefore remains controversial and the debate continues as to whether it simply provides distraction, altering the observed behaviour, or whether it produces true analgesia.

Opioids

Over the past 25 years opioids have become the mainstay of pain management for ventilated neonates. Opioids have been shown to improve ventilator synchrony and to reduce physiological instability and behavioural and hormonal stress responses (Dyke, Kohan, and Evans 1995; Guinsburg et al., 1998). Their use is limited, however, due to their significant dose-dependent side effects, which include respiratory depression, hypotension, urine retention and feed intolerance (Hall et al., 2005; Anand et al., 2004). A Cochrane review of the available evidence from randomised controlled trials concluded, however, that although some studies suggest that morphine causes a reduction in acute pain scores, there is at present insufficient evidence to recommend its routine clinical use in neonates (Bellu, de Waal and Zanini, 2008). Despite the widespread use of morphine, there is no consensus on the correct dosing: up to one hundredfold differences in starting doses have been reported (van den Anker, 2013). Opioids continue to be prescribed according to clinical judgement and consideration of the risks and benefits on an individual patient basis, while more conclusive evidence is still to emerge from clinical trials.

Non-pharmacological interventions

Numerous non-pharmacological therapies that use tactile, visual and oral stimuli have been implemented, with little or no side effects, in order to reduce mild to moderate pain in neonates undergoing minor clinical procedures (Pillai Riddell et al., 2011). 'Kangaroo care' or maternal skin-to-skin contact was introduced in the 1970s. This is a technique thought to stimulate the infant's proprioceptive system and encourages parental involvement in infant care. This non-pharmacological intervention has been shown to produce a significant reduction in pain scores during and after heel lance (Johnston et al., 2008; Johnston et al., 2003); in particular, it reduces crying and heart rate, and its superiority to the traditional swaddling of the infant has been demonstrated (Gray, Watt and Blass, 2000). Randomised controlled trials have shown that breastfeeding prior to or during heel lance or venepuncture also reduces crying time, tachycardia and overall pain scores calculated with the Premature Infant Pain Profile (Carbajal et al., 2003; Gray et al., 2002). Interestingly, expressed breast milk delivered by syringe does not produce the same desired effect (Skogsdal et al., 1997), but non-nutritive sucking of a pacifier does produce a reduction in pain scores during venepuncture and heel lance (Carbajal et al., 1999).

Long-term consequences of pain in early life

Providing effective pain management for infants is important not only to ensure that they receive the highest standard of acute clinical care, but to prevent the potential long-term neurodevelopmental consequences that may arise as a result of early exposure to painful interventions. The intense period of monitoring and the high number of noxious events, which form an essential part of neonatal care, occur at a time when the infant's nervous system is vulnerable. An extremely premature infant, who may be born as early as 23 weeks gestation, has an immature nervous system, which is undergoing rapid structural and functional development. During this period major changes in the brain cytoarchitecture are occurring – for example the growth and differentiation of subplate neurons; the aligning, orienting and layering of cortical neurons; elaboration of dendrites and axons; formation and selective pruning of neuronal processes and synapses; and the proliferation and differentiation of glial cells (Grunau, 2013; Miller and Ferriero, 2009). Increased neuronal excitation as a result of frequent painful clinical interventions may disrupt the development of subplate neurons and the formation of thalamocortical connections (Smyser et al., 2010) leading to long-term alterations in sensory and nociceptive processing (Fitzgerald and Walker, 2009).

Numerous studies have been conducted to determine whether exposure to a high burden of pain in early life leads to long-term changes in sensory or nociceptive processing. Premature infants who have been hospitalised for at least 40 days display greater nociceptive-specific brain activity in response to noxious stimulation when than age-matched term-born infants (Slater, Fabrizi, et al., 2010). Sensitisation to mechanical stimulation has been observed in behavioural and reflex responses, and this hypersensitivity has been shown to persist throughout the first year of life (Abdulkader et al., 2008). Studies of former preterm school-aged children and adolescents have also provided evidence of persistently altered sensory processing; these children displayed increased pain catastrophising and pain sensitivity when compared to term-born children (Hohmeister et al., 2009; Buskila et al., 2003).

The neurodevelopmental effects of early pain exposure do not appear to be restricted to the premature infant population. Term infants circumcised during the first five days of life have been shown to display increased pain-related behavioural responses to routine vaccination at 4–6 months in comparison to uncircumcised infants (Taddio et al., 1995). Similarly, major surgery in the first three months of life has been associated with increased pain sensitivity and greater analgesic requirements at the time of subsequent surgeries in the same or a different dermatome; this suggests that both spinal and supraspinal changes may occur following neonatal surgery (Peters et al., 2005).

The effect of early pain exposure on other sensory modalities has also been explored. Children previously hospitalised in NICU as neonates displayed greater perceptual sensitisation to thermal pain than healthy controls (Hermann et al., 2006). Children born extremely prematurely were also found to have generalised reduced thermal sensitivity (but not mechanical sensitivity) on quantitative sensory testing; this effect was further exaggerated in those who had undergone neonatal surgery and suggests centrally mediated alterations to nociceptive pathways – alterations that are related to the degree of injury (Walker et al., 2009).

In addition to functional and behavioural changes, greater cumulative exposure to painful procedures has been associated with reduced postnatal body and head growth, a reduction in cortical thickness, and altered brain microstructural development in premature infants (Brummelte et al., 2012; Vinall et al., 2012; Ranger et al., 2013). Brummelte and colleagues found a reduction in white matter and in subcortical grey matter, even after adjustment for clinical confounders such as illness severity, morphine exposure, brain injury and surgery (Brummelte et al., 2012). Repetitive pain in early life may also influence cognition. Doesburg and colleagues identified a relationship between neonatal skin-breaking procedures, altered functional brain activity, measured using electroencephalography (EEG) and impaired visual–perceptual ability in school-age children (Doesburg et al., 2013). Furthermore, repetitive invasive procedures during neonatal care may contribute to lower IQ scores in infants born extremely premature (Vinall et al., 2014). There is a growing body of empirical evidence demonstrating that early-life exposure to noxious insults is associated with significant changes in brain development and function, which emphasises the importance of minimising exposure to pain in early life and the urgent need for improvements in the management of pain.

Ethical implications of pain in neonates

Three decades ago it was emphatically believed that newborn infants did not feel pain. It was therefore considered ethically acceptable to submit them to surgical procedures without analgesic or anaesthetic agents (Anand and Hickey, 1987). Following the publication of the seminal work by Anand, Sippell and Aynsley-Green (1987), a flurry of research began to dispel this myth, and the medical community was obliged to reconsider standards of neonatal care. Studies provided evidence of the significant effects of analgesia and anaesthesia on morbidity and mortality, and evidence has continued to accumulate demonstrating the deleterious long-term neurodevelopmental consequences of untreated early pain experiences. Despite this progress, pain assessment and effective pain management remain unit-dependent and often physician-dependent. Any failure to acknowledge, assess and treat pain in neonates disregards the universal principles of beneficence and non-maleficence, contravenes the physician's duty of care and ignores the basic expectations of parents when they entrust physicians with the care of their newborn.

Why did we deny the existence of infant pain? Pain is, by definition, a subjective experience. The epistemological problem of 'other minds' and the view that behaviour may not be in itself sufficient to corroborate the presence of mentality – or, in this case, the presence of a conscious pain experience – raise a fundamental difficulty in the care of non-verbal patient populations such as newborns, infants, and brain-injured children and adults (Wilkinson, Savulescu and Slater, 2012). For these patients, methods of pain assessment cannot simply be extrapolated from adult practice; and they cannot rely on self-report either. Not until as recently as 2003 was the definition of pain given by the International Association for the Study of Pain (IASP) modified to clarify that 'the inability to communicate in no way negates the possibility that an individual is experiencing pain' (Merskey and Bogduk, 2014: 209). Some have argued that the immaturity of higher level processing in neonates does not allow for a conscious experience equivalent to that of an adult. In the past decade neurophysiological correlates of pain have been demonstrated in infants through techniques such as near-infrared spectroscopy, EEG and, more recently, functional magnetic resonance imaging (fMRI); and they suggest a greater degree of cortical processing than previously assumed (Slater et al., 2006; Slater, Worley et al., 2010; Goksan et al., 2015). It remains to be seen, however, how closely these correlates can be tied, if ever, to what we recognise as a conscious experience of pain. Although it has been suggested that lack of memory may amount to lack of pain, in light of the emerging evidence about the benefits of analgesia and the neurodevelopmental effects of untreated early pain experience, conscious experience and memory are irrelevant to whether pain should be treated. We would like to believe that some of the views discussed here are simply anachronistic; however, barriers to pain relief in neonates still exist today, the greatest being a fear of the potential risks associated with the administration of analgesic agents. As in any form of medical intervention, here too a fundamental balance should be found between risk and benefit and between efficacy and safety.

Pain research

We have progressed from an era of 'cultural blindness', in which the possibility of pain in neonates was denied, to an age in which we are striving to address the management of pain. If we are to satisfy our moral obligation to assess and treat pain effectively, it is essential that we continue to carry out research in the neonatal population so as to establish the best possible pain assessment techniques and most safe and effective analgesic interventions. Historically, it has been difficult to obtain ethical approval to perform clinical research on infants and children, which consequently has led to the use of unlicensed, potentially unsafe drugs and drug doses in clinical practice (Caldwell et al., 2004). Although infants were 'protected' from any potential adverse events that can be associated with clinical research, they were equally denied the benefits of evidence-based care and of appropriate and effective analgesia (Caldwell et al., 2004). Despite recent intensification of the research and a drive towards evidence-based practice, investigations into effective analgesia in neonates remain limited in comparison to other patient populations.

The birth of a child is an intensely emotional and stressful event for any parent, and even more so when the child is born prematurely or is acutely unwell. It is therefore essential to ensure that research is conducted only when informed parental consent is given and that parental wishes are fully respected (Golec et al., 2004). Exposing an infant to unnecessary clinical pain violates the fundamental medical principle of non-maleficence; this is why trials that aim to investigate analgesic efficacy typically rely on measuring the effect of interventions on clinically essential procedures. Experimental procedures that distress an infant should not be conducted purely for the purpose of research. However, some research studies require extra procedures to be performed or involve the administration of a placebo agent when there is no standard treatment for comparison and, understandably, some parents may find these risks unacceptable, regardless of the potential benefits for other infants in the future.

Research that seeks to establish the best methods of assessing and relieving pain in neonates is actively being undertaken. The greatest limitation at present is the lack of a sensitive and specific universal tool for pain assessment, without which it remains difficult to design clinical trials of analgesic therapies. Simultaneous recording of evoked changes in behaviour, physiology and brain activity will likely provide the best approach to understanding neonates' experience of pain and provide new insights into how best to assess and treat pain in this vulnerable patient population.

References

Abdulkader, H. M., Freer, Y., Garry, E. M., Fleetwood-Walker, S. M. and McIntosh, N. 2008. Prematurity and neonatal noxious events exert lasting effects on infant pain behaviour. *Early Human Development* 84 (6): 351–5.

Anand, K. J. and Hickey, P. R. 1987. Pain and its effects in the human neonate and fetus. *New England Journal of Medicine* 317 (21): 1321–9.

Anand, K. J. and the International Evidence-Based Group for Neonatal Pain. 2001. Consensus statement for the prevention and management of pain in the newborn. *Archives of Pediatrics and Adolescent Medicine* 155 (2): 173–80.

Anand, K. J., Sippell, W. G. and Aynsley-Green, A. 1987. Randomised trial of fentanyl anaesthesia in preterm babies undergoing surgery: Effects on the stress response. *Lancet* 1 (8527): 243–8.

Anand, K. J., Hall, R. W., Desai, N., Shephard B., Berggvist L. L., Young, T. E., Boyle, E. M., Carbajal, R., Bhutani, V. K., Moore, M. B., et al. 2004. Effects of morphine analgesia in ventilated preterm neonates: Primary outcomes from the NEOPAIN randomised trial. *Lancet* 363: 1673–82.

Bartocci, M., Bergqvist, L. L., Lagercrantz, H. and Anand, K. J. 2006. Pain activates cortical areas in the preterm newborn brain. *Pain* 122 (1–2): 109–17.

Bellu, R., de Waal, K. A. and Zanini, R. 2008. Opioids for neonates receiving mechanical ventilation. *Cochrane Database of Systematic Reviews* 1, CD004212. doi: 10.1002/14651858.CD004212.pub3.

Brummelte, S., Grunau, R. E., Chau, V., Poskitt, K. J., Brant, R. and Vinall, J. 2012. Procedural pain and brain development in premature newborns. *Annals of Neurology* 71 (3): 385–96.

Buskila, D., Neumann, L., Zmora, E., Feldman, M., Bolotin, A. and Press, J. 2003. Pain sensitivity in prematurely born adolescents. *Archives of Pediatrics and Adolescent Medicine* 157 (11): 1079–82.

Caldwell, P. H. Y., Murphy, S. B., Butow, P. N. and Craig, J. C. 2004. Clinical trials in children. *Lancet* 364 (9436): 803–11.

Carbajal, R. 1999. Analgesia with sugar nipples in the newborn [in French]. *Soins Pédiatrie/Puériculture* 187: 22–3.

Carbajal, R., Chauvet, X., Couderc, S., Olivier-Martin, M. 1999. Randomised trial of analgesic effects of sucrose, glucose, and pacifiers in term neonates. *British Medical Journal* 319 (7222): 1393–7.

Carbajal, R., Veerapen, S., Couderc, S., Jugie, M. and Ville, Y. 2003. Analgesic effect of breast feeding in term neonates: Randomised controlled trial. *British Medical Journal* 326 (7379): 1–5.

Carbajal, R., Rousset, A., Danan, C., Coquery, S., Nolent, P., Ducrocq, S., Saizou, C., Lapillonne, A., Granier, M., Durand, P. et al. 2008. Epidemiology and treatment of painful procedures in neonates in intensive care units. *Journal of the American Medical Association* 300 (1): 60–70.

Ceelie, I., de Wildt, S. N., van Dijk, M., van den Berg, M. M. J., van den Bosch, G. E., Duivenvoorden, H. J., de Leeuw, T. G., Mathôt, R., Knibbe, C. A. J. and Tibboel, D. 2013. Effect of intravenous paracetamol on postoperative morphine requirements in neonates and infants undergoing major noncardiac surgery: A randomized controlled trial. *Journal of the American Medical Association* 309 (2): 149–54.

Craig, K. D., Whitfield, M. F., Grunau, R. V., Linton, J. and Hadjistavropoulos, H. D. 1993. Pain in the preterm neonate: Behavioural and physiological indices. *Pain* 52 (3): 287–99.

Doesburg, S. M., Moiseev, A., Herdman, A. T., Ribary, U. and Grunau, R. E. 2013. Region-specific slowing of alpha oscillations is associated with visual–perceptual abilities in children born very preterm. *Frontiers in Human Neuroscience* 7 (791). doi: 10.3389/fnhum.2013.00791.

Dyke, M.P., Kohan, R. and Evans, S. 1995. Morphine increases synchronous ventilation in preterm infants. *Journal of Paediatrics and Child Health* 31 (3): 176–9.

Fabrizi, L., Slater, R., Worley, A., Meek, J., Boyd, S., Olhede, S. and Fitzgerald, M. 2011. A shift in sensory processing that enables the developing human brain to discriminate touch from pain. *Current Biology* 21 (18): 1552–8.

Fitzgerald, M. and Walker, S. M. 2009. Infant pain management: A developmental neurobiological approach. *Nature Clinical Practice Neurology* 5 (1): 35–50.

Goksan, S., Hartley, C., Emery, F., Cockrill, N., Poorun R., Moultrie, F., Rogers, R., Campbell, J., Sanders, M., Adams, E., et al. 2015. fMRI reveals neural activity overlap between adult and infant pain. *eLife* 4: e06356.

Golec, L., Gibbins, S., Dunn, M. S. and Hebert, P. 2004. Informed consent in the NICU setting: An ethically optimal model for research solicitation. *Journal of Perinatology* 24 (12): 783–91.

Gray, L., Watt, L. and Blass, E. M. 2000. Skin-to-skin contact is analgesic in healthy newborns. *Pediatrics* 105 (1): e14.

Gray, L., Miller, L. W., Philipp, B. L. and Blass, E. M. 2002. Breastfeeding is analgesic in healthy newborns. *Pediatrics* 109 (4): 590–3.

Grunau, R. E., 2013. Neonatal pain in very preterm infants: Long-term effects on brain, neurodevelopment and pain reactivity. *Rambam Maimonides Medical Journal* 4 (4): e0025.

Grunau, R. V., Johnston, C. C. and Craig, K. D. 1990. Neonatal facial and cry responses to invasive and non-invasive procedures. *Pain* 42 (3): 295–305.

Grunau, R. V., Whitfield, M. F. and Petrie, J. H. 1994a. Pain sensitivity and temperament in extremely low-birth-weight premature toddlers and preterm and full-term controls. *Pain* 58 (3): 341–6.

Grunau, R. V., Whitfield, M. F., Petrie, J. H. and Fryer, E. L. 1994b. Early pain experience, child and family factors, as precursors of somatization: A prospective study of extremely premature and fullterm children. *Pain* 56 (3): 353–9.

Guinsburg, R., Kopelman, B. I., Anand, K. J., de Almeida, M. F., Peres, C. de A. and Miyoshi, M. H. 1998. Physiological, hormonal, and behavioral responses to a single fentanyl dose in intubated and ventilated preterm neonates. *Journal of Pediatrics* 132 (6): 954–9.

Hall, R. W., Kronsberg, S. S., Barton, B. A., Kaiser, J. R., Anand, K. J. and NEOPAIN Trial Investigators Group. 2005. Morphine, hypotension, and adverse outcomes in preterm neonates: who's to blame? *Pediatrics* 115 (5): 1351–9.

Hartley, C. and Slater, R. 2014. Neurophysiological measures of nociceptive brain activity in the newborn infant: The next steps. *Acta Paediatrica* 103 (3): 238–42.

Hermann, C., Hohmeister, J., Demirakca, S., Zohsel, K. and Flor, H. 2006. Long-term alteration of pain sensitivity in school-aged children with early pain experiences. *Pain* 125 (3): 278–85.

Hohmeister, J., Demirakea, S., Zohsel, K., Flor, H. and Hermann, C. 2009. Responses to pain in school-aged children with experience in a neonatal intensive care unit: Cognitive aspects and maternal influences. *European Journal of Pain* 13 (1): 94–101.

Holsti, L., Grunau, R. E. and Shany, E. 2011. Assessing pain in preterm infants in the neonatal intensive care unit: Moving to a 'brain-oriented' approach. *Pain Management* 1 (2): 171–9.

Johnston, C. C., Stevens, B., Pinelli, J., Gibbins, S., Filion, F., Jack, A., Steele, S., Boyer, K. and Veilleux, A. 2003. Kangaroo care is effective in diminishing pain response in preterm neonates. *Archives of Pediatrics and Adolescent Medicine* 157 (11): 1084–8.

Johnston, C. C., Filion, F., Campbell-Yeo, M., Goulet, C., Bell, L., McNaughton, K., Byron, J., Aita, M., Finley, G. A. and Walker, C.-D. 2008. Kangaroo mother care diminishes pain from heel lance in very preterm neonates: A crossover trial. *BMC Pediatrics* 8 (13). doi: 10.1186/1471–2431–8–13.

Johnston, C. C., Stevens, B., Pinelli, J., Gibbins, S., Filion, F., Jack, A., Steele, S., Boyer, K., Veilleux, A. and Platt, R. 1999. Factors explaining lack of response to heel stick in preterm newborns. *Journal of Obstetric, Gynecologic, & Neonatal Nursing* 28 (6): 587–94.

Lee, G. Y. and Stevens, B. J. 2013. Neonatal and infant pain assessment. In *Oxford Textbook of Paediatric Pain*, ed. by P. J. McGrath, B. J. Stevens, S. M. Walker and W. T. Zempsky. Oxford: Oxford University Press, pp. 353–69.

Maxwell, L. G., Malavolta, C. P. and Fraga, M. V. 2013. Assessment of pain in the neonate. *Clinics in Perinatology* 40 (3): 457–69.

McKechnie, L. and Levene, M. 2008. Procedural pain guidelines for the newborn in the United Kingdom. *Journal of Perinatology* 28 (2): 107–11.

Merskey, H. and Bogduk, N., eds. 2014. Pain terms: A current list with definitions and notes on usage. In *Classification of Chronic Pain*, 2nd rev. edn. IASP Task Force on Taxonomy, 209–13. http://www.iasp-pain.org/PublicationsNews/Content.aspx?ItemNumber=1673 (accessed November 3, 2015).

Miller, S. P. and Ferriero, D. M. 2009. From selective vulnerability to connectivity: Insights from newborn brain imaging. *Trends in Neuroscience* 32 (9): 496–505.

Oberlander, T. F., Grunau, R. E., Whitfield, M. F., Fitzgerald, C. E., Pitfield, S. and Saul, P. 2000. Biobehavioral pain responses in former extremely low birth weight infants at four months' corrected age. *Pediatrics* 105 (1): e6.

Pacifici, G. M. and Allegaert, K. 2015. Clinical pharmacology of paracetamol in neonates: A review. *Current Therapeutic Research* 77: 24–40.

Parry, S. 2008. Acute pain in the neonate. *Anaesthesia and Intensive Care Medicine* 9 (4): 147–51.

Peters, J. W., Schouw, R., Anand, K. J., van Dijk, M., Duivenvoorden, H. J. and Tibboel, D. 2005. Does neonatal surgery lead to increased pain sensitivity in later childhood? *Pain* 114 (3): 444–54.

Pillai Riddell, R. R., Racine, N. M., Turcotte, K., Uman, L. S., Horton, R. E., Din Osmun, L., Ahola Kohut, S., Hillgrove Stuart, J., Stevens, B. and Gerwitz-Stern, A. 2011. Non-pharmacological management of infant and young child procedural pain. *Cochrane Database of Systematic Reviews* 10: CD006275.

Raeside, L. 2011. Physiological measures of assessing infant pain: A literature review. *British Journal of Nursing* 20 (21): 1370–6.

Ranger, M., Johnston, C. C. and Anand, K. J. 2007. Current controversies regarding pain assessment in neonates. *Seminars in Perinatology* 31 (5): 283–8.

Ranger, M., Chau, C., Garg, A., Woodward, T. S., Beg M. F., Bjornson B., Poskitt, K., Fitzpartick, K., Synnes, A. R., Miller S. P., et al. 2013. Neonatal pain-related stress predicts cortical thickness at age 7 years in children born very preterm. *PLoS One* 8 (10): e76702.

Roofthooft, D. W., Simons, S. H., Anand, K. J., Tibboel, D. and van Dijk, M. 2014. Eight years later, are we still hurting newborn infants? *Neonatology* 105 (3): 218–26.

Schmeling, D. J. and Coran, A. G. 1991. Hormonal and metabolic response to operative stress in the neonate. *Journal of Parenteral and Enteral Nutrition* 15 (2): 215–38.

Skogsdal, Y., Eriksson, M. and Schollin, J. 1997. Analgesia in newborns given oral glucose. *Acta Paediatrica* 86 (2): 217–20.

Slater, R., Cantarella, A., Franck, L., Meek, J. and Fitzgerald, M. 2008. How well do clinical pain assessment tools reflect pain in infants? *PLoS Medicine* 5 (6): e129. doi:10.1371/journal.pmed.0050129.

Slater, R., Fabrizi, L., Worley, A., Meek, J., Boyd, S. and Fitzgerald, M. 2010. Premature infants display increased noxious-evoked neuronal activity in the brain compared to healthy age-matched term-born infants. *Neuroimage* 52 (2): 583–9.

Slater, R., Cantarella, A., Gallella, S., Worley, A., Boyd, S., Meek, J. and Fitzgerald, M. 2006. Cortical pain responses in human infants. *Journal of Neuroscience* 26 (14): 3662–6.

Slater, R., Cantarella, A., Yoxen, J., Patten, D., Potts, H., Meek, J. and Fitzgerald, M. 2009. Latency to facial expression change following noxious stimulation in infants is dependent on postmenstrual age. *Pain* 2009. 146 (1–2): 177–82.

Slater, R., Cornelissen, L., Fabrizi, L., Patten, D., Yoxen, J., Worley, A., Boyd, S., Meek, J. and Fitzgerald M. 2010. Oral sucrose as an analgesic drug for procedural pain in newborn infants: A randomised controlled trial. *Lancet* 376 (9748): 1225–32.

Slater, R., Worley, A., Fabrizi, L., Roberts, S., Meek, J., Boyd, S. and Fitzgerald, M. 2010. Evoked potentials generated by noxious stimulation in the human infant brain. *European Journal of Pain* 14 (3): 321–6.

Smyser, C. D., Inder, T. E., Shimony, J. S., Hill, J. E., Degnan, A. J., Snyder, A. Z. and Neil, J. J. 2010. Longitudinal analysis of neural network development in preterm infants. *Cerebral Cortex* 20 (12): 2852–62.

Stevens, B., Johnston, C., Petryshen, P. and Taddio, A. 1996. Premature infant pain profile: Development and initial validation. *Clinical Journal of Pain* 12 (1): 13–22.

Stevens, B., Yamada, J., Lee, G. Y. and Ohlsson, A. 2013. Sucrose for analgesia in newborn infants undergoing painful procedures. *Cochrane Database of Systematic Reviews* 1. doi: 10.1002/14651858.CD001069.pub4.

Stevens, B., Gibbins, S., Yamada, J., Dionne, K., Lee, G., Johnston, C. and Taddio, A. 2014. The premature infant pain profile-revised (PIPP-R): Initial validation and feasibility. *Journal of Clinical Pain* 30 (3): 238–43.

Taddio, A., Shah, V. and Katz, J. 2009. Influence of repeated painful procedures and sucrose analgesia on the development of hyperalgesia in newborn infants. *Pain* 144 (1–2): 43–8.

Taddio, A., Goldbach, M., Ipp, M., Stevens, B. and Koren, G. 1995. Effect of neonatal circumcision on pain responses during vaccination in boys. *Lancet* 345 (8945): 291–2.

van den Anker, J. N. 2013. Treating pain in newborn infants: Navigating between Scylla and Charybdis. *Journal of Pediatrics* 163 (3): 618–9.

Vinall, J., Miller, S. P., Bjornson, B. H., Fitzpatrick, K. P. V., Poskitt, J. J., Brant, R., Synnes, A. R., Cepeda, I. L. and Grunau, R. E. 2014. Invasive procedures in preterm children: Brain and cognitive development at school age. *Pediatrics* 133 (3): 412–21.

Vinall, J., Miller S. P., Chau, V., Brummelte, S., Synnes, A. R. and Grunau, R. E. 2012. Neonatal pain in relation to postnatal growth in infants born very preterm. *Pain* 153 (7): 1374–81.

Walker, S. M., Frank, L. S., Fitzgerald, M., Myles, J., Stocks, J. and Marlow, N. 2009. Long-term impact of neonatal intensive care and surgery on somatosensory perception in children born extremely preterm. *Pain* 141(1–2): 79–87.

Wilkinson, D. J., Savulescu, J. and Slater, R. 2012. Sugaring the pill: Ethics and uncertainties in the use of sucrose for newborn infants. *Archives of Pediatrics and Adolescent Medicine* 166 (7): 629–33.

CHAPTER 12

How do people with autism spectrum disorders (ASD) experience pain?

Cecile Rattaz, Amandine Dubois and Amaria Baghdadli

Introduction

Since its original description by Leo Kanner in 1943, autism has been defined as a severe and life-long neurodevelopmental disorder, characterised by impairments in socialisation and communication and by a pattern of repetitive behaviours and interests (World Health Organization, 1993). Autism is a heterogeneous disorder according to its clinical presentation, aetiology, and degree of severity. It is associated with cognitive impairment in 50 per cent to 70 per cent of cases (Fombonne, 2003; Matson and Shoemaker, 2009). Because of this heterogeneity, the name 'autism spectrum disorders' (ASD) is now being commonly used. Current estimates are that autism and other ASDs occur in one out of 150 children, which is substantially more than was previously recognised (Fombonne, 2003, Baird et al., 2006).

People with autism and developmental disorders inevitably experience pain in their everyday life, just as other people do; but, for several reasons, they may also be confronted with painful situations on a more frequent basis (Dubois, Rattaz, et al., 2010). First, aberrant or challenging behaviour, such as aggression, self-injury, stereotyped movements and extreme tantrums are frequent and can be highly persistent in people with ASD (Emerson, 1995; Matson and Nebel-Schwalm, 2007; Baghdadli et al., 2008). When untreated, challenging behaviours may have deleterious effects on people's health: they can cause accidents, injuries, chronic pain, and so on. Lee et al. (2008) reported that children with autism, as well as children with other developmental disabilities, were about 2–3 times more likely to experience an injury that needs medical attention than unaffected control children. Breau and colleagues (Breau, Camfield, McGrath, et al., 2003; Breau, Camfield, Symons, et al., 2003) also showed that accidental injury was frequent in children with severe cognitive impairments. Secondly, medical comorbidities are common in this population. Epilepsy is associated with ASD in around 25 per cent of cases, and the risk is increased in people with lower IQ (Mouridsen et al., 2008; Amiet et al., 2008). Genetic syndromes such as tuberous sclerosis, fragile X, Down syndrome, or neurofibromatosis are largely more frequent in people with ASD than in the general population and require specific health care (for a review, see Zafeiriou, Verveiri and

An Introduction to Pain and its Relation to Nervous System Disorders, First Edition. Edited by Anna A. Battaglia.
© 2016 John Wiley & Sons, Ltd. Published 2016 by John Wiley & Sons, Ltd.

Vargiami, 2006). It has also been shown that gastro-intestinal problems such as constipation, diarrhea, or food selectivity were widespread in this population (Gillberg and Coleman, 1996; Coury, 2010; Molloy and Manning-Courtney, 2003). In a survey of children's health, respiratory, food and skin allergies were reported more frequently in parents of children with autism, who had higher mean physician visits over 12 months and were far more likely than control children to receive a long term medication (Gurney, McPheeters and Davis, 2006). Thirdly, people with ASD often are in poor health because of life conditions (medical treatment, prolonged inactivity) and lack of hygiene and prevention (for example dental care). According to Gurney and colleagues (2006), on average, parents report children with autism to be in fair or poor health far more frequently than do other parents. Studies about mortality in autism also show elevated death rates in this population; these rates are due to seizures, accidents and physical illnesses (Shavelle, Strauss and Pickett, 2001; Mouridsen et al., 2008).

To summarise, studies show that people with ASD are likely to experience pain in a large variety of contexts. They also suggest that children with the poorest abilities experienced the highest number of medical problems and, consequently, the greatest pain. For these reasons the assessment and treatment of pain in people with ASD is a major issue.

Pain and autism: Review of the literature

Little knowledge is available about pain in people with ASD. For a long time, such people have been described as less sensitive or insensitive to pain. The same belief existed a few decades ago regarding pain in newborns (Anand and Hickey, 1987) and pain in people with severe intellectual disabilities (Biersdorff, 1991: 1994). (To read more about pain in neonates and infants, see Chapter 11 in this volume.)

Several case studies report difficulties to detect the presence of pain in people with developmental disabilities. In some cases, individuals may not label their experience as 'pain' for days or weeks, during which they continue their usual routine. Through third-party reporting, Biersdorff (1994) described several cases of pain insensitivity or indifference in people with developmental disabilities where this condition manifested itself through an absence of reflexive withdrawal from a hot surface, a delay in the reported expression of pain or discomfort in the case of severe illnesses, diagnoses of painful conditions made on the basis of symptoms other than pain, or paradoxical use of pain behaviours. Those examples suggest that pain expression may be different in people with developmental disabilities, who display in some cases delayed, reduced or paradoxical reactions; but thus does not mean that pain behaviours are absent.

In people with ASD, unusual patterns of pain reactions are also described in parents' or clinicians' reports – for example behaviour modifications, or an expression of discomfort without localisation of the source of pain (Tordjman et al., 1999). The absence of nociceptive reflex or analgesic posture is also described in some individuals with ASD. For these reasons parents often experience great difficulties in finding out when their child is in pain and in identifying the source of pain. Case studies suggest that challenging behaviours are often related to unidentified and untreated pain in this population (Delaitre, 2012). These studies show that, in a functional approach (O'Neill et al., 1996), a careful and precise observation of the person's behaviour may help to identify the presence of pain, and, subsequently, to propose adapted treatments that can help reducing challenging behaviours.

The first experimental studies on the expression of pain in children with autism suggested that behavioural pain reactions were reduced (Tordjman et al., 1999; Gilbert-MacLeod et al., 2000). However, the authors noted the presence of self-injurious behaviours immediately after a painful experience, which indicated that children reacted to pain. Moreover, the results showed an increase in physiological reactions (heart rate and plasma β-endorphin level) during venepuncture (Tordjman et al., 1999, 2009). The authors interpreted these results as indicating a dissociation between observable reactions and physiological responses; and they argued for a different mode of pain expression rather than an insensitivity to pain. In a parental interview, parents of children with ASD described a lower pain reactivity than that found in controls (Militerni et al., 2000), but the authors noted that results must be considered with careful attention because only three categories were used (normal, low, very low pain reactivity).

Nader et al. (2004) compared pain reactivity during venepuncture in a group of 21 children with autism and in a non-impaired comparison group comprising 22 children. They showed that children with autism displayed a significant facial reaction to venepuncture, and that facial reactions in them were even stronger during needle injection than in the comparison group. It must be noted that both procedures differed (children with autism were bundled, while children in the control group were not) and that the purpose of venepuncture was different in the two groups (injection for children with autism, blood collection for the control group), and this calls for caution when considering differences between the groups.

In a recent study (Rattaz et al., 2013), we observed reactions during venepuncture in children with ASD and we compared them to the reactions of children with an intellectual disability and to the reactions of non-impaired control children. The three groups were matched on the basis of developmental age, as measured on the Vineland Adaptive Behavior Scale (Sparrow, Balla and Cicchetti, 1984). All children had a venepuncture for the purpose of a routine blood collection, and we ensured that the procedure was similar in the three groups. Results showed that children with ASD reacted to venepuncture by an increase in facial expressions and heart rate, as did children with an intellectual disability and non-impaired control children. As regards behavioural reactions, which were measured with the Non-communicating Children Pain Checklist (NCCPC) (Breau, Finley, et al., 2002; Breau, McGrath, et al., 2002), the pattern of reactions over time was slightly different among groups. Children with ASD tended to display more behavioural reactions than children in the two other groups, and their behavioural reactions remained high at the end of venepuncture. These results counter the idea of pain insensitivity in autism and suggest that the stress induced by venepuncture and, probably, by the whole context (hospital, nurses, lack of understanding, sensory stimuli, etc.) has a greater impact on children with ASD than on children with an intellectual disability or on control children.

Those results suggest that children with ASD display significant pain reactions in an acute medical situation. However, children's reactions during venepuncture cannot be generalised and made to encompass pain reactivity in other contexts. Children's reactions may be similar in an acute pain situation, but they may differ in prolonged or chronic pain, as was found in schizophrenic patients (Potvin et al., 2008; Bonnot et al., 2009; Lévesque et al., 2012).

Overall, the idea of global pain insensitivity in autism is no longer supported by clinical studies, even if an idiosyncratic expression of pain is noted in several cases. Studies rather argue for a different mode in pain expression; this would be due to several factors that will now be reviewed.

Main theories about pain in autism

Pain is a global phenomenon with neurochemical, sensory, affective, behavioural and cognitive components. Because of abnormalities in neurotransmitters, sensory processing disorders, and cognitive deficits, people with ASD may experience pain in a different way from ordinary people.

Neurochemical hypothesis

First, studies showed abnormalities in a wide array of neurotransmitters in ASD. One of the first neurochemical hypothesis to account for the symptoms associated with autistic disorder was related to dysfunction of the endogenous opioid system (Panksepp, 1979). The opioid theory was based on studies about reported therapeutic effects of opiate antagonists and reported elevated ß-endorphin levels in individuals with ASD (Bouvard et al., 1995; Cazzullo et al., 1999; Tordjman et al., 2009; Leboyer et al., 1992; Campbell et al., 1993; Gillberg, 1995). According to this theory, hyperfunction of the endogenous opioid system could explain some symptoms such as reduced socialisation, affective lability, stereotyped behaviours, self-injurious behaviours, and a reduced sensitivity to pain. However, the results of studies on ß-endorphin levels are inconsistent, and the large number of studies that have measured plasma or central ß-endorphin levels in autism didn't reach a consensus. Some studies showed, on the contrary, equal or lower levels of ß-endorphin in children with ASD (Ernst et al., 1993; Hunter et al., 2003; Nagamitsu et al., 1997; Sandman et al., 1991; Gillberg et al., 1990; Dettmer et al., 2007). Recently Tordjman et al. (2009) observed elevated levels of plasma ß-endorphin that were not associated with behavioural pain reactivity. They argue that, when measured in plasma, ß-endorphin should be considered a stress hormone and not an indicator of central opioid functioning. Moreover, the nature of the relationship between pain and self-injurious behaviours is not clear. Symons and Danov (2005) noted that pain may be the cause, but also the consequence of self-injurious behaviours, and Breau, Camfield, Symons, et al. (2003) suggested that chronic pain may lead to an increase in self-injurious behaviours in children with severe cognitive impairments. Consequently, self-injurious behaviours per se cannot be considered an indicator of hyperfunction of the endogenous opioid system. We can hypothesise that, on the contrary, untreated and chronic pain may result in self-injurious behaviours, which have a calming effect through opioid discharge. Overall, there is little support for the notion of analgesia due to a primary dysfunction of the endogenous opioids in autism (Lam, Aman and Arnold, 2006; Dubois, Rattaz, et al., 2010).

Apart from the opioid theory, studies showed a potential role of many neurotransmitters in modulating pain sensitivity. Serotonin has stimulated the greatest share of the neurochemical investigations carried out in autism, showing in particular a peripheral hyperserotonemia and a depressed central serotonergic functioning. More recently studies have been conducted on the cholinergic system, on oxytocin, and on amino acid neurotransmitters (for a review, see Lam et al., 2006 and Lévesque, Gaumond and Marchand, 2011), which may also be implicated in pain perception or modulation.

A dysfunction in inhibitory or excitatory endogenous modulation systems, which have been shown to be related to chronic pain conditions such as fibromyalgia (Julien et al., 2005), may also be involved. Studies on inhibitory systems have been led in patients with schizophrenia. The first results revealed no changes in diffuse inhibitory controls, but a lack of sensitisation (Potvin et al., 2008). They also evidenced a specific pain response profile in schizophrenic patients, with an elevated sensitivity to acute pain but a reduced sensitivity to prolonged pain (Lévesque et al., 2012). There are no such studies with ASD patients yet.

Overall, a wide array of neurochemical systems have been studied; it is still difficult to draw conclusions because of the heterogeneity of ASD and because of methodological limits. Further research is needed in this area – including studies on modulation systems, which are lacking in the field of autism.

Sensory processing disorders

Researchers reported that, in addition to impairments in social interaction and communication and to the restricted, repetitive and stereotyped pattern of behaviour, children with ASD respond to sensory stimuli differently from peers without disability (Tomchek and Dunn, 2007; Rogers, Hepburn and Wehner, 2003; Ben-Sasson et al., 2009; Wiggins et al., 2009). According to Baranek et al. (2006), the prevalence of sensory processing disorders is much more important in children with ASD and has a much higher rate (almost 70 per cent) than in children with intellectual disability (almost 40 per cent). Leekam et al. (2007) reported that over 90 per cent of children with ASD had sensory abnormalities, whereas two thirds of the children in the comparison group where affected by sensory symptoms. In a literature review, Rogers and Ozonoff (2005) report that sensory processing disorders are more frequent and more important in children with ASD than in non-impaired control children, but that they are also present in children with other developmental disabilities. Disturbances of sensory modulation are sometimes viewed as primary symptoms of autism, and in these authors' view poor social and communicative skills are consequences of poor modulation of the sensory input (Ornitz, 1989; Ayres, 1979). An enhanced perceptual functioning was also evidenced in the visual and auditory domains in people with ASD (Mottron and Burack, 2001; O'Riordan and Passetti, 2006). Sensory processing disorders are evidenced in parental reports through questionnaires, in experimental studies on physiological and behavioural reactions, or in the self-reports of high-functioning individuals with ASD. As pain and tactile reactivity partly share the same receptors and circuits, we will focus here on studies about tactile responsiveness in people with ASD.

As in other sensory modalities, tactile processing problems can be described as a high or low response to tactile stimuli, through either avoidance or seeking of such stimuli. The term 'tactile defensiveness' was used to describe avoidance of being touched, apparent discomfort in wearing certain clothes, or resistance to hair brushing and washing (Baranek, Foster and Berkson, 1997).

In autobiographical narratives, people with ASD reported lack of responsiveness or, at the opposite end, strong reactions to tactile input (Elwin et al., 2012). Grandin and Scariano (2005) reported an overresponsiveness to tactile stimuli; they noted for instance that some clothing textures could be experienced as painful and make them extremely anxious. Williams (1999) experienced touch and physical closeness as overwhelming. As regards proprioceptive and interoceptive stimuli, as well as pain, hyposensitivity is more commonly described in autobiographical accounts. People with Asperger syndrome or with high-functioning autism report indistinct registration of internal stimuli such as hunger or thirst, difficulties to identify and interpret stimuli, and a high pain threshold (Elwin et al., 2012; Gerland, 1997). Craving for specific stimulation is also described, for instance a need for deep pressure through a 'squeeze machine' (Grandin and Scariano, 2005).

Most studies about sensory processing in ASD were conducted through parental reports, using the Sensory Profile questionnaire (Dunn and Westman, 1997). Aberrant tactile processing is among the most commonly reported sensory symptoms in ASD (Rogers et al., 2003; Tomchek and Dunn, 2007). Tomchek and Dunn (2007) reported tactile sensitivity symptoms among 60 per cent of children with ASD in their sample (N = 171), the most frequently reported difficulty being to tolerate grooming and hygiene tasks. Kern et al. (2006) also described global sensory abnormalities in children and adults with ASD, including tactile processing. Wiggins et al. (2009) found that children with ASD had more tactile sensitivity than children with other developmental disorders.

In recent years a wide range of experimental studies using direct observation and psychophysics also evidenced specific responses to tactile stimuli in people with ASD. Pernon, Pry and Baghdadli (2007) found more intense facial reactions to a warm stimulus in children with ASD than in a comparison group. They also observed paradoxical reactions in these children, such as smiling when touching an unpleasant stimulus (roughness). These observations suggest that some aspects of tactile sensitivity may be differently perceived in people with ASD. In a psychophysical tactile study assessing thresholds and sensitivity through vibrotactile stimuli, Blakemore et al. (2006) showed lower tactile perceptual thresholds for high-frequency (200 Hz), but not for low-frequency (30 Hz) stimuli in adults with ASD. Tactile hypersensitivity in response to vibrotactile stimuli and thermal stimuli, but not to light touch, was also reported in adults with autism (Cascio et al., 2008). In a review on neurophysiologic findings, Marco et al. (2011) reported a high prevalence of atypical tactile behaviours in children with ASD. The same authors later showed that children with autism have diminished primary somatosensory cortical responses (Marco et al., 2012). Cascio et al. (2012) studied perceptual and neural responses to pleasant and unpleasant touch and found that the ASD group gave pleasant and rough textures more extreme average ratings than did controls. For the most unpleasant texture, the ASD group exhibited greater brain blood oxygenation level-dependent (BOLD) response than controls in affective somatosensory processing areas.

It is important to note that, from a developmental point of view, touch sensitivity does not appear to be solely dependent on age or cognitive level. Crane, Goddard and Pring (2009) showed that abnormalities in sensory processing were also found in adults with ASD. Congruently with autobiographical reports, parental reports and experimental studies show that symptoms such as discomfort with gentle touch can persist into adulthood (Blakemore et al., 2006; Leekam et al., 2007). Kern et al. (2006) evidenced a low threshold for touch that did not improve over age, as opposite to visual or auditory processing.

This field of research, which focuses on sensory processing disorders, is quite recent in the ASD domain and further studies need to be done – in particular studies using psychophysics. The first results evidence abnormalities in tactile processing, and we can hypothesise that pain perception could also be altered at some level in people with ASD. There is at the moment no psychophysical study on pain perception in people with ASD.

Cognitive development

The cognitive component of pain is involved at different levels and implies the ability to detect pain, understand its underlying cause, express one's emotions, communicate pain to others through verbal and non-verbal behaviours, and self-regulate through cognitive mechanisms (interpretation, attention orienting, distraction, etc.) or affective sharing (comfort seeking, social referencing, etc.).

Pain expression is likely to change due to maturation, life experience, and the development of sociocommunicative skills (Lilley, Craig and Grunau, 1997). Many studies have shown that pain expression changes during the first two years of age in normally developing children (Lewis and Thomas, 1990; Maikler, 1991; Craig et al., 1984). Later in development, the growing cognitive and verbal abilities allow the child to express pain in a more precise way. Under 24 months of age, children express their emotions mainly through wail and cry, with some words from 18 months ('ouch', 'boo-boo'). The verbal expression of pain appears at 24 months, and from 4–5 years children use more precise verbalisations (Dubois et al., 2008; Stanford, Chambers and Craig, 2006). Emotional control, which begins between the ages of three and six years (Saarni, 1999) also plays a role in pain expression, enabling children to express their feelings and inhibit some of their reactions (Craig, 1999). Then, from eight years of age, children develop the ability to grade internal states and to differentiate between the components of pain, such as intensity and unpleasantness (Goodenough et al., 1999).

Studies about pain expression in people with intellectual disability showed a relationship between pain expression and the level of cognitive development. Children with mild to moderate cognitive impairment seem to be able to communicate their pain verbally, whereas children with severe to profound cognitive impairment are described as having more 'indirect' pain reactions, less easily recognisable by the parents (Fanurik, 1999; Voepel-Lewis, Malviya and Tait, 2005). According to Hadden and von Baeyer (2002), parents of children with cerebral palsy reported that their children were able to express pain through a variety of behaviours; however, non-verbal children displayed certain idiosyncratic behaviours – such as laughing, grunting or self-injuring – that were not reported in verbal children. Dubois, Capdevila, et al. (2010) also showed the influence of the level of expressive communication on pain expression in children with an intellectual disability. In their study, children who had verbal abilities exhibited normative pain reactions, whereas non-verbal children produced particular pain reactions, with a low facial, vocal and motor reactivity and self-stimulating behaviours. These results suggest that pain expression is affected by the level of cognitive development, even if other factors – such as age and life experience – are also involved in pain expression (Zabalia, 2006).

For what concerns people with ASD, impairments in cognitive development may affect pain reactions at different levels. First, sociocommunicative disorders have an impact on pain expression; this impact is manifest in difficulties to use adequate verbal and non-verbal behaviours to communicate pain. Then, because of deficits in information processing, people with ASD may have difficulties to identify the source of pain and to react in an adequate way; and such difficulties are accompanied by a lack of protective or escape behaviours (Tordjman et al., 1999). Impairments in attention orienting may be involved (Lévesque et al., 2011), as pain processing is influenced by attention (Van Damme, Crombez and Eccleston, 2008; Villemure and Bushnell, 2002). Finally, deficits in the socioemotional domain have been largely studied in autism; these include difficulties regarding the ability to express and understand emotions, emotional control, and social referencing (Hobson, 1986; Kasari et al., 1990).

There are very few studies on developmental aspects of pain expression in people with ASD. In an exploratory study about reactions to an unpleasant stimulus, Pernon and Rattaz (2003), found that children with ASD did not decrease their reactions to a noxious stimulus from three years of age on, as control children did. Recently we also found no significant difference across developmental age in the reactions of children with ASD during venepuncture, whereas children in the control group displayed fewer facial and behavioural reactions from 30 months of age on (Rattaz et al., 2013). We hypothesised that, in an unpleasant or painful situation, typically developing children decreased their reactions due to the development of emotional control; this is not observed in children with ASD, who retain the same developmental level. These children have specific deficits regarding understanding and responding to social stimuli, and we can hypothesise that they have difficulties in inhibiting their behavioural reactions in this situation. These results are exploratory and further research needs to be done on the impact of developmental age and on the evolution of children's reactions over age.

How to assess pain in people with communicative and social deficits

The assessment of pain is a complex issue, which requires specific skills from the observer (empathy, objectivity, reliability), but also from the patient (ability to communicate, to understand questions, to share his/hers feelings and sensations, to express social signals). When the patient is able to communicate verbally and to quantify pain intensity, the assessment is quite easy to make and reliable. In people with ASD, who have difficulties in the social and communicative domains, pain assessment can be particularly challenging.

The International Association for the Study of Pain (IASP) stated that the inability to communicate verbally does not negate the possibility that an individual is experiencing pain and is in the need for an appropriate pain-relieving treatment (Merskey and Bogduk, 1994). This assertion concerns newborns, non-verbal young children, the elderly, and people with cognitive deficits, including ASD (for a detailed overview of pain in dementia, see Chapter 14 in this volume; for the role of cognitive impairment in the placebo and nocebo effects, see Chapter 15 in this volume). Difficulty in assessing pain has been cited as one of the primary reasons for infrequent and inadequate assessment and analgesia in people with cognitive impairments (Malviya et al., 2005), and many professionals don't use assessment tools to measure pain intensity and treatment efficacy (Kankkunen, Jänis and Vehviläinen-Julkunen, 2010; Breau, 2010).

Using validated tools to evaluate pain is therefore crucial – first, from a clinical point of view, because it allows to identify the presence of a painful sensation and to treat pain as a function of its nature and intensity. Then, since pain is defined as a subjective sensory and emotional experience, it is important to use assessment tools in order to have as unbiased an information as possible. Finally, for people who have communicative deficits, observing and measuring their behaviours through a specific scale is often the only way to identify the presence of pain, to assess its type, duration and intensity and to propose an appropriate treatment.

Self-report

The first method of evaluating pain is self-report through different tools: the Visual Analogue Scale (VAS) (Huskisson, 1974), the Faces Pain Scale – Revised (FPS-R) (Hicks et al., 2001), the block ordering or Poker Chip tool (Hester, Foster and Kristensen, 1990), the Eland Colour Scale (Eland, 1985). The ability to use such tools was studied in children with developmental disabilities. Zabalia, Jacquet and Breau (2005) conducted a study about children's abilities to evaluate pain in others (on pictures). They showed that children with low to moderate cognitive impairment were able to use a visual analogue scale (VAS) or the Faces Pain Scale and that they could describe verbally the type of pain (e.g., burning, pricking) in a comparable manner to control children of the same mental age. In children with low to moderate cognitive impairment, self-report can be helpful and reliable in some cases. In a study on 47 children with cognitive impairment hospitalised for surgery, Fanurik et al. (1998) showed that only 21 per cent (the ones with a low cognitive impairment) could use theses scales (block ordering tool, Faces Pain Scale). Benini et al. (2004) showed that children and adolescents with developmental delay could use a VAS during venipuncture, but that some tools had to be adapted. They proposed a simplified version of the Eland Colour Scale (larger and adapted to better distinguish the different parts of the body), of the block-ordering test (four cubes in sizes varying from 1 to 5 cm), and the Faces Pain Scale (four instead of nine facial expressions with more exaggerated somatic traits). These adapted tools proved to be easier to use in the self-assessment of pain in children with low to mild developmental delay. In people with severe or profound cognitive impairment, self-assessment was shown to be much more difficult or impossible (Defrin, Lotan and Pick, 2006; LaChapelle, Hadjistavropoulos and Craig, 1999). As regards the Faces Pain Scale, for example, Defrin et al. (2006) showed that adults with profound cognitive disability don't understand the meaning of the interval between two faces. Hadjistavropoulos et al. (1998) showed that an adapted VAS that provides added cues such as colour (McGrath et al., 1996) was useful with some elders who had mild to moderate dementia but not with seniors who had more severe cognitive impairments.

As Hadjistavropoulos and Craig (2002) pointed out, the usefulness of different assessment tools depends largely on the degree of developmental delay, and some people with modest cognitive impairment can provide self-report. Concerning people with ASD, there is only one study on self-report of pain at the time (Bandstra et al., 2012). In this study high-functioning children and adolescents with ASD were asked to rate, using the Faces Pain Scale – Revised and a numeric pain-rating scale, the amount of pain they would expect to feel in a series of hypothetical pain situations depicted in images (e.g., scraping knee on sidewalk). Results showed that children and adolescents with ASD were able to successfully use both of the self-report scales to rate pain. This study was done with high-functioning individuals, in other words individuals without cognitive impairment, and we can hypothesise that the self-assessment of pain is much more difficult for low-functioning individuals with ASD.

The self-assessment of pain has several cognitive prerequisites, for example numerical reasoning, language skills, general cognitive abilities (Stanford et al., 2006), which are impaired in a number of individuals with ASD. Individuals with ASD have difficulties in the spheres of self-recognition (Carmody and Lewis, 2012), symbolic development (Lam and Yeung, 2012) and emotional understanding (Hobson, 1986). They also have particularities in visual processing and tend to pay more attention to details than to the global form (Caron et al., 2006; Pelphrey et al., 2002). Those deficits may interfere with the ability to use self-assessment scales such as the Faces Pain Scale, the Eland Colour Scale, a VAS, and so on. As for individuals with developmental delay, these scales need to be adapted to the person's social and cognitive level. Another important point is to use tools or scales adapted to the person's communicative abilities. Visual communication strategies such as the picture exchange communication system (PECS) (Bondy and Frost, 1994) are commonly used with people with ASD. Such tools can be used to assess pain and help individuals with ASD to communicate pain to others. Finally, in the Bandstra et al. (2012) study, pain reports were made on images and not during a real painful event. We believe that the self-report of pain during a painful event is much more difficult and that in a number of cases a prior learning is essential – for example being able to use the pain assessment scales or tools before the painful event occurs.

Proxy report

Apart from self-report, hetero-assessment tools can be used to identify the presence of pain. Herr et al. (2006) recommend the parents' presence and participation when the individual is not able to communicate pain verbally. There is no pain assessment tool specifically designed for people with ASD, but some scales were validated for children with severe cognitive impairment (Symons, Shinde and Gilles, 2008): the Non-communicating Children's Pain Checklist (NCCPC) (Breau, Finley, et al., 2002; Breau, McGrath, et al., 2002), the Face Legs Activity Cry and Consolability – Revised (FLACC-R) (Malviya et al., 2006), the Pain Indicator for Communicatively Impaired Children (PICIC) (Stallard et al., 2002), the Douleur Enfant San Salvadour (DESS) (Collignon and Giusiano, 2001), the Checklist Pain Behaviour (CPB) (Terstegen et al., 2003; Duivenvoorden et al., 2006). In the validation studies, some of these scales (NCCPC and FLACC-R) have included a number of children with ASD within their population. The NCCPC scale (Breau, Finley, et al., 2002; Breau, McGrath, et al., 2002) was designed to evaluate postoperative and everyday pain in children with severe cognitive impairments. It is composed of 27 items divided into seven categories: vocal, social, facial, activity, body, limbs and physiological signs. The everyday pain version contains three more items, linked to eating and sleeping. The French validation study (Zabalia et al., 2011) included 15 per cent of children with autism. The FLACC-R Scale is a scale validated for postoperative pain in children with mild, moderate, severe or profound intellectual disability. Each of the five categories (face, legs, activity, cry and consolability) is scored from 0 to 2, and the revised form adds behaviours such as head banging, jerking, outbursts, grunting, breath holding, resisting care or comfort measures and so on. The validation study included 16 per cent of children with ASD, and this scale shows good pragmatic qualities and clinical utilities for clinicians (Voepel-Lewis, Malvyia and Tait, 2008). These two scales can be used to assess pain in children with autism, but one must keep in mind that they are not ASD-specific and may for this reason lack reliability and sensitivity.

Changes in behaviours can also be identified through functional assessment (O'Neill et al., 1996). This method, using an observational grid, aims at assessing and treating challenging behaviours and can be useful in the assessment of pain. In some cases it allows us to identify untreated sources of pain that can be the cause of challenging behaviours and to evidence links between painful reactions and self-injurious behaviours (Symons and Danov, 2005; Symons et al., 2009; Bosch et al., 1997). It must be noted that functional assessment can only be used for one person in a specific context and is particularly relevant in chronic pain and when health problems haven't been correctly treated.

Another important issue regarding hetero-assessment is the lack of concordance that can exist between two different observers. In typically developing children, studies showed the absence of inter-rater concordance when evaluating pain (Duignan and Dunn, 2008; Knutsson, Tibbelin and von Unge, 2006; Rajasagaram et al., 2009). As regards children with ASD, Nader et al. (2004) showed that there was a lack of concordance between parental reports of pain, made with the Faces Pain Scale, and the child's facial activity, as coded by a trained observer (CFCS) (Chambers et al., 1996). They also found an absence of relationship between the retrospective parental report of pain through the NCCPC scale (Breau, Finley, et al., 2002; Breau, McGrath, et al., 2002) and the behaviours observed during venipuncture. The children who were most reactive during venipuncture were those who had been identified by their parents as less sensitive and reactive to pain. The authors hypothesised that pain expression varied according to the environment (home vs hospital) and the type of painful stimulus (everyday pain vs acute pain linked to medical situations). In our study on pain in children with autism, we asked parents and nurses to rate the child's reactions during venipuncture from 0 (no pain) to 10 (extreme pain), on a VAS. Our results showed a lack of concordance between parents and nurses but a better concordance between both nurses (results are unpublished yet). Those results indicate that pain assessment is subjective and varies according to one's position as parent, nurse, observer, and so on.

In summary, pain assessment in people with ASD is a very complex and challenging issue. Self-reports of pain are limited on account of social and cognitive deficits or particularities, and proxy reports are almost as challenging as a result of the difficulty to assess pain in this population, the number of variables that may interfere (painful context, the observer's position, etc.) and the absence of specific and validated tools.

How to manage and treat pain in this population

A central prerequisite of pain management is the possibility to identify the presence of a painful event and to describe the type of pain in terms of location, intensity, duration and so on. The difficulties and challenges of pain assessment in people with ASD, described above, could partially explain why pain management is often insufficient in this population. However, there are several ways to alleviate pain in people with ASD, even if they are not routinely used and some adaptations need to be made.

As regards analgesic administration or sedation, some studies showed that children with cognitive impairments received fewer analgesics than children without cognitive impairments. Malviya et al. (2005) demonstrated a considerable discrepancy in pain management practices between children with and children without cognitive impairment. In their study, children with cognitive impairment undergoing spine fusion surgery received smaller opioid doses than children without cognitive impairment undergoing the same surgery. Koh et al. (2004) also found that children with cognitive impairment undergoing surgery received less opioid in the perioperative period than children without cognitive impairment; however, the amount and the type of analgesics were comparable across both groups in the postoperative period. On the other hand, sedation or general anaesthesia is sometimes used in people with developmental disabilities during medical exams that are not painful but are likely to provoke challenging behaviours, for example the magnetic resonance imaging (MRI). Overall, it may be safer to sedate some children, while others can cooperate if properly prepared; and the decision to use sedation or general anaesthesia must be evaluated in each case (Johnson and Rodriguez, 2013).

In our study on pain expression in children with ASD we observed that the use of a local anaesthetic (EMLA cream) or sedation through analgesic gas (N_2O/O_2 nitrous oxide/oxygen) during venipuncture was almost systematic in non-impaired control children but was proposed in less than half of children with ASD or intellectual disability (Rattaz et al., 2013). This suggests that health professionals may be less preoccupied by pain alleviation in this population, who is often considered to be less sensitive to pain. We can also hypothesise that, in a number of cases, EMLA cream and N_2O/O_2 are not appropriate for children with developmental disorders because they require the child's cooperation and understanding of the situation, which are often difficult to obtain for different reasons (challenging behaviours, sensory processing disorders, communication deficits, etc.). In the absence of prior learning, the use of a local anaesthetic or analgesic gas could have an adverse effect, increasing the person's anxiety during a painful situation. The use of anaesthetics must be adapted to the child with ASD, and for this purpose visual aids and training are of great interest (Cuvo et al., 2010). For example, describing venepuncture step by step, in the format of a social story, and including in this description the use of the local anaesthetic helps to improve social understanding and, consequently, to reduce anxiety (Thorne, 2007).

Non-pharmacological pain treatment, in particular behavioural techniques such as distracting, video modelling, or shaping, may be helpful for decreasing distress and improving compliance to medical procedures (Souders et al., 2002). For example, Shabani and Fisher (2006) used behavioural techniques (stimulus fading and differential reinforcement) to teach tolerance to blood drawing when checking glucose levels in a young adult with autism, mental retardation and diabetes. As regards sensory processing disorders, desensitisation techniques can help the person to tolerate an aversive stimulus. In their study, Cuvo et al. (2010) used a desensitisation procedure with a person who had difficulties tolerating the contact of the stethoscope on the chest. The procedure consisted in placing the stethoscope near the body and then on the chest for increasing durations of time. Another important point is to allow the person to know what to expect, in order to reduce anxiety and prevent challenging behaviours. Adapted communication methods can be used (for example, augmentative communication devices such as PECS; see Bondy and Frost, 1994) during and, if possible, a few days before the event, in order to help prepare the child.

Finally, the adaptation of care and the environment are very useful – especially in a hospital environment, where sensory overload could be overwhelming. Individuals with ASD can be disturbed by noises (medical equipment beeping to alert staff, doors opening, people screaming), by visual stimuli (flashing light, movements) and by tactile stimuli (textures, touch related to health care). A sensory-adapted environment can help to reduce anxiety and improve compliance during stress-provoking medical situations. For example, Shapiro et al. (2009) showed that children with developmental disability were more relaxed during dental care (behavioural and physiological measures) in a sensory-adapted environment (with adapted lighting, auditory and somatosensory stimuli). Nordahl et al. (2008) developed a protocol for acquiring MRI scans during natural sleep in children with autism (storybook, interview with parents, familiarisation visit with a simulated MRI scanner, auditory habituation kit, etc.). These adaptations allowed the authors to complete scanning without sedation in children with a wide range of autism severity and levels of functioning. Children with ASD also benefit from having a reliable daily routine, which is difficult to maintain in a hospital. However, there are general activities (meals, bedtimes) that the multidisciplinary team should strive to maintain (Scarpinato et al., 2010; Nordahl et al., 2008).

Finally, a very important point is to be able to assess the person's unique needs, and Johnson and Rodriguez (2013) noted the importance of partnering with parents in order to develop strategies that prevent challenging behaviours when a child with ASD is hospitalised. In order to guide care, Scarpinato et al. (2010) propose taking into account the patient's developmental level, the somatosensory disorders, the emotional disturbances (easily frustrated, moodiness, easily overstimulated) and the effective communication techniques, and to ask the parents or the caregivers what interventions have worked for this patient in the past.

Pain and autism: An integrated model

As described above, pain is a complex phenomenon and many factors are implicated in its expression. Hadjistavropoulos and Craig (2002) proposed to conceptualise pain from the perspective of a human communication model. In this model, the experience of an internal state (A) is encoded in an expressive behaviour (B), allowing the observer to draw inferences (C) about the nature of the sender's experience (Figure 12.1). Applied to pain in individuals with

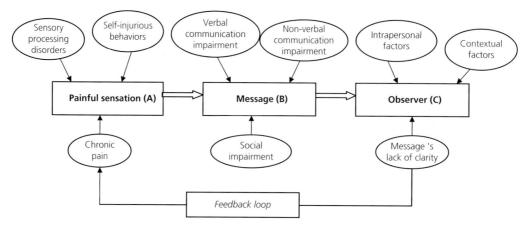

Figure 12.1 Pain expression in people with ASD: A communication model of pain. Source: Adapted from Hadjistavropoulos 2002.

autism, this model allows describing the different factors that may intervene in pain expression, assessment and treatment.

First, the painful sensation (A) can be altered as a result of sensory processing disorders (see above), but also because repeated and untreated pain has led to habituation and to a consequent decrease in reactions. This type of phenomenon has been described in the literature on premature newborns, for example by Johnston and Stevens (1996). Challenging behaviours such as self-injurious ones may also provoke pain or, through the endorphin mechanism, may decrease the pain perception threshold. Next, the message (B) expressed by the individual with ASD may lack clarity because of a sociocommunicative deficit, which affects verbal (ability to understand questions, to express and describe pain verbally) and non-verbal communication (behaviours such as pointing, facial expression, etc.). Social impairments also have an impact on pain expression in the form of difficulties regarding the expression of emotions, social referencing, comfort seeking, and so on. Finally, with respect to the observer (C), multiple factors may interfere with pain assessment. Some of them are intrapersonal or contextual, for example a lack of knowledge regarding pain expressions in people with disabilities, the observer's subjectivity as a function of his/her position (parent, nurse, etc.), or preconceived ideas about pain insensitivity in this population. The message's lack of clarity may also be involved: in some cases the relationship between the person's behaviour and the painful stimulus is hard to understand (for example, head banging caused by dental pain). A last element has to be integrated to the model: the feedback loop between observer (C) and sensation (A). If pain assessment is not correctly done because of several reasons linked to (A), (B) or (C) levels, then pain won't be treated and can become chronic, which can in turn affect the initial sensation (A).

This integrated model describes possible dysfunctions at the sensory, behavioural and cognitive levels; it allows for a better understanding of the difficulties encountered in pain assessment and treatment. In order to improve current practices, interventions should aim at the individuals' communicational abilities, the professionals' knowledge, the assessment tools, pain management techniques and environmental adaptations.

Conclusion

The belief in a reduced sensitivity to pain in people with ASD is no longer supported in the literature. However, people with ASD have deficits or disorders that could affect the neuro-chemical, sensory, affective, behavioural and cognitive components of pain. Moreover, there is a huge variability among persons, as a function of their profile and of the type of pain experienced (e.g., everyday pain vs pain related to medical care, acute vs chronic pain). For those reasons, in a number of cases pain assessment is very difficult for parents and professionals.

Overall, there is an urgent need to improve pain assessment and management in people with ASD and to develop tools and techniques adapted to each person's profile – as regards sociocommunicative abilities, intellectual level, sensory processing disorders and so on. More generally, there is a need to improve health care for persons with special needs, whose health-care associated conditions are still too often insufficiently diagnosed and treated.

References

Amiet, C., Gourfinkel-An, I., Bouzamondo, A., Tordjman, S., Baulac, M. Lechat, P., Mottron, L. and Cohen, D. 2008. Epilepsy in autism is associated with intellectual disability and gender: Evidence from a meta-analysis. *Biological Psychiatry* 64: 577–82.

Anand, K. and Hickey, P. 1987. Pain and its effect in the human neonate and fetus. *New England Journal of Medicine* 317: 1321–9.

Ayres, J. 1979. *Sensory Integration and the Child*. Los Angeles, CA: Western Psychological Services.

Baghdadli A., Picot M.-C., Pry R., Michelon, C., Bursztejn, C., Lazartigues, A. and Aussilloux, C. 2008. What factors are related to a negative outcome of self-injurious behaviour during childhood in pervasive developmental disorders? *Journal of Applied Research in Intellectual Disabilities* 21: 142–9.

Baird, G., Simonoff, E., Pickles, A., Chandler, S., Loucas, T., Meldrum, D. and Charman, T. 2006. Prevalence of disorders of the autism spectrum in a population cohort of children in South Thames: The special needs and autism project (SNAP). *The Lancet* 368 (9531), 210–15.

Bandstra, N., Johnson, S., Filliter, J. and Chambers, C. 2012. Self-reported and parent-reported pain for common painful events in high-functioning children and adolescents with autism spectrum disorder. *Clinical Journal of Pain* 28 (8): 715–21.

Baranek, G., Foster, L. and Berkson, H. 1997. Tactile defensiveness and stereotyped behaviors. *American Journal of Occupational Therapy* 51 (2): 91–5.

Baranek, G. T., David, F. J., Poe, M. D., Stone, W. L. and Watson, L. R. 2006. Sensory experiences questionnaire: Discriminating sensory features in young children with autism, developmental delays, and typical development. *Journal of Child Psychology and Psychiatry* 47: 591–601.

Benini, F., Trapanotto, M., Gobber, D., Agosto, C., Carli, G., Drigo, P., Eland, J. and Zacchello, F. 2004. Evaluating pain induced by venipuncture in pediatric patients with developmental delay. *Clinical Journal of Pain* 20 (3): 156–63.

Ben-Sasson, A., Hen, L., Fluss, R., Cermak, S., Engel-Yeger, B. and Gal, E. 2009. A meta-analysis of sensory modulation symptoms in individuals with autism spectrum disorders. *Journal of Autism and Developmental Disorders* 39: 1–11.

Biersdorff, K. 1991. Pain insensitivity and indifference: Alternative explanations for some medical catastrophes. *Mental Retardation* 29: 359–62.

Biersdorff, K. 1994. Incidence of significantly altered pain experience among individuals with developmental disabilities. *American Journal on Mental Retardation* 98 (5): 619–31.

Blakemore, S. J., Tavassoli, T., Calo, S., Thomas, R., Catmur, C., Frith, U. and Haggard, P. 2006. Tactile sensitivity in Asperger syndrome. *Brain and Cognition* 61: 5–13.

Bondy, A. and Frost, L. 1994. The Picture Exchange Communication System (PECS). *Focus on Autism and Other Developmental Disabilities* 9 (3): 1–19.

Bonnot, O., Anderson, G. M., Cohen, D., Willer, J. C. and Tordjman, S. 2009. Are patients with schizophrenia insensitive to pain? A reconsideration of the question. *The Clinical Journal of Pain* 25 (3): 244–52.

Bosch J., Van Dyke D. C., Smith S. M. and Poulton, S. 1997. Role of medical conditions in the exacerbation of self-injurious behavior: An exploratory study. *Mental Retardation* 35: 124–30.

Bouvard, M., Leboyer, M., Launay, J. M., Recasens, C., Plumet, M. H., Waller-Perotte, D., Tabuteau, F., Bondoux, D., Dugas, M., Lensing, P., et al. 1995. Low-dose naltrexone effects on plasma chemistries and clinical symptoms in autism: A double-blind, placebo-controlled study. *Psychiatry Research* 58: 191–201.

Breau, L. M. 2010. The science of pain measurement and the frustration of clinical pain assessment: Does an individualized numerical rating scale bridge the gap for children with intellectual disabilities? *Pain* 150 (2): 213–14.

Breau, L. M., Finley, G. A., McGrath, P. J. and Camfield, C. S. 2002a. Validation of the non-communicating children's pain checklist: Psostoperative version. *Anesthesiology* 96: 528–35.

Breau, L. M., McGrath, P. J., Camfield, C. S. and Finley, G. A. 2002b. Psychometric properties of the non-communicating children's pain checklist-revised. *Pain* 99 (1–2): 349–57.

Breau, L. M., Camfield, C. S., McGrath, P. J. and Finley G. A. 2003a. The incidence of pain in children with severe cognitive impairments. *Archives of Pediatrics & Adolescent Medicine* 157: 1219–26.

Breau, L. M., Camfield, C. S., Symons, F., Bodfish, J., MacKay, A., Finley, A. and McGrath, P. 2003b. Relation between pain and self-injurious behavior in non-verbal children with severe cognitive impairments. *The Journal of Pediatrics* 142: 498–503.

Campbell, M., Anderson, L. T., Small, A., Adams, P., Gonzalez, N. and Ersnt, M. 1993. Naltrexone in autistic children: Behavioral symptoms and attentional learning. *Journal of the American Academy of Child and Adolescent Psychiatry* 32: 1283–91.

Carmody, D. P. and Lewis, M. 2012. Self-representation in children with and without autism spectrum disorders. *Child Psychiatry and Human Development* 43 (2): 227–37.

Caron, M. J., Mottron, L., Berthiaume, C. and Dawson, M. 2006. Cognitive mechanisms, specificity and neural underpinnings of visuospatial peaks in autism. *Brain* 129 (7): 1789–802.

Cascio, C., McGlone, F., Folger, S., Tannan, V., Baranek, G., Pelphrey, K. and Essick, G. 2008. Tactile perception in adults with autism: A multidimensional psychophysical study. *Journal of Autism and Developmental Disorders* 38: 127–37.

Cascio, C., Moana-Filho, E., Guest, S., Nebel, M. B., Weisner, J., Baranek, G. and Essick, G. 2012. Perceptual and neural response to affective tactile texture stimulation in adults with autism spectrum disorders. *Autism Research* 5: 231–44.

Cazzullo, A. G., Musetti, M. C., Musetti, L., Bajo, S., Sacerdote, P. and Panerai, A. 1999. Beta-endorphin levels in peripheral blood mononuclear cells and long-term naltrexone treatment in autistic children. *European College of Neuropsychopharmacology* 9: 361–6.

Chambers, C., Cassidy, K., McGrath, P., Gilbert, C. and Craig, K. 1996. *Child Facial Action Coding System Revised Manual*. Halifax: Dalhousie University.

Collignon, P. and Giusiano, B. 2001. Validation of a pain evaluation scale for patients with severe cerebral palsy. *European Journal of Pain* 5 (4): 433–42.

Coury, D. 2010. Medical treatment of autism spectrum disorders. *Current Opinion in Neurology* 23: 131–6.

Craig, K. 1999. Emotional aspects of pain. In *Textbook of Pain*, ed. by P. Wall and D. Melzack. London: Churchill Livingstone, pp. 261–74.

Craig, K. D., McMahon, R., Morison, J. and Zaskow, C. 1984. Developmental changes in infant pain expression during immunization injections. *Social Science & Medicine* 19 (12): 1331–7.

Crane, L., Goddard, L. and Pring, L. 2009. Sensory processing in adults with autism spectrum disorders. *Autism* 13: 215–28.

Cuvo, A., Law Reagan, A., Ackerlund, J., Huckfeldt, R. and Kelly, C. 2010. Training children with autism to be compliant with a physical exam. *Research in Autism Spectrum Disorders* 4: 168–85.

Defrin, R., Lotan, M. and Pick, C. G. 2006. The evaluation of acute pain in individuals with cognitive impairment: A differential effect of the level of impairment. *Pain* 124 (3): 312–20.

Delaitre, L. 2012. Dépistage de la douleur chez une personne avec autisme. *Perspectives Psy* 51 (2): 152–61.

Dettmer, K., Hanna, D., Whetstone, P., Hansen, R. and Hammock, B. 2007. Autism and urinary exogenous neuropeptides: Development of an on-line SPE-HPLC-tandem mass spectrometry method to test the opioid excess theory. *Analytic and Bioanalytical Chemistry* 388: 1643–51.

Dubois, A., Bringuier, S., Capdevila, X. and Pry, R. 2008. Vocal and verbal expression of postoperative pain in preschoolers. *Pain Management Nursing* 8 (4): 160–5.

Dubois, A., Capdevila, X., Bringuier, S. and Pry, R. 2010a. Pain in children with an intellectual disability. *European Journal of Pain* 14 (6): 654–60.

Dubois, A., Rattaz, C., Pry, R. and Baghdadli, A. 2010b. Autisme et douleur: Analyse bibliographique. *Pain Research and Management* 15 (4): 245–53.

Duignan, M. and Dunn, V. 2008. Congruence of pain assessment between nurses and emergency department patients: A replication. *International Emergency Nursing* 16 (1): 23–8.

Duivenvoorden, H., Tibboel, D., Koot, H., van Dijk, M. and Peters, J. W. 2006. Pain assessment in profound cognitive impaired children using the Checklist Pain Behavior; is item reduction valid? *Pain* 126 (1–3): 147–54.

Dunn, W. and Westman, K. 1997. The sensory profile: The performance of a national sample of children without disabilities. *American Journal of Occupational Therapy* 51: 25–34.

Eland, J. 1985. The child who is hurting. *Seminary in Oncology Nursing* 1: 116–22.

Elwin, M., Ek, L., Schröder, E. and Kjellin, L. 2012. Autobiographical accounts of sensing in Asperger syndrome and high-functioning autism. *Archives of Psychiatric Nursing* 26 (5): 420–9.

Emerson, E. 1995. *Challenging Behaviour: Analysis and Intervention in People with Learning Disabilities*. Cambridge: Cambridge University Press.

Ernst, M., Devi, L., Silva, R. R., Gonzalez, N., Small, A., Malone, R. and Campbell, N. 1993. Plasma betaendorphin levels, naltrexone, and haloperidol in autistic children. *Psychopharmacology Bulletin* 29: 221–7.

Fanurik, D., Koh, J. L., Schmitz, M. L., Harrison, D. and Conrad, T. 1999. Children with cognitive impairment: Parent report of pain and coping. *Journal of Developmental and Behavioral Pediatrics* 20: 228–34.

Fombonne, E. 2003. Epidemiological surveys of autism and other pervasive developmental disorders: An update. *Journal of Autism and Developmental Disorders* 33: 365–82.

Gerland, G. 1997. *A Real Person: Life on the Outside*, trans. J. Tate. London: Souvenir Press.

Gillberg C. 1995. Endogeneous opioids and opiate antagonists in autism: Brief review of empirical findings and implications for clinicians. *Developmental Medicine and Child Neurology* 37: 239–45.

Gillberg, C. and Coleman, M. 1996. Autism and medical disorders: A review of the literature. *Developmental Medicine and Child Neurology* 38: 191–202.

Gillberg, C., Hagberg, B., Witt-Engerstöm, I. and Eriksson, I. 1990. CSF beta-endorphins in childhood neuropsychiatric disorders. *Brain Development* 12: 88–92.

Gilbert-MacLeod, C. A., Craig, K. D., Rocha, E. M. and Mathias, M. D. 2000. Everyday pain responses in children with and without developmental delays. *Journal of Pediatric Psychology* 25: 301–8.

Goodenough, B., Thomas, W., Champion, G., Perrott, D., Taplin, J., von Baeyer, C. and Ziegler, J. 1999. Unravelling age effects and sex differences in needle pain: Ratings of sensory intensity and unpleasantness of venipuncture pain by children and their parents. *Pain* 80: 179–90.

Grandin, T. and Scariano, M. 2005. *Emergence: Labeled Autistic*, 2nd edn. New York: Grand Central.

Gurney, J., McPheeters, M. and Davis, M. 2006. Parental report of health conditions and health care use among children with and without autism. *Archives of Pediatric and Adolescent Medicine* 160 (8): 825–30.

Hadden, K. L. and von Baeyer, C. L. 2002. Pain in children with cerebral palsy: Common triggers and expressive behaviors. *Pain* 99: 281–8.

Hadjistavropoulos, T. and Craig, K. D. 2002. A theorical framework for understanding self-report and observational measures of pain: A communication model. *Behaviour Research and Therapy* 40: 551–70.

Hadjistavropoulos, T., LaChapelle, D. L., MacLeod, F., Hale, C., O'Rourke, N. and Craig, K. 1998. Cognitive functioning and pain reactions in hospitalized frail elders. *Pain Research and Management* 3: 145–51.

Herr, K., Coyne, P. J., Key, T., Manworren, R., McCaffery, M., Merkel, S., Pelosi-Kelly, J. and Wild, L. 2006. Pain assessment in the nonverbal patient: Position statement with clinical practice recommendations. *Pain Management Nursing* 7: 44–52.

Hester, N., Foster, R. and Kristensen, K. 1990. Measurement of pain in children: Generalizability and validity of the Pain Ladder and the Poker Chip Tool. In *Pediatric Pain*, ed. by D. Tyler and E. Krane, special issue of *Advances in Pain Research and Therapy* 15. New York: Raven Press, pp. 79–84.

Hicks, C., von Baeyer, C., Spafford, P., van Korlaar, I. and Goodenough, B. 2001. The Faces Pain Scale – Revised: Toward a common metric in pediatric pain measurement. *Pain* 93: 173–83.

Hobson, R. P. 1986. The autistic child's appraisal of expressions of emotion. *Journal of Child Psychology and Psychiatry* 27 (3): 321–42.

Hunter, L. C., O'Hare, A., Herron, W. J., Fisher, L. and Jones, G. 2003. Opioid peptides and dipeptidyl peptidase in autism. *Developmental Medicine and Child Neurology* 45: 121–8.

Huskisson, E. 1974. Measurement of pain. *The Lancet* 9 (2): 1127–31.

Johnson, N. and Rodriguez, D. 2013. Children with autism spectrum disorder at a pediatric hospital: A systematic review of the literature. *Pediatric Nursing* 39 (3): 131–41.

Johnston, C. and Stevens, B. 1996. Experience in a neonatal intensive care unit affects pain response. *Pediatrics* 98: 925–30.

Julien, N., Goffaux, P., Arsenault, P. and Marchand, S. 2005. Widespread pain in fibromyalgia is related to a deficit of endogenous pain inhibition. *Pain* 114: 296–302.

Kankkunen, P., Jänis, P. and Vehviläinen-Julkunen, K. 2010. Pain assessment among non-communicating intellectually disabled people described by nursing staff. *Open Journal of Nursing* 24 (4): 55–9.

Kasari, C., Sigman, M., Mundy, P. and Yirmiya, N. 1990. Affective sharing in the context of joint attention interactions of normal, autistic, and mentally retarded children. *Journal of Autism and Developmental Disorders* 20 (1): 87–100.

Kern, J., Trivedi, M., Garver, C., Grannemann, B., Andrews, A., Savla, J., Johnson, D., Mehta, J., Schroeder, J. 2006. The pattern of sensory processing abnormalities in autism. *Autism* 10 (5): 480–94.

Knutsson, J., Tibbelin, A. and von Unge, M. 2006. Postoperative pain after paediatric adenoidectomy and differences between the pain scores made by the recovery room staff, the parent and the child. *Acta Otolaryngologica* 126 (10): 1079–83.

Koh, J., Fanurik, D., Harrison, R., Schmitz, M. and Norvell, D. 2004. Analgesia following surgery with and without cognitive impairment. *Pain* 111 (3): 239–44.

LaChapelle, D. L., Hadjistavropoulos, H. D. and Craig, K. D. 1999. Pain measurement in persons with intellectual disabilities. *Clinical Journal of Pain* 15 (1): 13–23.

Lam, Y. G. and Yeung, S. S. 2012. Cognitive deficits and symbolic play in preschoolers with autism. *Research in Autism Spectrum Disorders* 6 (1): 560–4.

Lam, K. S., Aman, M. G. and Arnold, L. E. 2006. Neurochemical correlates of autistic disorder: A review of the literature. *Research in Developmental Disabilities* 27: 254–89.

Leboyer, M., Bouvard, M., Launay, J. M., Tabuteau, F., Waller, D., Dugas, M., Kerdelhue, B., Lensing, P. and Panksepp, J. 1992. Brief report: A double-blind study of naltrexone in infantile autism. *Journal of Autism and Developmental Disorders* 22: 309–19.

Lee, L., Harrington, R. A., Chang, J. J. and Connors, S. L. 2008. Increased risk of injury in children with developmental disabilities. *Research in Developmental Disabilities* 29: 247–55.

Leekam, S., Nieto, C., Libby, S., Wing, L. and Gould, J. 2007. Describing the sensory abnormalities of children and adults with autism. *Journal of Autism and Developmental Disorders* 37 (5): 894–910.

Lévesque, M., Potvin, S., Marchand, S., Stip, E., Grignon, S., Lalonde, P., Lipp, O. and Goffaux, P. 2012. Pain perception in schizophrenia: Evidence of a specific pain response profile. *Pain Medicine* 13: 1571–9.

Lévesque, M., Gaumond, I. and Marchand, S. 2011. Douleur et autisme. *Douleur et Analgésie* 24: 165–70.

Lewis, M. and Thomas, D. 1990. Cortisol release in infants in response to inoculation. *Child Development* 61: 50–9.

Lilley, C. M., Craig, K. D. and Grunau R. V. 1997. The expression of pain in infants and toddlers: Developmental changes in facial action. *Pain* 72: 161–70.

Maikler, V. E. 1991. Effects of a skin refrigerant/anesthetic and age on the pain responses of infants receiving immunizations. *Research in Nursing and Health* 14: 397–403.

Malviya, S., Voepel-Lewis, T., Merkel, S. and Tait, A. R. 2005. Difficult pain assessment and lack of clinician knowledge are ongoing barriers to effective pain management in children with cognitive impairments. *Acute Pain* 7: 27–32.

Malviya, S., Voepel-Lewis, T., Burke, C., Merkel, S. and Tait, A. 2006. The revised FLACC observational pain tool: Improved reliability and validity for pain assessment in children with cognitive impairment. *Paediatric Anaesthesia* 16 (3): 258–65.

Marco, E. J., Hinkley, L. B., Hill, S. S. and Nagarajan, S. S. 2011. Sensory processing in autism: A review of neurophysiologic findings. *Pediatric Research* 69 (5): 48–54.

Marco, E., Khatibi, K., Hill, S., Siegel, B., Arroyo, M., Dowling, A., Neuhaus, J., Sherr, E., Hinkley, L. and Nagajaran, S. 2012. Children with autism show reduced somatosensory response: An MEG Study. *Autism Research* 5: 340–51.

Matson, J., and Shoemaker, M. 2009. Intellectual disability and its relationship to autism spectrum disorders. *Research in Developmental Disabilities* 30 (6): 1107–14.

Matson, J. and Nebel-Schwalm, M. 2007. Assessing challenging behaviors in children with autism spectrum disorders: A review. *Research in Developmental Disabilities* 28: 567–79.

McGrath, P. A., Seifert, C. E., Speechley, K. N., Booth, J., Stit, L. and Gibson, M. 1996. A new analogue scale for assessing children's pain. *Pain* 64: 435–43.

Merskey, H. and Bogduk, N., eds. 1994. Pain terms: A current list with definitions and notes on usage. In *Classification of Chronic Pain*, 2nd edn. Seatntle, WA: IASP Press, pp. 209–13.

Militerni, R., Bravaccio, C., Falco, C., Puglisi-Allegra, S., Pascucci, T. and Fico, C. 2000. Pain reactivity in children with autistic disorder. *Journal of Headache and Pain* 1: 53–6.

Molloy, C. and Manning-Courtney, P. 2003. Prevalence of chronic gastrointestinal symptoms in children with autism and autistic spectrum disorders. *Autism* 7 (2): 165–71.

Mottron, L. and Burack, J. 2001. Enhanced perceptual functioning in the development of autism, in *The Development of Autism: Perspectives from Theory and Research*, ed. by J. Burack, T. Charman, N. Yirmiya and P. Zelazo. Mahwah , NJ: Erlbaum, pp. 131–48.

Mouridsen, S., Bronnum-Hansen, H., Rich, B. and Isager, T. 2008. Mortality and causes of death in autism spectrum disorders: An update. *Autism* 12 (4): 403–14.

Nader, R., Oberlander, T. F., Chambers, C. T. and Craig, K. D. 2004. Expression of pain in children with autism. *Clinical Journal of Pain* 20: 88–97.

Nagamitsu S, Matsuishi T, Kisa, T., Komori, H., Miyasaki, M., Hashimoto, T., Yamashita, Y., Ohtaki, E. and Kato, H. 1997. CSF β-endorphin levels in patients with infantile autism. *Journal of Autism and Developmental Disorders* 27: 155–63.

Nordahl, C., Simon, T., Zierhut, C., Solomon, M., Rogers, S. and Amaral, D. 2008. Brief report: Methods for acquiring MRI data in very young children with autism without the use of sedation. *Journal of Autism and Developmental Disorders* 38: 1581–90.

O'Neill, R., Albin, R., Storey, R., Horner, R. and Sprague, J. 1996. *Functional Assessment and Program Development for Problem Behavior: A Practical Handbook*, 2nd edn. Pacific Grove, CA: Brooks/Cole.

O'Riordan, M. and Passetti, F. 2006. Discrimination in autism within different sensory modalities. *Journal of Autism and Developmental Disorders* 36: 665–75.

Ornitz, E. 1989. Autism at the interface between sensory processing and information processing. In *Autism: Nature, Diagnosis and Treatment*, ed. by G. Dawson. New York: Guilford, pp. 174–207.

Panksepp, J. 1979. A neurochemical theory of autism. *Trends in Neurosciences* 2: 174–7.

Pelphrey, K. A., Sasson, N. J., Reznick, J. S., Paul, G., Goldman, B. and Piven, J. 2002. Visual scanning of faces in autism. *Journal of Autism and Developmental Disorders* 32 (4): 249–61.

Pernon, E. and Rattaz, C. 2003. Les modes d'expression de la douleur chez l'enfant autiste: Étude comparée. *Devenir* 15: 263–77.

Pernon, E., Pry, R. and Baghdadli, A. 2007. Autism: Tactile perception and emotion. *Journal of Intellectual Disability Research* 51: 580–7.

Potvin, S., Stip, E., Tempier, A. et al. 2008. Pain perception in schizophrenia: No changes in diffuse noxious inhibitory controls (DNIC) but a lack of pain sensitization. *Journal of Psychiatric Research* 42: 1010–16.

Rajasagaram, U., Taylor, D. M., Braitberg, G., Pearsell, J. and Capp, B. 2009. Paediatric pain assessment: Differences between triage nurse, child and parent. *Journal of Paediatrics and Child Health* 45 (4): 199–203.

Rattaz, C., Dubois, A., Michelon, C., Viellard, M., Poinso, F. and Baghdadli, A. 2013. How do children with autism spectrum disorders express pain? A comparison with developmentally delayed and typically developing children. *Pain* 154 (10): 2007–13.

Rogers, S. J., Hepburn, S. and Wehner, E. 2003. Parent reports of sensory symptoms in toddlers with autism and those with other developmental disorders. *Journal of Autism and Developmental Disorders* 33: 631–42.

Rogers, S. J. and Ozonoff, S. 2005. Annotation: What do we know about sensory dysfunction in autism? A critical review of the empirical evidence. *Journal of Child Psychology and Psychiatry* 46: 1255–68.

Saarni, S. 1999. *The Development of Emotional Competence*. New York: Guilford.

Sandman, C. A., Barron, J. L., Chiez-DeMet, A. and DeMet, E. M. 1991. Brief report: Plasma β-endorphin and cortisol levels in autistic patients. *Journal of Autism and Developmental Disorders* 21: 83–7.

Scarpinato, N., Bradley, J., Kurbjun, K., Bateman, X., Holtzer, B. and Ely, B. 2010. Caring for the child with an autism spectrum disorder in the acute care setting. *Journal for Specialists in Pediatric Nursing* 15 (3): 244–54.

Shabani, D. B. and Fisher, W. W. 2006. Stimulus fading and differential reinforcement for the treatment of needle phobia in a youth with autism. *Journal of Applied Behavior Analysis* 39: 449–52.

Shapiro, M., Sgan-Cohen, H., Parush, S. and Melmed, R. 2009. Influence of adapted environment on the anxiety of medically treated children with developmental disability. *The Journal of Pediatrics* 154 (4): 546–50.

Shavelle, R., Strauss, D. and Pickett, J. 2001. Causes of death in autism. *Journal of Autism and Developmental Disorders* 31 (6): 569–76.

Souders, M. C., Freeman, K. G., DePaul, D. and Levy, S. E. 2002. Caring for children and adolescents with autism who require challenging procedures. *Pediatric Nursing* 28 (6): 555–62.

Sparrow, S., Balla, D. and Cicchetti, V. 1984. *Vineland Adaptive Behavior Scales*. Circle Pines, MN: American Guidance Service.

Stallard, P., Williams, L., Velleman, R., Lenton, S., McGrath, P. J. and Taylor, G. 2002. The development and evaluation of the pain indicator for communicatively impaired children (PICIC). *Pain* 98 (1–2): 145–9.

Stanford, E. A., Chambers, C. T. and Craig, K. D. 2006. The role of developmental factors in predicting young children's use of a self-report scale for pain. *Pain* 120 (1–2): 16–23.

Symons, F. and Danov, S. 2005. A prospective clinical analysis of pain behavior and self-injurious behavior. *Pain* 117 (3): 473–7.

Symons, F., Harper, V., McGrath, P. J., Breau, L. and Bodfish, J. 2009. Evidence of increased non-verbal behavioral signs of pain in adults with neurodevelopmental disorders and chronic self-injury. *Research on Developmental Disability* 30: 521–8.

Symons, F., Shinde, S. and Gilles, E. 2008. Perspectives on pain and intellectual disability. *Journal of Intellectual Disability Research* 52 (Pt 4): 275–86.

Terstegen, C., Koot, H., De Boer, J. and Tibboel, D. 2003. Measuring pain in children with cognitive impairment: Pain response to surgical procedures. *Pain* 103 (1–2): 187–98.

Thorne, A. 2007. Are you ready to give care to a child with autism? *Nursing* 37 (5): 59–61.

Tomchek, S. D. and Dunn, W. 2007. Sensory processing in children with and without autism: A comparative study using the short sensory profile. *American Journal of Occupational Therapy* 61: 190–200.

Tordjman, S., Antoine, C., Cohen, D. J., Gauvain-Piquard, A., Carlier, M., Roubertoux, P. and Ferrari, P. 1999. Study of the relationships between self-injurious behavior and pain reactivity in infantile autism. *Encephale* 25: 122–134.

Tordjman, S., Anderson, G. M., Botbol, M., Brailly-Tabard, S., Perez-Diaz, F., Graignic, R., Carlier, M., Schmit, G., Rolland, A. C., Bonnot, O., et al. 2009. Pain reactivity and plasma β-endorphin in children and adolescents with autistic disorder. *PLoS One* 4 (8): e5289.

Van Damme, S., Crombez, G. and Ecleston, C. 2008. Coping with pain: A motivational perspective. *Pain* 139 (1): 1–4.

Villemure, C. and Bushnell, M. C. 2002. Cognitive modulation of pain: How do attention and emotion influence pain processing? *Pain* 95 (3): 195–9.

Voepel-Lewis, T., Malviya, S. and Tait, A. R. 2005. Validity of parent ratings as proxy measures of pain in children with cognitive impairment. *Pain Management Nursing* 6: 168–74.

Voepel-Lewis, T., Malviya, S., Tait, A. R., Merkel, S., Foster, R., Krane, E. and Davis, P. 2008. A comparison of the clinical utility of pain assessment tools for children with cognitive impairment. *Anesthesia and Analgesia* 106 (1): 72–8.

Williams, D. 1999. *Nobody Nowhere: The Remarkable Autobiography of an Autistic Girl* 2nd edn. London: Jessica Kingsley.

Wiggins, L. Robbins, D., Bakeman, R. and Adamson, L. 2009. Brief report: Sensory abnormalities as distinguishing symptoms of autism spectrum disorders in young children. *Journal of Autism and Developmental Disorders* 39 (7): 1087–91.

World Health Organization. 1993. *ICD-10 Classification of Mental and Behavioral Disorders Diagnostic Criteria for Research*. Geneva: World Health Organization.

Zabalia, M. 2006. Pour une psychologie de l'enfant face à la douleur. *Enfance* 1: 5–19.

Zabalia, M., Jacquet, D. and Breau, L. M. 2005. Rôle du niveau verbal sur l'expression et l'évaluation de la douleur chez des sujets déficients intellectuels. *Douleur et Analgésie* 2: 65–70.

Zabalia, M., Breau, L. M., Wood, C., Lévêque, C., Hennequin, M., Villeneuve, E., Fall, E., Vallet, L., Grégoire, M. and Breau, G. 2011. Validation francophone de la grille d'évaluation de la douleur: Version postopératoire. *Canadian Journal of Anesthesia* 58 (11): 1016–23.

Zafeiriou, D, Verveiri, A. and Vargiami, E. 2006. Childhood autism and associated comorbidities. *Brain and Development* 29 (5): 257–72.

CHAPTER 13

Pain and depression: The janus factor of human suffering

Angela Iannitelli and Paola Tirassa

Wir wissen es ja schon, daß der Zusammenhang der verwickelten seelischen Probleme uns nötigt, jede Untersuchung unvollendet abzubrechen, bis ihr die Ergebnisse einer anderen zu Hilfe kommen können.

Sigmund Freud, Trauer und Melancholie[1]

Introduction

The Cartesian dualism between the *res extensa* (material substance) and the *res cogitans* (thinking thing) has governed biological research and medical science for centuries and has determined the difference between the physical and the mental dimension of human illness. Research on pain and in psychiatry, more than in any other field, has 'suffered' the consequences of the Cartesian dualism's relegation, over time, to the body and the soul (or mind).

Almost 50 per cent of psychiatric patients report that pain, fatigue or muscle pain, headache, abdominal and backache are associated with sadness or panic as primary manifestations of anxiety and depression. Depression is indeed being increasingly recognised as a comorbid disorder in patients with severe and chronic medical conditions and pain syndromes. In this context, mood disorders represent a paradigm, but also a framework, for an integrated view that allows us to improve our knowledge and management of pain.

To underline the existence of common pathways and mechanisms mediating pain and mood, antidepressants have been shown to be effective in pain symptom management in patients with neurodegenerative diseases, diabetic neuropathy, and fibromyalgia, as well as in some of the other psychiatric disorders.

The comorbidity of pain in depressive and anxiety disorders may lower therapeutic response, worsen psychological symptoms and be a predictor of the relapse of depression.

---

[1] 'As we already know, the interdependence of the complicated problems of the mind forces us to break off every enquiry before it is completed – till the outcome of some other enquiry can come to its assistance' (Freud, 1963: 258).

An Introduction to Pain and its Relation to Nervous System Disorders, First Edition. Edited by Anna A. Battaglia.
© 2016 John Wiley & Sons, Ltd. Published 2016 by John Wiley & Sons, Ltd.

Further, pain knowledge appears to be able to modify brain activity; this capacity transforms the 'illusion of pain' into a trigger factor or an indicator of depressive state.

Beside this, physical discomfort or pain without organic diagnosis – a phenomenon referred as medically unexplained pain or non-specific pain – represents a therapeutic challenge when patients attribute their pain to causes other than the mood disorders. Actually the attempt to identify the medical cause of somatic complains increases the patient's distress and dissatisfaction for the care received and increases the disease burden.

On the basis of the relevance of bodily perception and as the result of integrated efforts between neuroscientists and clinicians, a new concept of 'mind' comes to be construed. The psychoanalytic point of view, which overcomes the separate concepts of body and mind and defines pain as a psychic state expressed through localised bodily sensation rather than as a concept in itself, offers a further interpretation of pain and its attributes. This topic will be discussed in this chapter, which presents the research findings and the clinical evidence, both historical and recent, that have contributed to disclosing the association between mood disorders and pain, with particular focus on depression.

The ache and depression: A close encounter in the past centuries

Human suffering in its physical and psychological dimensions might be seen as a paradigm in the constant search for a meeting point between the soul (*psyche*) and the body (*soma*), which is dramatically symbolised by the history of depression. (For an overview of the psychology of pain and human suffering, see also Chapter 10 in this volume.)

The epistemology, the descriptive psychopathology, and the nosography of psychiatric disorders, including depression, reflect and express the social and cultural environment in which these disorders develop. This is particularly evident when looking at how the human pain, discomfort and unhappiness have been considered by clinicians and scientists in the past centuries, but also at the different definitions and symptoms reported by the nosographic systems developed in the past seventy years.

The ache between black bile and bipolar disorder

The first term used to identify the sad mood was 'melancholia', literally 'black bile' (*melaina chole* in Greek), introduced by 'Hippocrates' – that is, ancient Greek medical authors who wrote roughly between the mid-fifth and the late forth/early third century BC – and it was largely used until the twelfth century by French, Italians and Germans to indicate a state of sadness, dejection and despondency – and often fear, anger and delusions. The mental suffering was therefore due to bodily secretions such as black bile – a conception that revealed the supremacy of the physical causes on psychological pain, for which there was no concept yet.

It is thanks to the new scientific approach advocated by René Descartes (1596–1650) that the origin of mania and melancholia – the two faces of the mood disorders – was disconnected from the supernatural and linked to the body; but the concept of 'mental illness' still remained an oxymoron – a contradiction in terms – because the soul was considered unassailable and psychic disturbances were treated as physical diseases. Indeed, following Newton's theories, Nicholas Robinson (1697–1775) suggested that the pathological relaxation of nervous fibres was the primary cause of melancholia, while George Cheyne (1671–1743) argued that behavioural and mood disorders are the result of an impairment of the digestive or nervous system. Cheyne linked melancholia and physical pain, asserting that patients manifested an organic condition marked by vasal occlusions, nodules and scirrhuses, and functional deterioration. According to Herman Boerhaave (1668–1738), melancholia was dependent on the 'evaporation' of the liquid part of the blood and the consequent condensation of the solid parts (black juices) and that this whole process was due to a combination of biological and behavioural factors. This was the modern concept of multifactorial genesis of mood disorders.

The Enlightenment and Voltaire (1694–1778) attempted to refute Descartes' metaphysics by supporting empiricism. The empiricist approach, according to which knowledge comes from sensory experience, imposed a new perspective on the interpretation of human nature and opened the way to a more materialistic concept of mental illness.

To remain in the West: it was Johann Christian Reil (1759–1813), a German physician, physiologist and anatomist working at the Halle hospital, who formulated a proto-thought that integrated mental and physical pain; today his conception is applicable to the psychosomatic. He indicated the stress arising from the urban life and bureaucracy as a factor generating *nervous fragility*. According to Reil, illness resulted from an altered relationship between the soul, the body and the environment; thus the nervousness generated by hedonism, intellectualism and social lifestyle might favour the development of hypochondria, convulsions, the Sydenham's chorea, nervous breakdowns, and madness.

Vincenzo Chiarugi (1759–1820), an Italian physician who favoured the psychiatric hospital reform, was one of the first to study the influence of the body (the nervous system) on the mind. He designated the *sensorium commune* (common sense organ), localised in the medulla oblongata, as the place were body and soul meet. In other words, Chiarugi underlined the importance of the sensory and intellectual apparatus, *the sensorium* or 'organ of sense', trying to find a psychological answer to the body–mind dualism, and he identified madness (or insanity) as an *idiopathic affection* of this apparatus.

The term 'melancholia' was still used by Chiarugi, together with 'mania' and 'amentia' (dementia), and it was only thanks to Jean-Étienne-Dominique Esquirol (1772–1840) that the mental illness becomes a disease per se, separate from dementia, discernible on its own, and therefore approachable through scientific methodology. Before Esquirol, melancholia was only a descriptive diagnosis, but he introduced the idea that emotional disturbances rather than brain damage might be the cause of mental illness and he advocated new 'moral' treatment for patients.

In the middle of the nineteenth century the term 'depression' came to replace the vaguer term 'melancholia', and mania and melancholia merged into new types of insanity. In the 1854 Jean-Pierre Falret (1794–1870), a French psychiatrist, described a condition he called *la folie circulaire* (circular insanity) – which can be considered the first documentation of a bipolar affective disorder – and suggested the importance of hereditary and epidemiological factors in the development of the disease.

It was Emil Kraepelin (1856–1926) who, at the beginning of the twentieth century, finally defined the concept of maniac depressive insanity, which he described as a 'mixture' of exaltations and depressions or as a chain of alternating episodes of mania and depression. Thus depression can be present in a clinical picture that will be later identified as 'bipolar disorder'; but it also manifests itself as a specific pathology, marked by a lowering of mood tone, abulia, apathia, anahedonia, and disturbances of appetite and of sleep. Only recently did depression acquire transnosographic significance – but always in the context of mental suffering.

Thus, to conclude, depression conquered its nosographic space throughout the centuries, becoming a pathology; but this has been done at the expense of the body and its pain, which were not represented as part of a psychiatric disturbance. Nowadays great effort is made to develop an integrated concept that should include different aspects of human suffering, from neurobiology to pharmacology, and should be helpful for diagnosis and management.

The psychoanalytical theory of ache

This chapter's choice of focus on the psychoanalytic reading of depression and pain is due to the fact that psychoanalysis offers the best lens for understanding these phenomena: it treats painful depression and the depressive attribute of pain as two sides of the same coin. The Freudian study of pain offers at least three useful ways of attempting to understand the mechanisms by which pain develops: first, the build-up of a 'complementary series' between melancholia, mania and bereavement; secondly, the psychoanalytic view on the genesis of physical and psychic pain; and, finally, the psychoanalytic view on the difference between mental pain and anguish.

According to Sigmund Freud, melancholia, mania and bereavement are three clinical entities belonging to the same dimension: they can all be inserted in the same 'complemental series',[2] having as common denominator a heavy and painful loss, but using diverse strategies to face it.

Bereavement is a normal affective reaction to the loss of an object. The grief (or sorrow) arises from the disinvestment of an object, which is imposed by the real need to detach oneself from the object that has disappeared. Bereavement is a work whose goal is to withdraw the lost object from the suffering that has been placed upon it.

It is very interesting to note that the symbolic significance of Freud's description can be better understood by considering that the German word *Trauer*, 'bereavement', also means 'sorrow', and that its French synonym, *deuil*, derives from the medieval Latin *dolus*, which denotes grief and sorrow; *dolus* was also inherited in old Italian in the form *duolo* and used to indicate both pain and grief. The words used and their meaning are universal, as well as worth expressing. Mourning, therefore grief, Freud says, is 'the great enigma', especially because 'it seems so natural that the layman does not hesitate to declare it's obvious' (see Freud, 1915: 173–6).

Like bereavement, melancholia is characterised by an agonising and depressive mood, abulia, apathy, anhedonia, loss of the ability to love. But these symptoms, present already in mourning but more serious in this picture, are associated with reduced self-esteem, which can develop delusions of guilt. Even the term 'melancholia' was, as we have seen before, widely

---

[2] According to Freud, a complemental series is related to the diverse etiological factors – endogenous and exogenous – that can be distributed along a scale with two types of factors that vary in inverse proportion, and therefore complement each other.

used to refer to a 'black' or sad 'humour' or mood, and it became 'depression' since the 1850s. In Freud's work, melancholia is the only medical condition that has a special treatment, because it may be the subject of clinical observation and because this observation can be made with the help of the concepts of narcissism and the superego. In addition, in another complementary series, melancholia lies between what is normal and healing (where 'normality' means the model of mourning that operates in life) and mania, which is a clumsy attempt at healing. 'She is not a mother, not a lover, is the pain itself, is the pure idea, made of marble, of the worldwide pain, pain that obscures sooner or later every human life', writes the famous Italian novelist Antonio Fogazzaro (Fogazzaro, 1901). The anguish complex masters mania, which is also characterised by expanded mood, hyperactivity, and excessive joy. The normal face of mania is joy, jubilation, triumph, and by this the manic denies the grief and kills the lost object, thus losing it a second time.

For a narcissistic person, bodily pain would develop excessive investment in a body part, which would become increasingly painful in order to empty the ego (Freud, 1925). While this takes place, we observe a process of concentrating on the psychic representation of that painful part of the body. The transition from physical to psychic pain corresponds to the transformation from a narcissistic investment (cathexis) to an object investment. As stated by Freud, 'The intense cathexis of longing which is concentrated on the missed or lost object (a cathexis which steadily mounts up because it cannot be appeased) creates the same economic conditions as are created by the cathexis of pain which is concentrated on the injured part of the body' (Freud, 1925: 171).

Pain is the reaction to the loss of the object, while anguish is the reaction to the danger that this loss entails.

A necessary point of view on DSM-5: A new approach to somatic symptoms?

The *Diagnostic and Statistical Manual of Mental Disorders* (DSM) is a classification of mental disorders used by clinicians and researches in order to improve on common language and communicate the essential characteristics of the illness presented by patients. The DSM is published by the American Psychiatric Association (APA) and represents the work of hundreds of international experts on all aspects of mental health. These experts collect information from medical practitioners, hospital statistics, scientific publications and so on, in order to elaborate the criteria and facilitate the identification of symptoms in diverse clinical settings: inpatient, outpatient, partial hospital, consultation–liaison, clinical practice, private practice, and primary care. The DSM code is in line with the International Statistical Classification of Diseases and Related Health Problems (ICD) produced by the World Health Organization (WHO), although some incongruities might arise from non-synchronous revisions of the two publications.

The first DSM was first published in 1952 as a variant of the sixth edition of ICD and still reflected the need of the post-Second World War era to incorporate the psychophysiological, personality, and acute disorders of war veterans, as well as Adolf Meyer's psychobiological view that mental disorders represented 'reactions' to individual, social, and biological factors. This 'mental section' of ICD6 failed to gain international acceptance, and therefore the WHO commissioned a review from the English psychiatrist Erwin Stengel. Stengel identified in the difficulty to communication and in the lack of a common nomenclature the 'serious obstacles in the progress of psychiatry', and therefore the reason for the failure of the DSM. Actually Stengel's suggestions did not find much use in DSM-II – the edition published in 1968 – which was similar to the first edition (but at least the term 'reactions' was eliminated).

In 1974 Robert Spitzer and Joseph L. Fleiss demonstrated that DSM-II was an unreliable diagnostic tool, and in the same year it was decided to initiate a new review of the manual. Spitzer was selected and appointed chairman of the task force.

DSM-III was published in 1980 and was developed with the additional goal of providing a medical nomenclature for clinicians and researchers. Actually a number of important methodological innovations, including explicit diagnostic criteria, a multiaxial system, and a descriptive approach that attempted to be neutral with respect to aetiological theories were incorporated in the new version of the manual.

Six years later, in 1994, DSM-IV was published as a result of the great and collaborative effort of more than 1,000 individuals and numerous professional organisations to create a comprehensive review of scientific literature and to establish a firm empirical basis for nomenclature, classification and diagnostic sets.

The most recent formulation of the DSM dates from 2013. The preface of the fifth edition of DSM states that 'the current diagnostic criteria are the best available description of how mental disorders are expressed and can be recognised by trained clinicians' (American Psychiatric Association, 2013: xli).

Somatic symptoms and related disorders are a new category in DSM-5 – one that includes disorders with prominently somatic symptoms; thus the number of disorders and subcategories is reduced in order to avoid problematic overlaps. Indeed somatisation disorders, hypochondriasis, pain disorder and undifferentiated somatoform disorder have been removed, along with the three coded subdiagnoses specified by DSM-IV: pain disorder associated with psychological factors, pain disorder associated with both psychological factors and a general medical condition, and pain disorder associated with a general medical condition. 'Somatisation' is the tendency to perceive normal bodily sensations as unusually intense and disturbing.

The 'disorders' presented in the relevant chapter of DSM-5 – such as somatic symptom disorder, illness anxiety disorder, conversion disorder, psychological factors affecting other medical conditions, factitious disorder, other specified and unspecified somatic symptoms and related disorders – share a common feature: 'the prominence of somatic symptoms associated with significant distress and impairment' (American Psychiatric Association, 2013: 309). This category is based on a reorganisation of DSM-IV somatoform disorder diagnoses that is intended to make them more useful to non-psychiatric clinicians – given that people with these symptoms are commonly encountered in primary-care and other medical settings and less in psychiatric and other mental health settings.

It is very interesting that the diagnosis of somatic symptom disorder – the principal diagnostic category in this class – is made 'on the basis of positive symptoms and signs (distressing somatic symptoms plus abnormal thoughts, feelings, and behaviours in response to these symptoms) rather than the absence of a medical explanation for somatic complaints' (American Psychiatric Association, 2013: 309–310). The main characteristic in many individuals with this diagnosis is the way they present and interpret the somatic symptoms. People with chronic pain would be appropriately diagnosed as having somatic symptom disorder, with predominant pain. The admission of affective, cognitive, and behavioural components among the criteria for diagnosing this disorder provide a new reflection on the true clinical picture, which can be achieved by assessing the somatic complaints alone.

The distinction between somatic symptoms and medically unexplained symptoms (MUS) – diagnosed as 'conversion disorder' in the past – is therefore overcome. DSM-5 addresses this question by asserting that the lack of a medical cause does not automatically call for a mental disorder diagnosis and, conversely, the existence of a medical diagnosis does not exclude comorbidity with a mental disorder.

DSM-5 recognises that, beside the somatic symptoms frequently associated with psychological distress and psychopathology, symptoms with obscure or unclear causes can also be spontaneously manifested. But the comorbidity of pain with depression remains a critical point, due to the definition of subsequent case history, which can become more complex and severe, and the inefficacy of conventional therapies.

Therefore a question is still open: Is the reality of pain dependent on a medical diagnosis or, better, on an organic cause? Indeed, as revealed by the change of 'conversion disorder' into functional (or psychogenic) neurological symptom disorder, the DSM-5 does not go further; it shows the need (1) to refer to a medical pathophysiological explanation in order to understand the symptomatology; and (2) to be inside a strong epistemic context. Indeed, the DSM-5 puts somatic disturbances and pain into a multifactorial context in which genetic and biological vulnerability (e.g., increased sensitivity to pain) is taken into account. The contribution of early traumatic experiences, learning, and the environmental cultural–social context to the development of pain symptomatology is, however, still missing.

There is a further distance to be covered between the pain perceived by patients and the nosographic classification. Moreover, there is a prejudice to be overcome, in psychiatry as well as in medical science: pain is not an accessory to human life.

Chronic pain and depression

Depression is often comorbid in case histories characterised by pain; but the nature of the relationship between these two pathological components is not clear. Why does pain became chronic in certain psychopathological conditions? Is pain a trigger factor for depression, or is it its consequence? What are the crucial variables in this relationship, and how do they influence the insurgence of depression or the intensification of pain?

We will try to answer these questions in the following paragraphs; but the first suggestions, concerning the relationship between pain and depression, can arise from the evidence that depression and pain symptomatology are co-present in diverse conditions, which include diagnosed neurodegenerative and chronic diseases but also situations not supported by a medical diagnosis (see Table 13.1).

This implies the existence of a common substrate for pain and depression, which is apparently independent from diagnostic labels and settings. Further, although it is intuitively

Table 13.1 Co-expression of depression and pain symptomatology in diagnosed diseases and situations without medical diagnosis.

Painful conditions	Source	No. of patients	Setting	Patients with depression
Degenerative diseases				
Alzheimer's disease	Thielsher, Thielsher and Kostev (2013)	13.652	General practice	31.5%
Amyotrophic lateral sclerosis	Atassi et al. (2011)	127	Hospital	35%
Hungtinton's disease	Zarowitz, O'Shea and Nance (2014)	340	Nursing home	59.4%
Parkinson's disease	Thielsher, Thielsher and Kostev (2013)	8.403	General practice	33.4%
Chronic diseases				
Arthritis (or rheumatism)	Chen et al. (2012)	383	Community	29.63%
Chronic daily headache (transformed migraine included)	Juang et al. (2000)	261	Hospital	66%
Chronic low back pain	Subramanian et al. (2013)	436	Community	11.6%
Diabetic neuropathy	Koroschetz et al. (2011)	1623	Outpatient clinic	72.1%
Multiple sclerosis	Thielsher, Thielsher and Kostev (2013)	5.137	Primary care practice	34.7%
Phantom pain(?)	Desmond and MacLachlan (2006)	582	Community	32%
Rheumatoid arthritis	Dougados et al. (2014)	3.920	Rheumatology clinic	15%
Unexplained symptoms chronic syndromes				
Allodynia (in migraine)	Tietjen, Brandes and Peterlin (2009)	857	Outpatient clinic	45%
Chronic fatigue	Aggarwal et al. (2006)	173	Community	60%
Chronic orofacial pain	Aggarwal et al. (2006)	163	Community	44%
Chronic pelvic pain syndrome (women)	Silva et al. (2011)	147	Community	46.9%
Chronic prostatitis/pelvic pain syndrome	Clemens (2008)	174	Outpatient clinic	12%
Fibromyalgia	Koroschetz et al. (2011)	1434	Outpatient clinic	92.7%
Irritable bowel syndrome	Ladabaum et al. (2012)	141.295	Community	39%
Non-cardiac chest pain	Hocaoglu, Gulec, and Durmus (2008)	70	Clinical outpatient	21.4%
Somatoform pain disorder	Steinbrecher and Hiller (2011)	90	Primary care practice	27.7%

plausible that both pain and low mood can accompany a chronic pathological status, whether or not it is correct to indicate one as 'organic' and the other as 'reactive' is unclear; doing so could generate the false idea of something not pathologic or an 'exciting' illness, which would only be in the 'mind of patients'.

Actually, besides the neurobiological or pathophysiological mechanisms, pain and depression might share a common cultural preconception, which becomes a paradox in medicine: that human suffering is secondary to any other human pathological conditions. This paradox has long governed the management of pain and depression symptoms that occur in parallel with other signs – signs featuring in diagnosable pathologies – so that the terms 'somatisation' and 'hypochondria' have been largely used both as medical explanations and to justify the lack of interventions.

Today, although both physical and psychological pain (depressive symptoms) might be underestimated in old patients with Alzheimer's disease (see Chapter 15 in this volume), and also in young and adult patients with Parkinson's (see Chapter 17 in this volume) or with other chronic neurological or systemic diseases, the medical approach has changed. This is due, among other things, to evidence that treatments that aim to reduce a patient's discomfort and pain and to increase positive feelings have a great impact on the course of disease, but even more on that patient's quality of life.

Medically unexplained pain

A different trend is observable in those pain conditions that occur in the absence of a 'real organic cause' – both when diagnostic criteria have been established, as for example migraine, fibromyalgia or irritable bowel syndrome, and when persistent physical symptoms cannot be explained by medical illness or injury. As we have seen, this second group of cases are referred to as medically unexplained pain (MUS); some of them are reported in Table 13.1. They affect various organs and body parts, for example the head, the joints, the chest, the pelvis; and they may also be manifested through fatigue, breathlessness, sleep disturbances, and so on (Brown, 2007).

There is a large debate on the classification of MUSs. Indeed in general medicine they are indicated as somatic syndromes; but in psychiatry they are classified as somatoform and dissociative disorders in DSM-IV and as somatic symptom disorders in DSM-5. Thus, without a unified classification, a patient can be given a different diagnosis for the same basic problem, depending on the medical setting – a fact that generates confusion in both patients and doctors and increases the involvement of specialists, clinical tests and drug prescriptions; and these in turn dramatically increase the economic and social costs of therapy.

In the MUS group, migraine and low back pain (LBP) are two of the most widely diffused types of pain reported by patients of all ages, from childhood to old age. Hence we can consider them as prototypes when we investigate the pain/depression hypothesis.

Migraines are a chronic, disabling disease, which significantly reduces quality of life. The prevalence is 46 per cent for headache in general, 11 per cent for migraine, 42 per cent for tension-type headache and 3 per cent for chronic daily headache (Stovner et al., 2007). There is a high sex and age dependence component: migraines are three times more common in women than in men and show two picks of age prevalence, one during adolescence and the other from 25 to 55 years of age (Lipton and Bigal, 2005). Epidemiological studies also support the prevalence of headache and migraine in children and in abused or maltreated women, underlining the importance of early or chronic stress and of negative emotional experiences in the development of migraines (Tietjen and Peterlin, 2011); but they also indicate the comorbidity between migraine, post-traumatic stress disorders (Juang and Yang, 2014), and depression.

Further, patients with high levels of somatisation easily develop migraines and show high levels of stress and anxiety, while somatic symptoms are more often observable in headache patients with psychiatric comorbidity (Antonaci et al., 2011). But more significant data show that migraine-related disability is always correlated with the level of depressive symptoms (Yavuz et al., 2013); and migraines, especially in adolescents, might represent a high risk factor for severe depression (Pine, Cohen and Brook, 1996).

Shared aetiological factors and common determinants explain the co-occurrence of pain and depression (see below) and strengthen the non-coincidental association between these two pathologic entities, opening the door to questions regarding the time and the type of an efficacious treatment. Indeed mood and anxiety disorders – like chronic stress – facilitate the transition from episodic headaches into chronic daily headaches with a consequent headache-related disability (Scher, Lipton and Stewart, 2002; Borkum, 2010). Also, migraines can be a symptom of subsyndromic or hidden depression, and people with recurrent headaches are twice more likely than others to develop depression (Sheftell and Atlas, 2002). Clinicians should be alert, therefore, to psychological stress and somatic amplification when evaluating migraine patients, in order to prevent the transition of episodic headaches into chronic daily headaches and to help carry out an early diagnosis of depression.

In addition, as in depression, factors like consideration and reinforcement by doctors and the patient's consequent satisfaction (or lack of it) with the cure received, as well as with the efficacy of the acute treatment, might also influence the recurrence of headaches or migraines and contribute to their chronicisation (Lipton and Stewart, 1999). This fact reveals the role of expectations and motivations – therefore of the reward/aversive circuit in migraines – further supporting the link with mood disorders (Cahill, Cook and Pickens, 2014). In particular, aversive behaviours can be triggered by food or drink as well as by music or light; but, like patients with depression and anxiety, those who suffer from chronic migraines often avoid leisure or withdraw from social interactions, thus limiting even more their access to positive experiences and reinforcements; and this behaviour promotes the perpetuation of pain and depression (Smitherman, Kolivas and Bailey, 2013).

LBP is another significantly common MUS/chronic pain disorder: it ranks fifth on the list of causes of medical examination, so it is one of the main reasons why people go to consult a specialist. It involves the muscles and the bones of the back, but usually it is reported together with neck pain and shoulder pain, which are leading causes of inability to work and sick leave (Maniadakis and Gray, 2000). Polatin et al. (1993) found that depression is a common diagnosis in patients with chronic LBP, and Currie and Wang (2004) found, in a cross-sectional analysis, that LBP is a strong predictor of major depression. Moreover, a recent study showed that depression is most strongly related to LBP by comparison to other diseases (Shaw et al., 2010).

Adolescents also manifest LPB or generalised back, neck and shoulder pain with associated stress and depressive symptoms (Diepenmaat et al., 2006). In this population, passive coping strategies and fear–avoidance beliefs were found to be predictive of persistent disability, just like in adults (Ramond et al., 2011).

The chronicisation of LBP, like that of other pain syndromes, induces changes in brain activity that do not overlap with what happens in acute or subchronic conditions but encode for the emotional and mnemonic processing of physical stimuli and transform sensory perception in a heightened and more complex experience. As postulated by Hashmi et al.

(2013), this functional evidence supports the motivation decision model of pain (on which see below), according to which LBP inception initiates a cascade of emotionally driven learning events, which effectually reorganise the brain into a chronic pain site with distinct functional, anatomical and resting state properties.

Thus, as in migraine, the sharing anatomical and functional brain substrates and, even more, a *changed brain* might account for the comorbidity between LBP and mood disorder. Two dominant models have been used to explain the association between depression and LBP: the 'antecedent model', in which depression is responsible for the development and maintenance of chronic pain; and the 'consequence model', which implies that LBP elicits or increases depressive symptoms as a reaction to the challenge of current experiences of pain. Both the models indicate the existence of a reciprocal relationship between pain and depression, but recently the 'longitudinal model' suggests that depressive symptoms and LBP are risk factors across time for each other and that their directional relation is specific depending on time under treatment. In the case of tardy or missing treatment success, LBP can be interpreted as a threatening and uncontrollable event, and consequently, more depressive symptoms develop (Elfering, Käserand and Melloh, 2014).

This model reinforces the idea that it is of extreme importance, especially in primary-care settings or when one is treated by the family doctor, not to underestimate pain symptomatology, especially when no *organic cause* can be directly associated with it and when patients report increase in the frequency of pain attacks and discomfort. Targeted pharmacological, physical or psychological therapeutic strategies should be selected on the basis of the patient's past medical history and his/her individual response to treatment.

The concept of mental pain

Apart from pain as a physical perception, negative or unpleasant feelings are also 'pain'. Many terms have been used to indicate this pain of non-physical origins: mental pain, psychic pain, psychological pain, emptiness, psychache. But, although their use is widespread, there is still a big debate on separate definitions for pain. The main criticism is that any of these terms, and especially 'mental pain', indicates the existence of two entities – physical pain and emotional/ psychological pain – suggesting the Cartesian mind/ body dualism and consequently implying that mental and physical pain (illnesses) are polar opposites, that mental illnesses are not real or are of less importance, and that people with mental illnesses are responsible for their condition (Sykes, 2002).

As recently underlined by Tossani (2014), the fact that physical and mental pain can be conceptually distinguished does not imply their separate existence, and therefore the term 'mental pain' does not exclude the existence of real pain but integrates the physical and emotional–psychological symptoms, including the peripheral pain that otherwise would not be considered.

The importance of this integration is clear in conditions in which physical pain seems to arise from the imagination and from illusion – which occurs for example in phantom pain or in placebo/nocebo effects.

Phantom pain

In the nineteenth century, the Civil War surgeon Silas Weir Mitchell, describing the pain reported in the absent limb after amputation, wrote: 'thousands of spirit limbs were haunting as many good soldiers, every now and then tormenting them' (Mitchell, 1866: 90). Mitchell coined the term 'phantom pain' (PP), which is still used to indicate the pain or non-painful sensations experienced in a body part that no longer exists. Different parts of the body – the eyes, the teeth, the tongue, the nose, the breast, the penis, the bowel and the bladder, but most of all the limbs – are reported as phantoms after amputation, and tingling, throbbing, piercing, and pins and needles sensations are among the most commonly described types of pain (Subedi and Grossberg, 2011). Because of its similarity to hallucination, phantom pain was initially classified as a psychiatric illness, and although the idea that phantom pain 'is just in the head' of the patient – like a psychosomatic manifestation of a premorbid personality – may persist and contribute to the large divergence in the reported incidence and prevalence of this phenomenon, today phantom pain is conceived of as a syndrome with peripheral and central aetiology.

In fact personality traits characterised by passive coping styles and catastrophising behaviour can be associated with the development of PP (Hill, Niven and Knussen, 1995; Jensen et al., 2002), but it is more consistently the case that stress and depression significantly affect the onset and exacerbation of the pain episodes (Arena et al., 1990; Flor, 2002), placing phantom pain into the category of chronic 'mental' pain syndromes.

To take into account all the peripheral and central and all the psychocognitive determinants, a comprehensive model of phantom-limb pain has been recently proposed (Flor, 2002). On the basis of this model, pain-provoking experiences lived before amputation (cortical pain memory) are integrated with the input from the stump neuroma and the changes at the spinal levels (central sensitisation) to induce the reorganisation of the somatosensory cortex (SC) so that cortical neurons coding for pain are preferentially activated in the area refereed to the preexisting limb or organ. In parallel, areas mediating the affective and motivational aspects of pain, such as the insula and the anterior cingulate cortex (ACC) (Wei, Li and Zhuo, 1999), and areas related to the memory processing of emotions, like the frontal cortex (FC) and the parahippocampus (Vanneste, Joos and De Ridder, 2012), undergo plastic changes that contribute to the experience of phantom pain.

Beside pharmacological treatments, therapies addressed to reorganising somatosensory memories like those resulting from long-lasting preamputation pain and pain flashbacks, which are part of a traumatic memory, have been proposed to treat patients with phantom pain.

Ramachandran and Rogers-Ramachandran (1996) were the first to propose the application of mirror therapy to help phantom-limb pain: the patient watches the reflection of his/her intact limb moving in a mirror placed parasagittally between his/her arms or legs while simultaneously moving the phantom hand or foot in a manner similar to what s/he is observing, so that the virtual limb replaces the phantom limb. The therapeutical efficacy of this method is based on the theory of mirror neurons in the brain, whose activation during the mirror setting might remodulate the somatosensory input, which produces suppression of the protopathic pain perception in the phantom limb (Ramachandran and Rogers-Ramachandran, 2008). Mirror therapy has also been proposed as a method of treatment in fibromyalgia.

More recently, eye movement desensitisation and reprocessing (EMDR) – an intervention specifically aimed at processing unresolved memories from negative experiences, largely studied and applied in patients with posttraumatic stress disorder (PTSD) – has been indicated as treatment for PP (Schneider et al., 2008; Tinker and Wilson, 2005; De Roos et al., 2010). It is interesting that, in many cases, the EMDR pain treatment resulted not only in a reduction of pain, but also in an increase in a positive sense of self. Indeed, identified *negative cognitions* ('I'm weak', 'I'm stupid', 'I'm useless') *were replaced by an increased sense of self-efficacy and self-determination*, as demonstrated by Marshall Wilensky (2006). Since the activity of the limbic system represents the neurobiological correlate of EMDR (Pagani et al., 2012), these data reinforce the strong association between pain and mood and the existence of a common brain substrate.

Placebo–nocebo phenomena

As in the case of phantom pain, the reality of placebo analgesia and nocebo hyperalgesia has generated a great debate in neurobiological and psychological research and in medicine. Even now, the existence of placebo/nocebo phenomena represents one of the most challenging scientific and clinical questions around the concept of health and well-being and of their 'endogenous' control, and therefore it impacts on clinical trials and medical practice. (L. Colloca, A. Horin and D. Finniss present a detailed overview of placebo effects in Chapter 9.)

The first to introduce the term *placebo* in a medical context was the English physician John Haygarth (1740–1827), who investigated the efficacy of wand-like metal 'tractors' – patented by the American Elisha Perkins and used to draw out the body's pains – by comparing such an object with a wooden object indistinguishable to the eye from a Perkins tractor. Haygarth published his findings in a book titled *On the Imagination as a Cause and as a Cure of Disorders of the Body* (Haygarth, 1800). Although the experimental approach was rudimentary and based on a small number of patients, Haygarth's observations and interpretations anticipated the notion that expectations are important in shaping pain and pointed to the relevance of the doctor's credibility (or suggestions) to the success of therapeutic interventions ('famous doctors are often more successful than unknowns') (Haygarth, quoted in Wootton, 2006).

The idea of placebo as 'an inert agent generating an effect' has shifted today towards a more complex concept, of simulating an active therapy within a psychosocial context (Colloca and Benedetti, 2005). Thus the placebo response is driven by many different environmental factors – including the conditioning manifested by doctors or caregivers – and by personal factors that influence patients' expectations, desires, memories, and emotions (Price, Chung and Robinson, 2005).

Koyama et al. (2005) proposed that expectations for reduced pain produce a reduction of perceived pain, because the brain mechanisms supporting pain expectations powerfully interact with those that process afferent nociceptive information. Further, since the mental representation of impending pain also depends on information from past experiences, the brain regions associated with memory recall can affect the SC and alter the perception of pain and its intensity.

Functional and pharmacological studies demonstrated that the core mechanism of placebo analgesia and of the placebo effect in general is in the mesocorticolimbic circuit (also called 'reward circuit') and involves dopamine and opioid transmission (Gear, Aley and Levine, 1999). As we shall see, the same happens in the case of pain and mood. This suggests that placebo neurophysiology belongs to the physiological processes that mediate the interaction between environmental conditions and the corresponding physical and emotional responses of the individual and further support the reality of mental pain.

Neuroplasticity in pain, mood and depression

The International Association for the Study of Pain (IASP) defines pain as 'an unpleasant sensory and emotional experience associated with actual or potential tissue damage or described in term of such damage' (Bonica, 1979). In condensed form, this sentence takes into account the biological and psychological nature of pain, but also its relational and social aspects. Indeed pain is an experience (such as the knowledge gained from perception) that can be referred to as a real fact or as a fact imagined as being possible and reported through words in which the personal and private aspects of this experience – including the feelings, the discomfort and the cure received – are manifest.

In line with this, the brain regions involved in the experience of pain are not limited to the areas commonly activated by noxious stimuli – such as the thalamus, which receives the spinal afferents and sends projections to the ACC, the SC, the insula, the FC and the prefrontal cortex (PFC) – but include a modulatory system represented by the spinal projcetions to the nucleus accumbens (NAc), the amygdala (AMY) and the hypothalamus (HYP); the spinomesenchephalic tract and the periaqueductal grey (PAG); and the cerebellum (CBL) (Figure 13.1). The interconnection of these brain areas and their dynamic interaction with other brain structures more directly implicated in emotion and cognition, including the hippocampal area, represent a complex and multifaceted system through which the pain is perceived, controlled, and integrated with contextual – external and internal – stimuli (see Chapter 8 in this volume for a detailed and critical discussion of the 'pain matrix').

Examples of how this system works could come from our ordinary way to refer to pain: *if you think about it, you feel pretty bad; the massage was a little painful but what relief it was for the muscles; the anticipation of pain is worse than pain in itself, you feel better when coming home (hospitalised patients)*. These sentences express the fact that that the perception of pain, the intensity of pain, and even what it is considered as pain are the result of previous sensory, emotional and cognitive experiences. They also contain the idea that the motivation, the affective context and the emotional state can affect pain.

To reinforce this notion, the neuronal network for mental pain overlaps to some extent with brain regions involved in physical pain (see Chapter 8 in this volume). As recently proposed by Meerwijk, Ford and Weiss (2013) in a systematic review, the increased activation of the posterior cingulate cortex (PCC) and parahippocampal gyrus (PHCG) and the deactivation of the PFC reveal the arousal and appraisal components of the mental pain with respect to the pure physical stimulus, while the lesser involvement of the insula and AMY underlines the importance of duration and of the type of evoked emotions (e.g., grief, sadness, or fear). At variance with acute or sporadic physical pain, psychological pain is a lasting feeling that

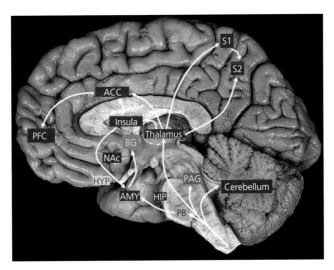

Figure 13.1 Pain and depression: Shared brain areas and pathology. Source: Blackburn-Munro 2001. Reproduced with permission of John Wiley & Sons.

takes time to be solved and easily generates frustration, sadness and a decreased well-being feeling: it is an enduring condition more similar to chronic pain and depression.

It is therefore not surprising that both physical and mental pain pathways overlap to a significant degree with those involved in the regulation of the mood and stress response, and consequently in depression. From an evolutionary point of view, pain, stress and mood might actually be seen as a highly conserved reward circuit that provides information on current conditions and therefore plays a primary role in assessing and responding to danger, whether from the external (e.g., a predator or social threat) or from the internal (e.g., infection, tissue damage) environment.

Normally, or rather in an 'hypothetically static' healthy condition, the perception and transmission of pain are invigilated at different levels from the relay medullary system, the PAG, and the limbic and supra-limbic structures, which tend to dump the body's signals in favour of increasing the attention paid to external events. Since we live in a dynamic emotional and social context, the pain-related brain areas can be activated in absence of a real pain stimulus, as occurs in healthy persons by imagining or anticipating a painful experience or by observing another person suffer, but also – as is discussed above – in phantom limb sensation and in placebo/nocebo effects. In this dynamic environment, chronic pain might represent an extreme point along a *continuum* in which the body and the brain (or mind) make an effort to maintain the equilibrium. Like depression, chronic pain remodels the brain, both directly, by influencing activity and plasticity in areas that regulate and control stress and mood, and indirectly, by affecting the neuronal networks of the peripheral brain.

In recent years, the notion that 'pain and mood change the brain' in terms of neuronal plasticity, connective rearranging and neuromodulation has replaced the idea of neurodegeneration in many chronic conditions, including pain syndromes and mood disorders, and has been proposed as an explanation for the grey or white matter alterations found in patient brains.

Moreover, the common ability of pain and mood to engage autonomic, neuroendocrine and neuroimmune components of the stress response and the evidence that depression and pain syndromes are characterised by a sympathetic/parasympathetic unbalanced, insufficient hypothalamus–pituitary–adrenal axis (HPAA) regulation and by an altered immune function (Robinson et al., 2009) have focused attention on the factors and mechanisms regulating the neuronal–immune–endocrine pathways (see Chapter 4 in this volume) – including hypothalamic hormones, cytokines, growth factors and neurotrophins.

Dysfunctional communication and neuroplastic changes – instead of irreversible damage – account for the production and maintenance of symptoms in pain and mood pathological conditions, and interaction with genetic vulnerabilities and environmental factors contribute to chronicity and comorbidity. More than common neuroanatomical and functional pathways, pain and depression share dysfunctional psychosomatic and somatopsychic communication, and because of this they can trigger and perpetuate each other, becoming the two faces of a multisystemic pathological condition.

Bringing about changes in one's perspective, brain changes in chronic pain conditions can also result in the alteration of personality traits; for example, traumatic brain injury can induce apathy, affective lability, and aggression (Schwarzbold et al., 2008), and even more radical changes in personality aspects, for instance self-transcendence (Urgesi et al., 2010).

Gustin et al. (2013) have recently demonstrated that the brain changes in chronic pain patients, but not in healthy controls, correlated with trait depression scores; this result supported the idea that acute pain induces subtle changes in depression states and trait values, which in turn can progress and become severe in chronic conditions. The evidence that an efficacious treatment for chronic pain also affects personality traits and depression (Mongini, Ibertis and Ferla, 1994; Naliboff et al., 1988) further supports the idea of a strict interaction between the subjective state and the manifestation or management of chronic pain.

On the other hand, psychiatric patients with functional illnesses such as depression (Fava et al., 2004), post-traumatic stress disorder (Otis, Keane and Kerns, 2003; Salomons et al., 2014), panic disorder, and anxiety (Fraenkel, Kindler and Melmed, 1996) show altered pain sensitivity. Patients with depression have significantly more frequent, more intense, and more unpleasant pain complaints than healthy control subjects (Lautenbacher et al., 1999). Further, psychological disorders predicted the onset of persistent pain disorder (Gureje, Simon and von Korff, 2001; Breslau et al., 2003), and psychological stress in subjects with a history of chronic pain predicted new episodes of pain (Croft et al., 1995).

It is worth noting that the discovery of the antidepressant action of the first psychoactive drugs came from accidental clinical observations. Chlorpromazine was synthesised in the 1950, in Paris, by Charpentier, who was actually looking for a compound analogous to promethazine but with greater sedative power. Chlorpromazine was initially used by Laborit in association with promethazine and peptidine, as a 'lithic cocktail' to potentiate anaesthesia in surgical patients, or with narcotics and barbiturates in patients manifesting untreatable pain crises. Only secondary because of its ability to calm but not to sedate, chlorpromazine was used in psychotic patients.

Imipramine was, in the late 1950s, the first tricyclic anti-depressant to be developed, and because of its antihistamine and anticholinergic proprieties it was initially applied in psychotic disorders such as schizophrenia; but it proved more efficacious in depressed patients and in the treatment of chronic pain, as reported below.

The discovery of the first psychoactive drugs reveals the multidimensionality of psychiatric pathology and, consequently, of its therapy – even more so as the pain is inscribed into the history of depression and psychopharmacological research (Bellantuono and Tansella, 1994).

Chronic pain and mood disorders as a maladaptation of the reward system

The ability to discriminate between pleasant and unpleasant stimuli and to associate with them a psychophysical (cold, satiation, fear, pain, etc.) or a behavioural (escape, avoidance, eating, fighting, etc.) response is crucial for survival and adaptation and actually represents a highly conserved mechanism during evolution. This mechanism is based on reward/punishment and reinforcement, so that an organism will strengthen a specific behaviour, producing pleasure or discomfort in response to a stimulus; and this kind of response is referred to as rewarding or aversive, respectively. The existence of a neuronal network in the brain that is able to detect the stimulus and to generate and regulate a specific response was identified in the 1950s by James Olds and Peter Milner, who demonstrated that the electrical stimulation of the septal area in rats produced a reinforcement similar to that of a natural reward stimulus, such as that produced by food or water.

Today the anatomical and functional characteristics of this network have been better defined with the help of pharmacological and brain imaging, both in humans and animals; and the mesocorticolimbic pathways are recognised as forming the brain's reward/aversion system (Wise, 2002). The core of the system is the ventral tegmental area (VTA) and its projections to the NAc, which works as a detector of rewarding/aversive stimuli, and sends the information to subcortical limbic areas like the AMY, the hippocampus (HIPP) and the HYP, and the CF and ACc for processing (Borsook et al., 2007).

Brain endogenous opioids and dopamine pathways have been largely reported as the principal mediators of the reward circuits, contributing to hedonic and motivational aspects, although other neurotrasmitters – including cis-3-aminocyclohexanecarboxylic acid (ACHchC) and gamma-aminobutyric acid (GABA) – that regulate the activity of NAc neurons and the release of dopamine can contribute to shape the behavioural response to rewarding or aversive stimuli (Fields, 2007; Bressan and Crippa, 2005; Navratilova et al., 2012).

Di Chiara and North (1992) proposed that opioid and dopamine pathways are associated with different aspects of reward – consumatory (opioid) and motivational–incentive (dopamine) respectively. This notion is, however, highly debated, since the role of opioids is also associated with the evaluation of the hedonic properties that make a reward desired and wanted, which means that opioids are involved in reward motivation and anticipation (Barbano and Cador, 2007). Since reward-related behaviours are very complex and require the processing of perceptions, memories, and expectations, the role that dopamine and opioids (or other neurotransmitters) might play can vary considerably, depending on the type of reward stimulus and on the internal (physical and mental) and external (environmental) conditions in which the stimulus occurs. For example, dopaminergic D2 receptors are crucial for the rewarding effects of opioids, but not for the development of physical opioid dependence or for the expression of some acute responses to morphine (Maldonado et al., 1997). This suggests a different contribution of dopamine to opioid-induced analgesia and addiction and to placebo analgesia (Tracey, 2010).

Independently of their relative contribution, dopamine and opioid pathways show their maximal cooperation and interdependence exactly in the rewarding aspect of pain in physiological and pathological conditions.

As mentioned before, brain changes in chronic pain can affect the emotional response to external stimuli, influence mood and favour the development of mood disorders. Bi-directionally similar brain changes can be responsible for the altered perception of objective versus subjective pain, impending sense of pain, and memory processing in psychiatric patients. The common core of these changes is located in brain areas that belong to the dopaminergic mesocorticolimbic pathway, which suggests that the endangerment of the reward/aversion process might explain the co-morbidity between pain syndrome and mood disorders (Borsook et al., 2007), but also the existence of MUS in healthy persons or in non-psychiatric patients (Gandhi, Becker and Schweinhardt, 2013).

To better understand the link between chronic pain and mood disorders, it is useful to think of pain and pleasure as the most potent aversive and reward stimuli for survival and therefore to postulate that seeking pleasure and avoiding pain represent two competitive motivations for the brain, which constantly need to be optimised and balanced. To achieve this goal, an individual has to engage in a decisional process that is based on evaluating the costs and benefits of various courses of action in terms of pain and pleasure. This process is unconscious and requires the integration of information about homeostatic state, sensory input, and possible threats or rewards.

According to the motivation decision model of pain (Fields, 2007), deciding is based on the assumption that anything that is potentially more important for survival (or for well-being) than pain exerts antinociceptive effects (pleasure). Therefore, when the reward (or the anticipation of reward) is more attractive than the impending pain, the pain response is suppressed.

This decision to 'not respond' engages antinociceptive mesocorticolimbic circuits (similarly to what occurs in placebo analgesia) and results in an interaction and balancing between their principal mediators: the endogenous opioids and phasic and tonic dopamine.[3] Simplifying the

---

[3] Phasic dopamine release is a transient dopamine release produced by the activation of dopamine neuron firing in response to behaviorally relevant stimuli. This large amplitude but brief pulse of dopamine is proposed to activate postsynaptic dopamine receptors but is rapidly removed from the synaptic space by fast, low-affinity/high-capacity re-uptake systems before it can trigger homeostatic responses. Tonic dopamine release is the release of dopamine from dopamine terminals activated presynaptically in a spike-independent manner by glutamate released from prefrontal cortical afferents. Dopamine released in this manner is proposed to underlie the background, steady-state level of extracellular dopamine in subcortical structures (Grace, 1991).

neurobiology of the process, opioids directly regulate the response to noxious stimuli through the descending pathways, by activating ON (responding) or OFF (not responding) cells in the rostral ventromedial medulla (RVM). The 'not respond' decision (OFF cells activation) induces an increase of phasic dopamine while suppressing the tonic release in the NAc, which results in pleasure and analgesia. At the same time, the enhancement of phasic dopamine in the NAc stimulates opioid release in the NAc, reinforcing analgesia, while increased levels of tonic dopamine down-regulate opioids and phasic dopamine, resulting in increased pain and decreased pleasure. The tonic dopamine levels are under the control of the corticostrial and limbic (hippocampal and amydgalic) afferents, so that the regulation of dopamine in the NAc is influenced by preexisting stimulus–reward associations such as those formed in the AMY, by spatial and contextual information derived from the hippocampus, or by the flexible processing of different behavioural strategies mediated by the PFC (Grace, 1991; Floresco, 2007; Leknes and Tracey, 2008). Stress also modulates the brain's dopamine/opioid balance, indicating that the activation of the HPAA axis can also influence the activity of the NAc, and therefore the processing of reward stimuli (Russo and Nestler, 2013). Based as it is on the brain's need to constantly rebalance the opioid/dopamine transmission in order to maintain homeostasis, the decision to 'respond', 'not respond' or 'overrespond' is susceptible of being affected in all the chronic conditions in which the continued presentation of the stimulus evokes lasting alterations of the central nervous system (CNS) – for instance chronic stress, chronic pain (Apkarian, Hashmi and Baliki, 2011) and neurological (Borsook, 2012) or psychiatric illness.

For example, prolonged stress and pain (Leitl et al., 2014), but also nerve injury (Alvarado et al., 2013) and neurodegeneration induce increase of tonic dopamine transmission and consequently reduce phasic dopamine. Reduced levels of brain phasic dopamine and opioids are also reported in the anhedonic state and in depression (Berridge and Robinson, 1998; Finan and Smith, 2013; Tye et al., 2013), while an increase of both tonic and phasic dopamine is associated with impulsive behaviour (Grace, 1991) and altered dopamine responsivity underlines schizophrenia.

Cerebral decrease of opioid receptor binding and dopaminergic deficit are observed in patients with neuropathic pain (Wood, 2006; Shaygan, Boger and Kroner-Herwig, 2013), rheumatoid arthritis (Nikolaus et al., 2013; Pinho de Oliveira Ribeiro et al., 2013), fibromyalgia (Wood et al., 2007), and migraines (Charbit, Akerman and Goadsby, 2010). This fact further supports the hypothesis that altered neurotransmitter homeostasis in the reward–aversion circuitry (RAC) can also be responsible for changed moods and depressive states (Barbano and Cador, 2007).

In terms of reward systems or the reward system, it is thus possible to speculate that the comorbidity between chronic pain and mood disorders is due to a lost ability to compensate through analgesia (pleasure) for aversive everyday events and therefore to not associate the 'not respond' decision to aversive stimuli or aversive expectations.

It is worth noticing that the mechanism of 'not responding' or 'overresponding', in case of impulsive or compulsive behaviour, is based mainly on the intrinsic regulation of dopamine/opioid transmission in the mesolimbic area but is also under the regulation of the cortical efferents. This evidence reinforces the 'unconscious' nature of the mechanism by indicating that the functions of the NAc facilitate the transformation of cognitive processes into meaningful patterns of behaviour without being involved in cognition per se (Floresco, 2007).

This aspect, together with the evidently overlapping pharmacological effects of analgesic and antidepressant drugs (on which see below), poses crucial questions as to therapeutic interventions and the need for a integrative medical practice. Even more importantly, it impacts on the body–mind concept and consequently on subjective and objective ideas of health and well-being.

Antidepressants as 'analgesic'

Pain can manifest itself as an autonomous symptom or in a symptomatologic constellation in several psychiatric disorders. Among these, depression not only shares with pain its aetiology and pathogenetic mechanisms, but its comorbidity with pain represents a complex challenge for clinicians. Depression can indeed be consequent upon no psychiatric case history with primary pain, or it can be a primary depressive state with prominent pain symptomatology, while paucisymptomatic pain might be present in subsyndromal or subthreshold depressive forms. Moreover, the pain/depressive dimension is observable in psychiatric conditions like psychosis, which have long been considered to manifest depression. The transnosographic use of certain pharmacological classes, first of all antidepressants – or, better still, a more correct dimensional use of them – is therefore necessary in order to treat patients.

In this context it is interesting to note that the prescription of antidepressants (ads) in the general population of industrialised countries has increased enormously in the past few years. Although the causes of this phenomenon are not yet well defined, it is possible to relate it to the overprescription of ads for psychiatric conditions on the one hand and to the prescription of ads for patients with non-psychiatric conditions on the other – the latter being done by general practitioners (GPs).

Paradoxically, the off-label uses or the uses in general medicine have revealed the analgesic activity of ads (Smith and Elliott, 2005; Bair and Sanderson, 2011; Chodosh et al., 2001), and now ads are emerging as the first line of agents in the treatment of chronic neuropathic pain (Finnerup et al., 2005).

Among the ads, those with the higher efficacy as analgesics are tricyclic antidepressants (TCAs), selective serotonin reuptake inhibitors (SSRI) and serotonin-norepinephrine reuptake inhibitors e (SNRI) (Stahl, 2002; Fava, 2003; Delgado, 2004), which mainly accentuate the effect of monoamine on the brain's endogenous pain-inhibiting systems in order to modulate pain transmission cells in the spinal cord (Max et al., 1992). Putative mechanisms of pain relief therefore include the inhibition of norepinephrine and/or serotonin re-uptake at synapse of the central descending pain control systems and the antagonism of NMDA receptors that mediates hyperalgesia and allodynia.

In addition, it has been demonstrated that the 5-HT1A serotonin receptor subtype – one of the 14 serotonin receptor subtypes known at the present – which is considered the preferential SSRI target, is also involved in pain and drug dependence; this explains the analgesic effects of this class of ads (Matsuda, 2013). Furthermore, since the activation of the serotonin system in the brain affects other neurotransmitters, including dopamine, noradrenaline, ach and glutamate in diverse cerebral areas of the mesolimbic system such as the FC, the hippocampus and the AMY, it is possible to speculate that ads might play a role in the reward mechanism and therefore may indirectly act on pain perception or motivation.

Trycyclic antidepressants (TCAs) have the longest track record in the treatment of depression and multiple pain conditions (Ansari, 2000; Bryson and Wilde, 1996). TCAs such as imipramine, amitriptyline, nortriptyline and clomipramine are generally recommended for trigeminal neuralgia, painful polyneuropathy and painful diabetic neuropathy (PDN) (Jensen et al., 2006). In postherpetic neuralgia, TCAs are alternative treatments of gabapentin or pregabalin (Tyring, 2007). In central post-stroke pain (Frese et al., 2006), migraine and tension headaches (Tomkins et al., 2001), amytriptiline is the second line of psychopharmacologic treatment. TCAs are also treatments in neuropathic pain after spinal cord injury (Rintala et al., 2007) and fibromyalgia (Goldenberg, 2007).

Apart from these ads' good efficacy in treatment, their use is limited by the high rates of adverse effects (tiredness, dry mouth) and several contraindications for patients with unstable angina, recent myocardial infarction, heart failure, ventricular arrhythmias history, significant conduction system disease and long QT syndrome. TCAs should be also used with caution in elderly patients, patients with glaucoma, orthostatic hypotension and diabetic uropathy (Mercier et al., 2013; Huang et al., 2013; Vinik and Casellini, 2013). For these reasons, many efforts have been made to develop other, more selective drugs (such as serotonin and noradrenalin inhibitor re-uptake) and drugs with only minor collateral effects.

The SSRIs had been the first class of ads to be ameliorated, and they include molecules like fluoxetine, paroxetine, citalopram, escitalopram, which specifically inhibit presynaptic re-uptake of serotonin but not of norepinephrine; and, unlike tricyclics, they lack the post-synaptic receptor-blocking effects and quinidine-like membrane stabilisation. However, only weak effects on pain were observed, and this class has not been licensed for the treatment of neuropathic pain (Vinik and Casellini, 2013).

SNRIs are a class of psychopharmacological drugs with the best efficacy on pain. Venlafaxine, duloxetine and milnacipran relieve pain by increasing the synaptic availability of 5-HT and NA in the descending pathways that inhibit pain impulses (Vinik and Casellini, 2013). Evidence from clinical trials demonstrated the efficacy of these drugs in the treatment of pain disorders, pain symptoms and depression (Gupta, Nihalani and Masand, 2007; Thor, Kirby and Viktrup, 2007) and showed that their analgesic effect is independent of antidepressant action (Perahia, Pritchett, et al., 2006; Perahia, Wang, et al., 2006).

The SNRIs also ameliorate the depressive symptomatology and the pain symptoms manifest in major depressive disorder (MMD) and in some subgroups, like women in the postmenstrual phase, elderly patients, elderly patients with arthritis, and sufferers of melancholic depression (Perahia, Wang, et al., 2006; Goldstein et al., 2004). Among patients with pain symptomatology, women with fibromyalgia, MMD, multisomatoform disorders, and trigeminal neuralgia are those who better respond to treatment with SNRI. Young people and adolescents with depressive and pain symptomatology can also be successfully treated with this class of ads (Robinson et al., 2009).

The pharmacological efficacy of ads in both pain and depressed patients indirectly demonstrated that somatic pain or generalised physical symptoms cannot be considered secondary to mood disorders (or vice-versa) and that therefore it is fundamental not to underrecognise or undertreat them.

Regardless of its origin and nosographic classification, pain as a symptom co-present in depression or in other psychiatric conditions creates the scientific and ethical necessity to find a treatment or a cure. Hence the teaching of and the training in the aetiopathogenesis, medical practice and management of pain are fundamental not only for psychiatry students (Elman, Zubieta and Borsook, 2011), but also for students in all the medical disciplines. The new generation of doctors must know that taking patients' pain seriously might be therapeutic in its own right and that relieving pain or curing it in all cases of comorbidity, psychiatric and non-psychiatric, is part of a medical mission to reduce discomfort and increase quality of life, all the more when pain is present as a primitive symptom.

References

Aggarwal, V. R., McBeth, J., Zakrzewska, J. M., Lunt, M. and Macfarlane, G. J. 2006. The epidemiology of chronic syndromes that are frequently unexplained: Do they have common associated factors? *International Journal of Epidemiology* 35: 468–76.

Alvarado, S., Tajerian, M., Millecamps, M., Suderman, M., Stone, L. S. and Szyf, M. 2013. Peripheral nerve injury is accompanied by chronic transcriptome-wide changes in the mouse prefrontal cortex. *Molecular Pain* 9: 21–32.

American Psychiatric Association. 2013. *Diagnostic and Statistical Manual of Mental Disorders: DSM-V*, 5th edn. Washington, DC: American Psychiatric Association.

Ansari, A. 2000. The efficacy of newer antidepressants in the treatment of chronic pain: A review of current literature. *Harvard Review of Psychiatry* 7: 257–77.

Antonaci, F., Nappi, G., Galli, F., Manzoni, G. C., Calabresi, P. and Costa, A. 2011. Migraine and psychiatric comorbidity: A review of clinical findings. *The Journal of Headache and Pain* 12: 115–25.

Apkarian, A. V., Hashmi, J. A. and Baliki, M. N. 2011. Pain and the brain: Specificity and plasticity of the brain in clinical chronic pain. *Pain* 152, S49–S64.

Arena, J. G., Sherman, R. A., Bruno, G. M. and Smith, J. D. 1990. The relationship between situational stress and phantom limb pain: Cross-lagged correlational data from six month pain logs. *Journal of Psychosomatic Research* 34: 71–7.

Atassi, N., Cook, A., Pineda, C. M., Yerramilli-Rao, P., Pulley, D. and Cudkowicz, M. 2011. Depression in amyotrophic lateral sclerosis. *Amyotrophic Lateral Sclerosis* 12: 109–12.

Bair, M. J. and Sanderson, T. R. 2011. Coanalgesics for chronic pain therapy: A narrative review. *Postgraduate Medicine* 123: 140–50.

Barbano, M. and Cador, M. 2007. Opioids for hedonic experience and dopamine to get ready for it. *Psychopharmacology (Berlin)* 191: 497–506.

Bellantuono, C. and Tansella, M. 1994. *Gli psicofarmaci nella pratica terapeutica*. Rome: Il Pensiero Scientifico.

Berridge, K. C. and Robinson, T. E. 1998. What is the role of dopamine in reward: Hedonic impact, reward learning, or incentive salience? *Brain Research Reviews* 28: 309–69.

Bonica, J. J. 1979. The need of a taxonomy. *Pain* 6: 247–8.

Borkum, J. M. 2010. Chronic headaches and the neurobiology of somatization. *Current Pain and Headache Reports* 14: 55–61.

Borsook, D. 2012. Neurological diseases and pain. *Brain* 135: 320–44.

Borsook, D., Becerra, L., Carlezon, W. A., Shaw, M., Renshaw, P., Elman, I. and Levine, J. 2007. Reward-aversion circuitry in analgesia and pain: Implications for psychiatric disorders. *European Journal of Pain* 11: 7–20.

Breslau, N., Lipton, R. B., Stewart, W. F., Schultz, L. R. and Welch, K. M. 2003. Comorbidity of migraine and depression: Investigating potential etiology and prognosis. *Neurology* 60: 1308–12.

Bressan, R. A. and Crippa, J. A. 2005. The role of dopamine in reward and pleasure behaviour: Review of data from preclinical research. *Acta Psychiatrica Scandinavica* 111 (Suppl. 427): 14–21.

Brown, R. J. 2007. Introduction to the special issue on medically unexplained symptoms: Background and future directions. *Clinical Psychology Review* 27: 769–80.

Bryson, H. M. and Wilde, M. I. 1996. Amitriptyline. A review of its pharmacological properties and therapeutic use in chronic pain states. *Drugs Aging* 8: 459–76.

Cahill, C. M., Cook, C. and Pickens, S. 2014. Migraine and reward system – or is it aversive? *Current Pain and Headache Reports* 18 (5): 410–19.

Charbit, A. R., Akerman, S. and Goadsby, P. J. 2010. Dopamine: What's new in migraine? *Current Opinion in Neurology* 23: 275–81.

Chen, X., Cheng, H., Huang, Y., Liu, Z. and Luo, X. 2012. Depression symptoms and chronic pain in the community population in Beijing, China. *Psychiatry Research* 200: 313–17.

Chodosh, J., Ferrell, B. A., Shekelle, P. G. and Wenger, N. S. 2001 Quality indicators for pain management in vulnerable elders. *Annals of Internal Medicine* 135: 731–5.

Clemens, J. Q. 2008. Male and female pelvic pain disorders: Is it all in their heads? *Journal of Urology* 179: 813–4.

Colloca, L. and Benedetti, F. 2005. Placebos and painkillers: Is mind as real as matter? *Nature Revews Neuroscience* 6: 545–52.

Croft, P. R., Papageorgiou, A. C., Ferry, S., Thomas, E., Jayson, M. I. and Silman, A. J. 1995. Psychologic distress and low back pain. Evidence from a prospective study in the general population. *Spine (Phila Pa 1976)*, 20: 2731–7.

Currie, S. R. and Wang, J. 2004. Chronic back pain and major depression in the general Canadian population. *Pain* 107: 54–60.

Delgado, P. L. 2004. Common pathways of depression and pain. *Journal of Clinical Psychiatry* 65 (Suppl. 12): S16–S19.

De Roos, C., Veenstra, A. C., de Jongh, A., den Hollander-Gijsman, M., van der Wee, N. J., Zitman, F. G. and van Rood, Y. R. 2010. Treatment of chronic phantom limb pain using a trauma-focused psychological approach. *Pain Research and Management* 15: 65–71.

Desmond, D. M. and MacLachlan, M. 2006. Affective distress and amputation-related pain among older men with long-term, traumatic limb amputations. *Journal of Pain and Symptom Menagement* 31: 362–8.

Di Chiara, G. and North, R. A. 1992. Neurobiology of opiate abuse. *Trends in Pharmacological Sciences* 13: 185–93.

Diepenmaat, A. C., van der Wal, M. F., de Vet, H. C. and Hirasing, R. A. 2006. Neck/shoulder, low back, and arm pain in relation to computer use, physical activity, stress, and depression among Dutch adolescents. *Pediatrics* 117: 412–16.

Dougados, M., Soubrier, M., Antunez, A. and Balint, P. 2014. Prevalence of comorbidities in rheumatoid arthritis and evaluation of their monitoring: Results of an international, cross-sectional study (COMORA). *Annals of the Rheumatic Diseases* 73: 62–8.

Elfering, A., Käser, A. and Melloh, M. 2014. Relationship between depressive symptoms and acute low back pain at first medical consultation, three and six weeks of primary care. *Psychology, Health & Medicine* 19: 235–46.

Elman, I., Zubieta, J. K. and Borsook, D. 2011. The missing p in psychiatric training: Why it is important to teach pain to psychiatrists. *Archives of General Psychiatry* 68: 12–20.

Fava, M. 2003. The role of the serotonergic and noradrenergic neurotransmitter systems in the treatment of psychological and physical symptoms of depression. *Journal of Clinical Psychiatry* 64 (Suppl. 13): S26–S29.

Fava, M., Mallinckrodt, C. H., Detke, M. J., Waltkin, J. G. and Wohlreich, M. M. 2004. The effect of duloxetine on painful physical symptoms in depressed patients: Do improvements in these symptoms result in higher remission rates? *Journal of Clinical Psychiatry* 65: 521–30.

Fields, H. L. 2007. Understanding how opioids contribute to reward and analgesia. *Regional Anesthesia and Pain Medicine* 32: 242–6.

Finan, P. H. and Smith, M. T. 2013. The comorbidity of insomnia, chronic pain, and depression: Dopamine as a putative mechanism. *Sleep Medicine Reviews* 17: 173–83.

Finnerup, N. B., Otto, M., McQuay, H. J., Jensen, T. S. and Sindrup, S. H. 2005. Algorithm for neuropathic pain treatment: An evidence-based proposal. *Pain* 118: 289–305.

Flor, H. 2002. Phantom-limb pain: Characteristics, causes, and treatment. *Lancet Neurology* 1: 182–9.

Floresco, S. B. 2007. Dopaminergic regulation of limbic–striatal interplay. *Journal of Psychiatry & Neuroscience* 32: 400–11.

Fogazzaro, A. 1901. *Il dolore nell'arte*. Milan: Baldini Castoldi.

Fraenkel, Y. M., Kindler, S. and Melmed, R. N. 1996. Differences in cognitions during chest pain of patients with panic disorder and ischemic heart disease. *Depression and Anxiety* 4: 217–22.

Frese, A., Husstedt, I. W., Ringelstein, E. B. and Evers, S. 2006. Pharmacologic treatment of central post-stroke pain. *Clinical Journal of Pain* 22: 252–60.

Freud, S. 1915. Caducità. In S. Freud, *Opere*, vol. 8. Turin: Bollati Boringhieri, pp. 174–5.

Freud, S. 1925. *Inibizione, sintomo e angoscia*. In S. Freud, *Opere*, vol. 10. Turin: Bollati Boringhieri.

Freud, S. 1963 [1917]. *Mourning and Melancholia*. In S. Freud, *The Standard Edition of the Complete Psychological Works*, vol. 14. London: Hogarth Press and the Institute of Psychoanalysis.

Gandhi, W., Becker, S. and Schweinhardt, P. 2013. Pain increases motivational drive to obtain reward, but does not affect associated hedonic responses: A behavioural study in healthy volunteers. *European Journal of Pain* 17: 1093–103.

Gear, R. W., Aley, K. O. and Levine, J. D. 1999. Pain-induced analgesia mediated by mesolimbic reward circuits. *Journal of Neuroscience* 19: 7175–81.

Goldenberg, D. L. 2007. Pharmacological treatment of fibromyalgia and other chronic musculoskeletal pain. *Best Practice & Research: Clinical Rheumatology* 21: 499–511.

Goldstein, D. J., Lu, Y., Detke, M. J., Hudson, J., Iyengar, S. and Demitrack, M. A. 2004. Effects of duloxetine on painful physical symptoms associated with depression. *Psychosomatics* 45: 17–28.

Grace, A. A. 1991. Phasic versus tonic dopamine release and the modulation of dopamine system responsivity: A hypothesis for the etiology of schizophrenia. *Neuroscience* 41: 1–24.

Gupta, S., Nihalani, N. and Masand, P. 2007. Duloxetine: Review of its pharmacology, and therapeutic use in depression and other psychiatric disorders. *Annals of Clinical Psychiatry* 19: 125–32.

Gureje, O., Simon, G. E. and von Korff, M. 2001. A cross-national study of the course of persistent pain in primary care. *Pain* 92: 195–200.

Gustin, S. M., Peck, C. C., Macey, P. M., Murray, G. M. and Henderson, L. A. 2013. Unraveling the effects of plasticity and pain on personality. *Journal of Pain* 14: 1642–52.

Haygarth, J. 1800. *On the Imagination, as a Cause and as a Cure of Disorders of the Body; Exemplified by Fictitious Tractors, and Epidemical Convulsions*. Bath: Crutwell.

Hashmi, J. A., Baliki, M. N., Huang, L., Baria, A. T., Torbey, S., Hermann, K. M., Schnitzer, T. J. and Apkarian, A. V. 2013. Shape shifting pain: Chronification of back pain shifts brain representation from nociceptive to emotional circuits. *Brain* 136: 2751–68.

Hill, A., Niven, C. A. and Knussen, C. 1995. The role of coping in adjustment to phantom limb pain. *Pain* 62: 79–86.

Hocaoglu, C., Gulec, M. Y. and Durmus, I. 2008. Psychiatric comorbidity in patients with chest pain without cardiac etiology. *Israel Journal of Psychiatry and Related Sciences* 45: 49–54.

Huang, X., Li, C., Luo, Y. L., Wang, B. and Ji, J. L. 2013. Efficacy of venlafaxine extended-release monotherapy for first-episode depression with painful physical symptoms. *Neuroreport* 24: 364–9.

Jensen, M. P., Ehde, D. M., Hoffman, A. J., Patterson, D. R., Czerniecki, J. M. and Robinson, L. R. 2002. Cognitions, coping and social environment predict adjustment to phantom limb pain. *Pain* 95: 133–42.

Jensen, T. S., Backonja, M. M., Hernandez Jimenez, S., Tesfaye, S., Valensi, P. and Ziegler, D. 2006. New perspectives on the management of diabetic peripheral neuropathic pain. *Diabetes & Vascular Disease Research* 3: 108–19.

Juang, K. D. and Yang, C. Y. 2014. Psychiatric comorbidity of chronic daily headache: Focus on traumatic experiences in childhood, post-traumatic stress disorder and suicidality. *Current Pain and Headache Reports* 18 (4). doi: 10.1007/s11916–014–0405–8 DOI:10.1007/s11916-014-0405-8 .

Juang, K. D., Wang, S. J., Fuh, J. L., Lu, S. R. and Su, T. P. 2000. Comorbidity of depressive and anxiety disorders in chronic daily headache and its subtypes. *Headache* 40: 818–23.

Koroschetz, J., Rehm, S. E., Gockel, U. and Brosz, M. 2011. Fibromyalgia and neuropathic pain-differences and similarities: A comparison of 3057 patients with diabetic painful neuropathy and fibromyalgia. *BMC Neurology* 11: 55–75.

Koyama, T., McHaffie J. G., Laurienti, P. J. and Coghill, R. C. 2005. The subjective experience of pain: Where expectations become reality. *Proceedings of the National Academy of Sciences of the United States of America* 102: 12950–5.

Ladabaum, U., Bovd, E., Zhao, W. K., Mannalithara, A., Sharabidze, A., Singh, G., Chung, E. and Levin, T. R. 2012. Diagnosis, comorbidities, and management of irritable bowel syndrome in patients in a large health maintenance organization. *Clinical Gastroenterlogy and Hepatology* 10: 37–45.

Lautenbacher, S., Spernal, J., Schreiber, W. and Krieg, J. C. 1999. Relationship between clinical pain complaints and pain sensitivity in patients with depression and panic disorder. *Psychosomatic Medicine* 61: 822–7.

Leitl, M. D., Onvani, S., Bowers, M. S., Cheng, K., Rice, K. C., Carlezon, W. A. Jr, Banks, M. L. and Negus, S. S. 2014. Pain-related depression of the mesolimbic dopamine system in rats: Expression, blockade by analgesics, and role of endogenous kappa-opioids. *Neuropsychopharmacology* 39: 614–24.

Leknes, S. and Tracey, I. 2008. A common neurobiology for pain and pleasure. *Nature Reviews Neuroscience* 9: 314–20.

Lipton, R. B. and Bigal, M. E. 2005. The epidemiology of migraine. *The American Journal of Medicine* 118 (Suppl. 1): 3S–10S.

Lipton, R. B. and Stewart, W. F. 1999. Acute migraine therapy: Do doctors understand what patients with migraine want from therapy? *Headache* 39 (Suppl. 2): S20–S26.

Maldonado, R., Saiardi, A., Valverde, O., Samad, T. A., Roques, B. P. and Borrelli, E. 1997. Absence of opiate rewarding effects in mice lacking dopamine D2 receptors. *Nature* 388: 586–9.

Maniadakis, N. and Gray, A. 2000. The economic burden of back pain in the UK. *Pain* 84: 95–103.

Matsuda, T. 2013. Neuropharmacologic studies on the brain serotonin1A receptor using the selective agonist osemozotan. *Biological & Pharmaceutical Bullettin* 36: 1871–82.

Max, M. B., Lynch, S. A., Muir, J., Shoaf, S. E., Smoller, B. and Dubner, R. 1992. Effects of desipramine, amitriptyline, and fluoxetine on pain in diabetic neuropathy. *New England Journal of Medicine* 326: 1250–6.

Meerwijk, E. L., Ford, J. M. and Weiss, S. J. 2013. Brain regions associated with psychological pain: Implications for a neural network and its relationship to physical pain. *Brain Imaging and Behavior* 7: 1–14.

Mercier, A., Auger-Aubin, I., Lebeau, J. P., Schuers, M., Boulet, P., Hermil, J. L., Van Royen, P. and Peremans, L. 2013. Evidence of prescription of antidepressants for non-psychiatric conditions in primary care: An analysis of guidelines and systematic reviews. *BMC Family Practice* 14: 55–64.

Mitchell, S. W. 1866. The case of George Dedlow. *Atlantic Monthly* 18 (105): 87–99.

Mongini, F., Ibertis, F. and Ferla, E. 1994. Personality characteristics before and after treatment of different head pain syndromes. *Cephalalgia* 14: 368–73.

Naliboff, B. D., McCreary, C. P., McArthur, D. L., Cohen, M. J. and Gottlieb, H. J. 1988. MMPI changes following behavioral treatment of chronic low back pain. *Pain* 35: 271–7.

Navratilova, E., Xie, J. Y. Okun, A., Qu, C., Eyde, N., Ci, S., Ossipov, M. H., King, T., Fields, H. L. and Porreca, F. 2012. Pain relief produces negative reinforcement through activation of mesolimbic reward-valuation circuitry. *Proceedings of the National Academy of Sciences of the United States of America* 109: 20709–13.

Nikolaus, S., Bode, C., Taal, E. and van de Laar, M. A. 2013. Fatigue and factors related to fatigue in rheumatoid arthritis: A systematic review. *Arthritis Care Research* 65: 1128–46.

Otis, J. D., Keane, T. M. and Kerns, R. D. 2003. An examination of the relationship between chronic pain and post-traumatic stress disorder. *Journal of Rehabilitation Research and Development* 40: 397–405.

Pagani, M., Di Lorenzo, G., Verardo, A. R., Nicolais, G., Monaco, L., Lauretti, G., Russo, R., Niolu, C., Ammaniti, M., Fernandez, I., et al. 2012. Neurobiological correlates of EMDR monitoring: An EEG study. *PLoS One* 7. doi: 10.1371/journal.pone.0045753.

Perahia, D. G., Pritchett, Y. L., Desaiah, D. and Raskin, J. 2006a. Efficacy of duloxetine in painful symptoms: An analgesic or antidepressant effect? *International Clinical Psychopharmacology* 21: 311–17.

Perahia, D. G., Wang, F., Mallinckrodt, C. H., Walker, D. J. and Detke, M. J. 2006b. Duloxetine in the treatment of major depressive disorder: A placebo- and paroxetine-controlled trial. *European Psychiatry* 21: 367–78.

Pine, D. S., Cohen, P. and Brook, J. 1996. The association between major depression and headache: Results of a longitudinal epidemiologic study in youth. *Journal of Child and Adolescent Psychopharmacology* 6: 153–64.

Pinho de Oliveira Ribeiro, N., Rafael de Mello Schier, A., Ornelas, A. C., Pinho de Oliveira, C. M., Nardi, A. E. and Silva, A. C. 2013. Anxiety, depression and suicidal ideation in patients with rheumatoid arthritis in use of methotrexate, hydroxychloroquine, leflunomide and biological drugs. *Comprehensive Psychiatry* 54: 1185–9.

Polatin, P. B., Kinney, R. K., Gatchel, R. J., Lillo, E. and Mayer, T. G. 1993. Psychiatric illness and chronic low-back pain: The mind and the spine – Which goes first? *Spine (Phila Pa 1976)*, 18: 66–71.

Price, D. D., Chung, S. K. and Robinson, M. E. 2005. Conditioning, expectation, and desire for relief in placebo analgesia. *Seminars In Pain Medicine* 3: 15–21.

Ramachandran, V. S. and Rogers-Ramachandran, D. 1996. Synaesthesia in phantom limbs induced with mirrors. *Proceedings, Biological Sciences* 263: 377–86.

Ramachandran, V. S. and Rogers-Ramachandran, D. 2008. Sensations referred to a patient's phantom arm from another subjects intact arm: Perceptual correlates of mirror neurons. *Medical Hypotheses* 70: 1233–4.

Ramond, A., Bouton, C., Richard, I., Roquelaure, Y., Baufreton, C., Legrand, E. and Huez, J. F. 2011. Psychosocial risk factors for chronic low back pain in primary care – a systematic review. *Family Practice* 28: 12–21.

Rintala, D. H., Holmes, S. A., Courtade, D., Fiess, R. N., Tastard, L. V. and Loubser, P. G. 2007. Comparison of the effectiveness of amitriptyline and gabapentin on chronic neuropathic pain in persons with spinal cord injury. *Archives of Physical Medicine and Rehabilitation* 88: 1547–60.

Robinson, M. J., Edwards, S. E., Iyengar, S., Bymaster, F., Clark, M. and Katon, W. 2009. Depression and pain. *Frontiers in Bioscience* 14: 5031–51.

Russo, S. J. and Nestler, E. J. 2013. The brain reward circuitry in mood disorders. *Nature Reviews Neuroscience* 14: 609–25.

Salomons, T. V., Moayedi, M., Erpelding, N. and Davis, K. D. 2014. A brief cognitive-behavioural intervention for pain reduces secondary hyperalgesia. *Pain* 22. doi: 10.1016/j.pain.

Scher, A. I., Lipton, R. B. and Stewart, W. 2002. Risk factors for chronic daily headache. *Current Pain and Headache Reports* 6: 486–91.

Schneider, J., Hofmann, A., Rost, C. and Shapiro, F. 2008. EMDR in the treatment of chronic phantom limb pain. *Pain Medicine* 9: 76–82.

Schwarzbold, M., Diaz, A., Martins, E. T. Rufino, A., Amante, L. N., Thais, M. E., Quevedo, J., Hohl, A., Linhares, M. N. and Walz, R. 2008. Psychiatric disorders and traumatic brain injury. *Journal of Neuropsychiatric Disease and Treatment* 4: 797–816.

Shaw, W. S., Means-Christensen, A. J., Slater, M. A., Webster, J. S., Patterson, T. L., Grant, I., Garfin, S. R., Wahlgren, D. R., Patel, S. and Atkinson, J. H. 2010. Psychiatric disorders and risk of transition to chronicity in men with first onset low back pain. *Pain Medicine* 11: 1391–400.

Shaygan, M., Boger, A. and Kroner-Herwig, B. 2013. Clinical features of chronic pain with neuropathic characteristics: A symptom-based assessment using the pain DETECT questionnaire. *European Journal of Pain* 17: 1529–38.

Sheftell, F. D. and Atlas, S. J. 2002. Migraine and psychiatric comorbidity: From theory and hypotheses to clinical application. *Headache* 42: 934–44.

Silva, G. P., Nascimento, A. L., Michelazzo, D., Alves Junior, F. F., Rocha, M. G., Silva, J. C., Reis, F. J., Nogueira, A. A. and Poli Neto, O. B. 2011. High prevalence of chronic pelvic pain in women in Ribeirão Preto, Brazil and direct association with abdominal surgery. *Clinics* 66: 1307–12.

Smith, B. H. and Elliott, A. M. 2005. Active self-management of chronic pain in the community. *Pain* 113: 249–50.

Smitherman, T. A., Kolivas, E. D. and Bailey, J. R. 2013. Panic disorder and migraine: Comorbidity, mechanisms, and clinical implications. *Headache* 53: 23–45.

Stahl, S. M. 2002. Does depression hurt? *The Journal of Clinical Psychiatry* 63: 273–4.

Steinbrecher, N. and Hiller, W. 2011. Course and prediction of somatoform disorder and medically unexplained symptoms in primary care. *General Hospital Psychiatry* 33: 318–26.

Stovner, L., Hagen, K., Jensen, R., Katsarava, Z., Lipton, R., Scher, A., Steiner, T. and Zwart, J. A. 2007. The global burden of headache: A documentation of headache prevalence and disability worldwide. *Cephalalgia* 27: 193–210.

Subedi, B. and Grossberg, G. T. 2011. Phantom limb pain: Mechanisms and treatment approaches. *Pain Research and Treatment*. doi: 10.1155/2011/864605.

Subramanian, M., Vaingankar, J. A., Abdin, E. and Chong, S. A. 2013. Psychiatric morbidity in pain conditions: Results from the Singapore Mental Health Study. *Pain Research and & Management* 18: 185–90.

Sykes, R. 2002. Physical or mental? A perspective on chronic fatigue syndrome. *Advances in Psychiatric Treatment* 8: 351–65.

Thielsher, C., Thielsher, S. and Kostev, K. 2013. The risk of developing depression when suffering from neurological diseases. *General Medical Science* 11. doi: 10.3205/000170.

Thor, K. B., Kirby, M. and Viktrup, L. 2007. Serotonin and noradrenaline involvement in urinary incontinence, depression and pain: Scientific basis for overlapping clinical efficacy from a single drug, duloxetine. *International Journal of Clinical Practice* 61: 1349–55.

Tietjen, G. E. and Peterlin, B. L. 2011. Childhood abuse and migraine: Epidemiology, sex differences, and potential mechanisms. *Headache* 51: 869–79.

Tietjen, G. E., Brandes, J. L., and Peterlin, L. 2009. Allodynia in migraine: Association with comorbid pain conditions. *Headache* 49: 1333–44.

Tinker, RH and Wilson, S. A. 2005. The phantom limb pain protocol. In *EMDR Solutions: Pathways to Healing*, ed. by R. Shapiro. New York: W. W. Norton, pp. 147–59.

Tomkins, G. E., Jackson, J. L., O'Malley, P. G., Balden, E. and Santoro, J. E. 2001. Treatment of chronic headache with antidepressants: A meta-analysis. *The American Journal of Medicine* 111: 54–63.

Tossani, E. 2014. Definition versus measurement of mental pain: A reply to Meerwijk and Weiss. *Psychothery and Psychosomatics* 83 (1). doi: 10.1159/000353394.

Tracey, I. 2010. Getting the pain you expect: Mechanisms of placebo, nocebo and reappraisal effects in humans. *Nature Medicine* 16: 1277–83.

Tye, K. M., Mirzabekov, J. J., Warden, M. R., Ferenczi, E. A., Tsai, H. C., Finkelstein, J., Kim, S. Y., Adhikari, A., Thompson, K. R., Andalman, A. S., et al. 2013. Dopamine neurons modulate neural encoding and expression of depression-related behaviour. *Nature* 493: 537–41.

Tyring, S. K. 2007. Management of herpes zoster and postherpetic neuralgia. *Journal of the American Academy of Dermatology* 57 (Suppl. 6): S136–S142.

Urgesi, C., Aglioti, S. M., Skrap, M. and Fabbro, F. 2010. The spiritual brain: Selective cortical lesions modulate human self-transcendence. *Neuron* 65: 309–19.

Vanneste, S., Joos, K. and De Ridder, D. 2012. Prefrontal cortex based sex differences in tinnitus perception: Same tinnitus intensity, same tinnitus distress, different mood. *PLoS One* 7. doi: 10.371/journal.pone.0031182.

Vinik, A. I. and Casellini, C. M. 2013. Guidelines in the management of diabetic nerve pain: Clinical utility of pregabalin. *Diabetes, Metabolic Syndrome and Obesity: Targets and Therapy* 6: 57–78.

Wei, F., Li, P. and Zhuo, M. 1999. Loss of synaptic depression in mammalian anterior cingulate cortex after amputation. *The Journal of Neuroscience* 19: 9346–54.

Wilensky M. 2006. Eye movement desensitization and reprocessing (EMDR) as a treatment for phantom limb pain. *Journal of Brief Therapy* 5: 31–44.

Wise, R. A. 2002. Brain reward circuitry: Insights from unsensed incentives. *Neuron* 36: 229–40.

Wood, P. B. 2006. Mesolimbic dopaminergic mechanisms and pain control. *Pain* 120: 230–4.

Wood, P. B., Schweinhardt, P., Jaeger, E., Dagher, A., Hakyemez, H., Rabiner, E. A., Bushnell, M. C. and Chizh, B. A. 2007. Fibromyalgia patients show an abnormal dopamine response to pain. *European Journal of Neuroscience* 25: 3576–82.

Wootton, D. 2006. *Bad Medicine: Doctors Doing Harm since Hippocrates.* Oxford: Oxford University Press.

Yavuz, B. G., Aydinlar, E. I., Dikmen, P. Y. and Incesu, C. 2013. Association between somatic amplification, anxiety, depression, stress and migraine. *The Journal of Headache and Pain* 14. doi: 10.1186/1129–2377–14–53.

Zarowitz, B. J., O'Shea, T. and Nance, M. 2014. Clinical, demographic, and pharmacologic features of nursing home residents with Huntington's disease. *Journal of the American Medical Directors Associations.* doi: 10.1016/j.jamda.

CHAPTER 14

Pain in multiple sclerosis: From classification to treatment

Claudio Solaro and Michele Messmer Uccelli

Introduction

Multiple sclerosis (MS) is an inflammatory, chronic, demyelinating disease of the central nervous system that results in motor, sensory and cognitive impairment. Other common symptoms are urinary dysfunction, visual impairment, fatigue, spasticity, incoordination, and pain (Paty and Ebers, 1999). Pain may be the most commonly treated symptom in MS, estimated to account for 30 per cent of all symptomatic treatment (Brichetto et al., 2003). MS typically occurs between the ages of 20 and 40 years, although diagnosis in children and in older adults does occur.

Pain in MS has been linked to quality of life (QOL) (Ehde et al., 2003), and there is an interdependent relationship between depression, fatigue and pain. A meta-analysis reported that the presence of pain is related to decreased employment (Shahrbanian et al., 2013), a result that confirms the importance of addressing MS comprehensively.

Classification of pain syndromes

Pain is defined as an 'unpleasant sensory and emotional experience associated with actual or potential tissue damage or described in terms of such damage' (Merskey and Bogduk, 1994: III, 210). Pain syndromes have been classified into nociceptive, somatic/visceral and neuropathic pain (Treede et al., 2008). Nociceptive pain occurs as an appropriate encoding of noxious stimuli and represents a physiological response, transmitted to the level of consciousness when nociceptors in bone, muscle, or any body tissue are activated, warning the organism of tissue damage and eliciting, in turn, coordinated reflexes and behavioural responses. Neuropatic pain arises directly from a lesion or from a disease that affects the somatosensory system; it has no biological advantage (such as warning) and causes distress (Treede et al., 2008). Clinical features include burning, dysesthetic, piercing pain, painful responses to non-painful stimuli (allodynia), and increased pain sensation when noxious stimuli are applied (hyperalgesia). Pain associated with MS is typically divided into four categories: continuous central neuropathic pain; intermittent

An Introduction to Pain and its Relation to Nervous System Disorders, First Edition. Edited by Anna A. Battaglia.
© 2016 John Wiley & Sons, Ltd. Published 2016 by John Wiley & Sons, Ltd.

central neuropathic pain (e.g., trigeminal neuralgia, Lhermitte's sign); musculoskeletal pain (e.g., painful tonic spasms, pain secondary to spasticity); and mixed neuropathic and non-neuropathic pain (e.g., headache) (O'Connor et al., 2008). More recently a mechanism-based classification of pain distinguished five types of pain: neuropathic, nociceptive, psychogenic, idiopathic and mixed pain (Truini et al., 2013). This classification is based on pathophysiology and treatment response more than the earlier one. Further, the use of the characterisation 'ongoing extremity pain' instead of 'dysaesthetic extremity pain' has been recommended, on the grounds that the latter is in contrast with the definition of sensory disturbances proposed by the International Association for the Study of Pain (IASP) (Merskey and Bogduk, 1994). Further, the International Headache Society has recently decided that the traditional term 'trigeminal neuralgia' be changed to 'painful trigeminal neuropathy attributed to multiple sclerosis plaque' (Headache Classification Committee of the International Headache Society, 2013). The present chapter applies this revised terminology.

Assessment

Due to the lack of a disease-specific scale, the assessment of pain in MS is currently limited to the use of generic pain instruments, generally, visual analogue and numerical rating scales. While these have some use in the subjective reporting of pain intensity and unpleasantness, they probably lack the level of depth sufficient to permit an accurate description of the complexity of neuropathic pain.

The DN4 questionnaire is a 10-item screening tool used to distinguish between neuropathic and non-neuropathic pain (a score of > 3 signifies the presence of neuropathic pain) (Bouhassira et al., 2005). In one study using the DN4 in 302 subjects with MS, 92 experienced pain (i.e., 30 per cent), and 42 of them (i.e., 13 per cent) experienced neuropathic pain. Fifteen subjects were diagnosed with ongoing extremity pain (Truini et al., 2012). Until recently the DN4 had not commonly been used in MS pain research, although its potential for differentiating pain syndromes could lead to better treatment outcomes.

Prevalence of pain in MS

A wide range of results from pain prevalence studies were reported over the past three decades. The inconsistencies maybe attributed to variations in the definitions and classifications applied, since there are no significant differences in prevalence results between data collection methods that used in-person interviews or self-administered questionnaires (Clifford and Trotter, 1984; Vermote, Ketalaer and Carltonn, 1986; Kassirer and Osterberg, 1987; Moulin, Foley and Ebers, 1988; Stenager et al., 1991; Warnell et al., 1991; Archibald et al., 1994; Indaco et al., 1994; Stenager, Knutsen and Jensen, 1995; Beiske et al., 2004; Svendsen, Jensen and Bach, 2003; Solaro et al., 2004; Österberg, Boivie and Thomas, 2005; Ehde et al., 2006; Hadjimichael et al., 2007; Piwko et al., 2007; Khan and Pallant, 2007; Boneschi et al., 2008; Douglas, Wollin and Windsor, 2008; Hirsh et al., 2009; Seixa et al., 2011; Truini et al., 2012). Table 14.1 provides details on sample sizes, selection and data collection

Table 14.1 Studies of the prevalence of pain in MS.

Study	N (males : females)	Selection method	Data collection	Prevalence (%)
Clifford and Trotter (1984)	317 (na)	Consecutive outpatients	Chart review	29
Vermote et al. (1986)	83 (na)	Rehabilitation inpatients	McGill Pain Questionnaire	54
Kassirer and Osterberg (1987)	28, 84% were wheelchair users (26:2)	Long-term outpatients	Specific questionnaire	75
Moulin et al. (1988)	159 (51:108)	Outpatients	Specific questionnaire plus interview	64
Stenager et al. (1991)	117 (52:65)	Random sample of inpatients	Specific questionnaire	65
Warnell (1991)	364 (115:249)	Outpatients	Specific questionnaire	64
Archibald et al. (1994)	85 (13:72)	Outpatients	Specific questionnaire	64
Indaco et al. (1994)	141 (61:80)	Outpatients	Structured interview	57 (excluding headache)
Stenager et al. (1995)	49 (22:27)	Follow-up of younger individuals from 1991 study	Structured interview	86
Beiske et al. (2004)	142 (47:95)	Outpatients	Interview	73.9 (pain and/ or sensory)
Ehde et al. (2003)	442 (111:331)	MS Society member database	Postal questionnaire	44
Svendsen et al. (2003)	627 with MS (214:413) vs 487 controls	Case–control study	Postal questionnaire	79.4 patients with MS vs 74.7 in controls
Solaro et al. (2004)	1,672 (520:1,152)	Random sample of outpatients	Structured interview	42.8
Österberg, Boivie and Thomas (2005)	364 (124:240)	Single-center registry	Postal questionnaire, telephone interview, in-person interview plus examination	57.5
Ehde et al. (2006)	180 (40:140)	Study pool	Postal questionnaire	66
Hadjimichael et al. (2007)	9,115 (2,188:6,927)	National patient registry	Postal questionnaire	74.5
Piwko et al. (2007)	297 (68:229)	Population based	Postal questionnaire	71
Khan and Pallant (2007)	94 (26:68)	Community based	Semi-structured interview	67
Boneschi et al. (2008)	428 (161:267)	Outpatients	Semi-structured interview	39.8
Douglas et al. (2008)	219 (40:179)	Community based	Structured interview	75
Hirsh et al. (2009)	2,994 (390:2,604)	MS population base	Postal questionnaire	91
Seixa et al. (2011)	85 (62 : 23)	MS population base	Interview	34
Truini et al. (2012)	302 (91/211)	Random sample of outpatients	Structured questionnaire	30

MS = multiple sclerosis; na = data not available.

methods, and prevalence results. In these studies pain has been inconsistently correlated with disability, disease course, disease duration, and age and has been reported to interfere with daily activities and employment and to influence quality of life negatively (Shahrbanian et al., 2013; Hirsh et al., 2009; Douglas et al., 2008; Khan and Pallant, 2007; Solaro and Tanganelli, 2004; Svendsen et al., 2003; Archibald et al., 1994). Overall, the prevalence of pain in MS is approximately 63 per cent. Pain itself is highly heterogeneous, not specifically linked to disease course, and can resolve over time in some subjects (Khan et al., 2013). For a systematic review and meta-analysis of the prevalence of pain in MS, one can consult Foley et al. (2013).

Pathophysiology of major pain symptoms in MS

Continuous central neuropathic pain

The pathology of central neuropathic pain continues to be inadequately defined. MS pain is thought to be a type of central pain due to demyelinating lesions in areas involved in pain perception.

Experimental autoimmune encephalomyelitis (EAE), an animal model of MS that is inducible in various rodent strains by immunisation with several myelin antigens, is widely used to better understand the pathophysiological features of the disease as well as the potential efficacy of medications, although its use in the study of pain is limited. The availability of an animal model could be an important step in understanding the mechanism that underlies pain in MS and could create an opportunity to test pharmacological therapies. Aicher et al. (2004) used proteolipid protein to induce active immunisation (chronic–relapsing form) and proteolipid protein-specific splenocytes for passive induction in Swiss/Jackson laboratory (SJL) mice. Nociceptive testing was done with withdrawal latencies to a radiant stimulus. In both models, hypoalgesia peaked prior to the peak of motor deficits, whereas during the chronic phase of the disease the mice developed hyperalgesia; this suggested that EAE may be a useful model for pain. In a chronic–relapsing model of EAE, Olechowski, Truong and Kerr (2009) studied pain sensitivity and found a significant decrease in elicited pain behaviour. This behaviour was found to involve the glutamatergic system, which suggested a potential mechanism underlying neuropathic pain. Thibault, Calvino and Pezet (2011) characterised sensory abnormalities, for example thermal and mechanical hyperalgesia, and showed a partial effect of medications (gabapentin, duloxetine, and tramadol) in two EAE models.

It has been hypothesised that MS is an acquired channelopathy (Waxman, 2001). Sodium channel Nav1.8, whose expression is normally restricted to the peripheral nervous system, is present in cerebellar Purkinje cells in a mouse model of MS. The ectopic expression of Nav1.8 contributes to symptom development in this model. Disregulated sodium channel expression on sensory fibres can lead to a functional change in axonal conduction, contributing to neuropathic pain (Waxman et al., 1999). Although the theory is intriguing and several hypotheses can be made regarding the mechanisms of action of sodium channel blockers such as antiepileptic drugs (e.g., cerbamazepine) on pain in MS, this remains an area that requires further study.

Different EAE models with genetic, clinical and histopathological heterogeneity show different profiles in pathological changes and sensory involvement – findings that may provide insight on the perception of pain in association with different types of MS (Lu et al., 2012). The field has gained limited insight into the pathophysiological mechanisms of neuropathic pain with the help of animal models, and of course replication in human subjects is unknown. Insights on the EAE model have recently been published by Khan and Smith (2014).

Treatment of central neuropathic pain

Treatment of central neuropathic pain in MS includes tricyclic antidepressants, antiepileptic medications, intrathecally administered baclofen, opioid analgesics, anaesthetic and antiarrhythmic agents and cannabinoids. An overview of published studies on medications for central neuropathic pain (excluding cannabinoids RTC) is given in Table 14.2.

Table 14.2 Studies of medications for central neuropathic pain in MS (excluding cannabinoid RTC).

Study	Design	Medication	No. subjects	Mean dose	Efficacy	No. subjects with AE
Breuer et al. (2007)	Double-blind, crossover	LMT	12	400 mg/day	No difference vs placebo	1
Cianchetti et al. (1999)	Open-label	LMT	15	400 mg/day	2/15 complete pr; 6/15 partial pr	0
Falah et al. (2011)	Double-blind	LEV	26	3000 mg/day	Significant pr in treatment group	12 (6 dropout)
Houtchens et al. (1997)	Open-label	GBP	25	600 mg/day	15/22 complete or partial pr	11/22 (5 dropout)
Kalman et al. (2002)	Single-blind,	morphine	14	0.67 mg/kg	pr	
Rog et al. (2007)	Open-label	CNB	63	9.6 spray daily [b]	Last week of rx NRS 2.9	30/34 (17/28 dropout)
Rossi et al. (2009)	Single-blind	LEV	12 vs 8 placebo	3000 mg/day	Significant pr	8/12 (1 dropout)
Solaro et al. (2009)	Open-label	PGB	16	154 mg/day	9/16 complete pr	3 dropout
Vollmer et al. (2013)	Double-blind	DUL	118 vs 121 placebo	60 mg/day	Significant difference in pr between groups	16 dropout

RTC = randomized controlled trial; b = Each spray delivered 2.7 mg of Δ9-tethrahydrocannabiol and 2.5 mg of cannabidiol; MS = multiple sclerosis; N = number; AE = adverse event; RX = treatment; NRS = numerical rating scale; CNB = cannabinoid; GBP = gabapentin; LEV = levetiracetam; LMT = lamotrigine; PGB = pregabalin; DUL = duloxetine; pr = pain relief/reduction.

Antidepressants

Tricyclic antidepressants such as amitriptyline, nortriptyline, and clomipramine, considered first-line drugs for treating central pain, act at the synaptic level, increasing serotoninergic and noradrenergic transmission, as well as on sodium channels. Their use in MS can be problematic due to common adverse effects that include drowsiness, dry mouth, constipation, urinary retention, and hypotension (Saarto and Wiffen, 2007). Currently there are no guidelines as to optimal dosing.

Duloxetine, a serotonine and norepinephrine reuptake inhibitor, has been used to treat pain in MS. In a six-week randomised trial, 239 subjects were administered daily 60 mg duloxtine (118 subjects) or placebo (121 subjects). The primary outcome was variation in mean pain intensity (0–10). Subjects receiving duloxetine showed a significant improvement (1.83 vs 1.01 points), but there was a higher frequency of adverse effects (13.6 vs 4.1 subjects), which included dizziness and somnolence (Vollmer et al., 2013).

Antiepileptic medications

Antiepileptic medications are also used in treating central neuropathic pain associated with MS, although these drugs have adverse effects that necessitate careful dosing vigilance and fine-tuning (Solaro et al., 2005). One example is carbamazepine, which is commonly used in MS despite its high incidence of adverse effects, which often result is non-adherence even at low doses. Adverse effects include drowsiness, vertigo, hypertension, bradycardia, rash, neutropenia, and abnormal liver function (Killian and Fromm, 2001).

Lamotrigine as an add-on drug has been shown to be partially successful in treating neuropathic pain when studied in a small group of patients, at a daily dose of up to 400 mg (Cianchetti et al., 1999). When used as the primary pain treatment, it does not have a clear effect on pain intensity. Lamotrigine seems to be relatively well tolerated by patients (Breuer et al., 2007).

Gabapentin has also been studied in small numbers of subjects and has shown to result in moderate to excellent pain relief at a mean dosage between 600 and 900 mg daily. However, approximately 50 per cent of the subjects studied experienced adverse effects that included somnolence and dyspepsia, and nearly 40 per cent of these (5/13 of the total number of subjects with adverse effects) discontinued treatment (Houtchens et al., 1997).

Levetiracetam, at 3,000 mg daily, has been shown to be effective for pain reduction in two small randomised placebo-controlled trials (Rossi et al., 2009; Falah et al., 2011). In one of these, levetiracetam was particularly effective for pain relief in patients with intense pain or without touch-evoked pain (Falah et al., 2011).

Pregabalin has been proposed for paroxysmal pain (Solaro, Boemker and Tanganelli, 2009). Adverse effects include dizziness and general malaise. Older subjects (over 60) treated with central nervous system (CNS)-targeted drugs such as benzodiazepine, baclofen, and tramadol have experienced delirium with low-dose pregabalin (75 mg daily) (Solaro and Tanganelli, 2009).

Overall, antiepileptic drugs for the treatment of central neuropathic pain in MS have demonstrated, at best, mediocre benefits. They require careful follow-up and fine-tuning of dosage, due to the frequent occurrence of intolerable adverse effects that often prevent patients from reaching the level of medication required for satisfactory pain relief (Solaro et al., 2005; Killian and Fromm, 2001; Rossi et al., 2009).

Intrathecally administered baclofen and opioid analgesics

Insufficient evidence exists for the use of baclofen or opioid analgesics for neuropathic pain associated with MS, although both have been reported in very small numbers of patients (4 baclofen, 9 baclofen plus morphine, 14 morphine alone) (Herman, D'Luzansky and Ippolito, 1992; Sadiq and Poopatana, 2007; Kalman et al., 2002). While subjects treated with a combination of baclofen plus morphine reported pain relief, the daily dose of morphine ranged from 210 to 9,500 mg; thus the possibility that this regimen merits further study was eliminated.

Cannabinoids

Unlike most other drugs, cannabinoids have been assessed for the treatment of pain in MS in double-blind randomised placebo-controlled clinical trials. Table 14.3 provides information on dosage, design, subjects and results from four trials (Svendsen et al., 2004; Rog et al., 2005; Zajicek et al. 2003; Langford et al., 2013; Rog, Nurmikko and Young, 2007). It is difficult to draw any conclusions from these reports due to inherent problems that include the small numbers of subjects in most cases, varying modes of administration (capsule vs oromucosal spray) and dosing, significant differences in treatment duration and other design peculiarities. The positive study on dronabinol reported that median spontaneous pain intensity was significantly lower in the treatment group than in the placebo group; but the sample consisted of 24 subjects (Svendsen et al., 2004). The largest study included 611 treated subjects, although pain was a secondary outcome and was limited to the subjects' impression of pain improvement after treatment, at the end of the study (Zajicek et al., 2003). Approximately half of the subjects felt that their pain had improved, while 20 per cent reported a worsening of pain with active treatment. A second large trial of 339 subjects found no significant difference on pain ratings between the treatment group and the placebo group, due to the large number of placebo responders (Langford et al., 2013).

Clearly these results are not sufficient for providing indications about any real capacity of cannabinoids to treat pain in MS. A further major shortcoming in cannabinoid research is that thus far no head-to-head trial comparing cannabinoids to other commonly used drugs for pain in MS has been conducted.

Painful trigeminal neuropathy attributed to multiple sclerosis plaque (trigeminal neuralgia)

Painful trigeminal neuropathy (PTN) attributed to multiple sclerosis plaque is ascribed to various causes (Da Silva et al., 2005) – for example, a plaque in the pons with or without vascular compression, or vascular compression of the nerve by an artery at the root entry zone (Lazar and Kirkpatrick, 1979; Gass et al., 1997; Meaney et al., 1995). One study reported 35 cases in which nerve compression was observed in 23 subjects (60 per cent) and T2-weighted imaging demonstrated a brainstem lesion in 26 subjects (74 per cent) (Broggi et al., 2004).

Table **14.3** Double-blind, randomized, placebo-controlled clinical trials on cannabinoids for neuropathic pain in MS.

Study	Agent/dosing	Mode of administration	Measure	No. subjects	Duration	Results
Svendsen et al., 2004	dronabinol/ max 10 mg	oral capsules	NRS of spontaneous pain intensity	24	3 week rx vs. placebo + 3 week washout + 3 week rx vs. placebo	median spontaneous pain intensity significantly lower during dronabinol rx than during placebo rx
Rog et al., 2005	Δ 9-THC:CBD/ self-titrated	oromucosol spray, add-on to analgesic rx	NRS-11	66	5 weeks	improved pain intensity in rx group NRS 3.8 vs. placebo NRS 5.0
Zajicek et al., 2003	Δ 9-THC vs. oral cannabis extract vs. placebo/ self-titrated	oral capsules	rating scale of perceived pain improvement (as secondary outcome)	611	15 weeks	approx half of rx subjects reported (general) pain improvment; 20% rx subjects reported worse pain
Langford et al., 2013	Δ 9-THC/ self-titrated	oromucosol spray, add-on to analgesi rx	NRS	339	14 weeks rx vs. placebo + 14 weeks open label	phase A: no significant difference on pain rating between groups
			NRS	53	4 weeks randomized to rx vs. placebo	phase B: 57% placebo failed rx vs. 24% rx group

RX = treatment; NRS = numerical rating scale; n = number.

It has been suggested that trigeminal evoked potentials are relatively accurate in indicating the site of lesions (Cruccu et al., 1990; Bergamaschi et al., 1994).

Treatment of painful trigeminal neuropathy attributed to multiple sclerosis plaque

PTN is estimated to occur in approximately 6 per cent of patients with MS (Putzki et al., 2009). Although relatively uncommon, this symptom is extremely painful and has a relevant impact on quality of life (Hooge and Redekop, 1995). Treatment of PTN consists primarily of antiepileptic medications acting on voltage-dependent sodium channels. Although there have been no randomised clinical trials, reports from small, open-label studies are available on carbamazepine (Ramsaransing, Zwanikken and De Keyser, 2000), lamotrogine (Leandri et al., 2000; Solaro et al., 1998), topiramate (Solaro et al., 2001), combination therapy (Solaro, Messmer Uccelli, Guglieri, et al., 2000), and misoprostol (DMKG 2003). Table 14.4 provides information on these reports. It is worth noting that, while carbamazepine is relatively effective in treating PTN, a particularly adverse effect is the reversible worsening of MS symptoms (Solaro et al., 2005; Ramsaransing et al., 2000), which requires careful monitoring and patient education. The other drugs appear to be useful in the small numbers of subjects studied and also require fine-tuning so as to balance benefit with adverse effects. Unfortunately the combined results do not provide adequate data necessary for evidence-based decisions on the pharmacological treatment of PTN in patients with MS.

Table 14.4 Studies of medications for painful trigeminal neuropathy attributed to multiple sclerosis plaque).

Study	Medication	No. subjects	Mean dose	No. subjects with pain reduction/ total subjects	No. subjects with AE
DMKG Study Group (2003)	MSP[a]	18	600 µg/day	14/18	4
Espir and Millac (1970)	CBZ	5	760 mg/day	4/5	1
Jorns et al. (2009)	LEV	10	320 mg/day	4/10	3
Khan (1998)	GBP	7	1400 mg/day	6/7 complete; 1/7 partial	NS
Leandri et al. (2000)	LMT	18	170 mg/day	17/18	1
Lunardi et al. (1997)	LMT	5	165 mg/day	5/5	NS
Pfau et al. (2012)	MSP	3	1000 µg/day	3/3	0
Ramsaransing et al. (2000)	CBZ + GBP	6	400 + 800 mg/ day	6/6	0
Reder and Arnason (1995)	MSP	7	570 µg/day	4/7 complete; 2/7 partial	0
Solaro et al. (1998)	GBP	6	900 mg/day	5/6	0
Solaro, Messmer Uccelli, Uccelli, et al. (2000)	LMT + GBP	5	170 mg + 780 mg/day	5/5	0
Solaro et al. (2001)	TPM	2	200 mg/day	2/2	NS
Zvartau-Hind et al. (2000)	TPM	6	100 mg bid	5/6	0

a = previous therapy maintained; AE = adverse event; bid = twice daily; CBZ = carbamazepine; DMKG = German Migraine and Headache Society; GBP = gabapentin; LMT = lamotrigine; MSP = misoprostol; NS = not specified; TPM = topiramate; LEV = levetiracetam.

Surgical interventions for painful trigeminal neuropathy attributed to multiple sclerosis plaque

For pharmacological-resistant PTN, invasive treatments can be considered. Glycerol injection (Linderoth and Hakanson, 1989; Kondziolka, Lunsford and Bissonette, 1994; Berk, Constantoyannis and Honey, 2003), radiofrequency lesioning (Broggi and Franzini, 1982), and radiosurgery (Rogers et al., 2002; Huang et al., 2002) ablate the retrogasserian ganglion in order to interrupt the trigeminal pathway; and they have been used in MS. These procedures can cause nerve damage that in rare cases can itself result in relevant adverse effects – such as hypoesthesia/hyperesthesia, decreased corneal reflex, transitory masticatory weakness, and hearing loss. Microvascular decompression of the trigeminal root may be effective in carefully selected subjects, although the recurrence of pain is relatively common, suggesting that combined mechanisms other than nerve compression have a role in the generation of PTN in MS (Broggi et al., 1999). In a study involving 96 subjects with MS and PTN who were followed for a mean period of approximately six years, initial pain relief was obtained in the majority of patients; recurrence of PTN was registered in 66 per cent of cases; and repeating the procedure, regardless of its type, was usually less effective than the initial intervention (Mohammad-Mohammadi et al., 2013).

Given these findings, surgical interventions for PTN in MS should be reserved for critical cases in which there is no acceptable alternative, obviously following careful patient education regarding outcome expectations (Mohammad-Mohammadi et al., 2013).

Lhermitte's sign

Lhermitte's sign is described as a painful, electric-like sensation that runs down the back and into the limbs and is elicited by flexing the neck. Approximately 40 per cent of patients with MS experience Lhermitte's sign at some point during the course of the disease. Although relatively common in MS, it is not limited to it. Treatment with low-dose carbamazepine is rarely necessary and is typically reserved for persistent cases (Al-Araji and Oger, 2005).

Musculoskeletal pain

Painful tonic spasms

Painful tonic spasms related to MS, present in approximately 11 per cent of subjects, are described as a cramping, pulling pain that most commonly occurs in the lower extremities, rarely in the upper limbs, more often at night, and is typically triggered by sensory stimuli (Solaro et al., 2004). This symptom is due to ectopic impulses resulting from demyelination and axonal damage, which determine a painful sensation at the level of the muscle (O'Connor et al., 2008).

Anecdotally, medications such as baclofen, benzodiazepines, gabapentin, carbamazepine and tiagabine, a selective inhibitor of the γ-aminobutyric acid transporter, are used for treating painful tonic spasms. Botulinum toxin has also been proposed for reducing painful spasms – an effect that is maintainable for at least 90 days, apparently with no noted adverse events (Restivo et al., 2003).

Overall the available data do not offer indications as to the most effective treatment of these various medications. This is an area where, due to the lack of evidence, clinical experience will probably continue to determine treatment choices.

Secondary pain

Pain in MS can be secondary to other characteristics of the disease, including other symptoms, particularly spasticity. Depression has also been shown to influence pain perception in patients with MS (Alschuler, Ehde and Jensen, 2013). Secondary pain can be experienced by patients with a more debilitating disease course, particularly those with impaired mobility. In these subjects, pain may be due to incorrect posture, improper use of technical aids, prolonged sitting, and the development of decubitus lesions in more extreme cases (Thompson, 1998). Given this, it is important not only that neuropathic pain symptoms be routinely evaluated in the clinical setting, but also that secondary-pain sources be included in the overall pain assessment and management.

Conclusions

- A clear correlation between pain and clinical variables has not been established. It seems likely that a correlation between disability and a more chronic course of MS is likely, while at the same time patients experience pain syndromes early in the disease course.
- The DN4 is the only currently available pain scale that is able to distinguish neuropathic from nociceptive pain. The next step should be assessing how best to use this tool in a clinical setting.
- For a better understanding of neuronal damage related to pain in MS, EAE studies should be expanded, since the initial suppositions are interesting but limited in scope and could in principle lead to new therapeutic strategies with a better understanding of the distinct mechanisms of pain in MS.
- Treatment of pain syndromes in MS continues to be generally based on clinical experience and on a scattering of small, open-label studies. There is insufficient evidence on any particular pain treatment necessary for developing treatment guidelines.

References

Aicher, S. A., Silverman, M. B., Winkler, C. W. and Bebo, B. F., Jr. 2004. Hyperalgesia in an animal model of multiple sclerosis. *Pain* 110: 560–70.

Al-Araji, A. H. and Oger, J. 2005. Reappraisal of Lhermitte's sign in multiple sclerosis. *Multiple Sclerosis* 11: 398–402.

Alschuler, K. N., Ehde, D. M. and Jensen, M. P. 2013. Co-occurring depression and pain in multiple sclerosis. *Physical Medicine and Rehabilitation Clinics of North America* 24 (4): 703–15.

Archibald, C. J., McGrath, P. J., Ritvo, P. G., Fisk, J. D., Bhan, V., Maxner, C. E., and Murray, T. J. 1994. Pain prevalence, severity and impact in a clinical sample of multiple sclerosis patients. *Pain* 58: 89–93.

Beiske, A. G., Pedersen, E. D., Czujko, B. and Myhr, K. M. 2004. Pain and sensory complaints in multiple sclerosis. *European Journal of Neurology* 1: 479–82.

Bergamaschi, R., Romani, A., Versino, M., Callieco, R., Gaspari, D., Citterio, A. and Cosi, V. 1994. Usefulness of trigeminal somatosensory evoked potentials to detect subclinical trigeminal impairment in multiple sclerosis patients. *Acta Neurologica Scandinavica* 89: 412–4.

Berk, C., Constantoyannis, C. and Honey, C. R. 2003. The treatment of trigeminal neuralgia in patients with multiple sclerosis using percutaneous radiofrequency rhizotomy. *Canadian Journal of Neurological Sciences* 30: 220–3.

Bouhassira, D., Attal, N., Alchaar, H., Boureau, F., Brochet, B., Bruxelle, J., Cunin, G., Fermanian, J., Ginies, P., Grun-Overdyking, A., et al. 2005. Comparison of pain syndromes associated with nervous or somatic lesions and development of a new neuropathic pain diagnostic questionnaire (DN4). *Pain* 114 (1–2): 29–36.

Breuer, B., Pappagallo, M., Knotkova, H., Guleyupoglu, N., Wallenstein, S., and Portenoy, R. K. 2007. Randomized double-blind, placebo-controlled two period crossover pilot trial of lamotrigine in patients with central pain due to multiple sclerosis. *Clinical Therapeutics* 29: 2022–30.

Brichetto, G., Messmer Uccelli, M., Mancardi, G. L. and Solaro, C. 2003. Symptomatic medication use in multiple clerosis. *Multiple Sclerosis* 9 (5): 458–60.

Broggi, G. and Franzini, A. 1982. Radiofrequency trigeminal rhizotomy in treatment of symptomatic non-neoplastic facial pain. *Journal of Neurosurgery* 57: 483–6.

Broggi, G., Ferroli, P., Franzini, A., Nazzi, V., Farina, L., La Mantia, L. and Milanese, C. 2004. Operative findings and outcomes of microvascular decompression for trigeminal neuralgia in 35 patients affected by multiple sclerosis. *Neurosurgery* 55 (4): 830–8. (Discussion: 838–9.)

Broggi, G., Ferroli, P., Franzini, A., Pluderi, M., La Mantia, L. and Milanese, C. 1999. Role of microvascular decompression in trigeminal neuralgia and multiple sclerosis. *Lancet* 354: 1878–9.

Cianchetti, C., Zuddas, A., Randazzo, A. P., Perra, L. and Marrosu, M. G. 1999. Lamotrigine adjunctive therapy in painful phenomena in MS: Preliminary observations. *Neurology* 53 (2): 433.

Clifford, D. B. and Trotter, J. L. 1984. Pain in multiple sclerosis. *Archives of Neurology* 41: 1270–2.

Cruccu, G., Leandri, M., Feliciani, M. and Manfredi, M. 1990. Idiopathic and symptomatic trigeminal pain. *Journal of Neurology, Neurosurgery, and Psychiatry* 53 (12): 1034–42.

Da Silva, C. J., Da Rocha, A. J., Mendes, M. F., Maja, A. C. M., Jr, Braga, F. T. and Tilbery, C. P. 2005. Trigeminal involvement in multiple sclerosis: Magnetic resonance imaging findings with clinical correlation in a series of patients. *Multiple Sclerosis* 11: 282–5.

DMKG Study Group. 2003. Misoprostol in the treatment of trigeminal neuralgia associated with multiple sclerosis. *Journal of Neurology* 250: 542–5.

Douglas, C., Wollin, J. and Windsor, C. 2008. Illness and demographic correlates of chronic pain among a community-based sample of people with multiple sclerosis. *Archives of Physical Medicine and Rehabilitation* 89: 1923–32.

Ehde, D. M., Gibbons, L.E., Chwastiak, L., Bombardier, C. H., Sullivan, M. D. and Kraft, G. H. 2003. Chronic pain in a large community sample of persons with multiple sclerosis. *Multiple Sclerosis* 6: 605–11.

Ehde, D. M., Osborne, T. L., Hanley, M. A., Jensen, M. P. and Kraft, G. H. 2006. The scope and nature of pain in persons with multiple sclerosis. *Multiple Sclerosis* 12: 629–38.

Espir, M. L. and Millac, P. 1970.Treatment of paroxysmal disorders in multiple sclerosis with carbamazepine (Tegretol). *Journal of Neurology, Neurosurgery, and Psychiatry* 33 (4): 528–31.

Falah, M., Madsen, C., Holbech, J. V. and Sindrup, S. H. 2011. A randomized, placebo-controlled trial of levetiracetam in central pain in multiple sclerosis. *European Journal of Pain* 1532–2149.

Foley, P. L., Vesterinen, H. M., Laird, B. J., Sena, E. S., Colvin, L. A., Chandran, S., MacLeod, M. R. and Fallon, M. T. 2013. Prevalence and natural history of pain in adults with multiple sclerosis: Systematic review and meta-analysis. *Pain* 154 (5): 632–42.

Gass, A., Kitchen, N., MacManus, D. G., Moseley, I. F., Hennerici, M. G. and Miller, D. H. 1997. Trigeminal neuralgia in patients with multiple sclerosis: Lesion localization with magnetic resonance imaging. *Neurology* 49 (4): 1142–4.

Hadjimichael, O., Kerns, R. D., Rizzo, M. A., Cutter, G. and Vollmer, T. 2007. Persistent pain and uncomfortable sensations in persons with multiple sclerosis. *Pain* 127: 35–41.

Headache Classification Committee of the International Headache Society (IHS). 2013. The International Classification of Headache Disorders, 3rd edition (beta version). *Cephalalgia* 33 (9): 629–808.

Herman, R. M., D'Luzansky, S. C. and Ippolito, R. 1992. Intrathecal baclofen suppresses central pain in patients with spinal lesions: A pilot study. *Clinical Journal of Pain* 8: 338–45.

Hirsh, A. T., Turner, A. P., Ehde, D. M. and Haselkorn, J. K. 2009. Prevalence and impact of pain in multiple sclerosis: Physical and psychological contributors. *Archives of Physical Medicine and Rehabilitation* 90: 646–51.

Hooge, J. P. and Redekop, W. K. 1995. Trigeminal neuralgia in multiple sclerosis. *Neurology* 45: 1294–6.

Houtchens, M. K., Richert, J. R., Sami, A. and Rose, J. W. 1997. Open label gabapentin treatment for pain in multiple sclerosis. *Multiple Sclerosis* 3: 250–3.

Huang, E., Teh, B. S., Zeck, O., Woo, S. Y., Lu, H. H.,Chiu, J. K., Butler, E. B., Gormley, W. B., and Carpenter, L. S. 2002. Gamma knife radiosurgery for treatment of trigeminal neuralgia in multiple sclerosis patients. *Stereotactic and Functional Neurosurgery* 79: 44–50.

Indaco, A., Iachetta, C., Nappi, C., Socci, L. and Carrieri, P. B. 1994. Chronic and acute pain syndromes in patients with multiple sclerosis. *Acta Neurologica* 16: 97–102.

Jorns, T. P., Johnston, A., and Zakrzewska, J. M. 2009. Pilot study to evaluate the efficacy and tolerability of levetiracetam (Keppra) in treatment of patients with trigeminal neuralgia. *European Journal of Neurology* 16: 740–44.

Kalman, S., Österbergn, A., Sörensenn, J., Boivie, J. and Bertler, A. 2002. Morphine responsiveness in a group of well-defined multiple sclerosis patients: A study with i.v. morphine. *European Journal of Pain* 6: 69–80.

Kassirer, M. R. and Osterberg, D. H. 1987. Pain in chronic multiple sclerosis. *Journal of Pain and Symptom Management* 2: 95–7.

Khan, F. and Pallant, J. 2007. Chronic pain in multiple sclerosis: Prevalence, characteristics and impact on quality of life in an Australian community cohort. *Journal of Pain* 8: 614–23.

Khan, N. and Smith, M. T. 2014. Multiple sclerosis-induced neuropathic pain: Pharmacological management and pathophysiological insights from rodent EAE models. *Inflammopharmacology* 22 (1): 1–22.

Khan, F., Amatya, B. and Kesserling, J. 2013. Longitudinal 7-year follow-up of chronic pain in persons with multiple sclerosis in the community. *Journal of Neurology* 260 (8): 2005–15.

Khan, O. A. 1998. Gabapentin relieves trigeminal neuralgia in multiple sclerosis patients. *Neurology* 51: 611–4.

Killian, J. M. and Fromm, G. H. 2001. Carbamazepine in the treatment of neuralgia: Use and side effects. *Archives of Neurology* 19: 129–36.

Kondziolka, D., Lunsford, L. D. and Bissonette, D. J. 1994. Long-term results after glycerol rhizotomy for multiple sclerosis-related trigeminal neuralgia. *Canadian Journal of Neurological Sciences* 21: 137–40.

Langford, R. M., Mares, J., Novotna, A., Vachova, M., Novakova, I., Notcutt, W. and Ratcliffe, S. 2013. A double-blind, randomized, placebo-controlled, parallel-group study of THC/CBD oromucosal spray in combination with the existing treatment regimen, in the relief of central neuropathic pain in patients with multiple sclerosis. *Journal of Neurology* 260 (4): 984–97.

Lazar, M. L. and Kirkpatrick, J. B. 1979. Trigeminal neuralgia and multiple sclerosis: Demonstration of the plaque in an operative case. *Neurosurgery* 5: 711–7.

Leandri, M., Lundardi, G., Inglese, M., Messmer-Uccelli, M., Mancardi, G. L., Gottlieb, A., and Solaro, C. 2000. Lamotrigine in trigeminal neuralgia secondary to multiple sclerosis. *Journal of Neurology* 247: 556–8.

Linderoth, B. and Hakanson, S. 1989. Paroxysmal facial pain in disseminated sclerosis treated by retrogasserian glycerol injection. *Acta Neurologica Scandinavica* 80: 341–6.

Lu, J., Kurejova, M., Wirotanseng, L. N., Linker, R. A., Kuner, R. and Tappe-Theodor, A. 2012. Pain in experimental autoimmune encephalitis: A comparative study between different mouse models. *Journal of Neuroinflammation* 9: 233.

Lunardi, G., Leandri, M., Albano, C., Cultrera, S., Fracassi, M., Rubino,V., and Favale, E. 1997. Clinical effectiveness of lamotrigine and plasma levels in essential and symptomatic trigeminal neuralgia. *Neurology* 48: 1714–7.

Martinelli Boneschi, F., Colombo, B., Annovazzi, P., Martinelli, V., Bernasconi, L., Solaro, C. and Comi, G. 2008. Lifetime and actual prevalence of pain and headache in multiple sclerosis. *Multiple Sclerosis* 14: 514–21.

Meaney, J. F. M., Watt, J. M. G., Eldridge, P. R., Whitehouse, G. H., Wells, J. C. D. and Miles, J. B. 1995. Association between trigeminal neuralgia and multiple sclerosis: Role of magnetic resonance imaging. *Journal of Neurology, Neurosurgery, and Psychiatry* 59: 253–9.

Merskey, H. and Bogduk, N. 1994. *Classification of Chronic Pain*, 2nd edn. Seattle, WA: IASP Task Force on Taxonomy, IASP Press.

Mohammad-Mohammadi, A., Recinos, P. F., Lee, J. H., Elson, P. and Barnett, G. H. 2013. Surgical outcomes of trigeminal neuralgia in patients with multiple sclerosis. *Neurosurgery* 73 (6): 941–50.

Moulin, D. E., Foley, K. M. and Ebers, G. C. 1988. Pain syndromes in multiple sclerosis. *Neurology* 38: 1830–4.

O'Connor, A. B., Schwid, S. R., Herrmann, D. N., Markman, J. D. and Dworkin, R. H. 2008. Pain associated with multiple sclerosis: Systematic review and proposed classification. *Pain* 137: 96–111.

Österberg, A., Boivie, J. and Thomas, K. A. 2005. Central pain in multiple sclerosis: Prevalence and clinical characteristics. *European Journal of Pain* 9: 531–42.

Olechowski, C. J., Truong, J. J. and Kerr, B. J. 2009. Neuropathic pain behaviours in a chronic-relapsing model of experimental autoimmune encephalomyelitis (EAE).S141 (1–2): 156–64.

Paty, D. W. and Ebers, G. C., eds. 1999. *Multiple Sclerosis*. Philadelphia: F. A. Davis.

Pfau, G., Brinkers, M., Treuheit, T., Kretzschmar, M., Sentürk, M., and Hachenberg, T. 2012. Misoprostol as a therapeutic option for trigeminal neuralgia in patients with multiple sclerosis. *Pain Medicine* 13: 1377–8.

Piwko, C., Desjardins, O. B., Bereza, B. G., Machado, M., Jaszewski, B., Freedman, M. S., Einarson, T. R. and Iskedjian, M. 2007. Pain due to multiple sclerosis: Analysis of the prevalence and economic burden in Canada. *Pain Research & Management* 12: 259–65.

Putzki, N., Pfriem, A., Limmroth, V., Yaldizli, O., Tettenborn, B., Diener, H. C. and Katsarava, Z. 2009. Prevalence of migraine, tension-type headache and trigeminal neuralgia in multiple sclerosis. *European Journal of Neurology* 16 (2): 262–7.

Ramsaransing, G., Zwanikken, C. and De Keyser, J. 2000. Worsening of symptoms of multiple sclerosis associated with carbamazepine. *British Medical Journal* 320: 1113.

Reder, A. T. and Arnason, B. 1995. Trigeminal neuralgia in multiple sclerosis relieved by a prostaglandin E analogue. *Neurology* 45: 1097–100.

Restivo, D. A., Tinazzi, M., Patti, F., Palmeri, A. and Maimone, D. 2003. Botulinum toxin treatment of painful tonic spasms in multiple sclerosis. *Neurology* 61: 719–20.

Rog, D. J., Nurmikko, T. J., Friede, T. and Young, C. 2005. Randomized, controlled trial of cannabis-based medicine in central pain in multiple sclerosis. *Neurology* 65 (6): 812–9.

Rog, D. J., Nurmikko, T. J. and Young, C. A. 2007. Oromucosal delta9-tetrahydrocannabinol/ cannabidiol for neuropathic pain associated with multiple sclerosis: An uncontrolled, open-label, 2-year extension trial. *Clinical Therapeutics* 29: 2068–79.

Rogers, C. L., Shetter, A. G., Ponce, F. A., Fiedler, J. A., Smith, K. A. and Speiser, B. L. 2002. Gamma knife radiosurgery for trigeminal neuralgia associated with multiple sclerosis. *Journal of Neurosurgery* 97 (Suppl. 5): 529–32.

Rossi, S., Mataluni, G., Codecà, C., Fiore, S., Buttari, F., Musells, A., Castelli, M., Bernardi, G. and Centonze, D. 2009. Effects of levetiracetam on chronic pain in multiple sclerosis: Results of a pilot, randomized placebo-controlled study. *European Journal of Neurology* 16: 360–6.

Saarto, T. and Wiffen, P. J. 2007. Antidepressants for neuropathic pain. *Cochrane Database of Systematic Reviews* 4. doi: 10.1002/14651858.CD005454.pub2.

Sadiq, S. and Poopatana, C. 2007. Intrathecal basclofen and morphine in multiple sclerosis patients with severe pain and spasticity. *Journal of Neurology* 254: 1464–5.

Seixa, D., Vasco Galhardo, M. J., Guimarães, J. and Lima, L. 2011. Pain in Portuguese patients with multiple sclerosis. *Frontiers in Neurology* 2 (20).

Shahrbanian, S., Auais, M., Duquette, P., Anderson, K. and Mayo, N. E. 2013. Does pain in individuals with multiple sclerosis affect employment? A systematic review and meta-analysis. *Pain Research & Management* 18 (5): e94–e100.

Solaro, C. and Tanganelli, P. 2004. Tiagabine for treating painful tonic spasms in multiple sclerosis: A pilot study. *Journal of Neurology, Neurosurgery, and Psychiatry* 75 (341).

Solaro, C. and Tanganelli, P. 2009. Acute delirium in patients with multiple sclerosis treated with pregabalin. *Clinical Neuropharmacology* 32: 236–7.

Solaro, C., Boemker, M. and Tanganelli, P. 2009. Pregabalin for treating paroxysmal symptoms in multiple sclerosis: A pilot study. *Journal of Neurology* 256: 1773–4.

Solaro, C., Brichetto, G., Battaglia, M. A., Messmer Uccelli, M. and Mancardi, G. L. 2005. Antiepileptic medications in multiple sclerosis: Adverse effects in a three-year follow-up study. *Neurological Sciences* 25: 307–10.

Solaro, C., Messmer Uccelli, M., Brichetto, G., Gasperini, C. and Mancardi, G. L. 2001. Topiramate relieves idiopathic and symptomatic trigeminal neuralgia. *Journal of Pain and Symptom Management* 21: 367–8.

Solaro, C., Messmer Uccelli, M., Guglieri, P., Uccelli, A. and Mancardi, G. L. 2000a. Gabapentin is effective in treating nocturnal painful spasms in multiple sclerosis. *Multiple Sclerosis* 6: 192–3.

Solaro, C., Messmer Uccelli, M., Uccelli, A., Leandri, M. and Mancardi, G. L. 2000b. Low dose gabapentin combined with either lamotrigine or carbamazepine can be useful therapies for trigeminal neuralgia in multiple sclerosis. *European Neurology* 44: 45–8.

Solaro, C., Lunardi, G. L., Capello, E., Inglese, M., Uccelli Messmer, M., Uccelli, A. and Mancardi, G. L. 1998. An open-label trial of gabapentin treatment of paroxysmal symptoms in multiple sclerosis patients. *Neurology* 51: 609–11.

Solaro, C., Brichetto, G., Amato, M. P., Cocco, E., Colombo, B., D'Aleo, G., Gasperini, C., Ghezzi, A., Merinelli, V., Milanese, C., et al. 2004. The prevalence of pain in multiple sclerosis: A multicenter cross-sectional study. *Neurology* 63: 919–21.

Stenager, E., Knutsen, L. and Jensen, K. 1991. Acute and chronic pain syndromes in multiple sclerosis. *Acta Neurologica Scandinavica* 84: 197–200.

Stenager, E., Knudsen, L. and Jensen, K. 1995. Acute and chronic pain syndromes in multiple sclerosis. *Italian Journal of Neurological Sciences* 16: 629–32.

Svendsen, K. B., Jensen, T. S. and Bach, F. W. 2004. Does the cannabinoid dronabinol reduce central pain in multiple sclerosis? Randomised double blind placebo controlled crossover trial. *British Medical Journal* 329 (7460). doi: 10.1136/bmj.38149.566979.

Svendsen, K. B., Jensen, T. S., Overvad, K., Hansen, H. J., Koch-Henriksen, N., and Bach, F. W. 2003. Pain in patients with multiple sclerosis: A population-based study. *Archives of Neurology* 60: 1089–94.

Thibault, K., Calvino, B. and Pezet, S. 2011. Characterisation of sensory abnormalities observed in an animal model of multiple sclerosis: A behavioural and pharmacological study. *European Journal of Pain* 15 (3): 231. e1–231.e16.

Thompson, A. J. 1998. Symptomatic treatment in multiple sclerosis. *Current Opinion in Neurology* 11: 305–9.

Treede, R. D., Jensen, T. S., Campbell, J. N., Cruccu, G., Dostrovsky, J. O., Griffin, J. W., Hansson, P., Hughes, R., Nurmikko, T. and Serra, J. 2008. Neuropathic pain: Redefinition and a grading system for clinical and research purposes. *Neurology* 70: 1630–5.

Truini, A., Galeotti, F., La Cesa, S., Di Rezze, S., Biasiotta, A., Di Stefano, G., Tinelli, E., Millefiorini, E., Gatti, A. and Cruccu, G. 2012. Mechanisms of pain in multiple sclerosis: A combined clinical and neurophysiological study. *Pain* 153: 2048–54.

Truini A., Barbanti, P., Pozzilli, C. and Cruccu, G. 2013. A mechanism-based classification of pain in multiple sclerosis. *Journal of Neurology* 260: 351–367A.

Vermote, R., Ketalaer, P. and Carltonn, H. 1986. Pain in multiple sclerosis patients: A prospective study using the McGill pain questionnaire. *Clinical Neurology and Neurosurgery* 88: 87–93.

Vollmer, T. L., Robinson, M. J., Risser, R. C. and Malcolm, S. K. 2013. A randomized, double-blind, placebo-controlled trial of duloxetine for the treatment of pain in patients with multiple sclerosis. *Pain Practice* 14 (8): 732–44.

Warnell, P. 1991. The pain experience of a multiple sclerosis population: A descriptive study. *Axone* 13 (1): 26–8.

Waxman, S. G. 2001. Acquired channelopathies in nerve injury and MS. *Neurology* 56 (12): 1621–7.

Waxman, S. G., Dib-Hajj, S., Cummins, T. R. and Black, J. A. 1999. Sodium channels and pain. *Proceedings of the National Academy of Sciences of the United Staes of America* 96 (14): 7635–9.

Zajicek, J., Fox, P., Sanders, H., Wright, D., Vickery, J., Nunn, A., Thompson, A. and UK MS Research Group. 2003. Cannabinoids for treatment of spasticity and other symptoms related to multiple sclerosis (CAMS study): Multicentre randomised placebo-controlled trial. *Lancet* 362: 1517–26.

Zvartau-Hind, M., Din, M. U., Gilan, A., Lisak, R. P. and Khan, O. A. 2000. Topiramate relieves refractory trigeminal neuralgia in MS patients. *Neurology* 55: 1587–8.

CHAPTER 15

Pain perception in dementia

Miriam Kunz and Stefan Lautenbacher

Introduction

There is an estimated 35 million people with dementia across the world. 'Dementia' designates a broad category of brain neurodegenerative diseases that are accompanied by long-term loss of cognitive ability in memory functioning, attention, executive functioning, orientation, language and other cognitive domains, as well as by changes in mood and behaviour. These cognitive and affective changes increase along the course of dementia and severely impact a person's daily functioning. The most common type of dementia is Alzheimer's disease. Other common forms are vascular dementia, frontotemporal dementia and Lewy body dementia. Given that prevalence rates of dementia are tightly linked to aging, demographic changes in the coming decades and the increasingly aging population will lead to a substantial growth in the number of people affected by dementia and in the scale of the challenge of providing appropriate treatment and care for all of them. Pain presents a great challenge in the treatment of patients with dementia. The prevalence of pain, and especially of chronic pain, is strongly related to age, hitting the oldest population the hardest: prevalence rates above the age of 85 years are of 70 per cent (Helme and Gibson, 2001). Given these circumstances, it is clear that pain is probably very common among people suffering from dementia. However, because such patients (especially those in the advanced stages of the disease) are often unable to use self-report to communicate their suffering, their pain is often overlooked, and therefore remains untreated (Scherder et al., 2009). For a long time the topic of pain in dementia has been overlooked; but in the last two decades clinicians and scientists have started to conduct research on the prevalence (Takai et al., 2010) and assessment (Hadjistavropoulos et al., 2007; Herr, 2011) of pain in older people with dementia and on the challenges related to its treatment in this population and have stressed the severe consequences of undetected and untreated pain (AGS Panel, 2009; Gloth, 2011). Persistent pain has been associated with a progressive decline in functional and mental capacity (Moriatry, McGuire and Finn, 2011), in social interaction (Lin et al., 2011), in quality of life (Cipher and Clifford, 2004), in appetite (Bosley et al., 2004) – as well as with sleep (Giron et al., 2002) and behavioural disturbances

An Introduction to Pain and its Relation to Nervous System Disorders, First Edition. Edited by Anna A. Battaglia.
© 2016 John Wiley & Sons, Ltd. Published 2016 by John Wiley & Sons, Ltd.

such as agitation, depression and anxiety (Husebo et al., 2011). Given these severe consequences, it is a pressing issue to improve our understanding of how dementia impacts pain processing and how pain can be assessed and treated in this vulnerable patient group.

On the basis of previous reviews made in our group (e.g., Husebo et al., 2012), we give here an overview of the literature on pain and dementia – including of basic research that investigates how pain processing and pain responses are affected by the disease. Our overview also summarises pain assessment and treatment recommendations for these individuals.

Experimental perspective

When research on pain and dementia started in the late 1990s, some believed that patients with dementia reported less pain simply because their neurodegenerative decline made them less sensitive to pain. In order to investigate how the pain system itself might be altered across the evolution of dementia, experimental studies are necessary. So far, there have been several studies that used experimental pain in patients with dementia (these were mostly patients with Alzheimer´s disease) and assessed changes in pain thresholds, pain tolerance, pain reflexes and supra-threshold responses. Applying experimental pain has the advantage of making it possible to gain complete control over the noxious stimulus and, in the process, to disentangle that stimulus from the assessable pain responses. Experimental pain is usually induced by means of thermal, electrical or pressure stimulation.

In an assessment of pain thresholds, each individual has to indicate when a stimulus starts to become painful. When carrying out this process in patients with dementia and in healthy controls, most studies found no differences in threshold levels between the two groups (Benedetti et al., 1999; Benedetti et al., 2004; Gibson et al., 2001; Jensen-Dahm et al., 2014; Lints-Martindale et al., 2007). Likewise, patients with dementia and controls were found to rate supra-threshold pain intensities – that is, situatons where stimuli were lying above the pain threshold – as equally painful (Cole et al., 2006; Jensen-Dahm et al., 2014; Kunz et al., 2007; Kunz et al., 2008). When pain tolerance is under investigation, each individual has to indicate when a stimulus starts to become unbearably painful. Here the findings were more contradictory: there was evidence for increased (Benedetti et al., 1999), decreased and unchanged tolerance levels in patients with dementia (Jensen-Dahm et al., 2014). An interesting approach has been chosen by Jensen-Dahm et al. (2014), who reported not only on changes in pain psychophysics in patients with dementia but also (in parallel) on the psychometric quality of their data. Their findings clearly suggest that the pain threshold, pain tolerance and supra-threshold pain ratings were as reliable in patients with mild degrees of dementia as they were in healthy controls. Thus we can conclude from these studies that sensitivity to low and moderate noxious stimuli seems unchanged in patients with dementia, but it remains unclear how pain tolerance is affected.

Some authors have argued that it might be more appropriate to assess pain responses that are independent of the patients' ability to give self-reports of pain, in view of the decline in their cognitive abilities (Kunz et al., 2007). Such a response is the nociceptive flexion reflex (NFR/RIII) – a physiological, polysynaptic reflex allowing for noxious stimuli to elicit an appropriate withdrawal response. The NFR threshold has been shown to be highly correlated with the subjective pain threshold (at least in cognitively unimpaired individuals); in consequence, it is believed to have the potential to serve as a non-verbal assessment tool for pain sensitivity. This reflex has been assessed in one study, and its threshold was found to be significantly decreased in dementia; this could point to an increase in pain processing, at least at the spinal level (Kunz et al., 2008). Moreover, when looking at brain responses to pain in patients with dementia, functional magnetic resonance imaging (fMRI) studies showed that brain activity in response to noxious stimulation is preserved and even elevated in patients with dementia by comparison to healthy controls (Cole et al., 2006). Another type of response that is independent of the patients' ability to give self-reports of pain is the autonomic response (e.g., heart rate, skin conductance). Studies showed a decline in autonomic responsiveness in patients with dementia (Kunz et al., 2008; Rainero et al., 2000). However, given that dementia is known to be accompanied by dysfunctions of the autonomic system even at its earliest stages, the autonomic response might not be a valid pain indicator in this patient group.

In conclusion, although the empirical findings on pain processing in patients with dementia are partly contradictory, most of them seem to suggest that the processing of experimentally induced pain is not diminished in patients with mild to moderate forms of dementia. When intepreting these findings, one has to keep in mind, however, that they come from, and are representative of, patients with only mild to moderate degrees of cognitive impairments. Most of the patients included in the studies were still able to use language in order to provide pain thresholds and pain tolerance estimates, as well as self-report ratings of their pain – an ability that declines rigorously in moderate to severe stages of dementia. We therefore do not know whether or how pain sensitivity might be altered in later stages of the disease. Due to ethical considerations, however, it is difficult to apply experimental pain induction procedures on patients in these severe stages of dementia.

Clinical perspective

Pain prevalence in older individuals with and without dementia

On account of their advanced age, individuals with dementia often suffer from multiple morbidities associated with pain. However, the exact pain prevalence in dementia is unknown, due to the lack of self-report in this patient group. Thus we will report prevalence rates from elderly individuals with and without dementia. Between 45 per cent to 83 per cent of the patients living in nursing homes experience acute or chronic pain, particularly those with moderate to severe dementia. Most of them (about 94 per cent) suffer from persistent pain (3–6 months or more: Miro et al., 2007), which is often located in the musculoskeletal system (Grimby et al., 1999). Chronic musculoskeletal pain affects over 100 million people in Europe and is by far the most common limiting factor that affects activities of the ageing population. Musculoskeletal pain increases the risk of reduced mobility, disability and muscle weakness and reduces the health-related quality of life (Woolf et al., 2004). But pain problems are not related only to movement. About 40 per cent of elderly individuals experience pain in internal organs, head and skin, which is more challenging to quantify (Husebo et al., 2008). Elderly patients with visceral painful conditions are far more likely than younger adults to present atypical pain responses (Helme and Gibson, 2001). Peptic ulcers, intestinal obstruction, and peritonitis are other visceral conditions, often with reduced or absent abdominal complaints (Helme and Gibson, 2001), and about 45 per cent of older persons with appendicitis do not have typical lower-right quadrant pain as a presenting symptom, by comparison with 5 per cent of younger adults (Wroblewski and Mikulowski, 1991). Living in a nursing home, 53 per cent of the elderly are at risk of developing a pressure ulcer (Horn et al., 2002), and skin diseases found in 95 per cent of the patients were described as constituting one of the most prevalent health problems (Black et al., 2006). Pain in connection with genito-urinary infections is quite often described in elderly patients. Catheter-associated urinary tract infection is the most common nosocomial infection; it accounts for more than 1 million cases every year in American hospitals and nursing homes (NHs) (Tambyah and Maki, 2000). Moreover, orofacial pain also has an incidence rate that increases with age (Koopman et al., 2009; Lobbezoo, Weijenberg and Scherder, 2011); around 30 per cent of the institutionalised elderly have experienced acute dental pain during the preceding year (Gluhak et al., 2010).

In conclusion, pain prevalence rates are high in elderly individuals and there is no reason to believe that these prevalence rates should substantially differ between those with and those without dementia.

Assessment of pain in individuals with dementia

Given the high prevalence of pain in the elderly, a proper assessment of pain by onlookers such as health-care professionals or family members is a prerequisite to successful pain treatment. The task of judging a sufferer's pain appropriately is a very complex one, given that pain is a very personal, private and subjective experience. Whenever older adults in pain also have severe cognitive impairment – which goes along with the loss of language and abstract thinking – the task of appreciating their pain becomes even more challenging. This is the case because one of the most important cues that judges rely on when assessing pain is the patient's self-report; but this self-report is often missing in cognitively impaired patients (Kappesser, Williams and Prkachin, 2006). Accordingly, older adults with dementia, particularly those in advanced stages of the disease, are at a very high risk for being underdiagnosed, and consequently undertreated for pain. Against this background, immense effort has been invested in the development of behavioural pain assessment tools that do not rely on the individual's capacity to provide a self-report of pain (e.g., Herr, Bjoro and Decker, 2006; Zwakhalen et al., 2006). These assessment tools are based on observations of typical behaviour in the patients that might be related to pain, such as vocalisations (e.g., moaning), facial expressions (e.g., grimacing), and body movements (e.g., defense).

There is also strong evidence that behaviours like agitation, pacing or resisting care are related to present pain problems. As stated above, a panel on Persistent Pain in Older Persons convened by the American Geriatric Society (AGS) has recommended that, as a prerequisite to appropriate pain treatment, a comprehensive, disease-specific and individual assessment of a patient's typical pain behaviour be carried out by using a validated pain assessment tool (AGS Panel, 1998). However, the recommendations of the AGS Panel are based on experience with older adults without dementia. This is of key importance, because in dementia symptoms attributed to neuropsychiatric disturbances may overlap with indicators of pain and thus can make interpretation quite challenging (Herr et al., 2006).

Self-report on pain

As stated before, the validity of self-report ratings might be questionable in patients with dementia, given the decline in language ability throughout the course of this disease. However, self-report might still be an appropriate method for pain assessment in its early stages, when the patient is still able to recognise and verbalise pain (Corbett et al., 2012). Nevertheless, studies that aimed to assess the performance of self-assessment scales (verbal and visual scales, the Faces Pain Scale) found that, whereas patients with mild to moderate degrees of impairment demonstrated comprehension of at least one scale, the comprehension rate decreased drastically in those with moderate to severe impairments (Kaasalainen and Crook, 2004; Pautex et al., 2006). Comprehension was defined as the ability to explain the scale's use and to indicate correctly on it positions for 'no pain' and for 'extreme pain'. Thus the self-report on pain seem to be still valid in mild forms of dementia. However, when the degree of cognitive impairment increases, the self-report seems to become a more and more invalid assessment tool (see Figure 15.1). Clinicians should be aware of this and should not interpret the absence of self-reported pain in a severely demented patient as a valid indicator of a pain-free state.

The facial expression

In the last two decades, more than 30 pain assessment instruments for older persons with dementia have been developed, tested, and reviewed in the literature. Most of these instruments are based on the idea that the patients' acute or chronic pain experience is communicated through changes in facial expression, vocalisation, and body movements (see Figure 15.1). Especially the *facial expression* seems to be one of the most promising non-verbal

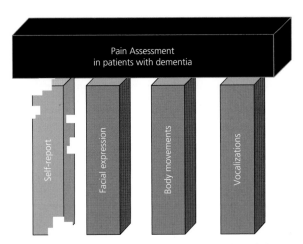

Figure 15.1 Pain assessment in patients with dementia should mostly rely on behavioural categories that are believed to be indicative of the presence of pain (facial expression, body movements and vocalisation), and not so much on self-reports on pain.

pain indicators, given that items relating to it are included in all pain assessment instruments. Interestingly, one of the first instruments developed that try to assess non-verbal behaviour is the Facial Action Coding System (FACS), which focuses exclusively on facial expressions (Ekman and Friesen, 1978). The FACS is based on the anatomical analysis of visible facial movements, which are categorised as action units. Using the FACS, it has been shown that there is a small subset of facial movements that occur in the context of pain; these include narrowing the eye aperture, contracting the eyebrows, lifting the upper lip and closing the eyes (Prkachin, 1992). Patients with dementia show the same subset of facial movements in response to pain as non-demented elderly individuals (Kunz et al., 2007). These findings are very promising: they clearly suggest that the face encodes the experience of pain in a specific way and that this specific encoding does not change throughout dementia.

It is, however, important to point out that, despite evidence that there is a small subset of pain-indicative facial movements, this does not imply the existence of only one facial expression of pain – a uniform and unique 'face', which can be observed at all times and in all individuals. Instead the frequencies of the occurrence of these key movements during pain usually range from 10 per cent to 60 per cent. Therefore the likelihood that all four key facial movements will occur simultaneously – in other words, the likelihood that an individual experiencing pain will display the complete subset or the 'prototypical expression of pain' – is very low. Rather individuals display only parts of this subset, sometimes even blending it with a limited range of other facial movements (e.g., smiling: Kunz, Prkachin and Lautenbacher, 2013). We have recently shown that it might be more helpful to differentiate between *at least three* different facial activity patterns that are displayed in the context of pain and that combine, or are made up of, various facial movements (Kunz and Lautenbacher, 2014). These patterns were (1) a tightening of the muscles surrounding the eyes, with furrowed brows and wrinkled nose; (2) furrowed brows, with a tightening of the muscles surrounding the eyes; and (3) an opened mouth, with a tightening of the muscles surrounding the eyes. These different facial activity patterns all have one facial movement in common, namely the tightening of the muscles surrounding the eyes. This facial movement is indeed the most frequent, and hence possibly the most important movement that occurs during pain.

Pain behaviour rating scales

Observational pain behaviour-rating scales have been developed mostly on the basis of literature reviews and interviews with nursing staff, and – as stated before – they include observational items related to facial expressions, vocalisation, and body movements (see Figure 15.1; Cohen-Mansfield, 2006). Most of these items are assessed by the rater according to frequency, intensity, or presence and absence. The different observational scales differ to a certain extent with regard to the types of items included, the interpretation of the total pain intensity, the scoring method, and the instructions for staff training. In addition, different systems of scoring the presumed pain intensity make the use of these scales challenging and sometimes only suitable for research purposes.

Usually these observational scales are filled out when the patient is at rest (after some minutes of observation); but sometimes patients are observed during activities of daily life (ADL). Recent findings clearly suggest that the observation of the patient at rest may not disclose the pain, especially chronic pain conditions, and it is now recommended that movement-related pains are better disclosed during ADLs. This idea has already been incorporated into at least four pain behaviour-rating scales, and these scales include instructions for spontaneous or guided movements during the pain observation process (Lefebvre-Chapiro, 2001; Snow et al., 2004; Husebo et al., 2010; Nygaard and Jarland, 2005).

Most of these scales are easy to use but require training and time for proper administration. In the past few years doctors, nurses and other caregivers have been involved in the validation processes. Evaluations of the most promising observational pain assessment scales in patients with dementia indicate that, although a number of tools demonstrate some promising potential, the validity of those scales is still not satisfactory (Villanueva, 2003). A further limitation is that some instruments are not validated for English or are validated for one language only. Another concern is related to the scoring system. To produce a total score, most instruments add up the number of observed behaviours. This means that a higher number of behaviours indicates more pain. However, a patient may be in severe pain even if this pain is manifested only in a few items, for instance in connection with moving a leg or an arm, and the rest of the body is pain-free. Some patients with Parkinson disease, for instance, or in a severely reduced condition would not be able to express enough behaviour to be judged to be in pain. Others, such as patients with Huntington's chorea, may express too much pain even when there is none.

In conclusion, as the severity of dementia is increasing, caregivers should rely less on self-report ratings and more on behavioural indicators of pain. Especially the facial expression seems to be a promising pain indicator. Patients with dementia display the same types of facial movements in response to pain as everyone else; the pain typicalness of their facial expression is not reduced. This implies that facial expressions of pain have the potential to serve as an alternative pain indicator in patients with dementia. With regard to pain behaviour-rating scales, several scales are available. However, the process of validating these scales has only just started. Although first results are promising, future studies are needed to define the most appropriate pain behaviour items – the ones that are able to discriminate between pain behaviours and behaviours related to other aspects and unmet needs in patients with dementia.

Pain treatment for individuals with dementia

Pharmacological management of chronic pain in older persons may be challenging. In 2009, with support from the American Pain Society and the American Academy of Pain Medicine, the AGS Panel revised its previous recommendations on pain management in older adults. The new guidelines include key approaches for safer opioid prescriptions in older adults. However, the current treatment recommendation guidelines, which aim to update the evidence base of the 2002 guidelines, are not yet evidence-based and do not consider the treatment of pain in patients with severe dementia, who often are excluded from randomised clinical trials and pain treatment studies (Corbett et al., 2012).

Although pharmacological treatment with analgesics is the most common form of pain treatment in older individuals, the use of non-pharmacological and alternative treatment should also be considered, given that these treatments have less adverse side effects (Ballard et al., 2011).

Prevalence studies of analgesic drug use in dementia

Traditionally, medications with analgesic effects are classified into three groups: periphery analgesics, such as acetaminophen; non-steroidal anti-inflammatory drugs (NSAIDs); and opioid agents. Even when pain is assessed and recognised, pain management often falls short of prescription recommendations. This seems to be a worldwide challenge, documented by studies from the Netherlands (e.g., Achterberg et al., 2007), Belgium (Elseviers, Vander Stichele and Van Bortel, 2010), England (Closs, Barr and Briggs, 2004), Norway (Nygaard and Jarland, 2005), the United States (Won et al., 2003) or China (Tse, Pun and Benzie, 2005).

In a study of 551 nursing-home residents in North Carolina, only around 50 per cent of the individuals with severe cognitive impairment received pain medications, by comparison with 80 per cent of the cognitively intact cohort ($P < 0.001$), despite a similar rate of pain-related conditions in both groups (Richards and Scott, 2002). Notably, cognitively impaired persons were more likely to be given pain medications 'as needed', while their peers had regularly scheduled analgesics. 'As needed' drug regimens are particularly inappropriate for individuals with moderate to severe cognitive impairment, who are unable to verbally communicate the need for pain relief. A study on the knowledge and beliefs of health-care professionals caring for older adults with dementia in a nursing-home setting found that a large number of the resident professionals thought that patients should only receive analgesics 'when necessary' rather than on a fixed schedule (Cramer et al., 2000).

In addition to the reduced analgesic treatment in patients with dementia, the analgesic effect itself might be reduced in these patients. There is evidence that patients with dementia do not experience a placebo effect (Benedetti et al., 2006). Losing the ability to modulate pain via endogenous placebo-related mechanisms could mean that patients with dementia need a higher dose of analgesic medications for pain relief to be achieved (Benedetti et al., 2006; for a comprehensive overview of this aspect, see Chapter 16 in this volume).

But time is changing. An increasing awareness of pain management in patients with dementia has been recently reported in the Swedish National Study of Aging and Care – Kungsholmen (Haasum et al., 2011). The study analysed the use of analgesics and psychotropics in 2,610 participants aged over 65 and found that 46 per cent of the patients with dementia used at least one analgesic drug, by comparison with 25 per cent of those without dementia. Persons with dementia reported pain less frequently, but the prevalence of pain-related diagnoses was similar to that found in persons without dementia. These findings are very promising, since they clearly suggest that the research findings of the last decades – which reported the undertreatment of pain in dementia – have already impacted the clinical practice and have led to an intensification of pain management in this frail patient group.

In conclusion, after many decades of reduced analgesic treatment for patients with dementia, health-care professionals seem to be better aware of the need for pain management in this patient group. Given that patients with dementia are still excluded from high-quality randomised-controlled analgesic trials, it is difficult to give recommendations for their pharmacological treatment with analgesics. We do know, however, that dementia is accompanied by a loss of placebo effects, which might mean that higher dosages of analgesics are required to treat pain.

Overall conclusion

Advanced age is associated with an increased prevalence of pain and of dementia. Given that the ability to report about pain depends on the patient's memory, expectation, and insight, patients with dementia (who are often limited in these respects) are at high risk for being underdiagnosed and for remaining untreated for pain. In patients with dementia, pain that goes unnoticed may contribute to increasing behavioural disturbances, like agitation and aggression. International recommendations to assess pain have been published and an impressive number of pain behaviour-rating scales have been developed and tested. However, the validity of these scales is still unsatisfactory. It is, moreover, an alarming fact that patients with dementia are still excluded from high-quality trials of pain treatment in older individuals. This underlines the high need for research as well as for excellent concepts for the implementation of pain assessment and pain treatment in elderly individuals with dementia.

Acknowledgement

We acknowledge financing support from the COST program (European Cooperation in the field of Scientific and Technical Research) for COST TD 1005 (Pain Assessment in Patients with Impaired Cognition, especially Dementia).

References

Achterberg, W. P., Pot, A. M., Scherder, E. J. and Ribbe, M. W. 2007. Pain in the nursing home: Assessment and treatment on different types of care wards. *Journal of Pain and Symptom Management* 34: 480–7.

AGS Panel. 1998. The management of chronic pain in older persons. *Journal of American Geriatric Society* 46: 635–51.

AGS Panel. 2009. Pharmacological Management of Persistent Pain in Older Persons. *Journal of American Geriatric Society* 57: 1331–6.

Ballard, C., Smith, J., Husebo, B., Aarsland, D. and Corbett, A. 2011. The role of pain treatment in managing the behavioural and psychological symptoms of dementia (BPSD) *International Journal of Paliativel Nursing* 17: 420–3.

Benedetti, F., Arduino, C., Costa, S., Vighetti, S., Tarenzi, L., Rainero, I. and Asteggiano, G. 2006. Loss of expectation-related mechanisms in Alzheimer's disease makes analgesic therapies less effective. *Pain* 121: 133–44.

Benedetti, F., Arduino, C., Vighetti, S., Asteggiano, G., Tarenzi, L. and Rainero, I. 2004. Pain reactivity in Alzheimer patients with different degrees of cognitive impairment and brain electrical activity deterioration. *Pain* 111: 22–9.

Benedetti, F., Vighetti, S., Ricco, C., Lagna, E., Bergamasco, B., Pinessi, L. and Rainero, I. 1999. Pain threshold and tolerance in Alzheimer's disease. *Pain* 80: 377–82.

Black, B. S., Finucane, T., Baker, A., Loreck, D., Blass, D., Fogarty, L., Philips, H., Hovanec, L., Steele c. and Rabins, P. V. 2006. Health problems and correlates of pain in nursing home residents with advanced dementia. *Alzheimer Disease and Associated Disorders* 20: 283–90.

Bosley, B. N., Weiner, D. K., Rudy, T. E. and Granieri, E. 2004. Is chronic nonmalignant pain associated with decreased appetite in older adults? Preliminary evidence. *Journal of the American Geriatric Society* 52: 247–51.

Cipher, D. J. and Clifford, R. A. 2004. Dementia, pain, depression, behavioral disturbances, and ADLs: Toward a comprehensive conceptualization of quality of life in long-term care. *International Journal of Geriatric Psychiatry* 19: 741–8.

Closs, S. J., Barr, B. and Briggs, M. 2004. Cognitive status and analgesic provision in nursing home residents. *British Journal of General Practice* 54: 919–21.

Cohen-Mansfield, J. 2006. Pain Assessment in Noncommunicative Elderly Persons: PAINE. *Clinical Journal of Pain* 22: 569–75.

Cole, L. J., Farrell, M. J., Duff, E. P., Barber, J. B., Egan, G. F., Gibson, S. J. 2006. Pain sensitivity and Fmri pain-related brain activity in Alzheimer's disease. *Brain* 129: 2957–65.

Corbett, A., Husebo, B. S., Malcangio, M., Staniland, A., Cohen-Mansfield, J., Aarsland, D. and Ballard, C. 2012. Assessment and treatment of pain in people with dementia Assessment, diagnosis and treatment of pain in people with dementia. *Nature Reviews Neurology* 8: 264–74.

Cramer, G. W., Galer, B. S., Mendelson, M. A. and Thomson G. D. 2000. A drug use evaluation of selected opioid and nonopioid analgesics in the nursing facility setting. *Journal of the American Geriatric Society* 48: 398–404.

Ekman, P. and Friesen, W. V. 1978. *Facial action coding system: Investigator's guide*. Palo Alto, CA: Consulting Psychologists Press.

Elseviers, M. M., Vander Stichele, R. R. and Van Bortel, L. 2010. Drug utilization in Belgian nursing homes: Impact of residents' and institutional characteristics. *Pharmacoepidemiology and Drug Safety* 190: 1041–8.

Gibson, S. J., Voukelatos, X., Ames, D., Flicker, L. and Helme, R. D. 2001. An examination of pain perception and cerebral event-related potentials following carbon dioxide laser stimulation in patients with Alzheimer's disease and age-matched control volunteers. *Pain Research and Management* 6: 126–32.

Giron, M. S. T., Forsell, Y., Bernsten, C., Thorslund, M., Winblad, B. and Fastbom, J. 2002. Sleep problems in a very old population: Drug use and clinical correlates. *Journals of Gerontology A-Biological Science and Medical Science* 57: 236–40.

Gloth, F. M. 2011. Pharmacological management of persistent pain in older persons: Focus on opioids and nonopioids. *Journal of Pain* 12: 14–20.

Gluhak, C., Arnetzl, G. V., Kirmeier, R., Jakse, N. and Arnetzl, G. 2010. Oral status among seniors in nine nursing homes in Styria, Austria. *Gerontology* 27: 47–52.

Grimby, C., Fastbom, J., Forsell, Y., Thorslund, M., Claesson, C. B. and Winblad, B. 1999. Musculoskeletal pain and analgesic therapy in a very old population. *Archives of Gerontology and Geriatrics* 29: 29–43.

Haasum, Y., Fastbom, J., Fratiglioni, L., Kåreholt, I. and Johnell, K. 2011. Pain treatment in elderly persons with and without dementia a population-based study of institutionalized and home-dwelling elderly. *Drugs and Aging* 28: 283–93.

Hadjistavropoulos, T., Herr, K., Turk, D. C., Fine, P. G., Dworkin, R. H., Helme, R., Jackson, K., Parmelee, P. A., Rudy, T. E., Lynn B. B., et al. 2007. An interdisciplinary expert consensus statement on assessment of pain in older persons. *Clinical Journal of Pain* 23: S1–S43.

Helme, R. D. and Gibson, S. J. 2001. The epidemiology of pain in elderly people. *Clinics in Geriatric Medicine* 17: 417–31.

Herr, K. 2011. Pain assessment strategies in older patients. *Journal of Pain* 12: S3–S13.

Herr, K., Bjoro, K. and Decker, S. 2006. Tools for assessment of pain in nonverbal older adults with dementia: A state-of-the-science review. *Journal of Pain and Symptom Management* 31: 170–92.

Horn, S. D., Bender, S. A., Bergstrom, N., Cook, A. S., Ferguson, M. L., Rimmasch, H. L., Sharkey, S. S., mout, R. J., Taler, G. A. and Voss A. C. 2002. Description of the National Pressure Ulcer Long-Term Care Study. *Journal of the American Geriatric Society* 50: 1816–25.

Husebo, B. S., Ballard, C., Sandvik, R., Nilsen, O. B. and Aarsland, D. 2011. Efficacy of treating pain to reduce behavioural disturbances in residents of nursing homes with dementia: Cluster randomised clinical trial. *British Medical Journal* 343: 1–10.

Husebo, B. S., Kunz, M., Achterberg, W., Lobbezoo, F., Kappesser, J., Tudose, C., Strand, L. I. and Lautenbacher, S. 2012. Pain assessment and treatment challenges in patients with dementia. *Zeitschrift fuer Neuropsychologie* 23: 237–46.

Husebo, B. S., Strand, L. I., Moe-Nilssen, R., Husebo, S. B. and Ljunggren, A. E. 2010. Pain in older persons with severe dementia: Psychometric properties of the Mobilization–Observation–Behaviour–Intensity–Dementia (MOBID-2) Pain Scale in a clinical setting. *Scandinavian Journal of Caring Sciences* 24: 380–91.

Husebo, B. S., Strand, L. I., Moe-Nilssen, R., Husebo, S. B., Aarsland, D. and Ljunggren, A. E. 2008. Who suffers most? Dementia and pain in nursing home patients: A cross-sectional study. *Journal of the American Directors Association* 9: 427–33.

Jensen-Dahm, C., Werner, M. U., Dahl, J. B., Jensen, T. S., Ballegaard, M., Hejl, A. M. and Waldemar, G. 2014. Quantitative sensory testing and pain tolerance in patients with mild to moderate Alzheimer's disease compared to healthy control. *Pain* 155: 1439–45.

Kaasalainen, S. and Crook, J. 2004. An exploration of seniors' ability to report pain. *Clinical Nursing Research* 13: 199–215.

Kappesser, J., Williams, A. C. and Prkachin, K. M. 2006. Testing two accounts of pain underestimation. *Pain* 124: 109–16.

Koopman, J. S., Dieleman, J. P., Huygen, F. J., de Mos, M., Martin, C. G. and Sturkenboom, M. C. 2009. Incidence of facial pain in the general population. *Pain* 147: 122–7.

Kunz, M. and Lautenbacher, S. 2014. The faces of pain: A cluster analysis of individual differences in facial activity patterns of pain. *European Journal of Pain* 6: 813–23.

Kunz, M., Prkachin, K. and Lautenbacher, S. 2013. Smiling in pain: Explorations of its social motives. *Pain Research and Treatment*. doi: 10.11555/2013/128093.

Kunz, M., Mylius, V., Scharmann, S., Schepelman, K. and Lautenbacher, S. 2008. Influence of dementia on multiple components of pain. *European Journal of Pain* 13: 317–25.

Kunz, M., Scharmann, S., Hemmeter, U., Schepelman, K. and Lautenbacher, S. 2007. The facial expression of pain in patients with dementia. *Pain* 133: 221–8.

Lefebvre-Chapiro, S. 2001. The Doloplus 2 scale: Evaluating pain in the elderly. *European Journal of Palliative Care* 8: 191–4.

Lin, P. C., Lin, L. C., Shyu, Y. I. L. and Hua, M. S. 2011. Predictors of pain in nursing home residents with dementia: A cross-sectional study. *Journal of Clinical Nursing* 20: 1849–57.

Lints-Martindale, A., Hadjistavropoulos, T., Barber, B. and Gibson, S. 2007. A psychophysical investigation of the facial action coding system as an index of pain variability among older adults with and without Alzheimer's disease. *Pain Medicine* 8: 678–89.

Lobbezoo, F., Weijenberg, R. A. F. and Scherder, E. J. A. 2011. Topical review: Orofacial pain in dementia patients: A diagnostic challenge. *Journal of Orofacial Pain* 25: 6–14.

Miro, J., Paredes, S., Rull, M., Queral, R., Miralles, R., Nieto, R., Huguet, A. and Baos, J. 2007. Pain in older adults: A prevalence study in the Mediterranean region of Catalonia. *European Journal of Pain* 11: 83–92.

Moriarty, O. M., McGuire, B. E. and Finn, D. P. 2011. The effect of pain on cognitive function: A review of clinical and preclinical research. *Progress in Neurobiology* 93: 385–404.

Nygaard, H. A. and Jarland, M. 2005. Are nursing home patients with dementia diagnosis at increased risk for inadequate pain treatment? *International Journal of Geriatric Psychiatry* 20: 730–7.

Pautex, S., Michon, A., Guedira, M., Emond, H., Lous, P. L., Samaras, D., Michel, J. P., Hermann, F., Panteleimon, G. and Gold, G. 2006. Pain in severe dementia: Self-assessment or observational scales? *Journal of the American Geriatric Society* 54: 1040–5.

Prkachin, K. M. 1992. The consistency of facial expressions of pain: A comparison across modalities. *Pain* 51: 297–306.

Rainero, I., Vighetti, S., Bergamasco, B., Pinessi, L. and Benedetti, F. 2000. Autonomic responses and pain perception in Alzheimers's disease. *European Journal of Pain* 4: 267–74.

Richards, S. C. M. and Scott, D. L. 2002. Prescribed exercise in people with fibromyalgia: Parallel group randomised controlled trial. *British Medical Journal* 325: 185–7.

Scherder, E., Herr, K., Pickering, G., Gibson, S., Benedetti, F. and Lautenbacher, S. 2009. Pain in dementia. *Pain* 145: 276–768.

Snow, A. L., Weber, J. B., O'Malley, K. J., Cody, M., Beck, C., Bruera, E., Ashton, C. and Kunik, M. E. 2004. NOPPAIN: A nursing assistant-administered pain assessment instrument for use in dementia. *Dementia and Geriatric Cognitive Disorders* 17: 240–6.

Takai, Y., Yamamoto-Mitani, N., Okamoto, Y., Koyama, K. and Honda, A. 2010. Literature review of pain prevalence among older residents of nursing homes. *Pain Management Nursing* 11: 209–23.

Tambyah, P. A. and Maki, D. G. 2000. Catheter-associated urinary tract infection is rarely symptomatic: A prospective study of 1497 catheterized patients. *Archives of Internal Medicine* 160: 678–82.

Tse, M. M. Y., Pun, S. P. Y. and Benzie, I. F. F. 2005. Pain relief strategies used by older people with chronic pain: An exploratory survey for planning patient-centred intervention. *Journal of Clinical Nursing* 14: 315–20.

Villanueva, M. R. 2003. Pain assessment for the dementing elderly (PADE): Reliability and validity of a new measure. *Journal of the American Medical Directors Association* 4: 1–8.

Won, A. B., Lapane, K., Vallow, S., Schein, J., Morris, J. N. and Lipsitz, L. A. 2003. Persistent nonmalignant pain and analgesic prescribing practices in elderly nursing home residents. *Journal of the American Geriatric Society* 51: S193–S194.

Woolf, A. D., Zeidler, H., Haglund, U., Carr, A. J., Chaussade, S., Cucinotta, D. E. E. A., Veale, D. J. and Martin-Mola, E. 2004. Musculoskeletal pain in Europe: Its impact and a comparison of population and medical perceptions of treatment in eight European countries. *Annals of the Rheumatic Diseases* 63: 342–7.

Wroblewski, M. and Mikulowski, P. 1991. Peritonitis in geriatric inpatients. *Age and Ageing* 20: 90–4.

Zwakhalen, S. M., Hamers, J. P., Abu-Saad, H. H. and Berger, M. P. 2006. Pain in elderly people with severe dementia: A systematic review of behavioural pain assessment tools. *BMC Geriatrics* 6 (3).

CHAPTER 16

The role of cognitive impairment in the placebo and nocebo effects

Martina Amanzio

Introduction

The *psychosocial context* surrounding the patient and the *psychobiological model* offer interesting perspectives from which to study the placebo response. Some authors use the term *placebo response* to mean any type of improvement in a placebo group under clinical trial, even if that improvement is related to statistical artefacts such as sampling bias and regression to the mean, or to the natural history of a clinical condition. Importantly, the term 'response' should only be reserved for an *active neurobiological process* that occurs as a result of a dummy treatment. Indeed, the changes in brain activity that were related to the psychosocial context in the form of a procedure designed to elicit placebo analgesia allowed us to demonstrate, through multiple studies, how the effect we measured is *real*. The interesting aspect of studying the placebo effect is that the specific effects of a drug are removed in order to collect the effects of the sensory stimuli, meaning the psychosocial context. The open–hidden paradigm allowed us to differentiate the specific effects of an active drug from those of the psychosocial context.

Contextual information leading to placebo responses arises either from *conscious expectancies* about the anticipated effects of a treatment or from prior learning in the form of *conditioning* through active treatments. The context surrounding the administration of a placebo that produces the placebo effect may prompt individuals to expect improvement and positive outcomes. (For a comprehensive discussion of placebo effects, see Chapter 9 in this volume.) Alternatively, contextual information can lead individuals to expect a worsening of symptoms; indeed changes in the negative direction are registered as part of the *nocebo effect*. One way of studying the nocebo effects of a medical treatment is to analyse the findings of randomised double-blind placebo-controlled trials (RCTs).

The importance of these studies is given by the fact that they attempt to maximise the placebo components of therapies and hence minimise the worsening of symptoms so as to avoid nocebo effects and the subsequent discontinuation of medication intake; moreover, assessing situations where the loss of placebo mechanisms may require increased therapeutic dosage represents another crucial aspect. These aspects become even more important in

An Introduction to Pain and its Relation to Nervous System Disorders, First Edition. Edited by Anna A. Battaglia.
© 2016 John Wiley & Sons, Ltd. Published 2016 by John Wiley & Sons, Ltd.

patients with cognitive impairment, who may exhibit a disruption of the expectancy-evoked placebo–analgesia network and impaired sensory and affective responses to pain.

In this chapter we attempt to answer specific questions about patients with cognitive impairment, even though at present we have only a limited understanding of the factors that influence patients' response to pain and to placebo treatment as well as of the underlying mechanisms. In particular, we analyse the responses to pain and placebo analgesia in patients with major neurocognitive disorders, and we pay special attention to the cerebral network at the basis of these phenomena. We conclude by addressing two questions on nocebo effects that have important and far-reaching implications for clinical trial design and for medical practice: (1) Do patients with cognitive impairment experience nocebo effects differently from the general population? (2) Are the level of cognitive impairment and the diagnosis (that is, a diagnosis of mild cognitive impairment versus one of Alzheimer's disease) related to negative symptom outcome?

Pain perception in patients with major neurocognitive disorders

Pain is a complex and subjective perceptual experience, incorporating sensory–discriminative, affective–motivational and cognitive–evaluative dimensions (Melzack and Casey, 1968). These aspects of the pain experience are subserved by distinct supraspinal pathways, which have been identified as a *pain matrix* (Derbyshire et al., 1997). The pain matrix consists of two parallel subsystems: the lateral, which is sensory–discriminatory, and the medial, which relates to affective–cognitive evaluation (Apkarian et al., 2005; Ingvar, 1999; Tracey and Mantyh, 2007). In general, the lateral pain subsystem consists of spinothalamic tract neurons that ascend via the ventroposterior lateral thalamus onto the primary and secondary somatosensory cortices (S1 and S2), and these in turn code the location, intensity, and quality of the sensation. Another major pathway, at the level of the medulla, ascends to the medial thalamus, the hypothalamic nuclei, the limbic regions including the cingulate cortex, the insula cortex (IC) and the prefrontal areas (Price, 2000; Rainville, 2002; Vogt, 2005), all of which are known to be involved in controlling emotion, arousal and attention. It is proposed that this medial pain pathway mediates the unpleasant, affective dimensions of pain and the motivation to escape from the noxious event (Price, 2000; Treede et al., 1999). The prefrontal cortex has a role in the cognitive–evaluative component; it has also been assigned a role of integration between the sensory–discriminative and the motivational–affective dimensions of pain (George et al., 1994; Ketter et al., 1993). Other pain matrix components are the posterior parietal cortex (Andersen, 1987; Coghill et al., 1994) and brainstem areas such as the periaqueductal gray (PAG) and the reticular formation (Kelly, 1985). Although this differentiation of the pain matrix into two subsystems makes its description easier, the two parts act in an integrated manner to produce the overall experience of pain (Price, 2000).

A further element that complicates pain perception studies is the presence of neurodegenerative diseases such as major neurocognitive disorders. It has been suggested that the experience of pain may be diminished in Alzheimer's disease (AD) on account of disease-related changes to brain regions involved in the transmission and processing of noxious input (Scherder, Sergeant and Swaab, 2003). Given the high prevalence of both cognitive impairment and chronic pain, it is likely that many older adults with AD have chronic or persistent pain. In mild to moderate stages of AD, patients may be unable to indicate pain perception through verbal or behavioural reports of pain (Feldt, Ryden and Miles, 1998; Ferrell, Ferrell and Rivera, 1995; Tsai et al., 2008).

As AD progresses to more severe stages, people lose the ability to communicate verbally, and this leaves them at great risk of experiencing untreated pain. M. Kunz and S. Lautenbacher examine in detail the complexities involved in pain perception in major neurocognitive disorders; moreover, they provide an exhaustive discussion of how pain is clinically assessed in such people (see Chapter 15 in this volume). Indeed trying to interpret the complex relationship between AD pathology, brain activation and pain perception is a task of great challenge.

An interesting part of this task will be to discover what pain-related central nervous system (CNS) processing may be altered by the changes that occur in primary neurocognitive disorders and to analyse it. Psychophysiological measures represent an interesting perspective from which to investigate the dissociation between sensory and affective–motivational aspects of experimental pain, also taking into consideration the rostral pain network (Devinsky, Morrell and Vogt, 1995). As reported by Monroe et al. (2012), some researchers have suggested that an additional rostral (limbic) network may be responsible for the behavioural expression of pain (Devinsky et al., 1995). The rostral pain system overlaps with several components of the medial pain network and consists of specific nuclei in the amygdala, periaqueductal gray (PAG), orbitofrontal cortex, anterior cingulate cortex (ACC), anterior insular cortex (Devinsky et al., 1995), striatum (Chudler and Dong, 1995; Devinsky et al., 1995), thalamus and hypothalamus (Sewards and Sewards, 2002). The striatum is a key structure in the rostral pain network that is not generally associated with either the lateral or the medial pain network (Apkarian et al., 2005; Treede et al., 1999). AD pathology studies have shown that structures in the rostral pain network such as the amygdala (Vogt et al., 1990), the orbitofrontal cortex, the insula (Tekin et al., 2001), the PAG (Parvizi, Van Hoesen and Damasio, 2000) and the striatum (Selden, Mesulam and Geula, 1994) develop neurofibrillary tangles and neuritic plaques. Damage in these areas is associated with altered behavioural responses to pain.

Neurodegenerative changes that occur in AD selectively target components of the medial pain system, in particular the medial thalamic nuclei, the hypothalamus, the cingulate and insular cortex. On the other hand, the cerebral regions comprising the lateral pain pathway are relatively well preserved, which suggests that the disease may compromise the affective–motivational component or dimension of pain more than its sensory intensity (Braak and Braak, 1991; Scherder and Bouma, 1997; Scherder, 2000; Scherder et al., 2003). In line with with these neuropathological findings, the pain threshold has been found to be normal in AD patients (Benedetti et al., 1999; Benedetti et al., 2004; Gibson et al., 2001), but tolerance to experimentally induced pain is increased, in parallel with cognitive deterioration (Benedetti et al., 1999). In frontotemporal dementia (FTD), a higher prevalence of loss of awareness of pain than in other kinds of major neurocognitive disorders was reported by caregivers; this possibly indicates the deterioration of the medial system (Bathgate et al., 2001). Among FTD patients, those displaying semantic dementia showed more frequently an increased responsiveness to sensory stimuli, including pain (Snowden et al., 2001). More recently, FTD patients have been shown to have a higher pain threshold and pain tolerance – as measured through electrical stimulation – than age-matched healthy controls (Carlino et al., 2010), which demonstrates a disruption in sensory–discriminative and affective aspects of pain sensation.

The study by Cole and co-workers (2006) is of interest to our purposes. They examined the pain matrix in a group of control subjects and AD patients by using psychophysiological measurements during moderately painful stimulation. In particular, they used a group-by-region approach to functional magnetic resonance imaging (fMRI) data analysis, aiming to directly compare pain-induced blood oxygenation level-dependent (BOLD) activity in regions of interest (ROIs) located in medial and lateral pain pathways. The analysis of fMRI data revealed a common network of pain-related activity, both groups showing increased activity in both the medial and the lateral pain systems during noxious versus innocuous pressure stimulation. It is important to underline that the factors contributing to mixed findings in psychophysical and neurophysiological studies of pain in people with cognitive impairment include study design, cognitive ability of participants, and acute versus chronic pain conditions.

The medial pathway ROI was located in the anterior midcingulate cortex (aMCC). This region of the cingulate cortex, which corresponds with Brodmann area (BA) 24, represents the terminal point of the majority of nociceptive inputs from the medial intralaminar thalamic nuclei (Vogt, 2005), and is consistently activated in studies of functional brain imaging of pain (Farrell, Laird and Egan, 2005; Vogt, 2005). ROIs were also generated for the anterior, middle and posterior subregions of the IC. On the other hand, the lateral pain-pathway ROIs were identified in S1 and S2.

Interestingly, by comparison with controls, patients showed significantly greater pain-related activity in several pain-pathway regions, including the aMCC, the ventrolateral and dorsomedial thalamus, AI, SI and SII, and frontal regions – including the dorsolateral prefrontal cortex (DLPFC). Evidence from pain-imaging research suggests that the DLPFC plays a role in the cognitive modulation of pain (Bornhovd et al., 2002; Lorenz et al., 2002; Gracely et al., 2004; Graff-Guerrero et al., 2005; Pariente et al., 2005; Zubieta et al., 2005). The cognitive–evaluative component of pain refers to the meaning of the pain experience and its possible consequences. The anticipatory response is part of a neuropsychological mechanism that is impaired in AD patients (Amanzio et al., 2011; Amanzio, Vase, et al., 2013); in particular, an analysis of context in terms of cognitive flexibility, prospective memory, monitoring and attention abilities appears to be involved in situations in which the painful experience should be integrated.

Indeed prolonged activation in the pain pathway – along with increased activity in cognitive processing regions such as the DLPFC, observed in the study by Cole et al. (2006) – indicates that the cognitive integration of the painful experience is altered in AD patients. In particular, the pain experience demands attention by its very nature, and the DLPFC is known to be involved in working memory processing and attention (Amanzio et al., 2011; Barch et al., 1997; MacDonald et al., 2000; Mesulam, 1998; Smith and Jonides, 1999).

However, in situations where the threat of tissue damage is low, as it is in an experimental pain environment (the experimental setting in the study by Cole et al., 2006), the propensity to escape from the stimulus is necessarily suppressed. Under such circumstances, it is more adaptive to disengage attention from the experience of pain, as the authors observed in the control group. In particular, the attenuation in pain-evoked activity in the control group of subjects can be interpreted as an accurate appraisal of the pain related to the experimental setting, followed by a consequent disengagement of attention from the sensation.

The patient group also had greater pain-related activity across the motor network, including M1, the premotor cortex, the supplementary motor cortex (SMA) and the cerebellum. This increase in motor activity suggests an inadequate inhibition of withdrawal responses and provides further evidence of impaired cognitive integration of the pain experience.

The impact of Alzheimer's disease on the functional connectivity between brain regions underlying pain perception was investigated by Cole (2011). Indeed the synchronicity or functional connectivity between pain-processing regions of the brain can be assessed by determining the degree of temporal correlation between regional responses to noxious stimulation. Some earlier studies investigating neurodegenerative diseases demonstrated increased regional brain activity during cognitive task performance, possibly related to the recruitment of a larger neural population in compensation for the impaired functional integration between the components of the neural system that underlies task performance (Amanzio et al., 2011; Grady et al., 2003; Thiruvady et al., 2007). Considering the pain, the increased regional pain-related brain activity observed by Cole et al. (2006) may also reflect compensatory activation in AD, as a consequence of a neurodegenerative impairment of the functional integration between the components of the pain-processing network. A subsequent study by Cole et al. (2011) also goes in this direction, further suggesting an interplay between pain and cognitive processes in patients with AD. In particular, the researchers found, in patients with AD by comparison with controls, an enhanced functional connectivity between the DLPFC and the aMCC, the periaqueductal gray, the thalamus, the hypothalamus and several motor areas, providing further evidence that the supraspinal processing of painful input is intact.

These studies revealed the special role played by attentional recourses that are allocated to perceptual processes on the basis of the salience of the incoming information and of the relevance of this information for prioritised goals (Legrain et al., 2009). Bottom-up attentional processes have been related mainly to the AI and MCC and to the *salience network*. Once a stimulus has been detected as being salient, as in the case of painful stimuli, the AI activates the cognitive control network (Sridharan, Levitin and Menon, 2008) and thereby facilitates task-related information processing. In this way painful stimuli will have preferential access to the brain's attentional and working memory resources (Menon and Uddin, 2010). The AI also decreases activity in the *default mode network* (DMN) (Sridharan et al., 2008), which comprises the ventromedial prefrontal cortex (VMPFC) and the cingulate cortex and shows decreased activation during sensory or cognitive processing. Although the relevance of DMN modulation for selective attention is less well understood, there is evidence to the effect that failure of this DMN regulation through the AI leads to inefficient cognitive control (Bonnelle et al., 2012). The results obtained by Cole et al. (2011) appear to be in line with these findings. Indeed their AD patients showed a heightened functional connectivity between the anterior insula and the DMN.

In the following section we consider the role of expectation in placebo analgesia in patients with major neurocognitive disorders.

Placebo analgesia in patients with major neurocognitive disorders

The majority of studies investigating placebo responses have been conducted in the realm of pain. In particular, since pain is an interoceptive modality, it can be quantitatively manipulated in the laboratory. The placebo analgesia (PA) network has been substantiated by several studies that used different experimental procedures (fake analgesic creams, sham acupuncture, and others). Researchers have used brain-imaging tools such as fMRI, positron emission tomography (PET), and cerebral blood flow (rCBF) to study changes in neural activity in experimental placebo paradigms. We performed a coordinate-based activation likelihood estimation (ALE) meta-analysis in order to search for the cortical areas involved in placebo analgesia in human experimental pain models (Amanzio, Benedetti, et al., 2013). Focusing on changes in brain activity that take place during the expectation phase, we observed an increased activity associated with PA in areas such as the ACC and the PAG. Interestingly, an area selectively activated in the PA expectation component and not present when we analysed the PA during noxious stimulation was BA10, which demonstrates that anticipatory responses of the DLPFC may well be part of a more general mechanism, which is not confined to pain. The crucial relevance of the functional and structural integrity of cognitive–evaluative areas such as the prefrontal cortex for the downstream circuitry is emphasised by experimental data: the temporary functional lesion of the prefrontal cortex through repetitive transcranial magnetic stimulation was associated with a reduction or complete loss of the placebo analgesic response (Krummenacher et al., 2010). The results of our analysis also demonstrated how the areas engaged by PA in human experimental pain may be part of a general circuit underlying the voluntary regulation of affective responses (Bechara et al., 1997; Damasio, 1996; Petrovic et al., 2005; Wager et al., 2011).

The crucial role of expectation in the therapeutic outcome is best illustrated in the so-called open–hidden drug paradigms. In these paradigms identical concentrations of drugs are administered in two different conditions: an open condition, where the patient is aware of the medication and the latter is given by a health-care provider who also announces the intended treatment outcome (e.g. analgesia); and a hidden condition, where the patient is unaware of the medication and the latter is administered through a computer-controlled infusion. This allows the pharmacodynamic effect of the treatment (the hidden treatment) to be dissociated from the additional benefit of the psychosocial context in which the treatment is given. The difference between the two outcomes after the administration of the expected and unexpected therapy can be seen as the 'placebo' (psychological) component, even though no placebo has been given. These studies reveal that psychosocial factors can considerably enhance the analgesic effect of a drug (Levine and Gordon, 1984).

A new appreciation of the role played by neuropsychological differences in response to a placebo or pharmacological treatment opens an exciting possibility for the care of patients with cognitive impairment: it promises to optimise their outcome. Although there is an important involvement of classical conditioning in the generation of PA, the expectation of relief in terms of a therapeutic benefit has important consequences. However, how expectations translate into basic cognitive and behavioural processes is an intriguing matter, which has yet to be resolved. Specific questions need to be answered in order for us to gain a deeper understanding of some crucial aspects, given the few results obtained in patients with cognitive impairment. *What happens in cognitively impaired patients when an open–hidden paradigm is applied? Do they experience placebo effects differently from the general population? What role do cognitive factors play? What is the impact of the disruption of prefrontal cortex circuitries?*

To date, only one study has considered the topic of PA in patients with major neurocognitive disorders (Benedetti et al., 2006). Benedetti et al. studied AD patients at time zero (T0) and after one year. Patients were treated with either open (expected) or hidden (unexpected) local lidocaine after venipuncture. In the open condition, which was used to study the effect of the psychosocial context, the drug was applied on a tape, in full view of the patients, while they were told that the pain would subside in a few minutes. In the hidden condition, the same dose of the drug was administered but the patient was completely unaware that a local anaesthetic had being applied (in this case the psychological context was removed). Just before and 15 minutes after the application of lidocaine, the patients rated their pain according to a numerical rating scale (NRS) ranging from 0 (= no pain) to 10 (= unbearable pain) (see Figure 16.1).

The study was performed on 28 patients suffering from AD. At T0 there was a significant difference between the open and the hidden applications of lidocaine. In fact the open application induced a significantly larger decrease in pain than the hidden application. This effect was smaller one year later (see the graph on the left in Figure 16.2). The authors also studied 10 more AD patients. They observed a significant difference between the O and the H applications of lidocaine at the initial stage of the disease and after one year (see the graph on the right in Figure 16.2). Patients in both groups were also analysed by means of electroencephalography (EEG) mutual information analysis.

The greatest reduction in the open–hidden difference was seen in the 28 AD patients, and was due to the reduction in the effects of the open lidocaine for pain (on the left in Figure 16.2). Using EEG mutual information analysis, the authors observed that those 28 patients with reduced prefrontal connectivity with the rest of the brain displayed the lowest effects of

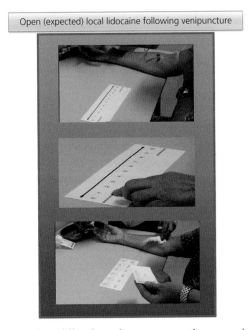

Open (expected) local lidocaine following venipuncture

Figure 16.1 The open–hidden paradigm differs from the presence or the removal of the psychosocial context.

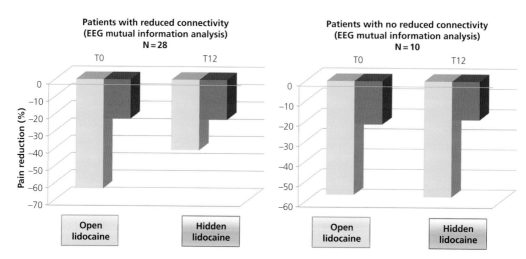

Figure 16.2 Histograms of the percentage of pain reduction in the open and hidden lidocaine trials that consider patients with reduced connectivity (graph on the left) and patients with no reduction in connectivity (graph on the right). Source: Benedetti 2006.

lidocaine and the lowest frontal assessment battery (FAB) scores. On the other hand, in the 10 patients who did not show reduced connectivity there was a significant difference between the O and the H applications of lidocaine at T0 and after one year (on the right in Figure 16.2).

Although this study opens up an important field of research on people with major neuro-cognitive disorders, demonstrating that the impairment of cognitive functions may disrupt the psychosocial component of treatment, further studies will be important to clarify the role of executive dysfunction in PA taking into account the NIA-Alzheimer's Association criteria (McKhann et al., 2011) and new neuropsychological studies on patients with AD (Amanzio et al., 2011; Amanzio, Vase, et al., 2013).

Adverse events caused by placebo treatments in patients with minor or major neurocognitive disorders: A way to study the nocebo effect

In clinical trials, nocebo effects are best illustrated as the development of side effects in placebo groups. In these studies patients know that they may receive an active treatment or a placebo and are accordingly informed by the investigator about the possible side effects they may experience after taking the active drug. This generates in subjects a negative expectancy that can lead to a negative outcome, in particular the appearance of specific side effects.

The development of side effects after placebo intake has been reported for several medical conditions. Many patients in the placebo groups of these clinical trials discontinued the treatment because of symptoms that were attributed to the medication. Notably, the side-effect profiles of placebo groups reflect the expected side effects of the drug, which also led to a similar rate of non-adherence to treatment (Amanzio, 2011, 2015). This has been shown for trials investigating migraine treatments, in which the placebo groups of non-steroidal anti-inflammatory drug (NSAID) trials reported more gastrointestinal symptoms, whereas the placebo groups of anticonvulsant trials reported specific side effects, namely anorexia, memory

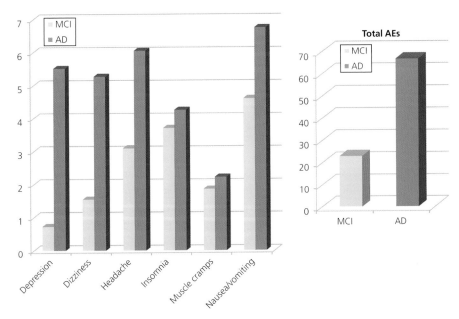

Figure 16.3 Histograms of the percentage of AEs in the placebo arms of mild cognitive impairment (MCI; in blue) and Alzheimer's disease (AD; in red) trials considering specific adverse events (AEs; graph on the left) and the total percentage of AEs (graph on the right). Source: Adapted from Amanzio 2012.

difficulties, paraesthesia and upper respiratory tract infection. In particular, as well as underlining the surprising similarities between the active-drug group and the corresponding placebo group, our study also revealed a similar rate of non-adherence to treatment when considering triptan–placebo versus NSAID–placebo (see Amanzio et al., 2009; Amanzio, 2011).

This final part of the chapter addresses the nocebo effect in the context of clinical trials mainly for people with cognitive impairment due to minor or major neurocognitive disorders. The systematic review approach to studying rates of adverse events (AEs) reported in the placebo arms of double-blind RCTs provides an interesting perspective from which to consider the phenomenon of the nocebo effect (see in particular Amanzio et al., 2009; Rief et al., 2009; Amanzio, Benedetti and Vase, 2012). Observations from these clinical trials indicate the role of the patients' expectations in the development of nocebo effects as part of the side effects registered in the placebo groups. Interestingly, non-pharmacodynamic factors such as expectation can trigger distressing symptoms just by themselves; this kind of effect is called the 'nocebo' phenomenon (Kennedy, 1961; Hahn, 1997). In a study we found the level of cognitive impairment to be a crucial factor in the occurrence of the AEs reported by patients with major neurocognitive disorders in the placebo arms of donepezil trials (Amanzio et al., 2012). Active treatment with donepezil induces cholinergic AEs such as nausea, vomiting, diarrhea, muscle cramps and dizziness (Waldemar et al., 2007); the same AEs were observed in the placebo arm of the trials (Amanzio et al., 2012).

In Figure 16.3 the graph on the left shows significant comparisons between the two groups of patients for specific AEs such as depression, dizziness, headache, insomnia, muscle cramps and nausea vomiting. The graph on the right shows that AD patients experienced a significantly higher number of total AEs than patients with mild cognitive impairment. This is the first study to show that patients who are in a more advanced stage of disease may be at greater risk of developing AEs. This important finding must be borne in mind if we wish to improve adherence to drug treatment.

The phenomenon we observed may be related to a greater presence of somatic comorbidity, which predisposes AD patients to experience emotional distress in the form of physical symptoms. Another possible interpretation is that AD patients are frailer and therefore more susceptible to AEs than subjects with mild cognitive impairment.

In particular, the knowledge of taking a medication and the anxiety associated with the clinical course of the disease may lead patients to monitor AEs more closely and thus amplify their perception of somatic symptoms. In previous studies somatisation has been associated with nocebo symptoms and with the occurrence of AEs in relation to active treatments. In daily clinical care we have observed that the presence of AEs such as tachycardia, dyspnea, or sweating in patients with high levels of anxiety is often a somatic manifestation of anxiety. Anxious patients monitor somatic changes more closely than the control population and amplify their symptoms accordingly. Although most of the studies that we considered here do not include a structured assessment of somatisation and mood changes, previous results have shown a clear association between cognitive status and somatic comorbidity in patients with AD (Doraiswamy et al., 2002). The second hypothesis is given by the so-called frailty syndrome (Fried et al., 2001), which is described in terms of a significant sense of loss related to a greater perception of fatigue, a higher incidence of somatic disturbances, the presence of more AEs to analgesic drugs, and increased vulnerability to stressful events. Future studies will be essential to clarify the role of somatisation and the frailty syndrome in the AEs experienced by AD patients.

In clinical trials as well as in clinical practice, the ability to anticipate who will develop side effects and to estimate the actual extent and strength of nocebo responses is crucial, since these unspecific side effects distress the patients, add burden to their illness, increase the costs of their care and lead to non-adherence to treatment and subsequent discontinuance. Healthcare specialists should be aware that every interaction with the patient has the potential to induce negative treatment outcomes. Consequently, the information given to patients before they receive medical treatment should be of particular relevance, with a view to reducing the risk of drug discontinuation.

References

Amanzio, M. 2015. Nocebo effects and psychotropic drug action. *Expert Review of Clinical Pharmacology* 8: 159–61.

Amanzio, M. 2011. Do we need a new procedure for the assessment of adverse events in anti-migraine clinical trials? *Recent Patents on CNS Drug Discovery* 6: 41–7.

Amanzio, M., Benedetti, F. and Vase, L. 2012. A systematic review of adverse events in the placebo arm of donepezil trials: The role of cognitive impairment. *International Psychogeriatrics* 4: 1–10.

Amanzio, M., Latini Corazzini, L., Vase, L. and Benedetti, F. 2009. A systematic review of adverse events in placebo groups of anti-migraine clinical trials. *Pain* 146: 261–9.

Amanzio, M., Benedetti, F., Porro, C. A., Palermo, S. and Cauda, F. 2013a. Activation Likelihood Estimation meta-analysis of brain correlates of placebo analgesia in human experimental pain. *Human Brain Mapping* 34: 738–52.

Amanzio, M., Vase, L., Leotta, D., Miceli, R., Palermo, S. and Geminiani G. 2013b. Impaired awareness of deficits in Alzheimer's disease: The role of everyday executive dysfunction. *Journal of the International Neuropsychological Society* 19: 63–72.

Amanzio, M., Torta, D. M., Sacco, K., Cauda, F., D'Agata, F., Duca, S. and Geminiani, G. C. 2011. Unawareness of deficits in Alzheimer's disease: Role of the cingulate cortex. *Brain* 134: 1061–76.

American Psychiatric Association. 2013. *Diagnostic and Statistical Manual of Mental Disorders*, 5th edn. Washington, DC: American Psychiatric Association.

Andersen, R. A. 1987. Inferior parietal lobule function in spatial perception and visuomotor integration. In V. B. Mountcastle, F. Plum and R. Geiger, *Handbook of Physiology 1: The Nervous System*, vol. 5: *Higher Functions of the Brain* (Part 2). Bethesda, MD: American Physiological Society, pp. 483–518.

Apkarian, V. A., Bushnell, C. M., Treede R. and Zubieta, J. 2005. Human brain mechanisms of pain perception and regulation in health and disease. *European Journal of Pain* 9: 463–84.

Barch, D. M., Braver, T. S., Nystrom, L. E., Forman, S. D., Noll, D. C. and Cohen, J. D. 1997. Dissociating working memory from task difficulty in human prefrontal cortex. *Neuropsychologia* 35: 1373–80.

Bathgate, D., Snowden, J. S., Varma, A., Blackshaw, A. and Neary, D. 2001. Behaviour in frontotemporal dementia, Alzheimer's disease and vascular dementia. *Acta Neurologica Scandinavica* 103: 367–78.

Bechara, A., Damasio, H., Tranel, D. and Damasio, A. R. 1997. Deciding advantageously before knowing the advantageous strategy. *Science* 275: 1293–5.

Benedetti, F., Arduino, C., Vighetti, S., Asteggiano, G., Tarenzi, L. and Rainero, I. 2004. Pain reactivity in Alzheimer patients with different degrees of cognitive impairment and brain electrical activity deterioration. *Pain* 111: 22–9.

Benedetti, F., Arduino, C., Costa, S., Vighetti, S., Tarenzi L., Rainero, I. and Asteggiano, G. 2006. Loss of expectation-related mechanisms in Alzheimer's disease makes analgesic therapies less effective. *Pain* 121: 133–44.

Benedetti, F., Vighetti, S., Ricco, C., Lagna, E., Bergamasco, B., Pinessi, L. and Rainero, I. 1999. Pain threshold and tolerance in Alzheimer's disease. *Pain* 80: 377–82.

Bonnelle, V., Ham, T. E., Leech, R., Kinnunen, K. M., Mehta, M. A., Greenwood, R. J. and Sharp, D. J. 2012. Salience network integrity predicts default mode network function after traumatic brain injury. *Proceedings of the National Academy of Sciences of the United States of America* 109: 4690–5.

Bornhovd, K., Quante, M., Glauche, V., Bromm, B., Weiller, C. and Buchel, C. 2002. Painful stimuli evoke different stimulus-response functions in the amygdala, prefrontal, insula and somatosensory cortex: A single-trial fMRI study. *Brain* 125: 1326–36.

Braak, H. and Braak, E. 1991. Neuropathological stageing of Alzheimer-related changes. *Acta Neuropathologica* 82: 239–59.

Carlino, E., Benedetti, F., Rainero, I., Asteggiano, G., Cappa, G., Tarenzi, L., Vighetti, S. and Pollo, A. 2010. Pain perception and tolerance in patients with frontotemporal dementia. *Pain* 151: 783–9.

Chudler, E. H. and Dong, W. K. 1995. The role of the basal ganglia in nociception and pain. *Pain* 60: 3–38.

Coghill, R. C., Talbot, J. D., Evans, A. C., Meyer, E., Giedde, A., Bushnell, M. C. and Duncana, G. H. 1994. Distributed processing of pain and vibration by the human brain. *The Journal of Neuroscience* 14: 4095–108.

Cole, L. J., Farrell, M. J., Duff, E. P., Barber, J. B., Egan, G. F. and Gibson, S. J. 2006. Pain sensitivity and fMRI pain-related brain activity in Alzheimer's disease. *Brain* 129: 2957–65.

Cole, L. J., Gavrilescu, M., Johnston, L. A., Gibson, S. J., Farrell, M. J. and Egan, G. F. 2011. The impact of Alzheimer's disease on the functional connectivity between brain regions underlying pain perception. *European Journal of Pain* 15: 568.e1–568.e11.

Damasio, A. R. 1996. The somatic marker hypothesis and the possible functions of the prefrontal cortex. *Philosophical Transactions of the Royal Society, Series B: Biological Sciences* 351: 1413–20.

Derbyshire, S. W., Jones, A. K., Gyulai, F., Clark, S., Townsend, D. and Firestone L. L. 1997. Pain processing during three levels of noxious stimulation produces differential patterns of central activity. *Pain* 73: 431–45.

Devinsky, O., Morrell, M. and Vogt, B. 1995. Contributions of the anterior cingulate cortex to behavior. *Brain* 118: 279–306.

Doraiswamy, M., Leon, J., Cummings, J. L., Martin, D. and Neumann, P. J. 2002. Prevalence and impact of medical comorbidity in Alzheimer's disease. *The Journals of Gerontology Series A, Biological Sciences and Medical Sciences* 57: 73–7.

Farrell, M. J., Laird, A. R. and Egan, G. F. 2005. Brain activity associated with painfully hot stimuli applied to the upper limb: A meta-analysis. *Human Brain Mapping* 25: 129–39.

Feldt, K. S., Ryden, M. B. and Miles, S. 1998. Treatment of pain in cognitively impaired compared with cognitively intact older patients with hip-fracture. *Journal of the American Geriatrics Society* 46: 1079–85.

Ferrell, B. A., Ferrell, B. R. and Rivera, L. 1995. Pain in cognitively impaired nursing home patients. *The Journal of Pain and Symptom Management* 10: 591–8.

Fried, L. P., Tangen, C. M., Walston, J., Newman, A. B., Hirsch, C., Gottdiener, J., Seeman, T., Tracy, R., Kop, W. J., Burke, G., McBurnie, M. A. and Cardiovascular Health Study Collaborative Research Group. 2001. Frailty in older adults: Evidence for a phenotype. *The Journals of Gerontology Series A: Biological Sciences and Medical Sciences* 56 (3): M146–56.

George, M. S., Ketter, T. A., Parekh, P. I., Rosinsky, N., Ring, H., Casey, B. J., Trimble M. R., Horowitz B., Herscovitch, P. and Post, R. M. 1994. Regional brain activity with selecting a response despite interference: An H2 15O PET study of the Stroop and an emotional Stroop. *Human Brain Mapping* 1: 194–209.

Gibson, S. J., Voukelatos, X., Ames, D., Flicker, L. and Helme, R. D. 2001. An examination of pain perception and cerebral event-related potentials following carbon dioxide laser stimulation in patients with Alzheimer's disease and age-matched control volunteers. *Pain Research & Management* 6: 126–32.

Gracely, R. H., Geisser, M. E., Giesecke, T., Grant, M. A., Petzke, F., Williams, D. A. and Clauw, D. J. 2004. Pain catastrophizing and neural responses to pain among persons with fibromyalgia. *Brain* 127: 835–43.

Grady, C. L., McIntosh, A. R., Beig, S., Keightley, M. L., Burian, H. and Black, S. E. 2003. Evidence from functional neuroimaging of a compensatory prefrontal network in Alzheimer's disease. *Journal of Neuroscience* 23: 986–93.

Graff-Guerrero, A., Gonzalez-Olvera, J., Fresan, A., Gomez-Martin, D., Mendez-Nunez, J. C. and Pellicer, F. 2005. Repetitive transcranial magnetic stimulation of dorsolateral prefrontal cortex increases tolerance to human experimental pain. Brain research. *Cognitive Brain Research* 25: 153–60.

Hahn, R. A. 1997. The nocebo phenomenon: Scope and foundations. In *The Placebo Effect: An Interdisciplinary Exploration*, ed. by A. Harrington. Cambridge, MA: Harvard University Press, pp. 56–76.

Ingvar, M. 1999. Pain and functional imaging. *Philosophical Transactions of the Royal Society of London, Series B: Biological Sciences* 354: 1347–58.

Kelly, D. D. 1985. Central representations of pain and analgesia. In *Principles of Neural Science*, ed. by E. R. Kandell and J. H. Schwartz, 2nd edn. New York: Elsevier, pp. 330–43.

Kennedy, W. P. 1961. The nocebo reaction. *Medical World* 95: 203–5.

Ketter, T. A., Andreason, P. J., George, M. S., Herscovich, P. and Post, R. M. 1993. Paralimbic rCBF increases during procaine-induced psychosensory and emotional experiences. *Biological Psychiatry* 33 (107): 66A.

Krummenacher, P., Candia, V., Folkers, G., Schedlowski, M. and Schonbachler, G. 2010. Prefrontal cortex modulates placebo analgesia. *Pain* 148: 368–74.

Legrain, V., Damme, S. V., Eccleston, C., Davis, K. D., Seminowicz, D. A. and Crombez, G. 2009. A neurocognitive model of attention to pain: Behavioral and neuroimaging evidence. *Pain* 144: 230–2.

Levine, J. D. and Gordon, N. C. 1984. Influence of the method of drug administration on analgesic response. *Nature* 312: 755–6.

Lorenz, J., Cross, D., Minoshima, S., Morrow, T., Paulson, P. and Casey, K. 2002. A unique representation of heat allodynia in the human brain. *Neuron* 35: 383–93.

MacDonald, A. W., Cohen, J. D., Stenger, V. A. and Carter, C. S. 2000. Dissociating the role of the dorsolateral prefrontal and anterior cingulate cortex in cognitive control. *Science* 288: 1835–8.

McKhann, G. M., Knopman, D. S., Chertkow, H., Hyman, B. T., Jack, C. R., Kawas, C. H., Klunk, W. E., Koroshetz, W. J., Manly, J. J., Mayeux, R., Mohs, R. C., Morris, J. C., Rossor, M. N., Scheltens, P. Carillo, M. C. Thies, B., Weintraub, S. and Phelps, C. H. 2011. The diagnosis of dementia due to Alzheimer's disease: Recommendations from the National Institute on Aging-Alzheimer's Association workgroups on diagnostic guidelines for Alzheimer's disease. *Alzheimer's & Dementia: The Journal of the Alzheimer's Association* 7: 263–9.

Melzack, R. and Casey, K. L. 1968. Sensory, motivational, and central control determinants of pain. In *The Skin Senses*, ed. by D. R. Kenshalo. Springfield, IL: C. C. Thomas, pp. 423–39.

Menon,V. and Uddin, L. Q. 2010. Saliency, switching, attention and control: A network model of insula function. *Brain Structure & Function* 214: 655–67.

Mesulam, M. M. 1998. From sensation to cognition. *Brain* 121: 1013–52.

Monroe, T. B., Gore, J.C, Chen, L. M., Mion, L. C. and Cowan, R. L. 2012. Pain in people with Alzheimer disease: Potential applications for psychophysical and neurophysiological research. *Journal of Geriatric Psychiatry and Neurology* 25: 240–55.

Pariente, J., White, P., Frackowiak, R. S. and Lewith, G. 2005. Expectancy and belief modulate the neuronal substrates of pain treated by acupuncture. *Neuroimage* 25: 1161–7.

Parvizi, J., Van Hoesen, G. W. and Damasio, A. 2000. Selective pathological changes of the periaqueductal gray matter in Alzheimer's disease. *Annals of Neurology* 48: 344–53.

Petrovic, P., Dietrich, T., Fransson, P., Andersson, J., Carlsson, K. and Ingvar, M. 2005. Placebo in emotional processing: Induced expectations of anxiety relief activate a generalized modulatory network. *Neuron* 46: 957–69.

Price, D. D. 2000. Psychological and neural mechanisms of affective dimension of pain. *Science* 288: 1769–72.

Rainero, I., Vighetti, S., Bergamasco, B., Pinessi, L. and Benedetti, F. 2000. Autonomic responses and pain perception in Alzheimer's disease. *European Journal of Pain* 4: 267–74.

Rainville, P. 2002. Brain mechanisms of pain affect and pain modulation. *Current Opinion Neurobiology* 12: 195–204.

Rief, W., Nestoriuc, Y., von Lilienfeld-Toal, A., Dogan, I., Schreiber, F., Hofmann, S. G., Barsky, A. J. and Avorn, J. 2009. Differences in adverse effect reporting in placebo groups in SSRI and tricyclic antidepressant trials: A systematic review and meta-analysis. *Drug Safety* 32: 1041–56.

Scherder, E. J. 2000. Low use of analgesics in Alzheimer's disease: Possible mechanisms. *Psychiatry* 63: 1–12.

Scherder, E. J. and Bouma, A. 1997. Is decreased use of analgesics in Alzheimer disease due to a change in the affective component of pain? *Alzheimer Disease and Associated Disorders* 11: 171–4.

Scherder, E. J., Sergeant, J. A. and Swaab, D. F. 2003. Pain processing in dementia and its relation to neuropathology. *Lancet Neurology* 2: 677–86.

Selden, N., Mesulam, M. M. and Geula, C. 1994. Human striatum: The distribution of neurofibrillary tangles in Alzheimer's disease. *Brain Research* 648: 327–31.

Sewards, T. V. and Sewards, M. A. 2002. The medial pain system: Neural representations of the motivational aspect of pain. *The Brain Research Bulletin* 59: 163–180.

Smith, E. E. and Jonides, J. 1999. Storage and executive processes in the frontal lobes. *Science* 283: 1657–61.

Snowden, S. J., Bathgate, D., Varma, A., Blackshaw, A., Gibbons, Z. C. and Neary, D. 2001. Distinct behavioural profiles in frontotemporal dementia and semantic dementia. *The Journal of Neurology, Neurosurgery, and Psychiatry* 70: 323–32.

Sridharan, D., Levitin, D. J. and Menon, V. 2008. A critical role for the right fronto-insular cortex in switching between central executive and default mode networks. *Proceedings of the National Academy of Sciences of the United States of America* 105: 12569–74.

Tekin, S., Mega, M. S., Masterman, D. M., Chow, T., Garakian, J., Vinters, H. V. and Cummings, J. L. 2001. Orbitofrontal and anterior cingulate cortex neurofibrillary tangle burden is associated with agitation in Alzheimer disease. *Annals of Neurology* 49: 355–61.

Thiruvady, D. R., Georgiou-Karistianis, N., Egan, G. F., Ray, S., Sritharan, A., Farrow, M., Churchyard, A., Chua, P., Bradshaw, J. L., Brawn, T. L. and Cunnington, R. 2007. Functional connectivity of the prefrontal cortex in Huntington's disease. *Journal of Neurology, Neurosurgery, & Psychiatry* 78: 127–33.

Tracey, I. and Mantyh, P. W. 2007. The cerebral signature for pain perception and its modulation. *Neuron* 55: 377–91.

Treede, R. D., Kenshalo, D. R., Gracely, R. H. and Jones, A. K. 1999. The cortical representation of pain. *Pain* 79: 105–11.

Tsai, P. F., Beck, C., Richards, K. C., Phillips, L., Roberson, P. K. and Evans, J. 2008. The pain behaviors for osteo-arthritis instrument for cognitively impaired elders (PBOICIE). *Research in Gerontological Nursing* 1: 116–22.

Vogt, B. A. 2005. Pain and emotion interactions in subregions of the cingulate gyrus. *Nature Reviews Neuroscience* 6: 533–44.

Vogt, L. J. K., Hyman, B. T., Van Hoesen, G. W. and Damasio, A. R. 1990. Pathological alterations in the amygdala in Alzheimer's disease. *Neuroscience* 37: 377–85.

Wager, T. D., Atlas, L. Y., Leotti, L. A. and Rilling, J. K. 2011. Predicting individual differences in placebo analgesia: Contributions of brain activity during anticipation and pain experience. *Journal of Neuroscience* 31: 439–52.

Waldemar, G., Dubois, B., Emre, M., Georges, J., McKeith, I. G., Rossor, M., Scheltens, P., Tariska P. and Winblad, B. 2007. Recommendations for the diagnosis and management of Alzheimer's disease and other disorders associated with dementia: EFNS guidelines. *European Journal of Neurology* 14, e1–e26.

Zubieta, J. K., Bueller, J. A., Jackson, L. R., Scott, D. J., Xu, Y., Koeppe, R. A., Nichols, T. E. and Stohler, C.S. 2005. Placebo effects mediated by endogenous opioid activity on mu-opioid receptors. *Journal of Neuroscience* 25: 7754–62.

CHAPTER 17

An overview of pain in Parkinson's disease

Panagiotis Zis, Elisaveta Sokolov and Kallol Ray Chaudhuri

Introduction

> We can ignore even pleasure. But pain insists upon being attended to. God whispers to us in our pleasures, speaks in our conscience, but shouts in our pains.
>
> *Lewis, 1944: 90–1*

Pain, one of the sensory symptoms in Parkinson's disease (PD) (Patel, Jankovic and Hallett, 2014), is a frequent yet poorly understood non-motor symptom (NMS), sometimes severe enough to overshadow the motor symptoms of the disorder. James Parkinson recognised that painful symptoms can be an early and manifest symptom in PD, a fact well recognised now (Parkinson, 1817; Garcia-Ruiz, Chaudhuri and Martinez-Martin, 2014). However, pain can occur throughout all stages of PD and exhibits considerable heterogeneity (Wasner and Deuschl, 2012).

Despite the fact that pain is one of the most common non-motor symptoms with adverse effect on quality of life in people with PD (Hanagasi et al., 2011), it continues to be neglected and poorly managed.

Definition of pain

Pain is a universal experience and the human body's most valuable alerting system. According to International Association for the Study of Pain (IASP), pain is an unpleasant sensory and emotional experience associated with actual or potential tissue damage, or described in terms of such damage (International Association for the Study of Pain, 1986). It is recognised that pain is a subjective symptom and that each person forms an understanding of the word through his/her experiences related to injury in early life (International Association for the Study of Pain, 1986). The need for a systematic classification of pain has long been recognised, and IASP suggests five axes for the description of pain: anatomical location, the body system

---

An Introduction to Pain and its Relation to Nervous System Disorders, First Edition. Edited by Anna A. Battaglia.
© 2016 John Wiley & Sons, Ltd. Published 2016 by John Wiley & Sons, Ltd.

involved, temporal characteristics, intensity, time of onset, and aetiology (International Association for the Study of Pain, 1986; van Seventer et al., 2013). A broad categorisation of pain that is useful in clinical practice divides it into nociceptive and neuropathic pain. Nociceptive pain is the pain that arises from actual or threatened damage to non-neural tissue and is due to the activation of nociceptors, while neuropathic pain is defined as the pain caused by a lesion or disease of the somatosensory nervous system (Treede et al., 2008). Neuropathic pain can be further classified through investigations such as neurophysiological testing and neuroimaging (Cruccu et al., 2010; for a detailed overview of this topic, see M. Thakur and S. McMahon's chapter in this volume).

Patients with PD experience pain that may fit into all of the above categories. However, pain in PD may or may not be related to the underlying disease process. For instance, pain unrelated to PD can arise in PD from comorbidities such as osteoarthritis; it can also be classified as neuropathic or nociceptive. Further, it can be described in relation to the affected bodily parts or to pathophysiological aspects of PD such as motor fluctuations, dystonia, dyskinesias and central pain. In this chapter we will consider pain that can specifically be attributed to PD, either directly (primary pain) or indirectly (secondary pain); we will also discuss the recently described concept of 'unexplained' pain in PD.

Pathogenesis

Basic pathophysiology

As Parkinson's is a multifocal and multisystem degenerative and progressive disease, it can affect the pain process at multiple levels (Fil et al., 2013). Pain sensation usually starts with sensing by specialised nociceptors located at the nerve terminals of the primary afferent fibres. The pain signal is then transmitted to the dorsal horn of the spinal column and through the central nervous system (CNS), where it is processed and interpreted in the somatosensory cerebral cortex. Four physiologic processes are associated with pain: transduction, transmission, modulation and perception (McCaffery and Pasero, 1999). A. Todd gives an extensive overview of the role of the dorsal horn in the integration and modulation of the nociceptive information in Chapter 1 of this volume.

'Transduction' refers to the conversion of a stimulus into electrical activity in the peripheral terminals of nociceptor sensory fibres. 'Transmission' denotes the passage of action potentials from the peripheral terminal, along axons, to the central terminal of nociceptors in the CNS. Two phylogenetically distinct systems, the medial and the lateral pain systems, transmit noxious information to the thalamus and to multiple areas in the brain (Figure 17.1). 'Modulation' describes the augmentation or the suppression of sensory input. 'Perception' refers to the interpretation to a specific sensory experience. The pathogenesis of pain may involve any of these processes.

As shown in Figure 17.1, in PD several areas could be affected, and particularly the medial system – as is also evident from Braak's theory of the pathogenesis of this disease (Braak et al., 2003). In theory, pain could be a feature of PD from Braak's stage 2 to advanced PD. This would

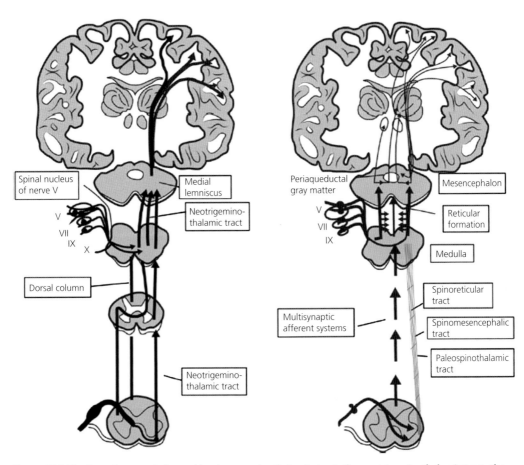

Figure 17.1 The lateral system is formed by the neospinothalamic tract, the neotrigeminothalamic tract, the cervical bundle and the beam of the dorsal horn, whose fibres terminate in the lateral thalamus, in the primary and secondary somatosensory areas, in the parietal operculum and in the insula. The medial system is mainly constituted by the palaeospinothalamic, spinomesencephalic, spinoreticular, spinoparabrachial hypothalamic and spinothalamic tract fibres, which terminate in the parabrachial nucleus, the locus coeruleus, the periaqueductal gray substance, the intralaminar and medial thalamic nuclei, the thalamic ventral caudal parvocellular nucleus and ventral caudal portae, the insula, the parietal operculum, the secondary somatosensory cortex, the amygdala and the hippocampus. Source: Fil 2013.

explain the emergence of pain as one of the 'premotor' non-motor markers of PD (Wolters and Braak, 2006), which is discussed below in more detail.

In a neurophysiologic study of central pain in PD it was shown that conduction along peripheral and central pain pathways is normal in PD patients with or without primary central pain. However, apart from signs of hyperalgesia, PD patients exhibited lack of habituation to sympathetic sudomotor responses to repetitive pain stimuli, which suggested an abnormal control of the effects of pain inputs on autonomic centres (Schestatsky et al., 2007). Interestingly, these abnormalities were attenuated by L-dopa, and this suggested that the dysfunction may occur in dopamine-dependent centres that regulate both the autonomic function and the inhibitory modulation of pain inputs (Schestatsky et al., 2007). This observation is consistent with the findings of a positron emission tomography (PET) study where it was shown that the pain threshold is lower in PD patients but returns to normal ranges after levodopa administration, and also that PD patients have higher pain-induced activation in nociceptive pathways, which can be reduced through levodopa (Brefel-Courbon et al., 2005).

Genetics

Among humans there is a marked variability in pain perception and in sensitivity and susceptibility to chronic painful conditions, and this variability is due to both genetic and environmental factors (Nielsen et al., 2008). Recent technological breakthroughs have considerably advanced the field of pain genetics in humans (Foulkes and Wood, 2008). Single nucleotide polymorphisms (SNPs) in several genes – such as COMT, OPRMI, GCH1, SCN9A and FAAH – have been implicated in pain sensitivity (Greenbaum et al., 2012). In Chapter 5 of this volume F. Denk and S. McMahon offer an excellent review of the genetic and epigenetic mechanisms that affect pain variability in humans.

Some of the genes mentioned above are involved in, or interact with, dopamine biochemical pathways that are impaired in PD. Genetic factors have been shown to be particularly important in early-onset PD; they may also influence the expression and perception of pain, which varies substantially among patients with PD. Recently single-nucleotide polymorphisms of the SCN9A and FAAH genes have been shown to be associated with the susceptibility to and expression of pain in these patients (Greenbaum et al., 2012).

Moreover, studies in non–PD patients with chronic pain have examined the association between variation in the COMT gene and the development of chronic pain. Having a low-activity polymorphism in COMT increases the likelihood of developing chronic pain. Of all the genetic polymorphisms that have been studied in the search for risk factors of chronic pain, polymorphisms in COMT – rs4818 single-nucleotide (SNP) polymorphisms – appear to be the most significant and are responsible for over 10 per cent of the variance in pain perception across the non PD population. People carrying lower activity COMT polymorphisms have lower pain thresholds and an almost three times higher chance of developing chronic pain. Studies addressing COMT polymorphism and pain in PD are currently underway (Diatchenko et al., 2007; Jarcho et al., 2012).

Pain as a premotor and presenting symptom in PD

A range of NMSs can precede the onset of motor symptoms in PD. Olfactory dysfunction and rapid eye movement behaviour disorder or constipation are possible biomarkers in the identification of an 'at-risk' population (Chaudhuri et al., 2011). In addition, sporadic reports link other NMSs, including central pain in PD, as possible premotor features (Metta, Rizos and Chaudhuri, 2010). This is in line with Braak's hypothesis about the pathogenesis of PD (Braak et al., 2003); and Wolters and Braak describe stage 2 (in Braak's terminology) as being linked to an expression of pain where the involvement of the coeruleus–subcoeruleus systems, lower raphe nuclei and gigantocellular reticular nucleus could be the causative factor (Wolters and Braak, 2006).

Pain may be the first symptom of the development of motor PD and has been increasingly recognised as a feature that can adversely affect quality of life in this disease (Todorova and Chaudhuri, 2013). Interestingly, in 15 per cent of the PD patients who participated in a recent retrospective study, pain was reported as the initial symptom (Lee et al., 2006). Eleven per cent of patients in this cohort suffered from shoulder pain before the onset of motor symptoms, and the side of the pain correlated with the side of the most severe motor symptoms. Interestingly, studies have shown that spinal cord lamina I neurons implicated in the pain pathway have α-synuclein deposits (Braak et al., 2007). In a study of the London PD brain bank, out of 433 cases examined, 21 per cent were reported to have presented a NMS first, 15 per cent being due to pain (O'Sullivan et al., 2008). The precise nature or type of pain that may be classified as early/premotor pain in PD remains to be characterised.

Epidemiology

Epidemiological studies estimate that the prevalence of pain in PD ranges from 30 per cent to 83 per cent, being significantly greater than in controls subjects (Ha and Jankovic, 2012). However, estimates vary greatly across studies. For instance, it has been reported that pain remains undeclared in 40.5 per cent of PD patients, and that up to 80 per cent of these patients experience chronic pain (Chaudhuri et al., 2010).

In a large systematic review, Broen and colleagues reviewed prevalence studies of pain in PD using the modified Quality Assessment of Diagnostic Accuracy Studies (QUADAS) tool (Broen et al., 2012) and found that prevalence frequency ranges from 40 per cent to 85 per cent, with a mean of 67.6 per cent. However, Chaudhuri and colleagues, in a controlled study of the holistic spectrum of non-motor symptoms in PD, showed that there is no difference in pain prevalence between PD and controls (Chaudhuri et al., 2006). Table 17.1 summarises the prevalence of pain in PD, as reported in clinical studies.

The variability of the definition of pain in PD, the methodology used and the non-uniformity of the patient population may account for variability in the reported prevalence. The outcome measures are also important for prevalence data, especially if the main outcome measure was not designed to assess pain. In addition, as will be discussed in detail below, pain in PD is heterogeneous and PD patients experience various kinds of pain to various degrees. One study demonstrated that 5 per cent of patients had three different types of pain and 24 per cent had two different types (Gallagher, Lees and Schrag, 2010). The PRIAMO study addressed NMSs in a holistic fashion, in a study of 1,072 patients across Italy (Barone et al., 2009). Pain was present in 60.9 per cent overall, rising from a prevalence of 50.9 per cent in untreated to 79.6 per cent in advanced (Hoehn and Yahr stage 4 or 5) PD patients.

Table 17.1 A list of studies addressing prevalence and incidence of pain in PD.

Author	Date	Study type	Pain type	Prevalence (%)
Goetz et al.	1986	Uncontrolled study	Pain directly related to PD.	46
Giuffrida et al.	2005	Uncontrolled study	All pain	67
Etchepare et al.	2006	Controlled study	Back pain	59 PD versus 23 controls
Chaudhuri et al.	2006	Controlled study	Unexplained pain	27.6 PD versus 30.2 controls
Lee et al.	2006	Uncontrolled study	All pain	85
Tinazzi et al.	2006	Uncontrolled study	All pain	40
Broetz et al.	2007	Controlled study	Back pain	74 PD versus 24 controls
Martinez-Martin et al.	2007	Uncontrolled study	All pain	28.8
Sullivan et al.	2007	Uncontrolled study	All pain	35
Defazio et al.	2008	Controlled study	All pain	69.9 PD versus 62.8 controls
Negre-Pages et al.	2008	Controlled study	All pain	61.8 PD chronic pain; twice more frequent in PD than in patients without PD
Stamey, Davidson and Jankovic	2008	Uncontrolled study	Shoulder pain	35
Ehrt, Larsen and Aarsland	2009	Controlled study	All pain	67 PD versus 39 controls
Barone et al.	2009	Uncontrolled study	All pain	60.9
Beiske et al.	2009	Uncontrolled study	All pain	83
Letro, Quagliato and Viana	2009	Uncontrolled study	Pain directly related to PD.	54
Madden and Hall	2010	Controlled study	Shoulder pain	80 PD versus 40 controls
Chaudhuri et al.	2010	Uncontrolled study	Unexplained pain	45.9
Hanagasi et al.	2011	Uncontrolled study	All pain	64.9
Santos-Garcia et al.	2011	Uncontrolled study	All pain	72.3

Source: adapted from Ha and Jankovic, 2012.

Systems of classification of pain in PD

There have been various proposals for categorising or classifying pain in PD. In 1976 Snider and colleagues presented a case series of 101 outpatients with parkinsonism, out of which 43 reported that they regularly experienced primary sensory symptoms, including burning sensations and spontaneous pain. The researchers classified the possible sources of these sensory symptoms into two broad categories: primary and secondary. Primary sensory symptoms were further categorised into symptoms arising from a peripheral nervous system malfunction (for example, chemical irritation and autonomic dysfunction, either through primary effect on sensation or through secondary effect via blood flow) or from a CNS pathophysiological process (for example, direct pathologic involvement of sensory pathways and release of sensory pathways from inhibition). Secondary sensory symptoms included not only abnormalities of joints, muscles and vessels but also rigidity, muscle cramps and dystonia (Snider et al., 1976).

In 1986 Quinn and colleagues proposed that pain associated with PD could be classified into four major categories: pain preceding diagnosis of PD, off-period pain (without dystonia) in patients with a fluctuating response to levodopa, painful dystonic spasms and peak-dose pain (Quinn et al., 1986).

However, the most commonly used classification system is the one proposed by Ford (1998, 2010). Ford offered five different categories of painful sensations in PD: musculoskeletal pain, radicular/neuropathic pain, pain associated to dystonia, central or primary pain, and pain associated to akathisia.

In 2009 Chaudhuri and Schapira proposed a classification of distinct pain syndromes in PD (Chaudhuri and Schapira, 2009). This classification consists of musculoskeletal pain, chronic pain (central or visceral), fluctuation-related pain, nocturnal pain, coat-hanger pain, orofacial pain, and peripheral limb or abdominal pain. This classification also categorises pain syndromes according to whether they respond to dopaminergic therapy. The first PD-specific pain scale (the Kings Parkinson Pain Scale) has been developed on the basis of this classification and its clinimetric validation study has been published (Chaudhuri et al., 2015).

In 2012 Del Sorbo and Albanese made a clear distinction between the categories 'primary' and 'secondary' in their taxonomy of PD-related pain. They use Ford's classification to divide secondary pain still further, into musculoskeletal pain, radicular/neuropathic pain, dystonia-related pain, and pain related to akathisia and to restless legs syndrome (Del Sorbo and Albanese, 2012).

Also in 2012, Wasner and Deuschl proposed a four-tier taxonomy of pain in PD. Tier 1 makes a distinction between pain related and pain unrelated to PD. Tier 2 further classifies pain into nociceptive, neuropathic and miscellaneous. Tier 3 continues to divide pain into musculoskeletal, visceral, cutaneous, peripheral and central, and tier 4 focuses on the specific structures involved and the pathology of these forms (Wasner and Deuschl, 2012).

Description of pain in PD

Primary or central pain

Primary pain in PD is assumed to be of central origin and to occur as a direct consequence of the disease itself, not from the disease's complications (Del Sorbo and Albanese, 2012). The origin of primary central pain in PD is believed to be in the dysfunction of dopaminergic-dependent autonomic centres, which regulate autonomic function and the inhibitory control of pain (Schestatsky et al., 2007). This dysfunction in basal ganglia probably alters the sensory processing of nociceptive inputs (Wasner and Deuschl, 2012; Beiske et al. Truini, Frontoni and Cruccu, 2013). In addition, the coeruleus and subcoeruleus systems, the lower raphe nuclei and the gigantocellular reticular nucleus are possibly implicated too, as described above (Braak et al., 2003; Wolters and Braak, 2006).

Unlike 'classic' central pain, central PD pain has no association with the clinically evident sensory deficits that reflect afferent pathway damage. The available data argue against the currently accepted criteria for defining neuropathic pain, given that the diagnosis requires a sensory deficit. However, primary pain presents many 'neuropathic' characteristics; thus its spectrum includes unexplained painful burning, formication, stabbing, aching, itching, tingling sensations that occur in undefined and peculiar body regions, or a vague overall sensation of tension and discomfort (Snider et al., 1976). There are several reports of unusual pain syndromes involving the face, head, pharynx, epigastrium, abdomen, pelvis, rectum, and genitalia.

Pain of central origin may have a relentless, obsessional and distressing quality that overshadows the motor symptoms. Furthermore, some patients with motor fluctuations experience fluctuations of painful sensations that may have a distinctly autonomic or visceral character (Del Sorbo and Albanese, 2012). Endogenous pain in PD is also associated with increased sensitivity to some specific painful stimuli (hyperalgesia) (Djaldetti et al., 2004).

Secondary pain
Musculoskeletal pain

Pain of musculoskeletal origin has been long described in PD, and it is the most frequent type of pain reported by patients (Sophie and Ford, 2012). Musculoskeletal pain comes from problems in the muscles or in the skeleton, or in both (Del Sorbo and Albanese, 2012), and it is a nociceptive pain that usually originates from abnormal posture, rigidity, and painful dystonia, or even from akinesia – because of motor fluctuations (Truini et al., 2013). Therefore musculoskeletal pain covers pain with muscular, joint and postural aetiologies, for instance muscle cramps. Muscle cramps or tightness in patients with PD can affect any part of the body, but they appear more typically in the neck, arm, paraspinal or calf muscles, while joint pains occur most frequently in the shoulder, hip, knee, and ankle (Fil et al., 2013; Goetz et al., 1986).

Radicular pain

Pain and discomfort that are well localised along the sensory distribution (dermatome) of a nerve or nerve root are described as radicular or neuritic pain (Ford, 2010). By comparison with the general population, patients with PD suffer more frequently from radicular pain (Wasner and Deuschl, 2012; Broetz et al., 2007). This higher frequency reflects the lumbar discal structure damage due to festination, kyphosis (Del Sorbo and Albanese, 2012), and immobility from rigidity and wheelchair or bed confinement (Sophie and Ford, 2012). Radicular pain may accompany musculoskeletal pain (Sophie and Ford, 2012). Radicular pain is a typical neuropathic pain; however, the paraesthetic sensations of coolness, numbness, or tingling may be mistakenly attributed to a central pain syndrome (Ford, 2010). Therefore further work of evaluating pain with his characteristics is necessary, as the pain could reveal a compressive root or nerve lesion.

Fluctuation related pain

Motor and non-motor fluctuations occur with PD progression. In relation to levodopa dosage, fluctuations can broadly be classified into three different patterns: 'wearing off' fluctuations (the duration of the beneficial effect of each levodopa dose becomes progressively shorter); delayed 'on'/no 'on' fluctuations (the interval between the intake of a levodopa dose and the subsequent effect is delayed or is totally absent); and random 'on–off' fluctuations. On the basis of the timing of their appearance in relation to the 'on–off' cycle of the patient, several

Table 17.2 Types of motor state and associated characteristics during levodopa treatment.

	Most commonly affected site	Abnormal involuntary movements	Associated pain
Dyskinesias in relation to ON/OFF cycle			
OFF period dystonia	Usually lower limbs and toes	Dystonic	+++
ON period	Extremities and trunk	Choreic or choreoathetoid	+
• *Peak dose dyskinesia*			
• *Square wave dyskinesia*			
Diphasic	Lower limbs	Dystonic or ballistic	+++

patterns of dyskinesia may occur. Table 17.2 summarises the different types of motor state and their associated characteristics in relation to pain.

Dyskinesia-related pain

The word δυσκινησία (*dyskinesia*) is an ancient Greek word literally meaning 'bad movement'. The general spectrum of dyskinesia includes a variety of movements, such as choreic, dystonic, athetotic, ballistic and myoclonic. In relation to levodopa dosage, fluctuations can broadly be classified into three different patterns: 'wearing off' fluctuations, delayed 'on' fluctuations and random 'on–off' fluctuations. Apart from the obvious motor implications, dyskinesias could be associated with pain, the strongest association being with the 'wearing off' state.

Dystonia-related pain

The word 'dystonic' comes from the ancient Greek word δυστονία (*dystonia*), literally meaning 'bad tone'. Dystonia is defined as a movement disorder characterised by sustained or intermittent muscle contractions and by involuntary, patterned, and often repetitive contractions of opposing muscles, which cause twisting movements or abnormal postures, or both. Dystonia is often initiated or worsened by voluntary action and associated with overflow muscle activation.

Dystonic spasms are among the most painful symptoms that a patient with PD may experience; and they tend to occur in the 'off' state or during the period leading up to an 'off' state. The dystonic spasms may be paroxysmal, spontaneous, or triggered by movement or activity. They may be brief, lasting for minutes; prolonged, lasting for hours; or even continuous, remaining unrelieved by treatment attempts (Ilson, Fahn and Côté, 1984).

Restlessness/akathisia related pain

Restlessness is a frequent complaint in PD (Ford, 2010). There are two distinct forms of restlessness that are associated with pain: akathisia and restless leg syndrome (RLS).

The term 'akathitic discomfort' (*akathisia*, 'incapacity to sit' in ancient Greek) is used to describe subjective inner restlessness or the painful impulse to move continually (Fil et al., 2013). Parkinsonian akathisia has long been observed in PD and must be distinguished from a need to move due to somatic urges, dyskinesia, anxiety, depression or claustrophobia. It has been suggested that akathisia results from a dopaminergic deficiency involving the mesocortical pathway, which originates in the ventral tegmental area and is known to appear in PD even in the premotor stage (Ford, 2010). The painful discomfort of akathisia primarily involves the lower limbs, can be severe and is relieved by motor activities, such as walking around the room or just getting out of bed. Akathisia is common in treated PD and its symptoms may mimic the clinical picture of restless legs syndrome, which makes it difficult to distinguish the two (Del Sorbo and Albanese, 2012).

RLS is a sensory-motor disturbance characterised by an urge to move and accompanied by uncomfortable and unpleasant sensations that worsen when at rest or during periods of inactivity; it peaks in the evening or at night, and is relieved by movement (Fil et al., 2013). RLS is common in PD and often associated with lower limb discomfort and pain. Pain may also be a dominant feature associated with periodic limb movement (PLM), which often accompanies RLS.

Atypical variants

Apart from the typical pain syndromes described above, there are other – atypical – variants. In their proposed classification system, Chaudhuri and Schapira describe nocturnal, orofacial, coat-hunger, and drug-induced pain (Chaudhuri and Schapira, 2009); and a new, specific, non-motor phenotype in PD – the unexplained lower limb pain (ULLP) – has more recently been described (Wallace and Chaudhuri, 2014).

Nocturnal pain is pain that may be related to RLS, but also to periodic limb movement. Moreover, during the night, PD patients may suffer from severe akinesia ('off' state), because of their levodopa dose timing.

Orofacial pain can involve the temporo-mandicular joint, can be related to bruxism, or can be part of a syndrome called 'burning mouth syndrome' (BMS). The IASP defines BMS as a distinctive nosological entity characterised by unremitting burning or a similar pain, which occurs in the absence of detectable mucosal changes and involves the tongue or other oral mucous membranes (International Association for the Study of Pain, 1994). Although BMS more often affects older individuals and postmenopausal women, it is also more common in patients with PD than in the general population (Clifford et al., 1998). Diminished endogenous dopamine and dysregulation of dopaminergic receptors in the nigrostriatal pathway have been implicated as a possible pathophysiologic mechanism for primary BMS (Jääskeläinen et al., 2001; Hagelberg et al., 2003).

Coat-hanger pain is pain located in suboccipital and paracervical regions, which are associated with an autonomic dysfunction. Coat-hanger pain is usually reported in patients suffering from orthostatic hypotension (Khurana, 2012). It is often ameliorated by lying down (or by other manoeuvres which increase blood pressure) and worsened in the head-up posture. Its exact aetiology is unknown; however, coat-hanger pain is thought to be caused by hypoperfusion in the relevant muscle groups. This hypothesis is consistent with the crampy and ischaemic nature of the described pain and with its relationship to posture.

Drug-induced pain is pain secondary to medication, arising from its side effects. Peripheral oedema-linked pain in the extremities is usually associated with treatment with dopamine agonists (Kleiner-Fisman and Fisman, 2007). A rare type of drug-induced pain is lower bowel pain associated with retroperitoneal fibrosis, which has also been described in association with ergot-derived dopamine receptor agonists (Kvernmo, Härtter and Burger, 2006).

ULLP has recently been proposed by Wallace and Chaudhuri (2014) as a specific non-motor phenotype in PD. From a cohort of 225 patients, Wallace and Chaudhuri have identified 22 with an unexplained lower limb pain that was described as persistent leg pain ranging from unilateral to bilateral, sometimes associated with whole-body pain; therefore the two researchers have proposed clinical criteria for the assessment of a syndrome of 'unexplained lower limb pain of PD'. This type of pain shows a male predominance and can occur in any age group, but usually is manifest in moderate to advanced PD patients. Patients describe it as internal, starting in the anterior proximal aspect of lower limb. Imaging of spine is normal, apart from age-related spondylotic changes. Although this pain is unrelated to non-motor fluctuations, a high association with non-motor symptom burden is evident (Wallace and Chaudhuri, 2014).

Table 17.3 summarises the characteristics of pain variants associated with PD and the respective responses to treatment, which is discussed in more detail below.

Assessment scales

Generic pain scales used to assess pain in PD include visual analogue scales (VASs), the Numerical Rating Scale (VRS), the Brief Pain Inventory (BPI) (Cleeland and Ryan, 1994), the McGill Pain Questionnaire (Melzack, 1975) and the DN4 Questionnaire (Bouhassira et al., 2005). The first pain scale specifically designed for PD patients (the Kings Parkinson Pain Scale) has now been validated by Chaudhuri and colleagues (2015). It is the first scale to cover pains related to PD over 14 items in seven domains, and it is based on the classification we described above. These seven domains are musculoskeletal pain (1 item), chronic pain (2 items), fluctuation-related pain (3 items), nocturnal pain (2 items), orofacial pain (3 items), radicular pain (1 item), oedema-related pain and lower abdominal pain (2 items). In each domain, pain is assessed along two axes; severity and frequency. The characteristics, the advantages and the disadvantages of these instruments are summarised in Table 17.4.

Treament

There are as yet no validated and evidence-based treatment algorithms for pain in PD. We provide here an evidence base for pain management strategies to be used in this disease.

Levodopa
Oral levodopa
Levodopa alters the pain threshold, as shown in a small study by Schesatsky and colleagues. They demonstrated that conduction along peripheral and central pain pathways is within normal limits in people with PD, regardless of whether or not these people experienced primary central pain; however, they failed to create habituation to repetitive pain stimulation. This suggests an atypical regulation of the autonomic centres that control sympathetic

Table 17.3 Characteristics of pain variants associated with PD and the respective response to treatment.

Pain in PD			Clinical characteristics	Management
Typical	Primary	Central	Burning, tingling or pruritus; relentless and bizarre in quality; poorly localizable	• Dopaminergic therapy (levodopa, dopamine agonists) but questionable effect • Opiates such as oxycodone • Anti-inflammatory agents, opioids, antiepileptics, tricyclic antidepressants, and atypical neuroleptics
	Secondary	Musculo-skeletal	Aching, cramping sensations experienced in muscles or articular/periarticular locations; exacerbation of pre/co-existing orthopaedic and rheumatological conditions	• Musculoskeletal examination, eventually rheumatological/orthopedic evaluation • Physical therapy and occupational therapy • Medical therapy: dopaminergic therapy (for parkinsonian rigidity and akinesia); anti-inflammatory and analgesic drugs (for rheumatological and orthopedic conditions). • Surgical therapy: orthopedic joint surgery if indicated
		Radicular	Localized to root or nerve territory; radicular pain may accompany musculoskeletal pain (in particular spinal disease)	• Neurological examination, eventually electrophysiological and imaging investigations • Physical and occupational therapy • Medical therapy: opioids analgesic, non-steroidal anti-inflammatory drugs, also in combination • Surgical therapy: decompressive surgery if indicated
	Fluctuation-related	Dyskinesia-related	Associated with dystonic (forced, sustained muscular contractions)	• Treat on off flcutiations: continuous drug delivery strategies (rotigotine, levodopa and apormophie infusion)
		Dystonia-related	posturing of limbs; may involve facial and pharyngeal musculature	• injections of botulinum toxin (dystonic limbs), baclofen
	Restlessness-related	Akathisia	Subjective sense of restlessness, often accompanied by an urge to move;	• Dopaminergic therapy (levodopa, dopamine agonists), opioids, clozapine
		Restless leg syndrome	Subjective sense of restlessness, often accompanied by an urge to move. However, patients with restless legs syndrome are able to control voluntary movements	• Lifestyle changes and activities: decreased use of caffeine, alcohol, and tobacco. • Eventual supplements to correct deficiencies in iron, folate, and magnesium • A program of moderate exercise and massaging the legs • Dopaminergic therapy (dopamine agonists, levodopa); benzodiazepines, opioids, anticonvulsants

Atypical			
	Nocturnal	May be related to restless leg syndrome or periodic limb movement	• Dopaminergic therapy (levodopa, dopamine agonists)
	Orofacial	Temporo-mandibular joint pain	
		Bruxism-related	
		Burning mouth syndrome	Unremitting oral burning or similar pain in the absence of detectable mucosal changes, involving the tongue or other oral mucous membranes	• Dopaminergic therapy (levodopa, dopamine agonists)
	Drug induced	Peripheral oedema linked	• Medication adjustment
		Lower bowel	
	Unexplained lower limb		• Opioids

Table 17.4 Characteristics, advantages and disadvantages of pain assessment scales.

Assessment Tool	PD Specific	Rater	Brief Description	Advantages	Disadvantages
Visual Analog Scale	No	Self-administered	Measures pain intensity. The patients make a cross on a 10 cm line. The score is obtained by measuring the distance between the line start and the cross.	Simple, easy to use, well-validated	Only allows for pain intensity evaluation. May be difficult to score in severely disabled or demented patients.
Visual Numerical Scale	No	Self-administered	Measures pain intensity. The patients provide verbal description is possible with this scale but not with VAS.	Simple, easy to use, well-validated	Only allows for pain intensity evaluation.
BPI (short form)	No	Self-administered	9 items. Designed to capture two dimensions of pain: severity and interference. BPI also intends to capture two components of interference – activity and affect (emotions).	Simple, easy to use, well validated. Linguistically adapted and validated in many languages. It does not only assess pain intensity, but also assesses the interferences of pain with daily living.	May be difficult to score in severely disabled or demented patients.
McGill pain questionnaire (short form)	No	Self-administered	4 subscales, 22 items. Designed to assess characteristics of pain (and related symptoms) and the correspondent intensity on each one	Simple, easy to use, covers many domains and combines intensity and different characteristics, including neuropathic pain.	Latest form (SF-MPQ-2) has not been used widely and its psychometric properties remain under-investigated. In addition, it is not validated in no other language but English and Persian. May be difficult to score in demented patients.
DN4	No	Clinician-administered	4 categories, 10 items. First 7 items are completed during the interview (pain characteristics and accompanied symptoms) and last 3 during the clinical examination (signs, presence of allodynia)	Simple, easy to use, well validated, linguistically adapted and validated in many languages. It is probably the best tool to discriminate neuropathic pain from nociceptive pain, which may alter the therapeutic management. It is clinician administered and therefore even in patients with advanced motor PD the questionnaire can be completed.	It does not assess intensity, but only pain type. May be difficult to score in demented patients.
Kings PD Pain questionnaire [55]	Yes	Clinician-administered	7 domains, specific to PD pain, 14 items. Severity and frequency is assessed on each item.	Non applicable	Non applicable

sudomotor responses. This dysfunction may occur in centres dependent on dopamine, which regulates the autonomic function and the inhibitory modulation of pain inputs, and it can be at least in part attenuated by levodopa (Schestatsky et al., 2007). In addition, Brefel-Courbon et al. (2005) showed that, in people with PD, levodopa raised the objective pain threshold as assessed by the nociceptive flexion reflex threshold. Some studies demonstrated that dopamine exhibited antinociceptive effects through the modulation of D2 subtype receptors at the spinal level (see Liu, Qiao and Dafny, 1992). Gerdelat-Mas et al. (2007) pointed to evidence of a dopaminergic modulation of the objective pain threshold in PD patients. It was demonstrated that levodopa significantly raises this threshold in pain-free PD patients, but not in healthy subjects, and that this threshold is lower in patients with PD than in healthy subjects.

Zesiewicz et al. (2010) recently produced a report on the quality standards for the management of non-motor symptoms in PD. This gives an excellent overview of treatment options for the non-motor symptoms of PD and reviews the evidence base for the hypothesis that oral levodopa is effective in treating dystonic pain in PD.

Intrajejunal levodopa infusion

Current open-label and comparative data suggest that intrajejunal levodopa infusions are beneficial in PD for non-motor symptoms, including pain (NMS) and health-related quality of life, in addition to improving motor fluctuations and dyskinesia (Honig et al., 2009). A comparative study of 17 patients with advanced PD were treated with intrajejunal L-dopa infusion (IJL) and were compared with a matched group of nine patients not given IJL. The IJL-treated group demonstrated significant improvements on the Unified Parkinson's Disease Rating Scale (UPDRS), parts III and IV, on the Non-Motor Symptoms Scale (NMSS), total score, and in quality of life, whereas the control group demonstrated no difference along these parameters (including the NMSS miscellaneous domain, part of which is 'unexplained' pain) (Reddy et al., 2012). This study provides indirect evidence for the efficacy of intrajejunal levodopa in the treatment of pain in PD.

Dopamine agonists

Rotigotine patch

The Randomised Evaluation of the 24-hour Coverage: Efficacy of Rotigotine (RECOVER) study demonstrated improvements on a Likert pain scale (measuring any type of pain). These improvements were obtained through the use of a rotigotine patch (Trenkwalder et al., 2011). They were most probably due to the effect of rotigotine on non-motor fluctuation in PD. In a recent post hoc analysis of data from 267 patients (178 rotigotine, 89 placebo), it was shown that pain was improved in people with PD treated with rotigotine. However, this effect may be partly attributable to benefits in motor function and sleep disturbances (Kassubek et al., 2014). Moreover, a phase IV multicenter, multinational, double-blind placebo-controlled two-arm trial (DOLORES) on the effect of the rotigotine patch on PD-related pain has just been completed and the results are awaited (Schrag, Sauerbier and Chaudhuri, 2015).

Apomorphine

Apomorphine can be administered through injection as well as through subcutaneous infusion and can also be effective for pain in PD, particularly pain related to non-motor fluctuation. Martinez-Martin and colleagues demonstrated in a pilot study that the infusion of apomorphine led to significant improvements, as measured on parts III and IV of the UPDRS, on the Parkinson's Disease Quality of Life Questionnaire (PDQ-8), and on the NMSS, total score. This suggested an improvement in 'unexplained' pain in PD (Martinez-Martin et al., 2011). [79]. Apomorphine injection, often used as a 'rescue' therapy in this disease, is usually effective for dystonic pain seen during 'wearing off'-related motor fluctuation. The effect of apomorphine infusion on pain in PD is, however, controversial and, although there may be a beneficial effect on 'off'-related pain, Dellapina et al. (2011) suggested that apomorphine, in contrast to levodopa, has no significant effect on the pain threshold in people with PD. More specifically, they showed that neither subjective nor objective pain thresholds and pain activation profiles were modified by apomorphine by comparison with placebo in 25 PD patients.

Non-dopaminergic options
Duloxetine

Djaldetti et al. (2007) studied the effect of duloxetine on pain symptoms in PD. Thirteen of the twenty patients who took part in the study reported pain relief to varying extents. The mean VAS, Brief Pain Inventory, and Short-Form McGill Pain Questionnaire scores decreased significantly. There was no change in pain threshold after treatment.

Deep brain stimulation (DBS)

This has been shown to improve 'off' pain (Kim et al., 2008). The procedure has a beneficial effect, which lasts for at least 24 months. In some subjects a new musculoskeletal pain had developed by this time (Kim et al., 2012). Pallidal DBS has demonstrated improvements for dystonia during the 'off' period and for other sensory symptoms in patients with advanced PD (Figure 17.2; Loher et al., 2002).

Opioids

Opioids are a key to chronic pain management and have been effective in treating many types of pain, prolonged-release opioid formulations proving to be especially effective. However, the use of opioids may be associated with opioid-induced constipation, which can affect up to 80 per cent of the patients who receive opioid therapy (Bell et al., 2009). Opioids have been used in PD; in particular oxycodone prolonged release (OXN PR), which is indicated for the treatment of moderate to severe pain, in combination with a naloxone component (an opiate antagonist used as a solution for injection in the treatment of opioid overdose), appears promising. When administered orally, naloxone can reduce opioid-induced constipation due to local action in the gut. It has a high first-pass metabolism and, when given with OXN PR, it facilitates the laxative effect, without significant antagonism of the narcotic analgesic effect of oxycodone (Bell et al., 2009). On the basis of this observation, a multicentre, double-blind, randomised, placebo-controlled parallel group study in male and female subjects was designed, in order to assess the efficacy of OXN PR in controlling PD-related chronic severe pain (see

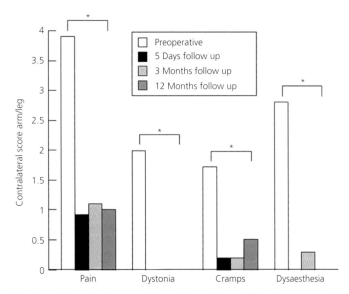

Figure 17.2 Effect of unilateral deep brain stimulation of the globus pallidus internus on 'off'-period dystonia, cramps, and sensory symptoms. Mean baseline scores and follow-up scores (percentage improvement compared with preoperative scores) during continuous pallidal stimulation are shown. Bar graphs display the unilateral scores of the arm and leg contralateral to the stimulation site. Source: Loher 2002.

https://clinicaltrials.gov/ct2/show/NCT01439100). The results of this study – the PANDA study – highlight the potential efficacy of OXN PR for patients with PD-related pain and might warrant further research on OXN PR in this setting.

A multidisciplinary approach

The role of multidisciplinary therapy in the management of pain in PD is paramount. In particular, specific approaches and strategies with physiotherapy can significantly improve the aching, cramping sensation associated with musculoskeletal pain as well as radicular pain. It may also help to receive input from occupational therapists and, occasionally, specific referrals to a pain team. Pain may be a dominant issue in the palliative stage of PD; in consequence, input from a palliative care team may also be important. A cross-sectional survey using a palliative care assessment tool, the Palliative Outcome Scale, reported that, out of the study's 82 patients with a diagnosis of idiopathic PD, multiple systems atrophy or progressive supranuclear palsy, over 80 per cent had pain (Saleem et al., 2013).

In conclusion, there are as yet no validated and evidence-based treatment algorithms for pain in PD. The PANDA study will become the first ever study to attempt a randomised placebo-controlled study of an active agent with the assessment of pain as a primary outcome measure; in addition, the validation of the first PD specific pain scale (Rizos et al., 2014) will aid future studies addressing pain in PD. Dopaminergic therapies can be particularly useful in alleviating a range of pains in PD – for example 'off'-related pain, aspects of central pain, night-time pain related to RLS and PLM, and pain of non-motor fluctuation. Furthermore, the role of opioids in managing central pain in PD is being increasingly recognised.

References

Barone, P., Antonini, A., Colosimo, C., Marconi, R., Morgante, L., Avarello, T. P., Bottacchi, E., Cannas, A., Ceravolo, G., Ceravolo, R., Cicarelli, G., Gaglio, R. M., Giglia, R. M., Iemolo, F., Manfredi, M., Meco, G., Nicoletti, A., Pederzoli, M., Petrone, A., Pisani, A., Pontieri, F. E., Quatrale, R., Ramat, S., Scala, R., Volpe, G., Zappulla, S., Bentivoglio, A. R., Stocchi, F., Trianni, G., Dotto, P. D. and PRIAMO study group. 2009. The PRIAMO study: A multicenter assessment of nonmotor symptoms and their impact on quality of life in Parkinson's disease. *Movement Disorders* 24 (11): 1641–9.

Beiske, A. G., Loge, J. H., Rønningen, A. and Svensson, E. 2009. Pain in Parkinson's disease: Prevalence and characteristics. *Pain* 141 (1–2): 173–7.

Bell, T. J., Panchal, S. J., Miaskowski, C., Bolge, S. C., Milanova, T. and Williamson, R. 2009. The prevalence, severity, and impact of opioid-induced bowel dysfunction: Results of a US and European Patient Survey (PROBE 1). *Pain Medicine* 10 (1): 35–42.

Bouhassira, D., Attal, N., Alchaar, H., Boureau, F., Brochet, B., Bruxelle, J., Cunin, G., Fermanian, J., Ginies, P., Grun-Overdyking, A., Jafari-Schluep, H., Lantéri-Minet, M., Laurent, B., Mick, G., Serrie, A., Valade and D., Vicaut, E. 2005. Comparison of pain syndromes associated with nervous or somatic lesions and development of a new neuropathic pain diagnostic questionnaire (DN4). *Pain* 114 (1–2): 29–36.

Braak, H., Sastre, M., Bohl, J. R., de Vos, R. A. and Del Tredici, K. 2007. Parkinson's disease: Lesions in dorsal horn layer, I., involvement of parasympathetic and sympathetic pre- and postganglionic neurons. *Acta Neuropathologica* 113 (4): 421–9.

Braak, H., Del Tredici, K., Rüb, U., de Vos, R. A., Jansen Steur, E. N. and Braak, E. 2003. Staging of brain pathology related to sporadic Parkinson's disease. *Neurobiology of Aging* 24 (2): 197–211.

Brefel-Courbon C1, Payoux, P., Thalamas, C., Ory, F., Quelven, I., Chollet, F., Montastruc, J. L. and Rascol, O. 2005. Effect of levodopa on pain threshold in Parkinson's disease: A clinical and positron emission tomography study. *Movement Disorders* 20 (12): 1557–63.

Broen, M. P., Braaksma, M. M., Patijn, J. and Weber, W. E. 2012. Prevalence of pain in Parkinson's disease: A systematic review using the modified QUADAS tool. *Movement Disorders* 27 (4): 480–4.

Broetz, D., Eichner, M., Gasser, T., Weller, M. and Steinbach, J. P. 2007. Radicular and nonradicular back pain in Parkinson's disease: A controlled study. *Movement Disorders* 22 (6): 853–6.

Chaudhuri, K. R. and Schapira, A. H. 2009. Non-motor symptoms of Parkinson's disease: Dopaminergic pathophysiology and treatment. *Lancet Neurology* 8 (5): 464–74.

Chaudhuri, K. R., Odin, P., Antonini, A. and Martinez-Martin, P. 2011. Parkinson's disease: The non-motor issues. *Parkinsonism & Related Disorders* 17 (10): 717–23.

Chaudhuri, K. R., Prieto-Jurcynska, C., Naidu, Y., Mitra, T., Frades-Payo, B., Tluk, S., Ruesmann, A., Odin, P., Macphee, G., Stochhi, F., et al. 2010. The nondeclaration of nonmotor symptoms of Parkinson's disease to health care professionals: An international study using the nonmotor symptoms questionnaire. *Movement Disorders* 25: 704–9.

Chaudhuri, R. K., Rizos, A., Trenkwalder, C., Rascol, O., Pal, S., Martino, D., Carroll, C., Paviour, D., Falup-Pecurariu, C., Kessel, B., Silverdale, M., Todorova, A., Sauerbier, A., Odin, P., Antonini, A. and Martinez-Martin, P., on behalf of EUROPAR and the IPMDS Non Motor PD Study Group. 2015. King's Parkinson's Disease Pain Scale, the first scale for pain in PD: An international validation. *Movement Disorders*. 30 (12): 1623–31.

Chaudhuri, K. R., Martinez-Martin, P., Schapira, A. H., Stocchi, F., Sethi, K., Odin, P., Brown, R. G., Koller, W., Barone, P., MacPhee, G., Kelly, L., Rabey, M., MacMahon, D., Thomas, S., Ondo, W., Rye, D., Forbes, A., Tluk, S., Dhawan, V., Bowron, A., Williams, A. J. and Olanow, C. W. 2006. International multicenter pilot study of the first comprehensive self-completed nonmotor symptoms questionnaire for Parkinson's disease: The NMSQuest study. *Movement Disorders* 21 (7): 916–23.

Cleeland, C. S. and Ryan, K. M. 1994. Pain assessment: Global use of the Brief Pain Inventory. *Annals, Academy of Medicine, Singapore* 23 (2): 129–38.s

Clifford, T. J., Warsi, M. J., Burnett, C. A. and Lamey, P. J. 1998. Burning mouth in Parkinson's disease sufferers. *Gerodontology* 15 (2): 73–8.

Cruccu, G., Sommer, C., Anand, P., Attal, N., Baron, R., Garcia-Larrea, L., Haanpaa, M., Jensen, T. S., Serra, J. and Treede, R. D. 2010. EFNS guidelines on neuropathic pain assessment: Revised 2009. *European Journal of Neurology* 17 (8): 1010–8.

Defazio, G., Berardelli, A., Fabbrini, G., Martino, D., Fincati, E., Fiaschi, A., Moretto, G., Abbruzzese, G., Marchese, R., Bonuccelli, U., Del Dotto, P., Barone, P., De Vivo, E., Albanese, A., Antonini, A., Canesi, M., Lopiano, L., Zibetti, M., Nappi, G., Martignoni, E., Lamberti, P. and Tinazzi, M. 2008. Pain as a nonmotor symptom of Parkinson disease: Evidence from a case-control study. *Archives of Neurology* 65 (9): 1191–4.

Del Sorbo, F. and Albanese, A. 2012. Clinical management of pain and fatigue in Parkinson's disease. *Parkinsonism & Related Disorders* 18 (Suppl. 1): S233–6.

Dellapina, E., Gerdelat-Mas, A., Ory-Magne, F., Pourcel, L., Galitzky, M., Calvas, F., Simonetta-Moreau, M., Thalamas, C., Payoux, P. and Brefel-Courbon, C. 2011. Apomorphine effect on pain threshold in Parkinson's disease: A clinical and positron emission tomography study. *Movement Disorders* 26 (1): 153–7.

Diatchenko, L., Nackley, A. G., Tchivileva, I. E., Shabalina, S. A. and Maixner, W. 2007. Genetic architecture of human pain perception. *Trends in Genetics* 23: 605–13.

Djaldetti, R., Yust-Katz, S., Kolianov, V., Melamed, E. and Dabby, R. 2007. The effect of duloxetine on primary pain symptoms in Parkinson disease. *Clinical Neuropharmacology* 30 (4): 201–5.

Djaldetti, R., Shifrin, A., Rogowski, Z., Sprecher, E., Melamed, E. and Yarnitsky, D. 2004. Quantitative measurement of pain sensation in patients with Parkinson disease. *Neurology* 62 (12): 2171–5.

Ehrt, U., Larsen, J. P. and Aarsland, D. 2009. Pain and its relationship to depression in Parkinson disease. *American Journal of Geriatric Psychiatry* 17 (4): 269–75.

Etchepare, F., Rozenberg, S., Mirault, T., Bonnet, A. M., Lecorre, C., Agid, Y., Bourgeois, P. and Fautrel, B. 2006. Back problems in Parkinson's disease: An underestimated problem. *Joint Bone Spine* 73 (3): 298–302.

Fil, A., Cano-de-la-Cuerda, R., Muñoz-Hellín, E., Vela, L., Ramiro-González, M. and Fernández-de-Las-Peñas, C. 2013. Pain in Parkinson disease: A review of the literature. *Parkinsonism & Related Disorders* 19 (3): 285–94.

Ford, B. 1998. Pain in Parkinson's disease. *Journal of Clinical Neuroscience* 5 (2): 63–72.

Ford, B. 2010. Pain in Parkinson's disease. *Movement Disorders* 25 (Suppl. 1): S98–10.

Foulkes, T. and Wood, J. N. 2008. Pain genes. *PLoS Genetics* 4 (7): e1000086.

Gallagher, D. A., Lees, A. J. and Schrag, A. 2010. What are the most important non-motor symptoms in patients with Parkinsons disease and are we missing them? *Movement Disorders* 25: 2493–500.

Garcia-Ruiz, P. J., Chaudhuri, K. R. and Martinez-Martin, P. 2014. Non-motor symptoms of Parkinson's disease A review … from the past. *Journal of the Neurological Sciences* 338 (1–2): 30–33.

Gerdelat-Mas, A., Simonetta-Moreau, M., Thalamas, C., Ory-Magne, F., Slaoui, T., Rascol, O. and Brefel-Courbon, C. 2007. Levodopa raises objective pain threshold in Parkinson's disease: A RIII reflex study. *Journal of Neurology, Neurosurgery, and Psychiatry* 78 (10): 1140–2.

Giuffrida, R., Vingerhoets, F. J., Bogousslavsky, J. and Ghika, J. 2005. Pain in Parkinson's disease. *Revue neurologique* (Paris) 161 (4): 407–18.

Goetz, C. G., Tanner, C. M., Levy, M., Wilson, R. S., Garron, D. C. 1986. sPain in Parkinson's disease. *Movement Disorders* 1 (1): 45–9.

Greenbaum, L., Tegeder, I., Barhum, Y., Melamed, E., Roditi, Y. and Djaldetti, R. 2012. Contribution of genetic variants to pain susceptibility in Parkinson disease. *European Journal of Pain* 16 (9): 1243–50.

Ha, A. D., Jankovic, J. 2012. Pain in Parkinson's Disease. *Movement Disorders* 27 (4): 485–91.

Hagelberg, N., Forssell, H., Rinne, J. O., Scheinin, H., Taiminen, T., Aalto, S., Luutonen, S., Någren, K. and Jääskeläinen, S. 2003. Striatal dopamine D1 and D2 receptors in burning mouth syndrome. *Pain* 101 (1–2): 149–54.

Hanagasi, H. A., Akat, S., Gurvit, H., Yazici, J. and Emre, M. 2011. Pain is common in Parkinson's disease. *Clinical Neurology and Neurosurgery* 113 (1): 11–3.

Honig, H., Antonini, A., Martinez-Martin, P., Forgacs, I., Faye, G. C., Fox, T., Fox, K., Mancini, F., Canesi, M., Odin, P. and Chaudhuri, K. R. 2009. Intrajejunal levodopa infusion in Parkinson's disease: A pilot multicenter study of effects on nonmotor symptoms and quality of life. *Movement Disorders* 24 (0): 1468–74.

Ilson, J., Fahn, S., Côté, L. 1984. Painful dystonic spasms in Parkinson's disease. *Advances in Neurology* 40: 395–8.

International Association for the Study of Pain (IASP) Subcommittee on Taxonomy. 1986. Descriptions of chronic pain syndromes and definitions of pain terms. In *Classification of Chronic Pain*, ed. by H. Merskey and N. Bogduk. Special issue of *Pain* (Suppl. 3): S1–226.

International Association for the Study of Pain (IASP) Subcommittee on Taxonomy. 1994. Relatively localized syndromes of the head and neck: Lesions of the ear, nose, and oral cavity. Part IIB of *Classification of Chronic Pain*, ed. by H. Merskey and N. Bogduk. Seattle, WA: IASP Press. At http://www.iasp-pain.org/files/Content/ ContentFolders/Publications2/FreeBooks/Classification-of-Chronic-Pain.pdf (accessed 5 October 2015).

Jääskeläinen, S. K., Rinne, J. O., Forssell, H., Tenovuo, O., Kaasinen, V., Sonninen, P. and Bergman, J. 2001. Role of the dopaminergic system in chronic pain: A fluorodopa-PET study. *Pain* 90 (3): 257–60.

Jarcho, J. M., Mayer, E. A., Jiang, Z. K., Feier, N. A. and London, E. D. 2012. Pain, affective symptoms, and cognitive deficits in patients with cerebral dopamine dysfunction. *Pain* 153: 744–54.

Kassubek, J., Chaudhuri, K. R., Zesiewicz, T., Surmann, E., Boroojerdi, B., Moran, K., Ghys, L. and Trenkwalder, C. 2014. Rotigotine transdermal system and evaluation of pain in patients with Parkinson's disease: A post hoc analysis of the RECOVER study. *BioMedCentral Neurology* 14 (1): 42.

Khurana, R. K. 2012. Coat-hanger ache in orthostatic hypotension. *Cephalalgia* 32 (10): 731–7.

Kim, H. J., Jeon, B. S., Lee, J. Y., Paek, S. H., Kim, D. G. 2012. The benefit of subthalamic deep brain stimulation for pain in Parkinson disease: A 2-year follow-up study. *Neurosurgery* 70 (1): 18–23. (Discussion: 23–4.)

Kim, H. J., Paek, S. H., Kim, J. Y., Lee, J. Y., Lim, Y. H., Kim, M. R., Kim, D. G. and Jeon, B. S. 2008. Chronic subthalamic deep brain stimulation improves pain in Parkinson disease. *Journal of Neurology* 255 (12): 1889–94.

Kleiner-Fisman, G. and Fisman, D. N. 2007. Risk factors for the development of pedal edema in patients using pramipexole. *Archives of Neurology* 64 (6): 820–4.

Kvernmo, T., Härtter, S., Burger, E. 2006. A review of the receptor-binding and pharmacokinetic properties of dopamine agonists. *Clinical Therapeutics* 28 (8): 1065–78.

Lee, M. A., Walker, R. W., Hildreth, T. J. and Prentice, W. M. 2006. A survey of pain in idiopathic Parkinson's disease. *Journal of Pain and Symptom Management* 32 (5): 462–9.

Letro, G. H., Quagliato, E. M. and Viana, M. A. 2009. Pain in Parkinson's disease. *Arquivos de Neuro-Psiquiatria* 67 (3A): 591–4.

Lewis, C. S. 1944. *The Problem of Pain*. New York: Macmillan.

Liu, Q. S., Qiao, J. T. and Dafny, N. 1992. D2 dopamine receptor involvement in spinal dopamine-produced antinociception. *Life Sciences* 51 (19): 1485–92.

Loher, T. J., Burgunder, J. M., Weber, S., Sommerhalder, R. and Krauss, J. K. 2002. Effect of chronic pallidal deep brain stimulation on off period dystonia and sensory symptoms in advanced Parkinson's disease. *Journal of Neurology, Neurosurgery, and Psychiatry* 73 (4): 395–9.

Madden, M. B. and Hall, D. A. 2010. Shoulder pain in Parkinson's disease: A case-control study. *Movement Disorders* 25 (8): 1105–6.

Martinez-Martin, P., Reddy, P., Antonini, A., Henriksen, T., Katzenschlager, R., Odin, P., Todorova, A., Naidu, Y., Tluk, S., Chandiramani, C., Martin, A. and Chaudhuri, K. R. 2011. Chronic subcutaneous infusion therapy with apomorphine in advanced Parkinson's disease compared to conventional therapy: A real-life study of non motor effect. *Journal of Parkinson's Disease* 1 (2): 197–203.

Martinez-Martin, P., Schapira, A. H., Stocchi, F., Sethi, K., Odin, P., MacPhee, G., Brown, R. G., Naidu, Y., Clayton, L., Abe, K., Tsuboi, Y., MacMahon, D., Barone, P., Rabey, M., Bonuccelli, U., Forbes, A., Breen, K., Tluk, S., Olanow, C. W., Thomas, S., Rye, D., Hand, A., Williams, A. J., Ondo, W. and Chaudhuri, K. R. 2007. Prevalence of nonmotor symptoms in Parkinson's disease in an international setting: Study using nonmotor symptoms questionnaire in 545 patients. *Movement Disorders* 22 (11): 1623–9.

McCaffery, M. and Pasero, C. (1999) *Pain: A Clinical Manual*. St Louis, MO: Mosby.

Melzack, R. 1975. The McGill Pain Questionnaire: Major properties and scoring methods. *Pain* 1: 277–99.

Metta, V., Rizos, A., Chaudhuri, K. R. 2010. Non-motor symptoms and Parkinson's disease. *Advances in Clinical Neuroscience & Rehabilitation* 10 (1): 19–21.

Nègre-Pagès, L., Regragui, W., Bouhassira, D., Grandjean, H. and Rascol, O. 2008. DoPaMiP Study Group. Chronic pain in Parkinson's disease: The cross-sectional French DoPaMiP survey. *Movement Disorders* 23 (10): 1361–9.

Nielsen, C. S., Stubhaug, A., Price, D. D., Vassend, O., Czajkowski, N. and Harris, J. R. 2008. Individual differences in pain sensitivity: Genetic and environmental contributions. *Pain* 136 (1–2): 21–9.

O'Sullivan, S. S., Williams, D. R., Gallagher, D. A., Massey, L. A., Silveira-Moriyama, L. and Lees, A. J. 2008. Nonmotor symptoms as presenting complaints in Parkinson's disease: A clinicopathological study. *Movement Disorders* 23 (1): 101–6.

Parkinson, J. 1817. *An Essay on the Shaking Palsy*. London: Printed by Whittingham and Rowland for Sherwood, Neely and Jones.

Patel, N., Jankovic, J. and Hallett, M. 2014. Sensory aspects of movement disorders. *Lancet Neurology* 13 (1): 100–12.

Quinn, N. P., Koller, W. C., Lang, A. E. and Marsden, C. D. 1986. Painful Parkinson's disease. *Lancet* 1 (8494): 1366–9.

Reddy, P., Martinez-Martin, P., Rizos, A., Martin, A., Faye, G. C., Forgacs, I., Odin, P., Antonini, A. and Chaudhuri, K. R. 2012. Intrajejunal levodopa versus conventional therapy in Parkinson disease: Motor and nonmotor effects. *Clinical Neuropharmacology* 35 (5): 205–7.

Saleem, T. Z., Higginson, I. J., Chaudhuri, K. R., Martin, A., Burman, R. and Leigh, P. N. 2013. Symptom prevalence, severity and palliative care needs assessment using the Palliative Outcome Scale: A cross-sectional study of patients with Parkinson's disease and related neurological conditions. *Palliative Medcine* 27 (8): 722–31.

Santos-García, D., Abella-Corral, J., Aneiros-Díaz, Á., Santos-Canelles, H., Llaneza-González, M. A. and Macías-Arribi, M. 2011. Pain in Parkinson's disease: Prevalence, characteristics, associated factors, and relation with other non motor symptoms, quality of life, autonomy, and caregiver burden. *Revista de neurologia* 52 (7): 385–93.

Schestatsky, P., Kumru, H., Valls-Solé, J., Valldeoriola, F., Marti, M. J., Tolosa, E. and Chaves, M. L. 2007. Neurophysiologic study of central pain in patients with Parkinson disease. *Neurology* 69 (23): 2162–9.

Schrag, A., Sauerbier, A. and Chaudhuri, K. R. 2015. New clinical trials for nonmotor manifestations of Parkinson's disease. *Movement Disorders* 30 (11): 1490–504.

Snider, S. R., Fahn, S., Isgreen, W. P. and Cote, L. J. 1976. Primary sensory symptoms in parkinsonism. *Neurology* 26 (5): 423–9

Sophie, M. and Ford, B. 2012. Management of pain in Parkinson's disease. *CNS Drugs* 26 (11): 937–48.

Stamey, W., Davidson, A. and Jankovic, J. 2008. Shoulder pain: A presenting symptom of Parkinson disease. *Journal of Clinical Rheumatology* 14 (4): 253–4.

Sullivan, K. L., Ward, C. L., Hauser, R. A. and Zesiewicz, T. A. 2007. Prevalence and treatment of non-motor symptoms in Parkinson's disease. *Parkinsonism & Related Disorders* 13 (8): 545.

Tinazzi, M., Del Vesco, C., Fincati, E., Ottaviani, S., Smania, N., Moretto, G., Fiaschi, A., Martino, D. and Defazio, G. 2006. Pain and motor complications in Parkinson's disease. *Journal of Neurology, Neurosurgery, and Psychiatry* 77 (7): 822–5.

Todorova, A. and Chaudhuri, K. R. 2013. Olfaction, pain and other sensory abnormalities in Parkinson's disease. In *Clinical Insights: Parkinson's Disease: Diagnosis, Motor Symptoms and Non-Motor Features*, ed. by Joseph Jankovic. London: Future Medicine Ltd, pp. 35–52.

Treede, R. D., Jensen, T. S., Campbell, J. N., Cruccu, G., Dostrovsky, J. O., Griffin, J. W., Hansson, P. and Hughes, R. 2008. Neuropathic pain: Redefinition and a grading system for clinical and research purposes. *Neurology* 70 (18): 1630–5.

Trenkwalder, C., Kies, B., Rudzinska, M., Fine, J., Nikl, J., Honczarenko, K., Dioszeghy, P., Hill, D., Anderson, T., Myllyla, V., Kassubek, J., Steiger, M., Zucconi, M., Tolosa, E., Poewe, W., Surmann, E., Whitesides, J., Boroojerdi, B. and Chaudhuri, K. R. 2011. Recover Study Group. Rotigotine effects on early morning motor function and sleep in Parkinson's disease: A double-blind, randomized, placebo-controlled study (RECOVER). *Movement Disorders* 26 (1): 90–9.

Trenkwalder, C., Chaudhuri, K. R., Martinez-Martin, P., Rascol, O., Ehret, R., Vališ, M., Sátori, M., Krygowska-Wajs, A., Marti, M. J., Reimer, K., et al 2015. Prolonged-release oxycodone-naloxone for treatment of severe pain in patients with Parkinson's disease (PANDA): A double-blind, randomised, placebo-controlled trial. *Lancet Neurolology*. doi: 10.1016/S1474-4422(15)00243-4.

Truini, A., Frontoni, M. and Cruccu, G. 2013. Parkinson's disease related pain: A review of recent findings. *Journal of Neurology* 260 (1): 330–4.

van Seventer, R., Vos, C., Giezeman, M., Meerding, W. J., Arnould, B., Regnault, A., van Eerd, M., Martin, C. and Huygen, F. 2013; Validation of the Dutch version of the DN4 Diagnostic Questionnaire for Neuropathic Pain. *Pain Practice* 13 (5): 390–8.

Wallace, V. C. and Chaudhuri, K. R. 2014. Unexplained lower limb pain in Parkinson's disease: A phenotypic variant of 'painful Parkinson's disease'. *Parkinsonism & Related Disorders* 20 (1): 122–4.

Wasner, G. and Deuschl, G. 2012. Pains in Parkinson disease: Many syndromes under one umbrella. *Nature Reviews Neurology* 8 (5): 284–94.

Wolters, E. C., Braak, H. 2006. Parkinson's disease: Premotor clinico-pathological correlations. *Journal of Neural Transmission* (Suppl. 70): 309–19.

Zesiewicz, T. A., Sullivan, K. L., Arnulf, I., Chaudhuri, K. R., Morgan, J. C., Gronseth, G. S., Miyasaki, J., Iverson, D. J. and Weiner, W. J. 2010. Practice parameter: Treatment of nonmotor symptoms of Parkinson disease: Report of the Quality Standards Subcommittee of the American Academy of Neurology. *Neurology* 74 (11): 924–31.

APPENDIX

Interviews with chronic pain patients

Transcript of Joan Hester's Interview with Her Patient Maeve

JH: I first met Maeve in 2004 and she still comes to see me at the Pain Clinic in King's College Hospital. Maeve had her third baby at the age of 37. It was a difficult birth with forceps delivery, traumatic and painful. Maeve was exhausted, tearful and angry with the doctors who had done this to her; she felt treated roughly. Her tear required stitches and at that moment the pain started radiating in the left side of the perineal region; she also had pins and needles in the same distribution. It was not nociceptive pain – that is, visceral pain – but neuropathic pain, nerve pain.

The problem was that the doctors couldn't see the pain. This is where the misunderstandings started: the pain can't be seen.

When she first attended the pain clinic at King's in 2004, Maeve had lost self-esteem and self-confidence and felt that no one believed in her symptoms, which were severe and interfered massively with her life.

The pain never went away. When I first examined her, Maeve had limited leg straightening, weaknesses in left quads and muscles on the lower left leg, allodynia all over the thigh, and also allodynia in the lower left leg. Pin test was sharp; increased cold perception very unpleasant. Nerve conduction studies in 2005 were normal, but this was because these studies only examine certain types of fibres and not the smaller C fibres. [*Addressing Maeve*]

JH: Tell us what you still feel.

M: Allodynia to cold is still there, very painful, from waist down, mainly affecting the left leg, with pins and needles, and at times also affecting the right leg.

JH: What about the nature of the pain?

M: It can vary from throbbing constantly in both legs to stabbing, as if there was a knife permanently in it, which you tolerate...

JH: What is the hardest to cope with?

M: The lack of energy, which is mostly unbearable – how pain sucks your energy; it is like having a parasite that sucks your energy all the time and is always there and sometimes sleeps. And it is all about how you manage it, how you cope with it.

JH: Is it worse at any time of the day?

M: Can be worse at any time, it comes in attacks or you can be pain-free and coping much better; it can be for a couple of weeks; in the summer it is much easier than in the winter; when I have an attack I become bed-bound for three or four days and I function at a very low level.

JH: How has all this affected your life?

M: I used to work full-time; I still work, now three days a week, but it continues to be a fight, because it is difficult for employers to understand – you don't want to tell them. I made an adaptation, sitting in a way that helps, keeping both legs up, keeping the weight away. I have now an equipment at work, but it has been a fight. They cannot see it [the pain], it does not have a proper label, so if I use a stick they can see it. But people still say things, and if you do not use it every single day of your life it means you are better. It's easier for people with a broken leg: the others see an end to it and they can understand you until you take away the plaster. But it is very difficult to understand long-term conditions.

JH: You had three children to bring up, you had to work, and you were in pain; what were your emotions like back then?

M: It's constant – it's there, still like that: it's like scales and you have a slide rule your head. I expected that to balance out, to stop somewhere in the middle; but it never has. It's very difficult to come to terms with it if pain changes are difficult to handle. The birth was mismanaged; I should have had a caesarian section.

JH: Litigation was a big issue.

M: It was very, very difficult, stressful and emotional. They [the doctors] could not fix it, so I went for compensation. I did it for others, so they don't go through the same things; it made no difference to me – money can't take away the pain. It took six years in total.

JH: What drugs do you take?

M: Tramadol, pregabalin, tryciclic antidepressants

JH: Do they help?

M: Not for an attack, when pain has broken; but over long periods of time they help me to function. I don't like to take them but there is no other choice. Doses vary.

JH: What else do you do?

M: Acupuncture every fortnight, it's very effective: not for stopping the pain, but it gives me more energy. I also do cognitive behavioural therapy (CBT), once every sixth week; it helps me to maintain a good level of functioning, to keep a positive outlook – and it's good having contact with someone who understands, you can feel very isolated.

JH: Have you ever thought about how life would have been without pain?

M: Difficult to think about this, it's a very disruptive thought.

JH: Thank you.

Transcript of Joan Hester's Interview with Her Patient Lorraine

JH: Lorraine had a whiplash injury in 1991. She was stationary in her car and was hit in the rear by a van at about 30mph. She felt an immediate neck pain that became worse the next day and gave her immense stiffness, so that she could not turn her head at all from side to side. After a few days the pain started to radiate into her arms, with tingling and pins and needles in the fingers.

The pain has never gone away completely. Physiotherapy has helped to relieve stiffness and has improved mobility of the neck. She has continued the treatment for 20 years; without it, her neck stiffens again. MRI and nerve conduction studies showed no significant abnormality. Medication has given her side effects but has never really relieved pain, though it does improve sleep and mood.

Lorraine attended a residential pain management programme, which she describes as the "best thing that ever happened." Over a three-week period she learned how to move her neck again, how to relax, how to pace her activities, and she understood how stress immediately tightens the muscles in her neck and shoulders. She learned when to use and when not to use medication, and the importance of carrying on with normal things in life. She couldn't return to work but instead studied for a degree course from home, limiting the time she spent at the computer to 30 minutes at a time.

She made herself socialise and meet up with friends, and after several months she dared to get into a car again and then to drive. After about five years she did return to work and she negotiated working one day a week from home, to ease the stress of travelling. Life goes on … but not as it was before the accident.

Accidents change lives and none of us knows how we would cope until we are forced into such a situation. Living with chronic pain is difficult and learning to cope with it requires time and patience. The hope is that, instead of pain ruling the person's life, the person will rule the pain. This is the goal of pain management. [*Addressing Lorraine*]

JH: In a way you two as my patients have been fortunate, as you are still receiving treatments; lots of my patients I see in my clinic will not have this benefit, which is really sad. You can tell them about your injury.

L: I had an accident 20 years ago: I was hit by a car. I've been since then in a constant state of pain in the upper back and neck. I went to the doctor, he said it was a whiplash injury, take painkillers, come back after two weeks; but I was still I pain, and this went on for a long time. Then they said: the pain must be in your mind. I felt quite helpless when I saw that they couldn't do anything for my pain … after 20 years, it is still a battle.

JH: Do you think a doctor will still say the same thing today if you had the injury now?

L: Maybe.

JH: I would hope not! The brain is influencing the pain, but the pain is very real. We know that 10 per cent of the patients with a whiplash injury go on to develop chronic pain; they send you to do an X-ray, do brain scans but they all look like there is no difference between you and those who have no pain. That's why patients and doctors argue. They get orthopaedic surgeons into court, but these see things in a very structural manner; maybe if there was a psychologist things could be different. Can you tell us: What have you done for your pain?

L: Physiotherapy with a pain management physio: this made a big difference to me.

JH: What is the main difference?

L: She understands the pain better, she can give me a strategy to cope with my pain in terms of pacing; she is also a bit of a psychiatrist – she knows all the things in my life and helps me to manage my life better, to cope with my pain. An ordinary physio will give you exercises to do at home, then will sign you off and you would have to cope with your pain on you own. My physio – she does all the body exercises and explains other strategies, especially do not overdo exercises.

JH: Is pacing difficult to achieve?

L: Very difficult, especially if you have children. At work you can be up or down in terms of pain days and a lot of people can't tell the pain is there.

JH: How much can you walk?

L: Not for many meters, muscles get very tight; I drive everywhere.

JH: You have a very understanding employer?

L: Yes. I also do hydrotherapy twice a week, so my muscles can function; I feel my muscles stiffen up as I get older.

JH: You say muscles still get stiffer after 20 years.

L: I can function with physio- and hydrotherapy; I am more mobile.

JH: Do you take tablets?

L: I had – I do now – as I get older, pain is increasingly harder to cope with: lamotrigine, codeine…

JH: Thank you. Remember, do not dismiss anyone in chronic pain.

Dr Joan Hester is lead clinician and consultant in pain medicine in the Anaesthetics Department, King's College Hospital, London and former president of the British Pain Society.

Index

An Introduction to Pain and its Relation to Nervous System Disorders, First Edition. Edited by Anna A. Battaglia.
© 2016 John Wiley & Sons, Ltd. Published 2016 by John Wiley & Sons, Ltd.